RIVER *of* INTERESTS

Water Management in South Florida and the Everglades, 1948-2010

Matthew C. Godfrey and Theodore Catton, Historians
Historical Research Associates, Inc.

July 2011

Library of Congress Cataloging-in-Publication Data

Godfrey, Matthew C.
  River of interests : water management in south Florida and the Everglades, 1948-2010 / Matthew C.
Godfrey and Theodore Catton.
     p. cm.
  Includes bibliographical references and index.
  ISBN 978-0-16-090134-8 (alk. paper)
  1. Restoration ecology--Florida--Everglades--History. 2. Wetland restoration--Florida--Everglades--
History. 3. Water-supply--Management--Florida--History. 4. Water-supply engineering--Florida--
Everglades--History. I. Catton, Theodore. II. United States. Army. Corps of Engineers. III. Title.
  QH105.F6G63 2012
  639.909759'39--dc23
                          2012012034

For sale by the Superintendent of Documents, U.S. Government Printing Office
Internet: bookstore.gpo.gov
Phone: toll free (866) 512-1800; DC area (202) 512-1800
Fax: (202) 512-2104
Mail: Stop IDCC, Washington, DC 20402-0001

ISBN 978-0-16-090134-8

CONTENTS

THE DESTRUCTION OF AMERICA'S EVERGLADES, a campaign to drain and develop "a worthless morass of venomous creatures," as described in Marjory Stoneman Douglas' book "The Everglades: River of Grass," took decades. The good news from the campaign was that the Corps of Engineers was very effective and successful in achieving the end. The bad news, many of the well- intended actions of that campaign achieved unintended consequences. Some consequences are reversible in total, or in part, to recreate conditions similar to those Mother Nature intended, but never returning the system to its once natural state. Other consequences are irreversible, such as large scale urbanization.

Having researched the Everglades intensively while at the U. S. Army War College in 2008-2009, I read and scoured through dozens of well-known resources and there is none better than Godfrey and Catton's "River of Interests," commissioned by the Corps in 2004. "River of Interests" is the best book I've found to date to provide the reader with a comprehensive understanding of the Everglades' story and journey in a factual, clear and candid manner. What was missing from "River of Interests" in 2009 was the last decade of the Everglades' story. The last decade was historic in that projects long -ago envisioned in the Comprehensive Everglades Restoration Plan (CERP) and other programs, began to break ground. Groundbreakings have been held for the Tamiami Trail bridge and roadway, Site 1 Impoundment Area, Picayune Strand Faka Union Canal Pump Station project, the Melaleuca Eradication Facility, and the Indian River Lagoon-South C-44 Reservoir and Stormwater Treatment Area.

Restoration of the Everglades is a war consisting of countless battles, in the past, present and future, spanning decades of fighting for and in the Everglades. It should come as no surprise that it will take decades of time, thousand of warriors, billions of dollars and tremendous cooperation, collaboration, communication, consensus and compromise to win.

The survival of the Everglades rests in the hands of the people who understand it, cherish it, and find ways to push through the battles in order to win the war.

The readability of "River of Interests" is phenomenal, providing readers with all they need to know about the history of the Everglades—from mankind's initial intervention in nature to the restoration efforts conducted through 2010. Read it to understand the past, for the good of the future.

Tremendous hope and optimism exists today for the Everglades. In November 2011, the state of Florida and the U.S. Army Corps of Engineers launched the most ambitious study ever, the Central Everglades Planning Project (CEPP). CEPP will no doubt contain the ways, means and methods to connect the heart of Everglades, Lake Okeechobee, to the greater southern Everglades. No effort for Everglades has ever been more important! Many blessings to those who dare to restore this gem for the greatness of our Nation and our world. Let it be your legacy!

– Alfred A. Pantano, Jr.
Colonel, U. S. Army
District Commander

In the spring of 2004, the U.S. Army Corps of Engineers commissioned Historical Research Associates, Inc. (HRA), to complete a study of the water resources system in South Florida (generally the region south of Orlando) from 1948 to 2000. This history was to include a discussion of all interests involved in water management—whether federal, state, or local—rather than just focusing on the history of the Corps' Central and Southern Florida Flood Control Project (C&SF Project), first instituted in 1948.

The years between 1948 and 2000 saw numerous changes within South Florida, including an explosion of agricultural and urban growth and the subsequent diminishment of ecological resources. Because of these factors, the Corps performed a restudy of the entire C&SF Project in the late 1990s, resulting in the authorization of the Comprehensive Everglades Restoration Plan (CERP) in 2000, a gigantic environmental restoration program intended to enlarge the "water pie" for agricultural, urban, and environmental interests in South Florida and to ensure the health of the ecosystem. In 2010, the Corps commissioned HRA to provide an epilogue addressing the first decade of CERP's implementation.

Water is the essential focus of this history, including its distribution, its quality, and its essentiality for life in South Florida. The following study highlights its importance in the region, as well as the problems that have developed between different interests fighting over the resource. Moreover, this report outlines the environmental transformation of the Corps, the events leading up to CERP, and the initial stages of that program. By doing so, it provides a needed perspective of how and why CERP was developed, and what problems, concerns, and interests informed water management in South Florida between 1948 and 2010.

ASR	Aquifer Storage and Recovery		LOTAC	Lake Okeechobee Technical Advisory Committee
BMPs	Best Management Practices		NARA-SE	National Archives and Records Administration Southeast Region, Atlanta, Georgia
C&SF Project	Central and Southern Florida Flood Control Project			
CEQ	President's Council on Environmental Quality		NARA II	National Archives and Records Administration, College Park, Maryland
CERP	Comprehensive Everglades Restoration Plan		NEPA	National Environmental Policy Act
CR-ENPA	Central Records, Everglades National Park Archives, Homestead, Florida		NOAA	National Oceanic and Atmospheric Administration
EAA	Everglades Agricultural Area		NPS	National Park Service
EDD	Everglades Drainage District		NRDC	Natural Resources Defense Council
EIS	Environmental Impact Statement		OMB	Office of Management and Budget
EPA	U.S. Environmental Protection Agency		PPB	Parts per billion
FAA	Federal Aviation Administration		REAL	Restoring the Everglades, An American Legacy Act
FCD	Central and Southern Florida Flood Control District		RG	Record Group
FRC	Federal Records Center, Atlanta, Georgia		SAM	Spatial Analysis Methodology
FSA	Florida State Archives, Tallahassee, Florida		SFWMD	South Florida Water Management District
FWS	U.S. Fish and Wildlife Service		SFWMDAR	South Florida Water Management District administrative records
FWSVBAR	U.S. Fish and Wildlife Service, Vero Beach, Florida, administrative records		STAs	Stormwater treatment areas
			SWIM	Surface Water Improvement and Management
GAO	U.S. General Accounting Office			
HQUSACE	Headquarters, U.S. Army Corps of Engineers		USGS	U.S. Geological Survey
			WRDA	Water Resources Development Act
IIF	Internal Improvement Fund			
JDAR	Jacksonville District administrative records			

In July 2004, the authors of this history arrived in South Florida, ready to research water management in the region since 1948. For the next several days, we traveled through large cities on the southeastern Florida coast, such as West Palm Beach and Miami, intrigued by the massive buildings and thronging crowds demarcating the area. At other times, we traversed agricultural fields lined by canals and levees (with such designations as "C-51" or "S-6")—regions dominated by sugarcane and its producers. Perhaps influenced by popular culture depictions, we concluded that both of these areas constituted the real South Florida.

Our perceptions changed on a muggy, hot day when we arrived at the National Audubon Society's Corkscrew Swamp Sanctuary, a remnant of the historical Everglades located just west of Naples, Florida. There, we spent several hours engulfed in an entirely different world, one characterized by sawgrass flooded by three-inch deep pools of water, custard apple trees, pileated woodpeckers, alligators, deer, lizards, and the continual buzz of insects. For 20 minutes we were drenched in a thunderstorm that disgorged heavy sheets of rain from black clouds, causing an eerie silence to envelop the swamp. After the storm passed, the swamp came to life once more, resulting in a cacophony of frogs, birds, and insects. Truly, we commented to each other, *this* was South Florida.

In the course of our research, however, we realized that both worlds, the natural and the populated, made up South Florida. Each influenced the other, often in profound ways. Explaining the history of water management in South Florida, we comprehended, required telling the tale of the co-existence of these two disparate landscapes and the conflicts they engendered.

First, one had to consider how the urban and agricultural side of South Florida developed during the last century. Although the state of Florida had implemented drainage schemes and other plans in the late 1800s and early 1900s to stimulate settlement and development of the land, the major catalyst allowing urban and agricultural interests to dominate the landscape was the creation of the Central and Southern Florida Flood Control Project (C&SF Project) in 1948 by the U.S. Army Corps of Engineers. Ever since its establishment, this project, and all of the structures that it produced, has dictated how water would be managed in South Florida, including who would get what amounts (and when) and how such volumes would be distributed.

Yet the story is not so simple. Instead, it is a complex tale of how different interests in South Florida staked a claim in water management and vigorously defended their positions. The Corps and its lo-

cal sponsor, the South Florida Water Management District (preceded by the Central and Southern Florida Flood Control District), has had to operate the C&SF Project among a multitude of interests since 1948, and such operation has required a balancing act between competing views. In the eyes of many environmental organizations, the Corps sometimes stumbled as it traversed this tightrope by placing urban and agricultural needs above ecological concerns. Although there was some legitimacy to this criticism, some of the environmental damage that occurred from water management policies also stemmed from the law of unintended consequences, which states that all human action, especially governmental, have unintended or unanticipated consequences. Colonel Terry Rice, former District Engineer of the Jacksonville District, put it another way, claiming that problems were caused by the Corps' "innocent ignorance." According to Rice, at times the Corps "really did not know what they were doing [to nature] by changing the hydrology, or [by] fixing the hydrology [for the benefit of humans]."[1]

Despite the perceived environmental stumbles, however, the Corps sometimes led the vanguard against ecological destruction, most notably in the late 1990s when it headed a restudy of the C&SF Project resulting in the Comprehensive Everglades Restoration Plan (CERP), one of the most massive environmental restoration projects ever undertaken. But the development of CERP itself begs the question of how the Corps got from point A (the implementation of the C&SF Project) to point B (the undoing of much of that work) and especially how competing interests influenced the Corps' trajectory. These latter issues form the major themes of this book.

Because the Corps' C&SF Project is the major driver of water management in the region, that project—and the Corps itself—serves as the main character of this story. The Jacksonville District is the branch of the Corps serving most of Florida, as well as Puerto Rico, the U.S. Virgin Islands, and the Suwanee, Withlacoochee, and Alapaha river drainages in southern Georgia. The District, which is one of five districts within the South Atlantic Division, traces its roots in Florida back to 1821, although an official district office was not established until 1884. As the second largest civil works district in the nation, it has a variety of responsibilities, including beach erosion control and hurricane protection, emergency response and recovery, flood control, navigation, environmental restoration, and regulatory permitting. Led by a District Engineer (usually a military officer), the District has nearly 800 employees, most of whom are civilians.[2]

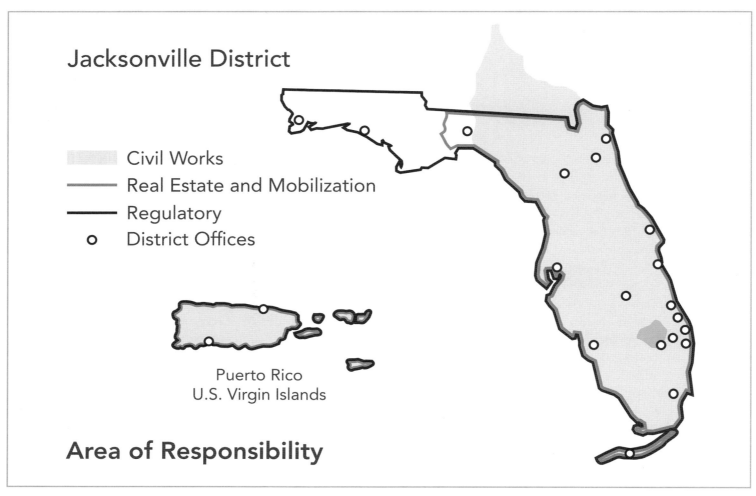

Boundaries and areas of responsibility of Jacksonville District. (U.S. Army Corps of Engineers, Jacksonville District)

In addition to the Corps, other characters play an important role in this history. The South Florida Water Management District (SFWMD), for example, frequently serves as the local sponsor for Jacksonville District projects. The descendant of the Central and Southern Florida Flood Control District, formed in 1949 to operate and maintain the C&SF Project, the district originally had responsibility only for flood control in South Florida. In 1972, however, the Florida legislature passed a water resources act that allowed for the establishment of five water management districts in Florida in 1977, one of which was the SFWMD. The SFWMD thereafter assumed responsibility for other issues pertaining to water management, such as supply and quality. Headquartered in West Palm Beach, the district, which is led by a governor-appointed board and an executive director, is responsible for the operation and maintenance of approximately 1,800 miles of canals. Its jurisdiction covers all or part of 16 counties in South Florida, including Miami-Dade, Broward, Palm Beach, Collier, Hendry, Lee, Martin, Monroe, and Okeechobee counties.[3]

Together with the Corps, the SFWMD, and other federal and state interests, such as the National Park Service and the U.S. Fish and Wildlife Service, the citizens of South Florida constitute the oth-

er main players in the water management saga. These individuals are diverse, creating an interesting political (and regional) dichotomy. Throughout the 1900s, for example, the southeastern coast of Florida became more and more urbanized, centering on Miami, a booming metropolis anchoring Miami-Dade County, one of the most populous counties in the United States with over 2.2 million people. Residents in Miami-Dade County, as well as in Palm Beach and Broward counties (areas containing the cities of West Palm Beach and Fort Lauderdale), range from senior citizens who moved from other parts of the United States to large populations of immigrants from Latin American and Caribbean countries, especially Cuba. This urbanized and socially diverse region sustains a strong environmental movement. Influential environmentalists from the region have included Marjory Stoneman Douglas, Arthur Marshall, and D. Robert "Bob" Graham, U.S. Senator from and governor of Florida. Yet having enough water for its population was also important to east coast residents, and their demands for water often superseded environmental needs.

Conversely, the southwestern part of South Florida, especially Okeechobee, Collier, Glades, and Hendry counties, is much more rural and racially homogeneous. Agriculture and tourism are the main components of the region's economy, an area including much of the remaining Everglades. Property rights and the right to pursue activities such as hunting, fishing, and frogging are generally more important to southwestern residents than preservation of areas for environmental purposes.

The sugar industry dominates a central region of South Florida, directly south of Lake Okeechobee. More racially diverse than southwestern Florida, in large part because of an influx of migrants from the West Indies working as laborers in the sugar fields, this region is largely controlled by various sugar companies, such as Flo-Sun, the U.S. Sugar Corporation, and the Sugar Cane Growers Cooperative of Florida. Although sugar has been raised in the area since at least the 1920s, it did not become a large influence until the 1960s. At that time, Florida's sugar industry became a major player in state and national politics because of the large sums of money it produced. The power of the sugar industry has led many urban environmentalists to claim that sugar unduly influences county, state, and federal governments on ecological issues, but sugar representatives respond with the same concerns about environmentalists.

The interplay of these diverse interests occurs against a unique geographical setting. Key to this ecosystem is its subtropical climate. The only area in the continental United States with a subtropical climate, South Florida has two seasons: wet and dry. The wet season, extending roughly from May through October, accounts for three-quarters of the average annual rainfall total of 60 inches, while the dry season, lasting from November through April, experiences little precipitation. The sun shines on an average of 70 percent of the time during the day, although it is common in the rainy season for afternoon thunderstorms to well up on a regular basis, and daily high temperatures range from 76 degrees in January to 92 degrees in July. The subtropical climate creates a large amount of humidity, averaging 75 percent annually. Hurricanes and fires also abound, serving a vital role in vegetation propagation.[4]

The climate influences how much water flows through South Florida, which is characterized by two inland ridges—one along the east coast and one to the west—forming a shallow bowl-like valley. A slight tilt in the bowl means that water drains in a southwesterly direction, but, before the beginnings of drainage and development in the late 1800s, this natural receptacle retained much of the large amounts of rainfall that cascaded to the ground. Supplementing this supply was a slow-moving flow of water emanating from the upper chain of lakes forming the headwaters of the Kissimmee River—Lakes Kissimmee, Tohopekaliga, Hatchineha, and Cypress, to name a few—located just south of present-day Orlando. Water from these lakes meandered down the twisting and turning Kissimmee to Lake Okeechobee, the second largest freshwater lake in the continental United States. Because the lake had no real outlet (the St. Lucie River began 20 miles to the east of the lake, flowing to the Atlantic Ocean, while the Caloosahatchee River started three miles west, running

to the Gulf of Mexico), it would overflow its southern rim during Florida's rainy season, spilling water into the South Florida valley. There, the liquid would begin a measured journey through a 60-mile wide region extending south from Lake Okeechobee to Florida Bay (off the southern tip of the peninsula), known as the Everglades. The extreme flatness of South Florida's landscape—the elevation drops less than 20 feet between Lake Okeechobee and Florida Bay—meant that the 100-mile journey through the Everglades was unhurried and leisurely. The slow-moving water essentially flooded the region, creating a rich habitat for flora and fauna, as well as layers of fertile peat and muck soils built on the basin's limestone bedrock.[5]

From the Everglades, water drained to the Gulf of Mexico through a series of open-water sloughs, including Shark River Slough. This channel took water southwest, dumping it in a region known as the Ten Thousand Islands, described by one scholar as "a bewildering green archipelago of mangrove keys at the edge of

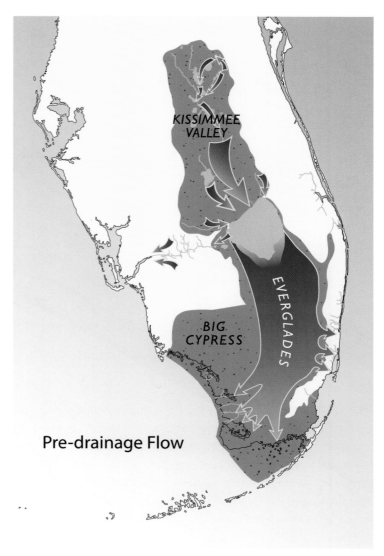

The historic drainage pattern of South Florida. (South Florida Water Management District)

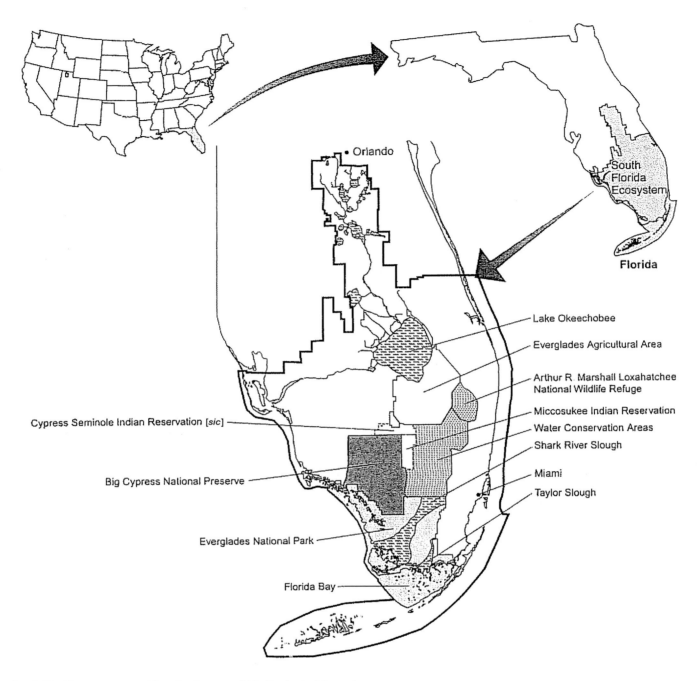

The South Florida ecosystem, with major features. (U.S. Geological Survey)

the Gulf."[6] Taylor Slough was the other major drainage, running southwest from a more easterly position into Florida Bay, located just south of Florida's southern tip. The Miami, New, and Hillsboro rivers also flowed through the Everglades, taking water east to Biscayne Bay. As these waterways deposited into the estuaries of the Ten Thousand Islands, Florida Bay, and Biscayne Bay, fresh water mixed with salt water, creating a habitat where shrimp, lobsters, and fish thrived.

Before Euro-American habitation of South Florida, the Everglades was thus "a complex system of plant life linked by water," in-cluding "expansive areas of sawgrass sloughs, wet prairies, cypress swamps, mangrove swamps, and coastal lagoons and bays."[7] It con-sisted of 2.9 million acres of land dominated by sawgrass and tree islands ("areas of slightly raised elevation covered by shrubs and woody vegetation"), housing a diversity of life, including cypress trees, pop ash, pines, buttonwood trees, mangroves, ferns, cabbage palm, orchids, sparrow hawks, red-cockaded woodpeckers, blue herons, egrets, roseate spoonbills, white ibises, otters, alligators, deer, and Florida panthers.[8] In the words of Marjory Stoneman Douglas, who bequeathed the term "river of grass" to the Everglades (playing off of *pahokee*, the Seminole word for the region meaning "grassy waters), "all these birds, insects, animals, reptiles, whispering, screaming, howling, croaking, fish in their kinds teeming, plants thrusting and struggling, life in its millions, its billion forms, the greatest concentration of living things on this continent, they made up the first Florida."[9]

Yet by the last quarter of the twentieth century, this diversity of life had largely ceased to exist, and the Everglades itself had shrunk to half its size. These conditions led to concerns about the C&SF Project's impact on the South Florida ecosystem, and, ultimately, to cries for dismantling the works. The following history of water management in South Florida since 1948 shows both the short-term value and the long-term pitfalls that the Corps' engineering of the South Florida environment has generated. In doing so, it focuses on the interaction of different interest groups, all with diverse stakes and perspectives, and how their conflicts and compromises influenced the direction that the Corps pursued.

Endnotes

[1] Quotation in Colonel Terry Rice interview by Brian Gridley, 8 March 2001, 3, Everglades Interview No. 4, Samuel Proctor Oral History Program, University of Florida, Gainesville, Florida. For more information on the law of unintended consequences, see Robert K. Merton, "The Unanticipated Consequences of Purposive Social Action," *American Sociological Review* 1 (December 1936): 894-904.

[2] U.S. Army Corps of Engineers, Jacksonville District, "Who We Are" < http://www.saj.usace.army.mil/cco/ who.htm> (10 February 2006).

[3] South Florida Water Management District, "Agency Overview" <http://www.sfwmd.gov/site /index.php?id=61> (10 February 2006).

[4] U.S. Army Corps of Engineers, "Overview: Central and Southern Florida Project Comprehensive Review Study, October 1998," 4, copy in File Restudy Feasibility, Box 1365, JDAR.

[5] For an in-depth discussion of the geology, hydrology, and biology of the Everglades before the 1800s, see David McCally, *The Everglades: An Environmental History* (Gainesville: University Press of Florida, 1999). Michael Grunwald also provides some information about the region before white settlement in *The Swamp: The Everglades, Florida, and the Politics of Paradise* (New York: Simon & Schuster, 2006), 9-23.

[6] Grunwald, *The Swamp*, 19.

[7] U.S. Army Corps of Engineers, "Overview: Central and Southern Florida Project Comprehensive Review Study, October 1998," 7, copy in File Restudy Feasibility, Box 1365, U.S. Army Corps of Engineers, Jacksonville District administrative records, Jacksonville, Florida [hereafter referred to as JDAR].

[8] Quotation in U.S. Department of the Interior, U.S. Geological Survey, *South Florida Ecosystem Program of the U.S. Geological Survey*, Fact Sheet FS-134-95 (Washington, D.C.: U.S. Geological Survey, 1995); see also McCally, *The Everglades*, 58-83.

[9] As quoted in U.S. Army Corps of Engineers, "Overview: Central and Southern Florida Project Comprehensive Review Study, October 1998," 6.

CHAPTER 1

Draining the Swamp: Development and the Beginning of Flood Control in South Florida, 1845-1947

IN THE MID-1800s, the Everglades, a region of water and sawgrass between Lake Okeechobee and the southern edge of Florida, percolated in Floridians' minds. What, they asked, was the purpose of this vast wetland? Was it destined to lay unoccupied, or were there measures they could take to make the area conducive to settlement? Unappreciative of the plethora of flora and fauna in the region, most Floridians could see only a wet swamp that had to be drained and seeded to crop before it could reach its full potential. Accordingly, throughout the late 1800s and the first decades of the twentieth century, Floridians, both privately and with state help, examined the possibility of draining the Everglades. Hamilton Disston and Napoleon Bonaparte Broward, for example, pursued drainage relentlessly, and railroads and land speculators marketed the dry land as an agricultural paradise. But problems appeared in the 1920s and 1930s: storms sporadically produced devastating floods, while flora and fauna dwindled because of the lack of water. Such problems required federal action; in 1930, the U.S. Army Corps of Engineers began a flood control project around Lake Okeechobee, and in the 1930s and 1940s, conservationists were able to secure protection for wildlife and vegetation through the creation of Everglades National Park. The state had sponsored drainage programs for much of the twentieth century, but by the mid-1940s, officials realized that federal help was necessary so that water in South Florida could be managed comprehensively.

Because this period of drainage, early flood control, and conservation laid the groundwork for the establishment of the Central and Southern Florida (C&SF) Flood Control Project in 1948 and for the subsequent water supply tensions prevalent throughout the rest of the century, it constitutes a critical era in the history of water management in South Florida. No flood control project or water supply scheme in the second half of the twentieth century began with a *tabula rasa*; instead, the Corps and other agencies had to construct projects in an environment that had already been extensively modified. In the words of historian George E. Buker, the Corps was "faced with correcting past mistakes."[1] By the time the Corps developed the C&SF Project, numerous political entities, including federal interests (the National Park Service and the U.S. Fish and Wildlife Service), state interests (the trustees of the Internal Improvement Fund), and local interests (boards of county commissioners) had already staked out their water terrain. Thus, the Corps would not only have to work within a manipulated and modified ecosystem, but also with existing political interests, each with a different perspective as to how water should be managed.[2]

Thousands of years before Americans had made any attempts to alter the South Florida environment, including the Everglades, native peoples had traversed the area, discovering ways to subsist and flourish within the soggy marshes. By the first years of the common era, three groups had settled in the Everglades area: the Calusa, who resided in a region that began north of the Caloosahatchee River and extended south through the Ten Thousand Islands to Cape Sable; the Mayaimi, who occupied the shores of Lake Okeechobee; and the Tekesta, who lived on the east coast beaches from Boca Raton south to Biscayne Bay and the keys.[3] By the time of Spanish contact in the early sixteenth century, the most dominant and populous group was the Calusa. This tribe, like the Mayaimi and the Tekesta, had learned how to use the Everglades, its water, and its resources in the most efficient ways. The groups subsisted mainly on food obtained in the freshwater and saltwater of the region, including cocoplum, sea

Ken Hughes' rendition of Pedro Menendez, the first Spaniard to explore South Florida extensively. (The Florida Memory Project, State Library and Archives of Florida)

grape, prickly pear, cabbage palm, and saw palmetto, as well as fish and game. They made clothes out of tree moss and palmetto strips, and employed conch shells as tools and drinking cups. They built houses using cabbage palm posts and palmetto, and applied fish oil to discourage mosquitoes and sandflies.[4]

Despite their knowledge of the land, the groups could not escape the problems that resulted from non-Indian settlement. In the early 1500s, Spanish explorers reached Florida, led by Juan Ponce de León in 1513. The first Spaniard to explore the region extensively was Pedro Menendez, who, in the 1560s, conducted investigations to try to find a waterway across the Florida peninsula to facilitate Spanish navigation to the Americas. By 1570, however, Spanish interest in South Florida had waned, mainly because no trans-peninsula waterway had been discovered. Yet non-Indians still influenced the region, and European diseases and slave raids decimated Indian populations. When Great Britain assumed authority over the area from 1763 to 1783, only 80 Calusa families remained, and they left with the Spanish. By the time the United States had gained official control over Florida in 1821, other Indian groups, including the Seminole,

an offshoot of the Creek in Georgia, had moved into the Everglades, and Americans spent a great amount of time and energy trying to remove them in the 1830s, 1840s, and 1850s.[5]

The Second Seminole War (1835-1842) and the Third Seminole War (1854-1855), for example, represented concerted campaigns by the United States to extricate the Seminole from the Everglades. Although these battles were characterized by one scholar as "America's first Vietnam," in that it was "a guerilla war of attrition, fought on unfamiliar, unforgiving terrain, against an underestimated, highly motivated enemy who often retreated but never quit," the expeditions provided numerous accounts and maps of the South Florida landscape, including the Ives map discussed below. Despite the colorful accounts of the landscape—or perhaps because of them, as most soldiers depicted the scenery as an "interminable, dreary waste of waters" infested with mosquitoes, snakes, and sawgrass—Floridians expressed little interest in the Everglades until the mid-1800s.[6]

This situation changed on 3 March 1845, when Congress allowed Florida to enter the United States as the 27th state in the

Union. Thereafter, the state's legislature, seeking new areas where people could settle, passed resolutions declaring that "there is a vast and extensive region, commonly termed the Everglades, in the southern section of this State, . . . which has hitherto been regarded as wholly valueless in consequence of being covered with water at stated periods of the year." The resolution asked Florida's representative and senators to "earnestly press upon" Congress to appoint "competent engineers to examine and survey the aforesaid region" in regard to the possibilities of drainage.[7] Buckingham Smith, an attorney from St. Augustine, Florida, received this appointment, and he submitted a report to the secretary of the treasury on 1 June 1848. In this document, Smith provided a detailed description of the Everglades landscape:

> The Everglades extend from the southern margin of Lake Okeechobee some 90 miles toward Cape Sable, the southern extremity of the peninsula of Florida, and are in width from 30 to 50 miles. They lie in a vast basin of lime rock. Their waters are entirely fresh, varying from 1 to 6 feet in depth. . . . As the Everglades extend southwardly from Lake Okeechobee they gradually decline and their waters move in the same course. They have their origin in the copious rains which fall in that latitude during the autumn and fall and in the overflow of Lake Okeechobee through swamps between it and the Everglades.[8]

Smith believed that in order to reclaim the Everglades, canals would have to be constructed from Lake Okeechobee to the Caloosahatchee and Loxahatchee rivers, thereby allowing the lake to drain into these rivers, lowering its water level and preventing it from sending water on its normal southward trek. Drains would also have to be placed at strategic locations "by which the waters accumulating from the rains may be conducted to the ocean or gulf." If such actions were not taken, Smith claimed, "the region south of the northern end of Lake Okeechobee will remain valueless for ages to come." But if drainage was implemented, the land could produce cotton, corn, rice, and tobacco, as well as lemons, limes, oranges, bananas, plantains, figs, olives, pineapple, and coconuts.[9] According to historian David McCally, Senator Westcott forwarded this report to the *Commercial Review of the South and West*, which "embraced Smith's conclusions and urged Congress to deed the Everglades to the State of Florida so that reclamation could begin."[10]

Congress listened to the *Commercial Review's* recommendation. In the Swamp Lands Act of 1850, Congress expanded the jurisdiction of an 1849 act granting swamp areas to the state of Louisiana, allowing the federal government to provide swamp and overflowed lands unfit for cultivation to other states as well.[11] Under the authority of this act, the federal government transferred title to more than 20 million acres to the state of Florida. In 1851 and 1855, the Florida legislature passed acts creating an Internal Improvement Fund (IIF), consisting of the land and the money obtained from land sales, and establishing a board of trustees to oversee the fund. This board, composed in part of the governor and his cabinet, essentially had authority over all state land sales and over all reclamation matters.[12]

In 1856, more information about the topographical features of South Florida was made available when Lieutenant J. C. Ives, a topographical engineer serving in the Third Seminole War, conducted a survey of the area and combined his data with other records produced in the 1840s by army officers traversing the region to produce a map of the "comparatively unknown region" south of Tampa Bay. The Department of War wanted the map to inform officers fighting the Seminole, but it became, in the words of Marjory Stoneman Douglas, "the first fine American map of the country."[13] Ives highlighted not only the Everglades, but other areas of South Florida, including Big Cypress Swamp and Lake Okeechobee, and he noted that the land was basically "a flat expanse, where the prairie of one day may at another be converted into a lake and where the lakes, rivers, swamps and hammocks" fluctuated as much as three feet at a time.[14]

Eager for a chance to promote the settlement of South Florida, the IIF began granting deals to railroad companies in which it would give the corporations land in return for completed rail lines. In this way, the IIF hoped to "open the interior and attract settlers, who would buy land and replenish the fund, which could perhaps be used to finance drainage ditches someday in the future."[15] After many railroads succumbed to financial difficulties in the post-Civil War era, the IIF essentially faced bankruptcy. Its situation worsened when Francis Vose, a New York metals manufacturer who had provided iron to railroad companies in Florida in return for state bonds, refused to accept the state's offer of 20 cents on the dollar for the bonds and sued the IIF instead. From that suit, an injunction was placed against the IIF's, preventing it from distributing any more land for discounted prices until Vose had been paid in full. Desperate for money, the IIF, under the leadership of Governor William D. Bloxham, began looking for new investors interested in obtaining land for reclamation purposes. In 1881, it found a candidate: Hamilton Disston.[16]

Disston was a 34-year-old entrepreneur from Philadelphia whose wealthy father owned a lucrative saw and file manufacturing company. First visiting Florida in 1877 on a fishing trip, Disston had been obsessed with draining the Everglades ever since. In 1881, Disston proposed to drain lands flooded by Kissimmee River and Lake Okeechobee waters by constructing a system of canals and ditches from Lake Okeechobee to the Caloosahatchee River, the St. Lucie River, and the Miami River, and by straightening and deepening the Kissimmee. This would convey water in the flooded Kissimmee basin to Lake Okeechobee, and the excess water would then be flushed out via the Caloosahatchee, St. Lucie, and Miami rivers,

Lieutenant J.C. Ives' military map of South Florida, 1856. (U.S. Army Corps of Engineers, Jacksonville District)

thereby lowering Lake Okeechobee's water level and allowing vast acreages of land to be cultivated. In exchange, the IIF would give Disston and his associates "one-half of all the reclaimed land already belonging to the state or later turned over by the federal government," as well as four million more acres for $1 million.[17] In September 1881, Disston's corporation, the Atlantic Gulf Coast Canal and Land Sales Company, began drainage operations.

By deepening and straightening the Kissimmee River, and by constructing canals connecting the various lakes that formed the headwaters of the river, Disston was able to drain portions of the area and sell it to cattle operators as grazing land in the 1880s. Disston's company also deepened the Caloosahatchee River and connected it to Lake Okeechobee through a linchpin canal. In addition, the corporation began a canal south of Lake Okeechobee, hoping to drain water into the Shark River, and started another east of the Kissimmee Valley toward the St. Johns River. To promote the reclaimed land, Disston produced advertising brochures, planned model cities, built hotels, settled families, and established agricultural enterprises such as sugar, rice, and peach cultivation. By the 1890s, however, Disston had overextended his operations, and the Panic of 1893 dealt a devastating blow to his finances. Banks began recalling loans and bonds became due. Faced with an increasingly precarious situation, Disston died on 30 April 1896, either through suicide or from a heart attack. Although his decade-long drainage effort reclaimed less than 100,000 acres, he left two legacies: first, he demonstrated conclusively the agricultural potential of the region through his experimental farms, and second, his connection of the Caloosahatchee River to Lake Okeechobee was "the first significant step in draining the Everglades."[18]

Meanwhile, the vision of canals and drainage lived on in other minds. John Westcott, for example, formed the Florida Coast Line Canal and Transportation Company in 1881 to build a canal from the mouth of the St. Johns River to Biscayne Bay. The enterprise received a boost in the 1890s when Henry L. Flagler, who became a millionaire with Standard Oil, formed the Florida East Coast Railroad to build a rail line from St. Augustine to Miami Beach. Flagler became interested in the canal project, perhaps because the company agreed to provide the railroad corporation with 270,000 acres of land it had obtained. However, even with Flagler's interest and resources, canal construction proceeded slowly, not reaching completion until 1912, although the construction of his railroad did precipitate South Florida's first settlement boom, leading to the establishment of West Palm Beach, Fort Lauderdale, and Miami.[19]

By the close of the nineteenth century, large-scale drainage and agricultural development of the Everglades, although attempted by many different parties, had not reached fruition. Despite the granting and sale of millions of acres of land in southern Florida to railroads and other corporations, successful reclamation lay in the future. An 1891 report written by H. W. Wiley of the U.S. De-

Hamilton Disston, the first to set up extensive drainage operations in South Florida. (The Florida Memory Project, State Library and Archives of Florida)

Early settlers to South Florida. (South Florida Water Management District)

partment of Agriculture observed that, although "the possibilities of bringing into successful cultivation the swamp lands of Florida have occupied the minds of capitalists for several years," large tracts remained inundated. Even those that had been drained were "still in the wild state, . . . no attempts having been made to fit them for cultivation."[20] Conditions were no better in 1903, leading Governor

William S. Jennings to compare Florida's drainage endeavors to "the man who undertook to lift himself," opining that the state was "almost as helpless."[21]

In the early 1900s, drainage schemes gained momentum, largely because of changing ideas about the human use of nature. The late 1800s and early 1900s saw the development of a conservation movement in the United States, characterized, in the words of historian Samuel P. Hays, by "rational planning to promote efficient development and use of all natural resources."[22] This movement expressed itself in several ways, including the formation of the U.S. Reclamation Service in 1902, and in the creation of national parks, which were conceived as areas to preserve pristine wilderness for the enjoyment of future generations. Other conservationists held that making wasteland productive was an excellent way to promote the efficient use of the nation's resources. The editors of *Collier's* magazine, for example, claimed that the terms "conservation" and "reclamation" meant not only the irrigation of dry land, but the draining of wetlands as well.[23] In Florida, these ideas, coupled with populist notions of the necessity of taking land from railroads and other large corporations to benefit small farmers, influenced state officials to implement drainage policies vigorously so that Everglades land could be used for agriculture.[24]

The drainage program was facilitated in 1903 when the federal government provided the IIF trustees with the patent to over two million acres of the Everglades, thereby ending several disputes over whether the state, railroad interests, or corporations were entitled to the land.[25] With this title secured, state officials actively implemented their own drainage program. Napoleon Bonaparte Broward, a Jacksonville jack-of-all-trades who had previously been employed as a steamboat captain, a sheriff, and a gunrunner, was especially active in promoting drainage.[26] In 1904, Broward entered Florida's race for governor, concerned that the state was relying too much on railroads and special interests to drain the land (and consequently was allowing these entities to accumulate large holdings and vast amounts of Florida wealth). During his campaign, Broward "carried his map of the Everglades from one end of the State to another, always crying in the hustings, 'Save and reclaim the people's land!'"[27] He pledged that, if elected, he would use state money to drain the land, financing the endeavor by selling the dry tracts for $5 to $20 an acre.[28]

After winning the election, Broward began to implement his promises, thereby inaugurating the first official state-sponsored drainage program. In May 1905, Broward gave a special message to the state legislature dealing exclusively with draining the Everglades. Insisting that it was the "duty" of the IIF trustees to drain Florida lands, he proposed that the state build a system of canals from Lake Okeechobee to the St. Lucie, St. Johns, and Caloosahatchee river basins, thereby allowing the lake's level to drop six feet. Such a scheme would allow large amounts of land, including three million acres held by private interests, to become productive. Broward also proposed that the state pass a constitutional amendment creating a

Napoleon Bonaparte Broward. (The Florida Memory Project, State Library and Archives of Florida)

drainage district that would collect taxes from private landowners "in proportion to benefits that the land will derive," thereby producing more money to be used in other drainage efforts.[29] The state legislature acted on Broward's recommendation, passing an act in 1903 that created the Everglades Drainage District (EDD) with boundaries roughly corresponding to the two million acres patented to the state in 1903.[30]

With the EDD in place, Broward ushered in an era of intensive state interest in drainage, including the construction of the New River Canal, running southeast from Lake Okeechobee to the New River near Fort Lauderdale. But in actuality, Broward accomplished relatively little; only 15 miles of canal were dug by the end of his term and the IIF fund had been depleted. Therefore, in December 1908, only a week before his term as governor ended, Broward convinced the IIF trustees to give Richard J. Bolles, a Colorado developer, 500,000 acres of land in exchange for $1 million. The trustees then proposed that most of this money be used to build five major canals—the North New River, South New River, Miami, Hillsboro, and Caloosahatchee. However, no studies had been completed on whether or not these waterways were practicable, resulting in a sale that "irrevocably committed the State of Florida to a specific drainage project even before the first engineering study regarding its feasibility appeared."[31]

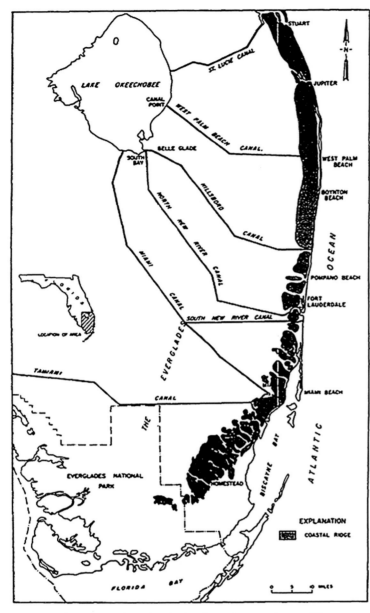

Location of major canals in South Florida. (U.S. Army Corps of Engineers, Jacksonville District.)

For the next several years, the state commissioned numerous engineering reports that revised the best methods to drain the land. These included the Wright Report (1909), which facilitated land speculation in South Florida based on low cost estimates of drainage schemes (which turned out to be faulty at best and fraudulent at worst); an Everglades Land Sales Company examination (1912) which recommended that Lake Okeechobee's water levels be regulated to facilitate drainage; and the Randolph Report (1913), which recommended the construction of a control canal from Lake Okeechobee to the St. Lucie River (the St. Lucie Canal) and that became "the master plan for all drainage work."[32] By the end of the 1920s, the major drainage canals were largely in place, consisting of the Caloosahatchee Canal, which ran from the western shore of

Lake Okeechobee to the Gulf of Mexico; the St. Lucie Canal, which extended from the eastern side of Lake Okeechobee to the Atlantic Ocean; and the West Palm Beach, Hillsboro, North New River, and Miami canals, which all ran from various points on the southern shore of Lake Okeechobee to the Atlantic Ocean.[33]

As these waterways were completed, agriculture developed in the region south of Lake Okeechobee. In the 1910s and 1920s, many new settlements appeared along the canals extending from Lake Okeechobee, including South Bay (on the North New River Canal), Lake Harbor (by the Miami Canal), Belle Glade (on the Hillsboro Canal), Pahokee (near the West Palm Beach Canal), and Moore Haven (on the southeast shore of Lake Okeechobee). By 1920, 23,000 people resided in the EDD. These numbers increased in the 1920s, in part because of better information about how to make Everglades soil productive and in part because of a growing demand for agricultural products. Perhaps even more important was the development of the sugar industry in the Everglades, started by the Southern Sugar Company in the 1920s and continued by Charles Stewart Mott, who rescued Southern Sugar from bankruptcy and reorganized it as the United States Sugar Corporation in 1931. Because of these efforts, cane sugar quickly became one of the predominant crops in the region.[34]

Yet even with the drainage works, flooding still occurred periodically in the Everglades region. After excessive rainfall in 1924, the EDD constructed a small dike around the southern end of Lake Okeechobee from Bascom Point to Moore Haven, the region's largest town.[35] Unfortunately, the barrier did not hold in 1926 when a hurricane swept over Moore Haven with winds between 130 to 150 miles an hour. Over 400 people were killed, approximately 1,200 had to be evacuated, and thousands of dollars of property damage occurred. Because of the devastation, the IIF trustees appointed an Everglades Engineering Board of Review in 1927 to examine the drainage program established by the Randolph Report, and to make additional recommendations about Everglades reclamation.[36]

The board, which consisted of Anson Marston (a prominent transportation engineer who had worked on the establishment of different highways), S. H. McCrory, and George B. Hills, spent two weeks examining drainage works, records, and data pertaining to reclamation. In its final report, published in May 1927, it stated that the Randolph Report's drainage plan had several fatal flaws, especially in terms of controlling floods. To correct the problems, the board recommended that the EDD complete and deepen the St. Lucie Canal as soon as possible (since its operation would have aided flood control efforts during the 1926 storm); that it enlarge the Caloosahatchee Canal; that Lake Okeechobee be controlled to a maximum and minimum level of 17 and 14 feet above mean low water (Punta Rosa datum, which the U.S. Coast and Geodetic Survey had determined to be 0.88 foot below mean sea level), respectively; and that a "greatly enlarged and highly safeguarded levee" be constructed on the south shores of Lake Okeechobee to protect the surrounding communities.[37]

The chances of the EDD implementing the board's suggestions were slim, however, because of continued financial problems.[38] Then, in 1928, another disaster struck the Lake Okeechobee region. In August and September, torrential rain fell in the area, causing the lake to reach a high level. On 16 September, another hurricane appeared, striking Florida at West Palm Beach and traveling northwest across Lake Okeechobee. Winds reached velocities of 135 miles per hour, causing wind tides and waves on the lake to exceed 29 feet in height on the southeastern shore. Unfortunately, the existing levees extended only 22 feet in elevation, causing water to pour over the dikes and into the streets of Belle Glade and other shore communities to depths of eight feet. The water ripped houses from their foundations and swept terrified residents to their deaths. By the time the hurricane moved on, it had killed over 2,000 people, most of them migrant black laborers.[39]

LAKE OKEECHOBEE AREA, FLORIDA. LIMITS OF EXTENSIVE FLOOD DAMAGE, 1926 AND 1928 HURRICANES
SCALE IN MILES

Area hit by the 1928 hurricane. (The Florida Memory Project, State Library and Archives of Florida)

Emerging from the disaster, residents called for help. But because of the financial difficulties of the EDD, and because it was unclear whether or not the EDD could properly operate for flood control instead of drainage, the state could do little to provide the desired flood protection. To rectify the situation, the state legislature created the Okeechobee Flood Control District in 1929, with boundaries including all of South Florida beginning at the northern shore of Lake Okeechobee, and directed it to construct flood control structures and to regulate Lake Okeechobee and the Caloosahatchee River to prevent damaging floods.[40]

To fulfill these missions, the Okeechobee district worked closely with the U.S. Army Corps of Engineers, which had already been making investigations as to what could be done to alleviate flooding from Lake Okeechobee. Since the early 1800s, the Corps had been the federal government's leading civil works agency, but most of its construction involved navigation projects on rivers and lakes. Until the 1930s, the federal government regarded flood control mainly as a local responsibility; not until 1936 would Congress recognize flood control as a proper federal activity nationwide, although it did pass a flood control act in 1917, allowing the construction of works on the Sacramento and Mississippi rivers.[41] Likewise, in 1928, Congress authorized the Corps to undertake an ambitious effort on the Lower Mississippi River, covering several states.[42] In 1924, U.S. Representative Herbert Drane, a Democrat from Florida, introduced a bill into Congress requesting that the Corps examine the Caloosahatchee River to ascertain whether deepening the channel could relieve flooding. Congress passed the act and provided $40,000, but the Corps, under the leadership of Chief of Engineers Major General Edgar Jadwin did not commence any work. After the hurricane passed, the Corps held public hearings at Pahokee and Moore Haven and completed its study, but found no justification for federal action. Nevertheless, Congress passed another bill requiring the Corps to investigate more comprehensively the problem of flood control in the region. After holding public hearings in communities around the lake, Jadwin recommended to Congress in April 1928 that the Corps take no flood control action until state and local resources had been exhausted. Jadwin believed that the plans already in place by the EDD, including enlargement and completion of the St. Lucie Canal, were sufficient. "If carried out," he promised, "they will provide for the control of floods in these areas with a reasonable factor of safety."[43]

After the devastation of the 1928 hurricane, Jadwin reexamined flood control possibilities around Lake Okeechobee, in part because Florida Governor John W. Martin and his cabinet sent a resolution to Congress asking that the federal government construct a high levee around the lake's southern shore. After considerable study by the Jacksonville District, Jadwin recommended that the Corps undertake a flood control and navigation program consisting of "a channel 6 feet deep and at least 80 feet wide from Lake Okeechobee to Fort Myers" (basically deepening the Caloosahatchee River to make it a second control canal); "the improvement of Taylor Creek to the extent of providing a channel 6 feet deep and 60 feet wide to Okeechobee [C]ity"; and the construction of levees along the south and north shores of the lake to heights of at least 31 feet. Jadwin estimated that the project would cost over $10 million, and he suggested that the state of Florida or other local interests provide 62.5 percent of that cost, not to exceed $6.74 million.[44]

Because of the expense of the Corps' proposal, the Okeechobee Flood Control District hired George B. Hills, one of the members of the 1927 Everglades Engineering Board of Review, to conduct an independent investigation of flood control. He recommended early in 1930 that Congress authorize a navigation and flood control project whereby the Corps, using the existing Caloosahatchee and St. Lucie canals, would build a waterway across Florida through the Everglades. At the same time, Congress requested that the Board of Engineers for Rivers and Harbors review Jadwin's 1929 report, and in March 1930, the board recommended that the levees be at least 34 feet above sea level and that instead of the $6 million contribution, the state provide $3.8 million and build at its own cost the north shore levee.[45]

In the spring of 1930, Congress passed a general river and harbor bill that included these provisions for flood control and navigation. Because many representatives were uneasy about the Corps implementing a flood control project, the House and Senate portrayed the program as primarily one that would improve navigation and provide only incidental flood protection. No matter how it was depicted, the plan, according to U.S. Senator Duncan Fletcher, would allow the Corps to make improvements to the St. Lucie Canal, to expand the levees along Lake Okeechobee's north and south shores, and to complete the "canalization" of the Caloosahatchee River. Fletcher believed that this would provide a "complete solution of the problems of adequate interstate navigation facilities and flood-control protection."[46]

Following this plan, the Corps built over 67 miles of dikes along Lake Okeechobee's south shore—later named the Hoover Dike after President Herbert Hoover—and another 15 miles of levees along the north shore near the city of Okeechobee. These were all constructed to handle crests of 32 to 35 feet in height. The Corps also performed the required deepening of the Caloosahatchee River, and by March 1938, the entire project was completed.[47] The Corps then assumed control of regulating the water level of Lake Okeechobee, maintaining a level between 14 and 17 feet through discharges to the St. Lucie Canal and the Caloosahatchee River.

Interesting, however, was the fact that in the 1930s, the U.S. Coast and Geodetic Survey, which had originally demarcated Lake Okeechobee's water levels in accordance with the Punta Rosa Datum (corresponding to the mean low water elevation of the Gulf of Mexico), discovered that the datum plane was not 0.88 foot below mean sea level, but was actually 1.44 feet below mean sea level.

Therefore, the original levee construction around Lake Okeechobee, which was supposed to have been 31 feet, was actually only 29.56 feet according to the National Geodetic Vertical Datum (NGVD) of 1929. Many continued to use the old Punta Rosa Datum plane for Lake Okeechobee (designating it as Lake Okeechobee Datum), even though the Corps had to convert the datum before designing any Lake Okeechobee project in order to avoid errors.[48] Regardless, by the end of the 1930s, the drainage system in southern Florida essentially consisted of the structures that enabled the Corps to regulate Lake Okeechobee; the four major drainage canals (West Palm Beach, Hillsboro, North New River, and Miami); and two canals connecting the four waterways (the Bolles and Cross canals).[49]

The success of drainage and flood control efforts, coupled with periods of drought, had detrimental effects on flora and fauna in the Everglades, emphasizing that proper amounts of water were essential to preserve the unique natural resources of the area. The region housed, among other things, orchids, mangroves, magnolia, cypress, mahogany, lignum vitae, rubber trees, egrets, cranes, herons, flamin-

A poster commemorating the construction of Hoover Dike. (South Florida Water Management District)

gos, spoonbills, alligators, turkeys, bear, deer, fox, wildcats, panthers, raccoons, and opossums. However, drainage, human settlement, and hunting slowly destroyed this rich diversity of life.[50] In the late 1800s, a flourishing plume trade brought hunters of all kinds to the Everglades, where they massacred thousands of egrets by invading rookeries.[51] The Florida state legislature passed a law in 1901 outlawing plume hunting, and the National Audubon Society, first formed in the 1880s, hired four game wardens to patrol the rookeries and enforce the law. Hunters did not welcome this supervision, and on 8 July 1905, Guy Bradley, one of the wardens, was murdered as he investigated a poaching incident, becoming America's first environmental martyr. This event led to laws "which strengthened bird protection and helped bring the significance of the Everglades to the American people."[52]

Drainage in South Florida only compounded the poaching destruction, as it enabled settlement to encroach on the Everglades. Recognizing the danger that human habitation posed, James Ingraham of the Florida East Coast Drainage and Sugar Company called for the preservation of Paradise Key, located in the Royal Palm area of the current Everglades National Park, in 1905. His efforts led Mary Barr Munroe of the Florida Federation of Women's Clubs to join the fight, and she, along with several scientists, including botanists David Fairchild and J. K. Small, advocated the creation of a Paradise Key reserve. Heeding these cries, the state established Royal Palm State Park in 1916.[53]

In the 1920s, Ernest Coe, a landscape architect from Connecticut who had moved to the Miami area, became the loudest voice for Everglades preservation. Coe had always been interested in nature, and he became entranced with the mangroves, the orchids, the giant royal palm trees, and other plants in the Everglades region, as well as the numerous bird rookeries and other wildlife. Coe claimed that these natural attributes justified the creation of a national park to preserve the unique ecology.[54] In promulgating these views, Coe was drawing on the ideas of many conservationists in the late 1800s and early 1900s who believed that the nation's natural wonders should be preserved as national parks for the enjoyment of future generations. Beginning with Yosemite and Yellowstone, Congress set aside vast tracts of land characterized by monumental scenery—huge mountain peaks, pristine vistas, waterfalls, canyons, and geysers—to protect these resources from exploitation and development, and in 1916, it created the National Park Service (NPS) to manage these areas.[55]

By the 1920s, some Americans had decided that national parks could also preserve plant and wildlife as well as scenery. Coe was one of these, and he began agitating for the creation of a national park to protect the ecology of the Everglades. In 1928, he formed the Tropic Everglades National Park Association and persuaded David Fairchild, a botanist with the U.S. Department of Agriculture, to serve as its first president. For the next several months, Coe, with the aid of the association, studied and mapped the area, conducting surveys by plane and boat. He brought his data to U.S. Senator Duncan U. Fletcher, a Democrat from Jacksonville, and in 1929, Fletcher ushered a bill through Congress authorizing an investigation of the Everglades as a possible national park.[56]

In 1930, an NPS committee, consisting of Director Horace Albright, Assistant Director Arno Cammerer, and Yellowstone National Park Superintendent Roger Toll, explored the Everglades on a four-day tour sponsored by the Tropic Everglades National Park Association. At the conclusion of this inquiry, the committee made a favorable report on the park's creation, and in December 1930, the secretary of the interior recommended that Congress establish a park constituting 2,000 square miles in Dade, Monroe, and Collier counties. However, Florida's congressional delegation had a difficult time passing a bill to create the park, mainly because many members of Congress could not understand why preservation of the area was necessary or important.[57]

The task became easier as more evidence mounted of how drainage and a lack of water affected plants and wildlife in the Everglades. In 1929, New York botanist John Kunkel Small had warned of the pending "extermination" of plants and wildlife in the Everglades because drainage facilitated fires that destroyed the soil. "Florida is being drained and burned to such an extent that it will soon become a desert!" he exclaimed.[58] Secretary of the Interior Ray Lyman Wilbur echoed these thoughts in 1933, stating that drainage prevented enough fresh water from reaching the Shark River and other waterways in South Florida, thus destroying "the most unique qualities" of the area.[59] John O'Reilly, a reporter for the *New York Herald Tribune*, also explained how the lack of water affected wildlife, noting that drainage had removed "a single block in the foundation on which the wild beauty and natural abundance of such a region is built." The evidence for this, he claimed, was "in the brown and dying vegetation; in the vast fires that have been eating plants and soil alike; [and] in the wholesale migration of birds and animals from a habitat which has been their home since before history." The solution, O'Reilly believed, was "to get the overflow of Lake Okeechobee directed back onto the Everglades," thereby reestablishing feeding grounds and allowing "thousands upon thousands of White Ibises and other water birds [to] return to their rookeries."[60]

The effects of drought on the land. (The Florida Memory Project, State Library and Archives of Florida)

White ibis. (The Florida Memory Project, State Library and Archives of Florida)

Influenced by these arguments, Congress passed an act in 1934 authorizing the creation of Everglades National Park. Heeding the report submitted by the NPS committee, this law recommended that an area of approximately 2,000 square miles be established as the Everglades National Park as soon as the state was able to transfer title to the lands to the United States.[61] This large area included much of Dade, Monroe, and Collier counties, including what would become known as the East Everglades area and islands in Florida Bay and the southern Gulf of Mexico. According to NPS Director Arno Cammerer, one of the main reasons for the establishment of the park was "so that the wild life may in fact be protected. . . . [T]he only hope the wild life has of surviving is to come under the protective wings of the National Park Service."[62]

Yet one group lost out in this effort to preserve Everglades flora and fauna: the Seminole Indians. The Seminole had originally been part of the Creek Confederacy. After the Yamasee War in the 1710s, a group of Creeks moved into northern Florida. After several years, those Creek that had not relocated began referring to the Florida Creek as *simanó·li*, meaning "wild" or "runaway." This term eventually morphed into "Seminole," the English term for this group. After a series of wars in the first half of the nineteenth century, the

United States removed the Seminole to southern Florida, establishing a reserve for the group in 1849 in Big Cypress Swamp, and most Seminole took up residence in either the swamp or the Everglades. When the Tamiami Trail was built in 1928, some families moved to areas surrounding the highway in order to conduct business with tourists.[63]

In 1917, the state of Florida created a reservation for the Seminole out of 99,000 acres of land in Monroe County. Likewise, in the early 1930s, the federal government consolidated several small areas of land into tracts set aside for the Seminole: Brighton (located to the northwest of Lake Okeechobee), Big Cypress (in the northeastern part of Big Cypress Swamp) and Dania (later called Hollywood, located near the eastern coast just south of Fort Lauderdale). Most Seminole ignored these reservations and continued to live wherever they wanted. Yet problems resulted in 1934 because the state reservation lay within the proposed boundaries of Everglades National Park. To resolve the situation, the state agreed to provide the federal government with the Seminole land in exchange for 104,800 acres in Broward and Palm Beach counties. This land lay north of the Tamiami Trail in the eastern part of Big Cypress Swamp.[64]

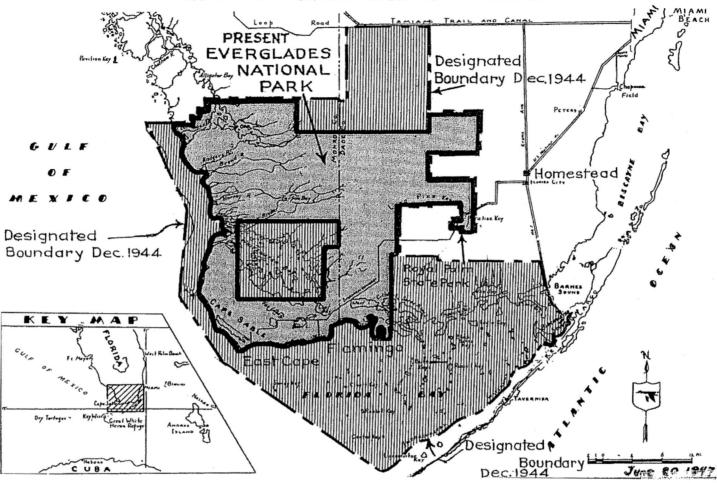

With the Seminole situation resolved, the state of Florida turned to the task of acquiring additional lands for the park, and it passed an enabling act allowing it to convey tracts to the United States as soon as it acquired them. But despite the best efforts of the Everglades National Park Association and the State Everglades National Park Commission (which had been created in 1935 to handle the land purchase and transfer issues), acquisition proceeded slowly.[65] One of the problems was that in the early 1940s oil was discovered in southern Florida, and the state began issuing oil and gas leases on the land it owned within the proposed park boundaries. By 1947, Humble Oil and Refining Company alone had produced 230,701 barrels of oil. This caused consternation among many conservationists; an article in *Natural History*, for example, lamented that "liquid death may ooze up from the bowels of the earth to spread its polluting destruction through the fresh water" and called for immediate action "to make certain that the production of oil entails a minimum

of damage to the numberless natural assets of this exotic wilderness."[66] Despite conservationists' concerns, drilling continued, and the NPS reported in the early 1940s that it "saw no way of establishing a national park for some time, since the area would be constantly subject to pressure for exploring and drilling for oil."[67]

In the meantime, wildlife and plants continued to be destroyed. In 1937 and 1938, Daniel Beard, a wildlife technician for the NPS, traversed the Everglades region and made observations about its flora and fauna and the effects of drainage on them.[68] Beard reported that before drainage began, "the park got the bulk of the western flow and some of the eastern flow that went through the Everglade Keys."[69] After the construction of the drainage canals, water entered the park only from the east. Drainage also lowered the water table, leading to the destruction of gator holes and the abandonment of large bird rookeries. According to A. E. Demaray, acting director of the NPS, Beard's main finding was that "changed water levels are in all probability fundamentally responsible for the depletion of characteristic plants and animals of the proposed park area." Based on these conclusions, Demaray proclaimed that "restoration of water levels is fundamental and must be accomplished if the area becomes a park. . . . Water is the basis for the unique features of southern Florida that make it of national park caliber."[70] The NPS therefore called for another extensive study of how drainage and flood control systems had affected the wildlife.

Meanwhile, the NPS participated in meetings in 1939 about saltwater intrusion and a shortage of drinking water for municipalities in South Florida. Although the conference focused on these issues, NPS representatives emphasized that state and federal interests should not deprive the Everglades of water in order to solve the problems. Continued inadequate water supplies, they stated, would "result in increasing the fire risk, decreasing soil building and destroying wildlife."[71] What was necessary, NPS officials declared, was the "restoration and maintenance of normal water conditions" in order to guarantee the "preservation and restoration of the national park character."[72]

An inspection of the Everglades in 1939 by Clifford C. Presnall, assistant chief of the NPS's Wildlife Division, reiterated the importance of water. Presnall reported that water levels were as much as three feet below normal and that some ditches were completely dry. He believed that "this lowering of the water table would not have been nearly so pronounced had there been no drainage canals." He blamed drainage for causing bird migrations and for decimating tree snail populations, thereby drastically reducing the number of Everglades kites. Drainage had also caused fire to become "unnaturally preponderant." Only the restoration of the "unhampered overflow from lake Okeechobee into the Everglades such as existed before the construction of dikes" would alleviate the situation, Presnall asserted, but he understood that the preponderance of agriculture south of Lake Okeechobee would make such a renewal difficult.[73]

In order to ensure that the animals and plants in the region had at least some form of protection, the state established a state wildlife refuge within the proposed park boundaries. Unfortunately, the designation did little to reduce the destruction, whether by drought or by poaching.[74] Therefore, on 6 December 1944, Congress passed an act allowing the secretary of the interior to accept "submerged land, or interests therein, subject to such reservations of oil, gas, or mineral rights" within the 2,000 square mile boundary, and to protect such land until the federal government could clear the mineral reservations.[75] The state then conveyed to the United States more than 850,000 acres of land within the proposed boundaries. One publication noted that the land consisted of three areas: Florida Bay; a 34-mile long and three-mile wide strip between Cape Sable and Lostman's River; and 400,000 acres from the Shark River to Royal Palm State Park and north to Forty Mile Bend on the Tamiami Trail, a highway constructed in the 1910s and 1920s from Miami to Fort

Myers and Tampa. Some of the lands not included were those in the Big Cypress region, those north of the Tamiami Trail, those located on the upper keys, and those which would become known as the East Everglades.[76] All of the deeded land was designated as the Everglades Wildlife Refuge, and the U.S. Fish and Wildlife Service was given administrative authority over it, with Daniel Beard as manager.[77]

Because of continuing difficulties with acquiring private land and with oil and gas rights, the state agreed in 1947 to the establishment of a "minimum" park, something that would at least get portions of the Everglades protected. This acreage, totaling 454,000 acres and corresponding roughly to the third section deeded to the United States in 1944, became Everglades National Park on 27 June 1947 when Secretary of the Interior J. A. Krug issued Order No. 2338.[78] Both park and state officials regarded this "minimum" park as only the beginning, noting that additional land to total 1,282,000 acres would "ultimately . . . be added to the park."[79] President Harry Truman officially dedicated the park on 6 December 1947, making it the first national park to be established not for its scenery but solely to protect its flora and fauna.[80] According to Acting Secretary of the Interior Warner W. Gardner, the establishment of the park only was a first step in its creation; more acreage would be added as it became available.[81]

August Burghard and Ernest Coe at the dedication of Everglades National Park. (The Florida Memory Project, State Library and Archives of Florida)

Everglades National Park advocates, as well as NPS personnel, were enthusiastic about the park's creation, believing that it was a step in the right direction for the preservation of the unique flora and fauna of southern Florida. However, because it was, in the words of Marjory Stoneman Douglas, "the only national park in which the wild-life, the crocodiles, the trees, the orchids, will be more important than the sheer geology of the country," it was essential that the flora and fauna had sufficient water.[82] Just two days before the creation of the park, NPS officials had reiterated that "this new national park is dependent to a large degree on the conservation and favorable distribution of the surface waters of the lower Everglades drainage basin." Therefore, "the restoration of natural conditions is the first requirement in any plan for bringing back many forms of wildlife which have been reduced to critical numbers." The NPS expressed its interest and concern "with any plans dealing with drainage, storage, and distribution of the waters of the lower Everglades," and believed that it was now an active player in any decisions involving this resource.[83]

In the 100 years following the state's declaration of interest in drainage, southern Florida had undergone vast transformations. Several canals had been built, and rivers flowing out of Lake Okeechobee had been channelized in order to control flooding from the lake and to remove water from the land. Settlement and agriculture had quickly followed the desiccation of land; the lower east coast of Florida's population had increased from 22,961 in 1900 to 228,454 in 1930, while cane sugar production had doubled between 1931 and 1941. Although the state had initiated drainage operations and implemented them for much of the first half of the twentieth century, it ultimately had to turn to the U.S. Army Corps of Engineers for flood control works. Yet all of these structures, whether for drainage or for flood control, had serious consequences for southern Florida's flora and fauna, especially in the Everglades. The federal government created Everglades National Park in 1947 to protect these resources, but the problem of ensuring that the park received adequate water remained. Many, including John H. Baker, executive director of the National Audubon Society, believed that the solution lay in "an intelligent water-control and land-use plan, backed by adequate legislative and administrative authority" and executed by "a qualified hydraulic engineer."[84] Whether one could be developed remained to be seen.

Endnotes

[1] George E. Buker, *Sun, Sand and Water: A History of the Jacksonville District, U.S. Army Corps of Engineers, 1821-1975* (Fort Belvoir, Va.: U.S. Army Corps of Engineers, 1981), 104.

[2] For examples of discussions of this period, see McCally, *The Everglades*; Nelson Manfred Blake, *Land Into Water—Water Into Land: A History of Water Management in Florida* (Tallahassee: University of Florida Presses, 1980); Lamar Johnson, *Beyond the Fourth Generation* (Gainesville: The University Presses of Florida, 1974); Junius Elmore Dovell, "A History of the Everglades of Florida," Ph.D. dissertation, University of North Carolina at Chapel Hill, 1947; and Christopher F. Meindl, "Past Perceptions of the Great American Wetland: Florida's Everglades during the Early Twentieth Century," *Environmental History* 5 (July 2000): 378-395.

[3] Marjory Stoneman Douglas, *The Everglades: River of Grass*, 50th anniversary edition (Sarasota, Fla.: Pineapple Press, 1997), 68.

[4] Douglas, *The Everglades*, 68-70; McCally, *The Everglades*, 39-53.

[5] McCally, *The Everglades*, 53-57, 59-60 (quotation on p. 56); see also Charlton W. Tebeau, *A History of Florida* (Coral Gables, Fla.: University of Miami Press, 1971), 19; Douglas, *The Everglades*, 185-188, 196-245; and Robert H. Keller and Michael F. Turek, *American Indians & National Parks* (Tucson: The University of Arizona Press, 1998), 217.

[6] Grunwald, *The Swamp*, 40-47.

[7] "Resolution by the Legislature of Florida," in Senate, *Everglades of Florida*, 62d Cong., 1st sess., 1911, S. Doc. 89, Serial 6108, 34-35.

[8] "Report of Buckingham Smith, Esq., on His Reconnaissance of the Everglades, 1848," in Senate, *Everglades of Florida*, 46.

[9] "Report of Buckingham Smith," 46-47, 49-50, 53.

[10] McCally, *The Everglades*, 88.

[11] Act of 28 September 1850, in Senate, *Everglades of Florida*, 67.

[12] Quotation in McCally, *The Everglades*, 88; see also "Acts of Florida Legislatures (1851-1855) Relating to the Everglades," in Senate, *Everglades of Florida*, 67-68; and "History of Drainage and Reclamation Work in the Everglades of Florida," in Senate, *Everglades of Florida*, 7.

[13] Douglas, *The Everglades*, 266.

[14] Lieutenant J. C. Ives, *Memoir to Accompany a Military Map of the Peninsula of Florida, South of Tampa Bay* (New York: M. B. Wynkoop, 1856), 5-7.

[15] Grunwald, *The Swamp*, 67.

[16] McCally, *The Everglades*, 89; Grunwald, *The Swamp*, 67-72.

[17] Quotation in Blake, *Land Into Water*, 75; see also McCally, *The Everglades*, 89; Grunwald, *The Swamp*, 85-87.

[18] Quotation in Ann Vileisis, *Discovering the Unknown Landscape: A History of America's Wetlands* (Washington, D.C.: Island Press, 1997), 136; see also Blake, *Land Into Water*, 81-83; McCally, *The Everglades*, 89; and Light and Dineen, "Water Control in the Everglades: A Historical Perspective," 53. The traditional historical accounts of Disston's death label it as a suicide, but more recent publications, including works by Joe Knetsch, a historian with the Florida Division of State Lands, and Michael Grunwald, a *Washington Post* reporter, attribute his death to heart problems, citing the coroner's report and several obituaries at the time. See Joe Knetsch, "Hamilton Disston and the Development of Florida," *Sunland Tribune* 24, no. 1 (1998): 5-19; Grunwald, *The Swamp*, 96.

[19] Blake, *Land Into Water*, 84-87; Grunwald, *The Swamp*, 99-109.

[20] "Report by Dr. H. W. Wiley, of the Bureau of Chemistry, United States Department of Agriculture, in 1891, on the Muck Lands of the Florida Peninsula," in Senate, *Everglades of Florida*, 73-74.

[21] "Message of Gov. W. S. Jennings to the Legislature of Florida Relative to Reclamation of Everglades," in Senate, *Everglades of Florida*, 84.

[22] Samuel P. Hays, *Conservation and the Gospel of Efficiency: The Progressive Conservation Movement, 1890-1920* (Cambridge: Harvard University Press, 1959; reprint, New York: Atheneum, 1969), 2; references are to the reprint edition. Hays' book is one of the best sources on the conservation movement in the late 1800s and early 1900s.

23 Introduction to Napoleon B. Broward, "Homes for Millions: Draining the Everglades," *Collier's* 44 (22 January 1910): 19.

24 Quotation in Blake, *Land Into Water*, 88; see also Meindl, "Past Perceptions of the Great American Wetland," 379; Vileisis, *Discovering the Unknown Landscape*, 113. For information about engineering studies and the Progressive influence, see Jeffrey Glenn Strickland, "The Origins of Everglades Drainage in the Progressive Era: Local, State and Federal Cooperation and Conflict," M.A. thesis, Florida Atlantic University, 1999, 69-81.

25 Quotations in W. Turner Wallis, "The History of Everglades Drainage and Its Present Status," *Soil Science Society of Florida Proceedings*, 4-A (1942): 31-32; see also Dovell, "A History of the Everglades of Florida," 178; Strickland, "The Origins of Everglades Drainage in the Progressive Era," 127; McCally, *The Everglades*, 88; and Blake, *Land Into Water*, 94.

26 For a complete biography of Broward, see Samuel Proctor, *Napoleon Bonaparte Broward: Florida's Fighting Democrat* (Gainesville: University of Florida Press, 1950).

27 Quotation in S. Mays Ball, "Reclaiming the Everglades: Reversing the Far Western Irrigation Problem," *Putnam's Magazine* 7 (April 1910): 798; see also Grunwald, *The Swamp*, 131.

28 N. P. Broward, "Draining the Everglades," *The Independent* 64 (25 June 1908): 1448; Blake, *Land Into Water*, 95-96; Vileisis, *Discovering the Unknown Landscape*, 136-137.

29 "Message of Gov. N. B. Broward to the Legislature of Florida Relative to Reclamation of Everglades," in Senate, *Everglades of Florida*, 99-109.

30 Blake, *Land Into Water*, 97; McCally, *The Everglades*, 92; Grunwald, *The Swamp*, 134-140.

31 Quotation in McCally, *The Everglades*, 93-94; see also Blake, *Land Into Water*, 105; Grunwald, *The Swamp*, 141-142. For more information about drainage under Broward's administration, see Proctor, *Napoleon Bonaparte Broward*, 216-224, 240-260

32 McCally, *The Everglades*, 114.

33 McCally, *The Everglades*, 95-98, 100-102, 109-110; Blake, *Land Into Water*, 107-112, 116-117, 120-121, 127-128; Meindl, "Past Perceptions of the Great American Wetland," 383-384. For information about the Wright Report, see "Report on the Drainage of the Everglades of Florida by J. O. Wright, Supervising Drainage Engineer," in Senate, *Everglades of Florida*, 140-180; and Aaron D. Purcell, "Plumb Lines, Politics, and Projections: The Florida Everglades and the Wright Report Controversy," *The Florida Historical Quarterly* 80 (Fall 2001): 161-197. For the Everglades Land Sales Company report, see Daniel W. Mead, Allen Hazen, and Leonard Metcalf, *Report on the Drainage of the Everglades of Florida* (Chicago: Board of Consulting Engineers, 1912). For the Randolph Report, see Senate, *Florida Everglades: Report of the Florida Everglades Engineering Commission to the Board of Commissioners of the Everglades Drainage District and the Trustees of the Internal Improvement Fund, State of Florida*, 63d Cong., 2d sess., 1913, S. Doc. 379, Serial 6574.

34 Blake, *Land Into Water*, 130-132; McCally, *The Everglades*, 121-125; Howard Sharp, "Farming the Muck Soil of the Everglades," *The Florida Grower* 32 (7 November 1925): 4; John A. Heitmann, "The Beginnings of Big Sugar in Florida, 1920-1945," *The Florida Historical Quarterly* 77 (Summer 1998): 44, 50-54; J. Carlyle Sitterson, *Sugar Country: The Cane Sugar Industry in the South, 1753-1950* (Lexington: Univeristy of Kentucky Press, 1953; reprint, Westport, Conn.: Greenwood Press, 1973), 361-370 (page references are to the reprint edition).

35 Robert Mykle, *Killer 'Cane: The Deadly Hurricane of 1928* (New York: Cooper Square Press, 2002), 34, 36.

36 McCally, *The Everglades*, 135; Blake, *Land Into Water*, 134-136; Grunwald, *The Swamp*, 186-189; Okeechobee Flood Control District, *A Report to the Board of Commissioners of Okeechobee Flood Control District on the Activities of the District and on Lake Okeechobee*, copy in Library, Jacksonville District, U.S. Army Corps of Engineers, Jacksonville, Florida.

37 Everglades Engineering Board of Review, *Report of Everglades Engineering Board of Review to Board of Commissioners of Everglades Drainage District* (Tallahassee, Fla.: T. J. Appleyard, 1927), 5-16, 52.

38 McCally, *The Everglades*, 139-140.

39 Blake, *Land Into Water*, 136; McCally, *The Everglades*, 139; Grunwald, *The Swamp*, 192-194; Edgar Jadwin, Major General, Chief of Engineers, to Hon. Wesley L. Jones, 31 January 1929, in Senate, *Caloosahatchee River and Lake Okeechobee Drainage Areas, Florida*, 70th Cong., 2d sess., 1929, S. Doc. 213, Serial 9000, 2. For a full discussion of the hurricane and its impacts, see Mykle, *Killer 'Cane* and Eliot Kleinberg, *Black Cloud: The Great Florida Hurricane of 1928* (New York: Carroll & Graf Publishers, 2003).

40 Okeechobee Flood Control District, *A Report to the Board of Commissioners of Okeechobee Flood Control District on the Activities of the District and on Lake Okeechobee*, 7-8; see also Senate Committee on Commerce, *Rivers and Harbors: Hearings Before the Committee on Commerce, United States Senate, Part 3*, 71st Cong., 2d sess., 1930, 322-323.

41 For more information on the 1917 act, see Matthew T. Pearcy, "A History of the Ransdell-Humphreys Flood Control Act of 1917," *Louisiana History* 41 (Spring 2000): 133-159.

42 Joseph L. Arnold, *The Evolution of the 1936 Flood Control Act* (Fort Belvoir, Va.: Office of History, United States Army Corps of Engineers, 1988), iii; Martin Reuss, *Designing the Bayous: The Control of Water in the Atchafalaya Basin, 1800-1995* (College Station: Texas A&M University Press, 2004), 121. Because the Flood Control Act of 1917 only applied to the lower Mississippi and the Sacramento river basins, Reuss contends that the 1936 Flood Control Act was "the real beginning of comprehensive federal flood control work." Martin Reuss, "Introduction," *The Flood Control Challenge: Past, Present, and Future*, Howard Rosen and Martin Reuss, eds. (Chicago: Public Works Historical Society, 1988), x-xi.

43 Quotation in Edgar Jadwin, Major General, Chief of Engineers, to The Secretary of War, April 2, 1928, in House, *Caloosahatchee River and Lake Okeechobee Drainage Areas, Florida*, 70th Cong., 1st sess., 1928, H. Doc. 215, Serial 8900, 5; see also McCally, *The Everglades*, 138-139; Blake, *Land Into Water*, 142-143; Grunwald, *The Swamp*, 197-199.

44 Quotations in Jadwin to Jones, 31 January 1929, in Senate, *Caloosahatchee River and Lake Okeechobee Drainage Areas*, 7; see also Blake, *Land Into Water*, 143-144; McCally, *The Everglades*, 139.

45 Quotation in Blake, *Land Into Water*, 146; see also Lytle Brown, Major General, Chief of Engineers, to Hon. Hiram W. Johnson, 15 March 1930, in Senate, *Caloosahatchee River and Lake Okeechobee Drainage Areas, Fla.*, 71st Cong., 2d sess., 1930, Serial 9219, 4-6; Okeechobee Flood Control District, *A Report to the Board of Commissioners of Okeechobee Flood Control District*, 8-9.

46 Quotation in Duncan U. Fletcher to Hon. Hiram W. Johnson, 13 May 1930, in Senate Committee on Commerce, *Rivers and Harbors: Hearings Before the Committee on Commerce, United States Senate, Seventy-First Congress, Second Session, Part 1*, 71st Cong., 2d sess., 1930, 203-205; see also Johnson, *Beyond the Fourth Generation*, 151.

47 Untitled document beginning "Q3: Hoover Dike Design," File Lake O, Box 745, JDAR; Okeechobee Flood Control District, *A Report to the Board of Commissioners of Okeechobee Flood Control District*, 22. The dike itself was not formally dedicated until 12 January 1961, when it was named after President Herbert Hoover. Kleinberg, *Black Cloud*, 198.

48 U.S. Army Corps of Engineers, Jacksonville District, *Central and Southern Florida Project for Flood Control and Other Purposes: Rule Curves*

and Key Operating Criteria, Master Regulation Manual, Volume 2, Part 1 (Jacksonville, Fla.: Department of the Army, Jacksonville District, Corps of Engineers, 1978), D-37A—D37B; U.S. Army Corps of Engineers, Jacksonville District, and South Florida Water Management District, "Central and Southern Florida Project, Comprehensive Everglades Restoration Plan: Topographic Technical Memorandum, Lake Okeechobee Watershed Project," Proj 01.2.6, March 2003, 1, copy at <www.evergladesplan.org> (8 March 2006).

[49] Light and Dineen, "Water Control in the Everglades: A Historical Perspective," 55.

[50] "Florida Fairyland," *Reader's Digest* 28 (June 1936): 32.

[51] Douglas, *The Everglades*, 279.

[52] Quotation in Cesar A. Becerra, "Birth of Everglades National Park," *South Florida History* 25-26 (Fall 1997/Winter 1998): 12. For more information on Bradley's death, see Stuart A. McIver, "Death of a Bird Warden," *South Florida History* 29 (Fall 2001): 20-27; Stuart A. McIver, *Death in the Everglades: The Murder of Guy Bradley, America's First Martyr to Environmentalism* (Gainesville: University Press of Florida, 2003).

[53] Becerra, "Birth of Everglades National Park," 12-14; Vileisis, *Discovering the Unknown Landscape*, 157-158; Grunwald, *The Swamp*, 170-171.

[54] Theodore Pratt, "Papa of the Everglades National Park," *The Saturday Evening Post* 220 (9 August 1947): 46, 49; Grunwald, *The Swamp*, 206-208.

[55] For more information on the development of the national park movement in America, see Alfred Runte, *National Parks: The American Experience*, 3rd edition (Lincoln: University of Nebraska Press, 1997).

[56] Act of 1 March 1929 (45 Stat. 1443); see also Duncan U. Fletcher to Dr. J. H. Paine, 10 July 1934, File Everglades H. R. 2837, Box 903, Entry 7, Record Group [RG] 79, Records of the National Park Service, National Archives and Records Administration II, College Park, Maryland [hereafter referred to as NARA II].

[57] See, for example, Fletcher to Paine, 10 July 1934; David Fairchild, "The Everglades National Park as an Introduction to the Tropics," at Library of Congress, "Reclaiming the Everglades: South Florida's Natural History, 1884-1934" <http://memory.loc.gov/ammem/award98/fmuhtml/everhome.html> (1 December 2004).

[58] John Kunkel Small, *From Eden to Sahara: Florida's Tragedy* (Lancaster, Penn.: The Science Press Printing Company, 1929), 7, 48.

[59] As quoted in "The Proposed Everglades National Park," *Science* 77 (17 February 1933): 185.

[60] John O'Reilly, "Wildlife Protection in South Florida," *Bird-Lore* 41 (May-June 1939): 130, 136, 138.

[61] Act of 30 May 1934 (48 Stat. 816).

[62] Quotation in Arno B. Cammerer, Director, Memorandum for Mr. Poole, Assistant Solicitor, 2 April 1934, File Everglades H.R. 2837, Box 903, Entry 7, RG 79, NARA II; see also Newton B. Drury, Director, to The Secretary, 10 May 1948, File 0-10 Laws and Legal Matters, Box 902, Entry 7, RG 79, NARA II.

[63] Buffalo Tiger and Harry A. Kersey, Jr., *Buffalo Tiger: A Life in the Everglades* (Lincoln: University of Nebraska Press, 2002), 7; William C. Sturtevant and Jessica R. Cattelino, "Florida Seminole and Miccosukee," in *Handbook of North American Indians*, ed. William Sturtevant, vol. 14, *Southeast*, ed. Raymond D. Fogelson (Washington, D.C.: Smithsonian Institution, 2004), 429-438; Keller and Turek, *American Indians and National Parks*, 217. For more information on the early history of the Seminole in the Everglades and Big Cypress Swamp, see Brent Richards Weisman, *Unconquered People: Florida's Seminole and Miccosukee Indians* (Gainesville: University Press of Florida, 1999), 66-89, 123-124.

[64] Sturtevant and Cattelino, "Florida Seminole and Miccosukee," 438; Harry A. Kersey, Jr., "The East Big Cypress Case, 1948-1987: Environmental Politics, Law, and Florida Seminole Tribal Sovereignty," *The Florida Historical Quarterly* 69 (April 1991): 457-458; Patricia R. Wickman, "The History of the Seminole People of Florida," *The Seminole Tribune*, copy at <http://www.semtribe.com/tribune/40anniversary/history.shtml> (5 January 2006).

[65] A. S. Houghton to Hon. Harold L. Ickes, Secretary of the Interior, 15 April 1938, File Everglades H.R. 2837, Box 903, Entry 7, RG 79, NARA II.

[66] Quotation in Kenneth D. Morrison, "Oil in the Everglades," *Natural History* 53 (June 1944): 282; see also Alfred Jackson Hanna and Kathryn Abbey Hanna, *Lake Okeechobee: Wellspring of the Everglades* (Indianapolis, Ind.: The Bobbs-Merrill Company, 1948), 345.

[67] Quotation in Department of the Interior, Information Service, National Park Service, For Immediate Release, n.d., File 714 Fishes, Box 919, Entry 7, RG 79, NARA II; see also Ernest F. Coe, Director, Everglades National Park Assn., Inc., to Honorable Harold L. Ickes, Secretary of the Interior, 21 April 1937, ibid.

[68] A. E. Demaray, Acting Director, to Mr. Abel Wolman, National Resources Committee, 12 June 1939, File E. G. 660-05 Water Supply Systems (Gen.), Box 918, Entry 7, RG 79, NARA II.

[69] Daniel Beard, "Wildlife Reconnaissance: Everglades National Park Project," October 1938, 46, 50, copy provided by Nancy Russell, Museum Curator, Everglades and Dry Tortugas National Parks, Homestead, Florida.

[70] Demaray to Wolman, 12 June 1939.

[71] O. B. Taylor, Regional Wildlife Technician, Memorandum for the Regional Director, Region I, 18 July 1939, File 801-02 Floods Everglades, Box 920, Entry 7, RG 79, NARA II.

[72] J. R. White, Acting Director, National Park Service, to C. G. Paulsen, Acting Chairman, Departmental Committee on Water Resources, 8 August 1939, File E. G. 660-05 Water Supply Systems (Gen.), Box 918, Entry 7, RG 79, NARA II.

[73] Clifford C. Presnall, "Wildlife Report on the Everglades National Park (Proposed)," 1-6; File 207 Paul Bartsch, Box 905, Entry 7, RG 79, NARA II.

[74] James O. Stevenson, Assistant in Charge, Section on National Park Wildlife, Memorandum for Mr. Ben Thompson, National Park Service, 26 June 1942, File 720-04 Everglades, Box 920, Entry 7, RG 79, NARA II.

[75] Act of 6 December 1944 (58 Stat. 794).

[76] Quotation in "The President's Report to You," *Audubon Magazine* 47 (January-February 1945): 46-47; see also Vileisis, *Discovering the Unknown Landscape*, 190-191.

[77] Newton B. Drury, Director, National Park Service, to The Secretary, 10 May 1948, File 0-10 Laws and Legal Matters, Box 902, Entry 7, RG 79, NARA II.

[78] United States Department of the Interior, Title 36—Parks and Forests, Chapter I—National Park Service, Department of the Interior, Order Establishing the Everglades National Park, Florida, in File Everglades National Park—Washington Liaison Office June 1947-July 1947, Box 900, Entry 7, RG 79, NARA II.

[79] "Everglades Becomes 28th National Park," National Park Service Advance Release, 20 June 1947, File Everglades National Park—Washington Liaison Office, June 1947-July 1947, Box 900, Entry 7, RG 79, NARA II.

[80] J. A. Krug, Secretary of the Interior, to Hon. Millard F. Caldwell, Governor of Florida, 2 April 1947, File Everglades National Park—1947, Box Labeled Everglades National Park 1950's, State Lands Records Vault, Division of State Lands, Florida Department of Environmental Protection, Marjory Stoneman Douglas Building, Tallahassee, Florida; Runte, *National Parks*, 108-109. For copies of speeches given during the dedication, see "Superintendent's Monthly Narrative Report for the month of December 1947 for

Everglades National Park," 2 January 194[8], File 207-02.3 Everglades Supt. Report, Box 906, Entry 7, RG 79, NARA II.

[81] Warner W. Gardner, Acting Secretary of the Interior, to Senator Butler, ca. 25 June 1947, File Everglades National Park—Washington Liaison Office June 1947-July 1947, Box 900, Entry 7, RG 79, NARA II. The Royal Palm State Park was included in the 454,000 acres.

[82] Douglas, *The Everglades*, 381.

[83] "Statement of the Interests of The National Park Service, United States Department of the Interior in the Water Resources of the Everglades," 18 June 1947, File Everglades National Park—Washington Liaison Office June 1947-July 1947, Box 900, Entry 7, RG 79, NARA II.

[84] John H. Baker, "Time Is Running Out on the Everglades," *Audubon Magazine* 45 (May-June 1943): 177.

CHAPTER 2   **Federal Intervention:** The Central and Southern Florida
Flood Control Project, 1948

SCHEMES TO DRAIN THE EVERGLADES in the first half of the twentieth century had created problems that few people foresaw, including the destruction of plant and wildlife in South Florida, a textbook example of the law of unintended consequences. Other problems resulted from soil subsidence, saltwater intrusion into freshwater wells, and fires raging in times of drought. The financial difficulties of the state of Florida and the Everglades Drainage District (EDD) precluded any local solutions to these problems. In addition, two hurricanes in 1947 caused devastating floods, destroying millions of dollars of property and cropland. These problems convinced state officials and other Floridians that it was time for drastic action, and they once again turned to the U.S. Army Corps of Engineers for help. The Corps proposed a comprehensive water control plan in 1948 that would curb floods and supply water for urban and agricultural interests, alleviating fires, soil subsidence, saltwater intrusion, and plant and wildlife damage in the process. Congress approved this plan in 1948, thereby creating the Central and Southern Florida Flood Control Project. Even though this project proposed an entire water control plan for Central and South Florida, it still left some people uneasy as to how it would address Florida's valuable fish, wildlife, and plant resources. Floridians generally lauded the establishment of the project, believing that it provided secure protection against future flooding, but there were fissures in this consensus that would eventually become gaping crevices.

By the 1930s, drainage had opened up numerous acres of land in South Florida to agriculture and settlement. But the removal of water had some unintended ecological consequences. For one thing, the muck soil exposed by drainage easily caught fire when it became too dry. These fires generally occurred underneath the surface and

produced heavy amounts of smoke, leading to rather bizarre scenes of trees with obliterated roots but no trunk damage sinking into the earth.[1] Such fires became numerous in the Everglades in the 1920s and 1930s; one periodical reported in 1931 that "there are areas in the glades . . . that have been burning underground for years."[2]

High rates of soil subsidence created other problems. The removal of water from the land oxidized bacteria in plant remains, thereby facilitating the complete decomposition of organic detritus. The subsequent soil loss sometimes amounted to as much as one inch per year. One observer claimed in 1942 that the city of Belle Glade was "six feet farther down than it was 25 years ago" and that Clewiston residents "add a new step to their front stoops every two or three years so they can reach the shrinking ground."[3] Drainage also caused saltwater from the Atlantic Ocean to intrude into freshwater wells because a loss of surface water allowed saltwater to flow into creeks during high tide and permeate the limestone strata underlying the banks. Because of this, by 1938, more than 1,000 wells moved inland by the city of Miami in the 1920s and 1930s had saltwater contamination.[4]

The prevalence of these problems, and the lack of state resources, led federal agencies, especially the Soil Conservation Service of the U.S. Department of Agriculture and the U.S. Geological Survey (USGS), to investigate solutions. The Soil Science Society of Florida, an organization formed by Florida scientists in 1939, aided these agencies in their efforts. At the first meeting of the society in 1939, R. V. Allison, its president and chief of the University of Florida's Everglades Agricultural Experiment Station, discussed soil and water problems. He explained that only in the last few years had scientists adequately understood "the duty of water and its relation to the soil as well as to the plant." Too much drainage had allowed subsidence to devastate soil levels, making it the most pressing soil problem affecting the Everglades. In order to curb subsidence, Allison proposed that a water control program be implemented that would flood uncultivated lands "as much of the year as possible" and consider the water needs of cultivated areas. Because Allison did not know what these needs were, he called for a "careful, exacting study" of "the handling of ground water" by cultivated tracts, "looking to economic plant response on the one hand and the best possible stabilization of the soil body on the other."[5] He also called on federal, state, and local officials to recognize that the Everglades hydrologic unit consisted of three elements: the Kissimmee River, which served as the watershed; Lake Okeechobee, which operated as the storage basin; and the Everglades itself, which was the overflow area.

Others in the Soil Science Society agreed with Allison's assessment. H. A. Bestor, a drainage engineer with the U.S. Sugar Corporation, stated in 1943 that an orderly plan for developing the Everglades needed to emphasize conservation of water over its disposal. Officials should institute "water control planning," Bestor continued, to preserve wildlife, to control soil subsidence and prevent muck burning, to utilize land for agriculture, and to preserve municipal water supplies. "All present land use," he concluded, "is challenged by lack of appreciation that conservation of its organic soils is vitally dependent on water control management."[6]

Meanwhile, the USGS, the Florida Geological Survey, and the Florida State Board of Conservation were conducting inquiries into saltwater intrusion and well contamination in South Florida. In 1939, the cities of Miami, Miami Beach, and Coral Gables, in conjunction with Dade County, entered into an agreement with the USGS to examine surface and well supplies in South Florida in order to receive information about how to prevent "a grave municipal water-supply shortage."[7] USGS scientists, including geologist Garald Parker, investigated the substrata of southern Florida and found that saltwater was entering the Biscayne Aquifer (the only source of fresh ground water in the Miami region) from below. The problem was that drainage had upset the natural balance between salt water and fresh water in the aquifer by lowering the groundwater table. To restore this equilibrium, Parker argued, freshwater tables had to be kept at 2.5 feet above sea level. The main way to ensure this was to build control dams at the mouths of canals draining Miami and its surroundings, and to establish a better water control plan for the region.[8]

Scientists, then, were well aware of the destruction that drainage was wreaking on natural resources, but the general public needed something more accessible to move them to action. Publications in national magazines such as *Collier's* and *Audubon* helped, but the biggest boost came in 1947 when a 57-year-old journalist named Marjory Stoneman Douglas published a book chronicling the destruction of the Everglades. Born in Minnesota in 1890 and raised in Massachusetts, Douglas moved to Miami in 1915 to join the staff of her father's newspaper, the *Miami Herald*. She quickly became involved with the Florida Federation of Women's Clubs, which was one of the area's major promoters of conservation, and focused many of her *Miami Herald* articles on creating a healthy urban environment through zoning, public parks, and tree planting. Douglas also championed the beauty and distinctiveness of the Everglades, and she joined the Tropic Everglades National Park Association soon after its formation. When her friend Hervey Allen, an editor at Rinehart Books, invited her to contribute to his series focusing on American rivers, she readily agreed, deciding to write about the Everglades, which the Seminole had called *pahokee*, meaning grassy waters. Using that word for her inspiration, she published *The Everglades: River of Grass* in the fall of 1947, and it soon became a bestseller.[9]

Using stunning and beautiful prose, Douglas painted a picture of the geological and ecological life of the Everglades, describing how, before drainage, water from Lake Okeechobee spilled over the lake's south rim, combined with rainwater, and became the "river of grass," flowing slowly, almost imperceptibly, southward, giving life to the disparate flora and fauna in the region. Douglas chronicled the different drainage programs that the state had instituted, as well as land development schemes and hurricanes that had influenced the area. Then, in the crowning chapter, she outlined how drainage was killing the Everglades:

> The endless acres of saw grass, brown as an enormous shadow where rain and lake water had once flowed, rustled dry. The birds flew high above them, the ibis, the egret, the heron beating steadily southward along drying watercourses to the last brackish pools. Fires that one night glittered along a narrow horizon the next day, before a racing wind, flashed crackling and roaring across the grassy world and flamed up in rolling columns of yellow smoke like pillars of dirty clouds. . . . But in all the creatures of these solitudes where the Tamiami Trail and the long canals stretched their thin lines, and in the hearts of the Indians, there was a sense of evil abroad, a restlessness, an anxiety that one passing rainfall could not change.[10]

To restore the beauty and natural conditions of the Everglades, Douglas argued, "a single plan of development and water control for the whole area, under the direction of a single engineer and his board" had to be instituted.[11] With that plan, the different water demands of disparate sections in South Florida then could be coordinated, and areas could be developed for water conservation. Ultimately, she concluded, the people of South Florida needed to cooperate with the federal government to develop this project.

Douglas's declaration of the necessity of federal involvement rang true to many Floridians observing the financial and administrative difficulties of the EDD. She referred to the period from 1931 to 1942 as "the era of utter confusion" in South Florida because of the financial straits of the EDD and the lack of a central authority in drainage matters. Florida's 1913 drainage law had authorized the establishment of subdrainage districts with their own taxing powers; when these districts were formed, they developed their "own plan of operation shaped to local desires."[12] By 1948, there were 12 of these districts covering approximately 100,000 acres of land.[13] Moreover, in 1931, the state legislature removed the governor and state officials from the EDD board, replacing them with five local members appointed by the governor. According to EDD engineer Lamar Johnson, this action "completely divorced" the district "from direct Tallahassee control," resulting in "non-payment of taxes, bond litigation, and little funds with which to operate for ten years."[14] Meanwhile, Douglas argued, cattle ranchers and vegetable farmers on lands surrounding Lake Okeechobee wanted the maintenance of a low water

level so that more agricultural land would be available, while residents of Broward and Dade counties desired a high level "to guard their own fields and their drying, over-used, city well-fields." These conditions produced "bad feeling, wrangling and confusion."[15]

After receiving financial help from the Reconstruction Finance Corporation, the EDD addressed some of the soil subsidence and other problems created by drainage, using studies conducted by the Soil Conservation Service and the Soil Science Society of Florida. In the early 1940s, the Soil Conservation Service had discovered that much of the land in southern Florida was unsuitable for agriculture because of an inadequate soil depth. The EDD's board wondered whether these tracts could be used for water conservation and storage, and asked engineers Turner Wallis and Lamar Johnson to work with a Soil Science Society committee to investigate the possibilities. In May 1944, this committee suggested that the EDD use public

Marjory Stoneman Douglas signing copies of *The Everglades: River of Grass*, 1947. (The Florida Memory Project, State Library and Archives of Florida)

lands as water conservation areas in order to improve wildlife and plant habitat and to stop soil subsidence and burning.[16]

Acting on these recommendations, Johnson, who became chief engineer of the EDD in 1946, drew up maps showing three possible water conservation areas in Palm Beach, Broward, and Dade counties, located mostly on acreage already owned by either the IIF trustees, the State Board of Education, the EDD, or the counties. The IIF trustees approved the plan, but the state legislature, influenced by a faction of landowners who wanted the land sold and the proceeds applied to the EDD's debt, mandated that voters in the three counties would have to approve the measure before any conservation areas could be created. "A popular referendum was usually considered a kiss of death at that period of Florida's history," Johnson later explained. "It looked dark for the future of the conservation areas at that point."[17]

In the meantime, settlement and economic development continued to increase in South Florida, especially around Lake Okeechobee and on the east coast. State officials noted in 1948 that there were "great tourist and business cities" along the coastal ridge of southern Florida, while agricultural communities clustered around the lake and on the western and northern side of the Kissimmee River Basin.[18] Cattle ranches and dairies proliferated in the Kissimmee and St. Johns basins; one estimate placed the number of cattle in these areas at 410,000 head. In addition, numerous farmers raised truck crops on the drained soil south of Lake Okeechobee, including beans, tomatoes, eggplant, cabbage, potatoes, and celery. The Corps reported that 160,000 acres in the Kissimmee River Basin and south of Lake Okeechobee were planted to truck crops in the 1945-1946 growing season, producing $67 million in vegetables. Citrus farms were also important, located from the Kissimmee Basin to Davie, southwest of Fort Lauderdale; approximately 268,000 acres of citrus groves existed in 1948. But cane sugar was the most significant crop in the Everglades. In 1934, Congress had passed the Sugar Act, which had divided cane sugar production into different quotas, thereby boosting prices. With this help, the U.S. Sugar Corporation and other companies planted 32,000 acres to cane, raising 873,000 tons of sugar in 1941. According to A. G. Matthews, chief engineer for the State of Florida's Division of Water Surveys and Research, the Everglades produced 2,330,232 tons of citrus fruits and vegetables from 1944 to 1946, as well as $11,764,000 worth of sugar and 120,000 head of beef.[19]

The high production of agricultural crops and the rising number of people living around Lake Okeechobee and along the east coast meant that any kind of storm similar to the 1926 and 1928 hurricanes would have a devastating impact on South Florida. But because of the levee that the Corps had built around Lake Okeechobee, and because of the existing drainage works, settlers felt secure from flooding, a feeling reinforced after drought hit the region in 1944 and 1945. According to Lamar Johnson, "the Everglades vegetation

and soil burned for months and the acrid smoke over Miami did not inspire the same sentimental emotions that the moon over that city does."[20] The Corps reported that between 1943 and 1946, "cattle died in the pastures of the Kissimmee Valley for lack of water; smoke from burning muck lands of the Everglades darkened the coastal cities; and salt water moved inland along drainage canals and through the underlying rock."[21]

But in the first months of 1947, rain began falling on the Everglades in large amounts. On 1 March, a storm dropped six inches of rain, while April and May also saw above average totals. The situation became severe in the summer, the apex of Florida's traditional rainy season (which usually lasts from June through October). As September approached and the rains continued, the ground in the Everglades became waterlogged and lake levels reached dangerous heights. Then, on 17 September, a hurricane hit Florida on the southwest coast, passing Lake Okeechobee on the west and dumping large amounts of rain on the upper Everglades, flooding most of the agricultural land south of Lake Okeechobee.[22] George Wedgworth, who would later become president of the Sugar Cane Growers Cooperative of Florida and whose parents were vegetable growers in the Everglades, related that his mother called him during the storm and told him, "This is the last call I'll make from this telephone because I'm leaving. . . . [W]e've got an inch or two of water over our oak floors and they're taking me out on a row boat."[23] Such conditions were prevalent throughout the region.

Before the area had a chance to recover from the devastation, another hurricane developed, moving into South Florida and the Atlantic Ocean by way of Fort Lauderdale. The amount of rainfall was

Damage caused by the 1947 hurricane. (The Florida Memory Project, State Library and Archives of Florida)

not as severe in the upper Everglades, but coastal cities received rain in large quantities, including six inches in two hours at Hialeah and nearly 15 inches at Fort Lauderdale in less than 24 hours. The EDD, under the direction of Johnson, kept its drainage canals open to discharge to the ocean as much of the floodwater in the agricultural area as it could, exacerbating coastal flooding. East coast residents charged Johnson with endangering their lives in order to please agricultural interests, but Johnson vehemently denied this, explaining that "the entire Everglades was flooded several feet and the flood was moving southward and eastward"—coastal cities would have been inundated regardless of the output from the drainage canals.[24]

Whoever was to blame, the hurricanes had devastating effects. Although the levee around Lake Okeechobee held, preventing the large numbers of deaths that occurred in 1926 and 1928, over 2,000 square miles of land south of the lake was covered by, in the words of U.S. Senator Spessard Holland, "an endless sheet of water anywhere from 6 to 7 feet deep down to a lesser depth." The Corps estimated that the storms caused $59 million in property damage throughout southern Florida, but Holland believed that the agency had "understated the actual figures."[25] The destruction shocked citizens of South Florida, both in the upper Everglades and in the coastal cities, and they demanded that something be done.[26]

Acting on these concerns were Florida's two U.S. senators, Holland (who became a member of the Senate Public Works Committee in 1947) and Claude Pepper. Both were Democrats; Holland had

Senator Claude Pepper. (The Florida Memory Project, State Library and Archives of Florida)

served as governor of Florida from 1941 to 1945, while Pepper had served in the Senate since 1936. After the September and October hurricanes, the two were inundated with resolutions and pleas from residents and city and county governments requesting more stringent water control. The Soroptimist Club of the Palm Beaches, for example, informed Holland that "it is clearer than ever that an overall Glades water control plan must be established" in South Florida.[27] The city commission of the city of Stuart was more specific, asking that a water plan provide not only for the control of flooding, but also "include conservation of fresh water."[28]

At the time, Corps leaders were already investigating flood control measures south of Lake Okeechobee, in part because of the flooding that occurred during the spring rains in 1947. With so many South Floridians in disparate parts of the region calling for flood control, however, Pepper became convinced that the Corps needed to expand its efforts. "The time has come when we have got to deal with the flood situation in the Peninsula of Florida, as a whole," he informed Lieutenant General R. A. Wheeler, Chief of Engineers, in October 1947. "It is all fundamentally one single problem and has got to be approached as a single problem with a single comprehensive program." Pepper therefore requested that the Corps "take steps . . . to formulate plans for a comprehensive flood-control program for the whole flooded area."[29] Wheeler agreed with Pepper, explaining that he had already implemented measures to begin "a comprehensive study of the entire flood problem of south Florida from the headwaters of the Kissimmee River to points south of Miami." The "urgent need" for a solution to the

Senator Spessard Holland. (The Florida Memory Project, State Library and Archives of Florida)

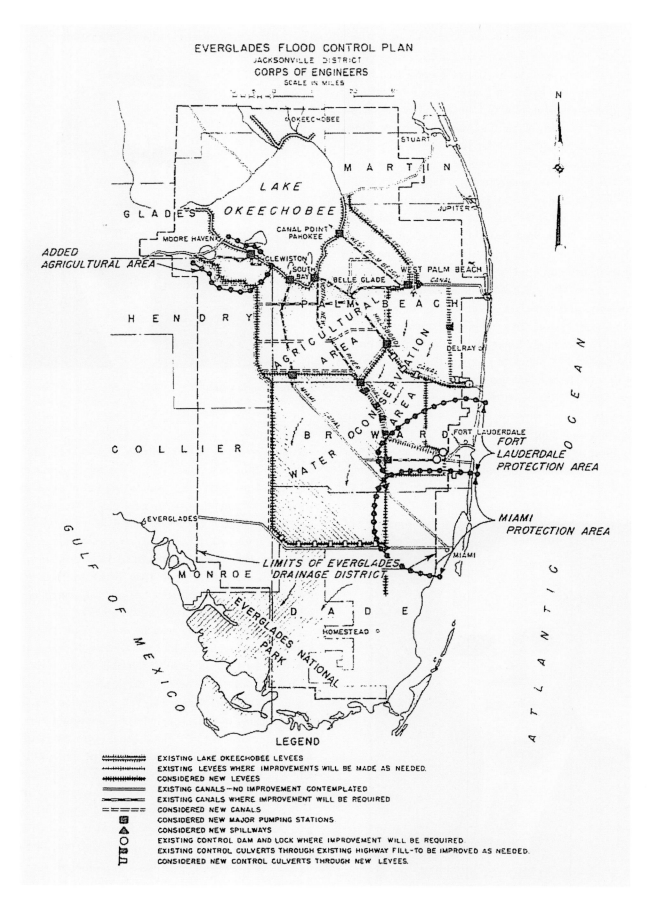

EVERGLADES FLOOD CONTROL PLAN
JACKSONVILLE DISTRICT
CORPS OF ENGINEERS
SCALE IN MILES

LEGEND

||||||| EXISTING LAKE OKEECHOBEE LEVEES
||||||| EXISTING LEVEES WHERE IMPROVEMENTS WILL BE MADE AS NEEDED.
||||||| CONSIDERED NEW LEVEES
===== EXISTING CANALS—NO IMPROVEMENT CONTEMPLATED
===== EXISTING CANALS WHERE IMPROVEMENT WILL BE REQUIRED
===== CONSIDERED NEW CANALS
⊞ CONSIDERED NEW MAJOR PUMPING STATIONS
▲ CONSIDERED NEW SPILLWAYS
○ EXISTING CONTROL DAM AND LOCK WHERE IMPROVEMENT WILL BE REQUIRED.
▯ EXISTING CONTROL CULVERTS THROUGH EXISTING HIGHWAY FILL-TO BE IMPROVED AS NEEDED.
▮ CONSIDERED NEW CONTROL CULVERTS THROUGH NEW LEVEES.

Map showing the Jacksonville District's initial comprehensive proposal, 1947. (Claude Pepper Collection, Claude Pepper Library, Florida State University, Tallahassee, Florida)

flood control problems, Wheeler noted, meant that the Jacksonville District would devote much of its resources to complete an over-all study and submit it to Congress "early in the coming calendar year." Wheeler told Pepper that the Corps already had enough congressional authorizations for "examinations, surveys and reports on individual streams and canals" to allow it to conduct a comprehensive study without additional legislation, meaning that the Corps could proceed immediately.[30]

In its preparation of the flood control plan, the Jacksonville District, led by District Engineer Colonel Willis Teale, held public hearings where local agencies and the general public relayed their wants and needs. Most of the comments in these meetings echoed Pepper's claims that uncoordinated local efforts in the past had failed to solve any of the region's water problems and that a comprehensive plan was "the only possible solution."[31] Listening to these concerns, Teale and the Jacksonville District formulated a plan recognizing that flood protection, drainage, and water control were all interrelated problems in South Florida. According to a Corps press release, the program

> contemplates the protection of 1,000 square miles of rich agricultural muck land immediately south of Lake Okeechobee, improvement of water control in large conservation areas outlined by local interests, and providing the coastal cities with protection from floodwaters from the Everglades by impounding such waters within water conservation areas, encircled by levees large enough to retain all of the water entering them during a period of severe rainfall such as has been experienced this year.

Teale explained that the Corps had developed this plan through ongoing field surveys and office studies, as well as through consultations with "various federal, state and local interests."[32] These included officers of the U.S. Sugar Corporation, the Palm Beach County Water Resources Board, Osceola Groves, and the Florida Division of Water Surveys and Research.[33]

Another important resource was the Soil Conservation Service. Because of the soil subsidence and muckburning problems in the Everglades, the Service had conducted numerous surveys, including those involving topography, subsurface rock strata, soil classification, and hydraulics. The studies, which were ongoing, had convinced Service officials that much of the Everglades could be "soundly developed for agricultural use," although specific areas, as explained above, were more fitting for water conservation.[34]

Essentially, Teale took the Corps' own studies of the Everglades and Lake Okeechobee and coupled them with the Soil Conservation Service's report in order to develop a flood control and water supply program that proposed to solve all of South Florida's flooding, saltwater intrusion, and soil subsidence concerns. On 2 November

1947—only a couple of weeks after the second hurricane had hit Florida and only five days after Pepper had requested a comprehensive plan—the Jacksonville District issued a press release delineating its preliminary proposal. This included the Corps' plan for the Everglades, as well as flood control structures within the Kissimmee and Upper St. Johns river basins (projects that were still tentative pending further District studies). In an innovative manner, given that the study of ecosystems and ecology had still not gained a wide audience in the United States, the District, influenced by the ideas of personnel in the Soil Conservation Service, declared that it would treat the whole area, from the Kissimmee headwaters to south of Miami, as "one watershed," or, essentially, as one ecosystem.[35]

The press release left no doubt that, despite some attempts to control soil subsidence and saltwater intrusion, the proposal was primarily a flood control plan that would protect the east coast and allow for "a sound program of expansion of agricultural activities." Yet the Corps also promised "improvement of conditions favorable to the propagation and maintenance of fish and wildlife within the conservation areas." Accordingly, the plan provided for the construction of levees and canals protecting and draining a 1,000 square mile area "suitable for long-term agricultural use," as well as structures discharging floodwater into water conservation areas for the protection of coastal cities such as Miami and West Palm Beach. The Corps still had to conduct economic studies of the plan, the press release explained, as well as more intensive surveys of the Central Florida region, but the core of the program was in place.[36]

Over the next several weeks, Teale, Holland, and Pepper held several public hearings with local interests to hear their comments about the plan. Although some flood control districts wanted an additional control canal to extend from Lake Okeechobee to alleviate high waters, few interests, if any, expressed any anxiety about the plan's effects on Everglades National Park.[37] Instead, most merely wanted something in place to safeguard South Florida from future floods. Pepper and Holland received numerous statements supporting the proposed program, and promised to keep in close contact with the Corps throughout the plan's preparation.[38]

Based on this feedback, Teale revised the tentative plan and issued the Jacksonville District's final report in December. Although not significantly different from the program delineated in the November press release, especially in its focus on flood prevention (which, of course, was what most Floridians wanted), the December version was more complex, delineating measures to relieve saltwater intrusion and water supply problems. Teale noted that the program would be executed over a wide area in Central and South Florida, including the Upper St. Johns River, the Kissimmee River Basin, Lake Okeechobee and its outlets, the Everglades itself (defined as a 40-mile-wide "grassy marsh" extending 100 miles from Lake Okeechobee to the South Florida coast), and coastal areas in Palm Beach, Broward, and Dade counties.[39]

According to Teale, who, again, was influenced by studies conducted by the Soil Conservation Service, the general problem affecting those areas was that drainage had "altered the natural balance between water and soil," causing "parched prairies and burning mucklands," saltwater intrusion, and flooding. A restoration of the "natural balance between soil and water" was necessary, and this could be accomplished through three means: flood control, water control, and water conservation.[40] Water conservation was especially key because the development of storage areas could prevent flooding and secure a more reliable supply for municipalities, agriculture, and the wildlife and plants within the Everglades. As later explained by Chief of Engineers Wheeler, the plan, which was for "flood protection, water control, and allied purposes," would eliminate flooding by removing water in wet seasons and storing it for use during dry periods. It would also control water levels to benefit agriculture and municipal water supplies.[41]

Recognizing that Everglades National Park had only been established the year before, Teale and the Jacksonville District also proclaimed that the "preservation of fish and wildlife" was an important element of the plan. Teale noted that South Florida had served in the past as "one of the greatest natural habitats for fish, birds, and game on the North American Continent," yet now, many of these species were "virtually extinct." The Corps had therefore consulted with the U.S. Fish and Wildlife Service to allow "full consideration" of fish and wildlife objectives in the comprehensive plan.[42]

In fulfill the various desired objectives, Teale made recommendations for each area covered by the project. For the Kissimmee River Basin, Teale proposed that the Corps turn several lakes into storage basins for flood control, conservation, and water supply, building levees and control structures around them. The Kissimmee River itself would also be enlarged. Teale suggested that projects be commenced in the Lake Okeechobee/Everglades area, including enlarging the St. Lucie Canal and the Caloosahatchee River for better water control, and extending the levee around the lake from the St. Lucie Canal northward, tying it into the north shore levee. He proposed that another levee be built on the northwestern shore of the lake, and possibly another outlet canal as well. To provide flood protection to agricultural lands in the upper Everglades, Teale recommended the construction of a levee around the 1,027-square-mile region, as well as "a canal network connected to eight pumping stations on the perimeter of the system."[43]

Following the EDD's suggestion, Teale also proposed that large parts of the Everglades be held as three water conservation areas, totaling 1,500 square miles in Dade, Broward, and Palm Beach counties.[44] As recommended by Teale, the conservation areas would be larger than those outlined by Johnson, but they would serve important functions. The pumping stations proposed for the agricultural areas, for example, could divert water into the conservation areas in times of excess rain, and could extract water in the same way during drought. Impounding water in the conservation areas would also prevent flooding in coastal cities, and the stored water could be used to "raise the ground-water table and improve water supply for the east-coast communities, ameliorate salt-water intrusion in the east-coast water supply well fields and streams, and benefit fish and wildlife in the Everglades." Teale proposed that the conservation areas be created by building levees from the West Palm Beach Canal south to the Tamiami Trail. The levee system would then follow the Tamiami Trail westward to the Collier County line, and then turn north where it would tie into the west rim levee blocking off the agricultural area south of Lake Okeechobee. Other levees would be built along the Hillsboro and North New River canals to divide the conservation areas into three sections.[45]

There were other features to the program, such as improving existing canals and building control structures on waterways within Dade County for flood control and to prevent saltwater intrusion, but these were the essential features of the Corps' plan. Yet the proposal was vague on how it would allow for fish and wildlife preservation, even though the Corps considered this an "important feature" of the project. Outside of the conservation areas, which would allow for the protection of "large parts of the Everglades" and the "preservation of wildlife," the plan offered no firm proposals for how the project could benefit fish and wildlife. Regardless, Teale estimated that the program would cost $208 million, with an annual operation and maintenance charge of $3.7 million. He recommended that local interests pay $29 million of the total cost, including 15 percent of all construction charges, and that the state establish an agency to coordinate the program locally.[46]

On 31 December 1947, South Atlantic Division Engineer Colonel Mason J. Young concurred with Teale's report, although he admitted that "since construction of the comprehensive project will take place over an extended period, many features of the plan will require further detailed study prior to the initiation of construction." He also emphasized that "if the coastal and Everglades sections of south Florida are to continue to prosper and develop, conservation of their water resources is as important and urgent" as flood control and drainage. Young foresaw increasing demands on water in South Florida, and he insisted that the Corps make adequate provision in the planning process to store water "to the maximum practicable limit."[47]

After gaining Young's approval, the Board of Engineers for Rivers and Harbors reviewed the report. During its deliberations, which included a public hearing on the plan, the board encountered some opposition; a few interests, such as the U.S. Sugar Corporation and the EDD, criticized parts of the plan. The board of commissioners of the EDD, for example, complained about the size of the water conservation areas, fearing that landowners in Palm Beach, Broward, and Dade counties would object to the impoundment of so much land. However, the EDD emphasized that it endorsed the program as a whole, believing that it was a sound plan for water management.[48]

Others voiced concern that the project would take too long to provide flood protection. These feelings were heightened in the first months of 1948 because of the continued saturation of the ground in South Florida and the high levels of Lake Okeechobee. To alleviate some of these concerns, the Board of Engineers recommended that initial construction begin with those structures that would protect the coastal cities and the agricultural area south of Lake Okeechobee, as well as whatever works were necessary to control the level of the lake. However, the board also suggested that the Jacksonville District examine the plan prior to construction "in cooperation with a responsible local or State agency" and make any feasible changes that did not "adversely [a]ffect the principal objectives of the plan."[49]

With the Board of Engineers' approval, the report went to Chief of Engineers Wheeler. Characterizing the plan as providing the works necessary "to prevent a repetition of recent destructive flooding" and "to stabilize the present agricultural economy of the region," Wheeler endorsed the project and recommended that it be presented to Congress.[50] He also suggested that Congress provide an appropriation of $70 million so that the Corps could begin the first phase.

Although the $208 million total cost of the project and the initial $70 million appropriation was a considerable sum of money, especially in the 1940s, it was not as much as the federal government had spent on other projects. In 1928, for example, Congress authorized $325 million for the Corps to conduct flood control efforts on the Mississippi River from Cairo, Illinois, south to the Gulf of Mexico. This was a much larger region than the South Florida flood control project would cover, but it still was a significant expense, especially considering that the total federal budget in 1930 only was $3.3 billion. According to historian Martin Reuss, "no other water project involved as great a percentage of the federal budget at the time of its authorization as did Mississippi valley flood control."[51] In comparison, the Corps asked Congress for less money for South Florida flood control, although, admittedly, the area was a great deal smaller and was confined to one state.

Before Congress received the report, the U.S. Department of the Interior submitted its comments on the plan. Assistant Secretary of the Interior William Warne explained that coordination with the Corps was essential because the project would affect the work of several Interior agencies, including the National Park Service (NPS), the Bureau of Indian Affairs, the USGS, and the U.S. Fish and Wildlife Service (FWS). Perhaps recognizing the plan's vagueness regarding fish and wildlife propagation, Warne noted that the NPS was especially concerned about how the project would affect Everglades National Park and its resources because the park was formed specifically to preserve flora and fauna. Because the park's dedication had occurred only recently, the NPS had not conducted any studies on the possible effects, whether beneficial or harmful, of the proposed plan. Its major worry, Warne explicated, was whether the Corps could guarantee an adequate water supply for the park, especially to prevent saltwater en-

80th Congress, 2d Session · - - - - House Document No. 643

COMPREHENSIVE REPORT ON CENTRAL AND SOUTHERN FLORIDA FOR FLOOD CONTROL AND OTHER PURPOSES

LETTER

FROM

THE SECRETARY OF THE ARMY

TRANSMITTING

A LETTER FROM THE CHIEF OF ENGINEERS, UNITED STATES ARMY, DATED FEBRUARY 19, 1948, SUBMITTING A REPORT, TOGETHER WITH ACCOMPANYING PAPERS AND ILLUSTRATIONS, ON A PRELIMINARY EXAMINATION AND SURVEY OF, AND A REVIEW OF REPORTS ON, RIVERS, LAKES, AND CANALS OF CENTRAL AND SOUTHERN FLORIDA FOR FLOOD CONTROL AND OTHER PURPOSES, MADE PURSUANT TO CONGRESSIONAL AUTHORIZATIONS

May 6, 1948.—Referred to the Committee on Public Works and ordered to be printed, with five illustrations

UNITED STATES
GOVERNMENT PRINTING OFFICE
WASHINGTON : 1949

90243

Cover of House Document 643.

croachment and "disastrous fires" in "the hazardous season between October and May." In its proposal, Warne explained, the Corps never discussed "what definite regulations would be promulgated to insure the release of such waters," nor did it outline what specific structures were needed to facilitate water releases to the park. Warne also recommended that park and Corps officials develop "the details of the plan" to guarantee that the park's "unique" ecological resources were preserved in their "natural state."[52]

The preliminary nature of data also tempered the FWS's overall commendation of the project. Its main conclusion was that if the project truly provided "adequate restoration and control of water levels in a large part of the Everglades," it would "generally improve" fish and wildlife conditions, especially if state or federal authorities operated the water conservation areas for fish and wildlife benefits.[53] But Warne emphasized that "loss of certain unique wildlife habitats" would result as well. The Corps' overall proposal "considers fish and wildlife as adequately as it can in light of the preliminary nature of the Service's findings," Warne explained, but the FWS still needed to coordinate closely with the Corps throughout project planning "to insure minimum damage to, and maximum benefits for, wildlife resources."[54]

The Corps responded to these concerns by assuring the Interior Department that it would remain in close contact with the pertinent agencies. Yet many Floridians were more concerned about receiving adequate flood control than they were with fish and wildlife issues, especially because saturated land and high water conditions in the spring of 1948 raised the specter of more flooding. According to Senator Holland, these conditions forced South Florida residents "to look ahead to next fall with great apprehension," leading Holland to place all of his efforts on getting the Corps project passed. "I shall continue to do everything in my power to get it enacted with the greatest possible speed," he declared, "and then to get the large appropriations which are required so that work can be started."[55]

Such assurances were comforting to South Florida residents, but some still decided to take matters into their own hands. One way that they tried to foster support for flood control was by putting together a book of photographs of the 1947 flood, a proposal first floated to Claude Pepper by the Atlantic and Gulf Canals Association, Inc. Fearing that Congress would not approve the necessary appropriations for the comprehensive program, the association recommended that it compile 150 photographs of flood conditions and publish "a booklet containing news stories from over the 11 counties with illustrations" that could be given to Florida's congressional delegation, the Corps, and "each member of the Congress."[56] The Palm Beach County Resources Development Board, the EDD, and the counties of Palm Beach, Broward, and Dade brought this idea to fruition, issuing a book that included a startling front-cover picture of a crying cow standing shoulder deep in water. The document soon became known as the "Weeping Cow" book, and, according to Lamar Johnson, it was "an indication of the concerted effort of the citizens of the area to promote the flood control project."[57]

Acting on these sentiments, and having remained in close contact with Corps officials as Teale's report made its way through the necessary channels, Holland and Pepper introduced a bill into the U.S. Senate in May 1948 to authorize the comprehensive water control project. The bill was referred to the Subcommittee on Flood Control and Improvement of Rivers and Harbors of the Committee on Public Works, and on 12 May 1948, the subcommittee began hearings. In order to expedite the authorization of the project, Florida's delegation presented a united front during the meetings, with Holland largely orchestrating the testimony that was presented. "The delegation is standing entirely together on this," Holland related. "Even those from the districts not directly affected are famil-

iar with the plight which is the plight of the State, and, we think, of the Nation."[58] Dwight L. Rogers, one of Florida's representatives to Congress, agreed. "There is absolutely no dissension," he declared. "We are all united, State, Federal, and everyone else down there."[59]

The testimony in the hearings almost solely focused on the flood control and water supply benefits were of the project. Almost all of the witnesses discussed the devastating damage of the 1947 flood and the necessity of preventing such a disaster from happening again. Moreover, the agricultural production of the region was emphasized repeatedly in order to convince senators that protection was necessary. There was little mention of the effects of the project on the South Florida ecosystem, outside of declarations that the water conservation areas would provide benefits to fish and wildlife. The only person in the hearings speaking solely as a representative of plant and wildlife interests was Eustace L. Adams, who represented the Dade County Conservation Council and the Florida Wildlife

TENTATIVE REPORT OF

FLOOD DAMAGE

FLORIDA EVERGLADES DRAINAGE DISTRICT

1947

Cover of the "Weeping Cow" book. (South Florida Water Management District)

Federation; no officials from Everglades National Park, the FWS, or even the Florida Game and Fresh Water Fish Commission testified. It is unclear why this oversight occurred, but District Engineer Colonel R. W. Pearson of the Jacksonville District later claimed that it stemmed from the lethargy of the interested agencies. He accused the Florida Game and Fresh Water Fish Commission, for example, of evincing "a considerable lack of interest in the project" during these formative stages.[60]

Regardless, the strong united front presented by Florida's congressional delegation convinced the Senate to include the project in its Flood Control Act of 1948, naming it the Central and Southern Florida Flood Control Project (C&SF Project). After some wrangling in the House of Representatives over the appropriation amount, the House passed the bill and President Truman signed it on 30 June, thereby authorizing $70 million to be expended on the first phase of the project.[61] This initial segment would include building levees and other flood control works to protect the east coast communities from flooding, and constructing structures to control Lake Okeechobee levels and to protect agriculture south of Lake Okeechobee. With the legislation passed, the next step was for the state to find a way to raise its share of the construction cost and to determine what local agency would cooperate with the Corps in the building and operation of the project. "It certainly is a source of joy to me that we have made a constructive start on the flood control program," Holland reported, "and I hope that we may continue to work with the complete unity which has manifested itself so far."[62]

The cooperation between the Corps, local and state agencies, and the federal government throughout the preparation of the flood control plan was remarkable, especially when compared to the development of a $325 million flood control project in 1927 and 1928 for the Mississippi River. That process was marked by political wrangling, jurisdictional disputes, and discord between Congress and President Calvin Coolidge. The development of South Florida's plan was not nearly as contentious for several reasons. For one, the Florida project involved only one state, rather than the multiple entities crossed by the Mississippi River. Among other things, this meant that Florida's plan did not receive the kind of national attention that the Mississippi River project garnered. For another, the overwhelming desire of most Floridians for some form of immediate flood protection necessitated that the Corps use all of the resources available to it in order to piece together a plan that could be passed as quickly as possible. Finally, Floridians were willing to pay part of the cost of the plan as necessitated by Congress, whereas local interests around the Mississippi River were more reluctant to share any costs.[63]

With the authorization of the C&SF Project, the state of Florida finally had a program that promised to eliminate the flood and water supply problems of South Florida. Because of the imbalance of water that drainage created, the region faced either too much water, as evidenced by the flood of 1947, or too little of the resource, as shown by the fires, soil subsidence, and saltwater intrusion problems that plagued the area. To resolve these issues, the Corps developed a plan that would prevent flooding in coastal cities and in the agricultural land south of Lake Okeechobee, while also providing conservation areas for water storage and fish and wildlife habitat. With almost universal approval in Florida, the plan seemed to be the solution to South Florida's water woes and the mechanism by which increased settlement in the area could occur.

Yet there were slight discolorations in this seemingly beautiful picture, blotches that in time would stain the entire canvas. It was clear both from the Corps' proposal and from testimony before Congress that, although fish and wildlife preservation was regarded as an "important feature" of the project, flood control and water supply were the biggest concerns of most Floridians. The U.S. Department of the Interior, the NPS, and the FWS all claimed that fish and wildlife preservation had to take a prominent position in the project's operation, but the vagueness of the plan on how it would aid fish and wildlife, coupled with the clamor for flood protection and water supply, virtually guaranteed that fish and wildlife interests would take a backseat. This made Interior officials nervous about the project, but the looming fear of flooding felt by most Floridians steamrolled these concerns and created a groundswell of support for the project that Congress could not ignore. Even Marjory Stoneman Douglas, who had decried the destruction of the Everglades, believed that the Corps was on the right track. Because of the project, she wrote, "the rich earth will be saved" and "the vast supply of wonderful water will be controlled and used to their utmost needs by the people of Florida and their unborn generations to come."[64] The ensuing decades would, in some fashion, fulfill her prediction, but, in the eyes of many critics, only by manipulating and damaging the already-imperiled and over-engineered flora and fauna of the Everglades.

Endnotes

[1] McCally, *The Everglades*, 142.

[2] John Chapman Hilder, "America's Last Wilderness: The Florida Everglades Are Going Dry," *World's Work* 60 (February 1931): 56.

[3] Robert McCormick, "Lavish Land," *Collier's* 110 (8 August 1942): 66.

[4] McCally, *The Everglades*, 145-146.

[5] Dr. R. V. Allison, "The Soil and Water Conservation Problem in the Everglades," *Soil Science Society of Florida Proceedings* 1 (1939): 35-42; McCally, *The Everglades*, 147. This was not the first time that Allison had advocated these ideas. He had declared as early as 1928 that overdrainage caused more problems than underdrainage, and had told the Florida State Horticultural Society that year that muck fires could be stopped by storing water on undrained areas. Dovell, "A History of the Everglades of Florida," 559-560.

[6] H. A. Bestor, "Reclamation Problems of Sub-Drainage Districts Adjacent to Lake Okeechobee," *Soil Science Society of Florida Proceedings* 5-A (1943): 164.

[7] Florida State Board of Conservation, *Fourth Biennial Report, Biennium Ending December 31, 1940* (Tallahassee: Florida State Board of Conservation, 1941), 55; see also McCally, *The Everglades*, 146.

[8] Garald G. Parker, G. E. Ferguson, S. K. Love, et al., *Water Resources of Southeastern Florida with Special Reference to the Geology and Ground Water of the Miami Area*, U.S. Geological Survey Water Supply Paper 1255 (Washington, D.C.: Government Printing Office, 1955); McCally, *The Everglades*, 146; Grunwald, *The Swamp*, 203-204.

[9] Jack E. Davis, "'Conservation is Now a Dead Word': Marjory Stoneman Douglas and the Transformation of American Environmentalism," *Environmental History* 8 (January 2003): 53-61. For an excellent biography of Douglas, see Jack E. Davis, *An Everglades Providence: Marjory Stoneman Douglas and the American Environmental Century* (Athens: The University of Georgia Press, 2009).

[10] Douglas, *The Everglades*, 349-350.

[11] Douglas, *The Everglades*, 383.

[12] Marjory Stoneman Douglas, "What Are They Doing To The Everglades?" 8, at Library of Congress, "Reclaiming the Everglades: South Florida's Natural History, 1884-1934" <http://memory.loc.gov/ammem/award98/fmuhtml/everhome.html> (3 December 2004).

[13] House, *Comprehensive Report on Central and Southern Florida for Flood Control and Other Purposes*, 80th Cong., 2d sess., 1948, H. Doc. 643, Serial 11243, 30.

[14] Johnson, *Beyond the Fourth Generation*, 85.

[15] Douglas, "What Are They Doing To The Everglades?" 8.

[16] Johnson, *Beyond the Fourth Generation*, 174-175; Blake, *Land Into Water*, 175.

[17] Quotation in Johnson, *Beyond the Fourth Generation*, 155-256, 177-179; see also Blake, *Land Into Water*, 175-176.

[18] A. G. Matthews, Chief Engineer, to Hon. George W. Malone, 19 May 1948, in Senate Committee on Public Works Subcommittee, *Rivers and Harbors—Flood Control Emergency Act: Hearings Before a Subcommittee of the Committee on Public Works, United States Senate*, 80th Cong., 2d sess., 1948, 269.

[19] Matthews to Malone, 19 May 1948; Hanna and Hanna, *Lake Okeechobee: Wellspring of the Everglades*, 278-304; House, *Comprehensive Report on Central and Southern Florida for Flood Control and Other Purposes*, 8, 19-21; Vileisis, *Discovering the Unknown Landscape*, 174.

[20] Johnson, *Beyond the Fourth Generation*, 139.

[21] House, *Comprehensive Report on Central and Southern Florida for Flood Control and Other Purposes*, 2.

[22] Johnson, *Beyond the Fourth Generation*, 135-136; Blake, *Land Into Water*, 176.

[23] George Wedgworth and Barbara Miedema interview by Matthew Godfrey, Belle Glade, Florida, 9 July 2004 [hereafter referred to as Wedgworth and Miedema interview].

[24] Quotation in Johnson, *Beyond the Fourth Generation*, 139-140; see also Blake, *Land Into Water*, 176-177; Grunwald, *The Swamp*, 219.

[25] Senate Committee on Public Works Subcommittee, *Rivers and Harbors—Flood Control Emergency Act*, 141-142.

[26] Johnson, *Beyond the Fourth Generation*, 145.

[27] Irene B. Burnham, Corresponding Secretary, Soroptimist Club of the Palm Beaches, to Spessard L. Holland, Senator, 12 November 1947, File Flood Control (Oct.-Dec. 1947), Box 287, Spessard L. Holland Papers. Manuscript Series 55, Special and Area Studies Collections, George A. Smathers Library (East), University of Florida, Gainesville, Florida [hereafter referred to as Holland Papers].

[28] Resolution No. 406, 12 November 1947, File Flood Control (Oct.-Dec. 1947), Box 287, Holland Papers.

[29] Claude Pepper to Lt General R. A. Wheeler, Chief of Engineers, 28 October 1947, Folder 11, Box 33, Series 201, U.S. Senate Correspondence, Claude Pepper Collection, Claude Pepper Library, Florida State University, Tallahassee, Florida [hereafter referred to as Pepper Collection].

[30] Lieutenant General R. A. Wheeler, Chief of Engineers, to Honorable Claude Pepper, 3 November 1947, Folder 11, Box 33, Series 201, Pepper Collection.

[31] House Committee on Public Works, *Central and Southern Florida Flood Control Project*, report prepared by the Library of Congress, 84th Congress, 2d session, 1956, Committee Print 23, 6.

[32] Press Release, Corps of Engineers, Jacksonville, Fla., District, 2 November 1947, Folder 11, Box 33, Series 201, Pepper Collection.

[33] See, for example, Colonel Willis E. Teale, District Engineer, to Honorable Claude Pepper, United States Senate, 22 April 1947, Folder 11, Box 33, Series 201, Pepper Collection.

[34] H. H. Bennett, Chief, Soil Conservation Service, United States Department of Agriculture, to Hon. Claude Pepper, United States Senate, 9 September 1947, Folder 11, Box 33, Series 201, Pepper Collection.

[35] Press Release, Corps of Engineers, Jacksonville, Fla., District, 2 November 1947.

[36] All quotations in Press Release, Corps of Engineers, Jacksonville, Fla., District, 2 November 1947.

[37] See Pelican Lake Sub-Drainage District, et al., to Honorable Claude Pepper, United States Senate, et al., 13 November 1947, Folder 11, Box 33, Series 201, Pepper Collection.

[38] See "Notes Covering Meeting Held at the El Comodoro Hotel, Miami, December 7, 10 A.M.—1 P.M.," Folder 11, Box 33, Series 201, Pepper Collection.

[39] Quotations in House, *Comprehensive Report on Central and Southern Florida for Flood Control and Other Purposes*, 15-18, 56-58; see also Hanna and Hanna, *Lake Okeechobee: Wellspring of the Everglades*, 352. For an example of a study conducted before 1947, see House, *Caloosahatchee River and Lake Okeechobee Drainage Areas, Florida (Side Channels)*, 79th Cong., 2d sess., 1947, H. Doc. 736, Serial 11059.

[40] House, *Comprehensive Report on Central and Southern Florida for Flood Control and Other Purposes*, 32-35.

[41] R. A. Wheeler, Lieutenant General, Chief of Engineers, to The Secretary of the Army, February 19, 1948, in House, *Comprehensive Report on Central and Southern Florida for Flood Control and Other Purposes*, 2. A paper written by Harold A. Scott, Chief of the Planning and Reports Branch of the Jacksonville District, in 1951, stated that the major purposes of the project were to protect the agricultural area and east coast communities from flooding. "Distribution of Water in the Central and Southern Florida Project," 26 September 1951, 2, South Florida Water Management District Reference Center, West Palm Beach, Florida.

[42] House, *Comprehensive Report on Central and Southern Florida for Flood Control and Other Purposes*, 36.

[43] House, *Comprehensive Report on Central and Southern Florida for Flood Control and Other Purposes*, 38-42.

[44] Despite Johnson's bleak opinion about Florida referenda, residents of Dade, Broward, and Palm Beach counties, convinced by the devastation of the floods, had ratified establishment of the water conservation areas late in 1947. Blake, *Land Into Water*, 177.

[45] House, *Comprehensive Report on Central and Southern Florida for Flood Control and Other Purposes*, 42-43.

[46] House, *Comprehensive Report on Central and Southern Florida for Flood Control and Other Purposes*, 44, 46-54 (quotation on p. 36).

[47] Colonel Mason J. Young, Corps of Engineers, Division Engineer, to the Chief of Engineers, United States Army, 31 December 1947, in House, *Comprehensive Report on Central and Southern Florida for Flood Control and Other Purposes*, 59-60.

[48] Board of Commissioners of Everglades Drainage District by W. D. Hilsabeck, Chairman, to Board of Engineers for Rivers and Harbors, 27 January 1948, File Flood Control Permanent (January 1948), Box 287, Holland Papers

[49] R. C. Crawford, Major General, Senior Member, to The Chief of Engineers, 9 February 1948, in House, *Comprehensive Report on Central and Southern Florida for Flood Control and Other Purposes*, 11-12.

[50] Wheeler to The Secretary of the Army, in House, *Comprehensive Report on Central and Southern Florida for Flood Control and Other Purposes*, 1-5.

[51] Reuss, *Designing the Bayous*, 111, 121.

[52] William G. Warne, Assistant Secretary of the Interior, to Lt. Gen. R. A. Wheeler, 13 April 1948, in House, *Comprehensive Report on Central and Southern Florida for Flood Control and Other Purposes*, vii. Likewise, in January, Daniel Beard, superintendent of the park, had conveyed to Teale his understanding that "future [Corps] studies will give full consideration to the interest of the Everglades National Park and that the water problem in the park area will be a subject of cooperative study." Beard to The District Engineer, 28 January 1948, File 1110-2-1150a (C&SF) Proj Genl—Flood Control (May 49-July 49), Box 8, Accession No. 077-01-0023, RG 77, Records of the Office of the Chief of Engineers, Federal Records Center, Atlanta, Georgia [hereafter referred to as FRC].

[53] United States Department of the Interior, Fish and Wildlife Service, Region 4, "A Preliminary Evaluation Report of the Effects on Fish and Wildlife Resources on the Everglade Drainage and Flood Control Project, Palm Beach, Broward, and Dade Counties, Florida," October 1947, 1, 18-19, copy in Library, Jacksonville District, U.S. Army Corps of Engineers, Jacksonville, Florida; see also Willis E. Teale, Colonel, Corps of Engineers, District Engineer, to The Regional Director, Fish and Wildlife Service, 21 November 1947, File A—Policy Action Taken by Corps of Engineers, Box 3, Entry 57A0179, Wetlands, 1944-1956, U.S. Fish and Wildlife Service, Atlanta Regional Office, Office of River Basin Studies, RG 22, Records of the U.S. Fish and Wildlife Service, National Archives and Records Administration Southeast Region, Atlanta, Georgia.

[54] Warne to Wheeler, 13 April 1948.

[55] Spessard L. Holland to Honorable Edwin A. Menninger, Publisher, The Stuart News, 10 March 1948, File Flood Control Permanent (Feb. 16-March 31, 1948), Box 287, Holland Papers. Apparently, the hurricanes had changed Holland's view of flood control. In 1947, EDD engineer Lamar Johnson had enlisted Holland's aid to get an appropriation for flood control. At that time, Johnson reported, Holland "figuratively jumped on me with both feet. He told me that I had no business in Washington and that I should take my problem to [Chief of Engineers] Colonel Jewett." Johnson, *Beyond the Fourth Generation*, 158.

[56] Atlantic Gulf Canals Association Inc. to Hon. Claude Pepper, Senate Office Building, 7 November 1947, Folder 11, Box 33, Series 201, Pepper Collection.

[57] Johnson, *Beyond the Fourth Generation*, 160.

[58] Senate Committee on Public Works Subcommittee, *Rivers and Harbors—Flood Control Emergency Act*, 144.

[59] Senate Committee on Public Works Subcommittee, *Rivers and Harbors—Flood Control Emergency Act*, 165.

[60] Quotation in Pearson to Colonel B. L. Robinson, Corps of Engineers, U.S. Army, South Atlantic Division, 21 August 1950, File 1110-2-1150a (C&SF) Proj Genl—Flood Control (May 50-Aug 50), Box 9, Accession No. 077-01-0023, RG 77, FRC; see also Senate Committee on Public Works Subcommittee, *Rivers and Harbors—Flood Control Emergency Act*, 183-184, 216-217, 249. Interestingly, the NPS initially proposed that the law authorizing the C&SF Project be amended to state that no work could occur that would affect Everglades National Park until the director of the NPS and the Chief of Engineers had established a mutually agreeable plan of operation. It dropped its request a few weeks later after Corps leaders assured it "that any plan of flood control will be taken up with us insofar as it may affect the Everglades National Park." "Chronological Documentation of National Park Service Efforts and Corps of Engineers Responsibility to Assure Everglades National Park of Fresh Water Supply from Central and Southern Florida Flood Control Project," 5, File CE-SE Central and Southern Florida FCP Everglades National Park Basic Data, U.S. Fish and Wildlife Service, Vero Beach administrative records [hereafter referred to as FWSVBAR].

[61] Act of 30 June 1948 (62 Stat. 1171, 1176).

[62] Spessard L. Holland to Honorable Ray Iverson, 23 June 1948, File Flood Control Permanent (May-June 1948), Box 287, Holland Papers.

[63] For more information on the development of the Mississippi River and Tributaries Project, see Charles A. Camillo and Matthew T. Pearcy, *Upon Their Shoulders: A History of the Mississippi River Commission from Its Inception Through the Advent of the Modern Mississippi River and Tributaries Project* (Vicksburg, Ms.: Mississippi River Commission, 2004), 141-172.

[64] Douglas, "What Are They Doing To The Everglades?" 12.

CHAPTER 3

Balancing Demands: Implementing the Central and
Southern Florida Flood Control Project, 1949–1960

THE INCLUSION OF the C&SF Project in the Flood Control Act of 1948 was the first step in the implementation of a water management program in South Florida. Throughout the 1950s, the state of Florida, the newly created Central and Southern Florida Flood Control District (FCD), and the U.S. Army Corps of Engineers worked together to construct and operate the project works. The Corps and the FCD attempted to coordinate the project with interested federal, state, and local agencies, but by the end of the 1950s, it was clear that these entities all had different views as to how water should be distributed in South Florida. Agriculturists wanted water for their crops, while rapidly growing urban interests demanded water as well. Everglades National Park and FWS officials, meanwhile, claimed that the Corps needed to provide them with enough water to preserve plants, fish, and wildlife in the Everglades and other areas. By the end of the 1950s, the collision of these different demands seemed inevitable.

In order for work to commence on the C&SF Project in the late 1940s, the state of Florida needed to raise around $3.25 million, its share of the construction cost of the first phase, as well as acquire the necessary lands and rights-of-way. Unfortunately, the federal law mandating these responsibilities (the Flood Control Act of 1948) was passed nine months before the state legislature was scheduled to meet, meaning that no action could be taken to fulfill these duties in 1948. In preparation for the 1949 legislative session, Governor Millard Caldwell established a committee to investigate what tax revenues could support the flood control plan, while other officials explored the creation of a new state agency to cooperate with the Corps in project implementation. The Okeechobee Flood Control District and the Everglades Drainage District (EDD) still existed, but the EDD did not have authority to operate for flood control and the

Okeechobee district had jurisdiction over a limited area. According to Lamar Johnson, engineer for the EDD, several individuals, including himself, drafted bills to establish a local cooperating agency. The EDD also kept in close contact with the Corps during this period, receiving and clarifying information pertaining to local cooperation, and compiling engineering data in preparation for the beginning of construction.[1]

In April 1949, the Florida state legislature convened, passing three bills that pertained to state involvement in the C&SF Project. The first established the FCD as the major local agency to coordinate with the Corps on the project, replacing the Okeechobee Flood Control District. The second provided for the abolishment of the EDD after it had paid off its bonds, giving its responsibilities to the FCD. The third was the state's General Appropriations Act, which included $3.25 million as the local share of the cost of the C&SF Project.[2]

The legislation authorizing the FCD established a five-member, non-salaried board, appointed by the governor for three-year overlapping terms, as the district's governing body. This board would have "full responsibility for the District's activities and interests."[3] One member of the board would become the executive director, who would serve with the executive staff, which included the heads of seven different divisions within the district: land, operation and maintenance, finance, engineering, public information and research, administration, and legal. Soon after the legislature created the FCD, the five appointed board members—Dave Turner, Fred Bartleson, Joe S. Earman, N. J. Hayes, and Lawrence Rogers—organized the district officially, designating Turner as executive director. The board also established its headquarters at West Palm Beach. As created, the FCD encompassed all or part of 17 counties in Central and South

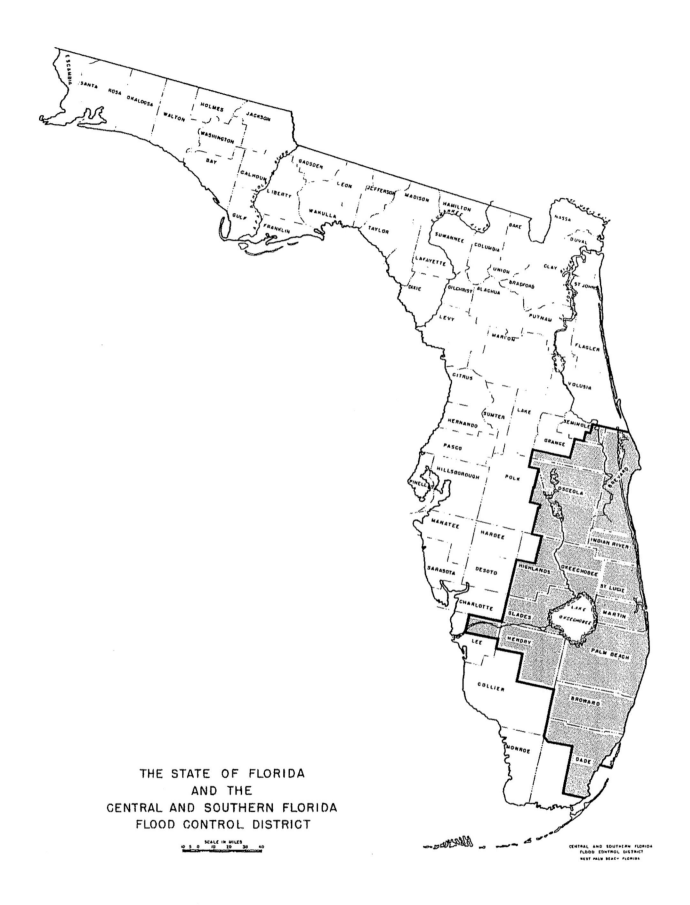

THE STATE OF FLORIDA
AND THE
CENTRAL AND SOUTHERN FLORIDA
FLOOD CONTROL DISTRICT

SCALE IN MILES

CENTRAL AND SOUTHERN FLORIDA
FLOOD CONTROL DISTRICT
WEST PALM BEACH FLORIDA

Boundaries of the Central and Southern Florida Flood Control District. Central and Southern Florida Flood Control District, *Facts about F.C.D.* (West Palm Beach, Fla.: Central and Southern Florida Flood Control District, 1955)

Florida, totaling 15,570 square miles. Its major responsibilities, according to a 1955 publication, was "cooperative participation in the advancement of studies design and construction" of the C&SF Project, as well as land acquisition, water control, and regulation once the system was developed.[4]

At a subsequent meeting attended by numerous state officials and legislators, W. Turner Wallis, appointed as chief engineer of the district, expounded on the FCD's functions. Essentially, he said, the FCD was "a cooperative agency between the State and the Federal Government and local interests in projects concerned with water conservation, flood and water control, and allied problems." John C. Stephens, a research project supervisor with the Soil Conservation Service of the U.S. Department of Agriculture, explained how the FCD coordinated with these interests. According to Stephens, the FCD held regular meetings with Corps engineers during the planning stages of C&SF project works, providing "basic data on economic, social, and physical factors essential to project development." The FCD received these data by "maintain[ing] close liaison with all agencies—Federal, State, and local—having an interest in problems of water conservation and control and natural resource developments."[5] These included the U.S. Department of Agriculture, the U.S. Geological Survey, the FWS, the Florida Geological Survey, the Florida Game and Fresh Water Fish Commission, and the State Board of Conservation, among others. The FCD also held meetings with land action groups, county commissioners, subdrainage districts, and landowners in order to understand what local interests wanted from the project, and then presented these views to the Corps. After the Corps made its final construction plans, the FCD reviewed the proposals before they were sent out to bid, and then it worked to obtain necessary property and rights-of-way for construction.[6]

In order to perform these functions, the FCD needed money from the state, including the funds necessary to cover the state's required contribution to the total cost of the project, and the financing to purchase lands and to provide operation and maintenance once the project was completed. The state legislature had created a flood control account in its general revenue fund, and had agreed to make occasional appropriations to the account, including the initial $3.25 million required for construction. Other charges, such as for right-of-way purchases and for operation and maintenance, would come from an ad valorem tax on all real and personal property in the FCD, whereby the amount paid would depend on the value of the property. This meant that landowners in Dade, Broward, and Palm Beach counties would be responsible for 95 percent of the total tax.[7]

Using the money provided by the state, as well as the federal appropriation, the Corps began its construction of the C&SF Project. According to the FCD, there were several major components to be completed in the first phase of the program. First, the Corps would build a levee from northwest Palm Beach County to the south of Dade County along the east coast, thereby preventing flooding from the Everglades to the coastal communities. Second, the Corps would modify control facilities and levees around Lake Okeechobee in order to create more water storage, and it would increase the discharge capacity from the lake in order to prevent flooding. Third, the Corps would create three water conservation areas in Palm Beach, Broward and Dade counties for water storage. Fourth, the Corps would construct canals, levees, and pumping stations to protect 700,000 acres of agriculture south of Lake Okeechobee in Palm Beach, Hendry, and Glades counties, known as the Everglades Agricultural Area (EAA). Fifth, the Corps would build canals and water control structures to handle drainage in Dade, Broward, Palm Beach, Martin, and St. Lucie counties.[8]

As this construction began, Corps representatives freely admitted that the C&SF Project as proposed in House Document 643 needed revising. Oscar Rawls, a spokesman for the Jacksonville District, informed state and local officials that because it had to produce a plan quickly, the Corps "in many instances" did not complete extensive studies of regional needs and instead relied on hasty estimates in its proposal. According to Rawls, the proposal was merely a quick report "stating the problems and in a preliminary sort of [way] an estimate of what the solution should be." The Jacksonville District thus had only "a plan that they would use for the basic frame work [sic] on which further and more complete planning would take place."[9] W. Turner Wallis, an engineer with the FCD, was even more blunt, stating that House Document 643 was "a hastily assembled document based on hydrological and agronomic data that even the most optimistic admitted was far from adequate."[10] More studies of the needs of Central and South Florida were necessary, and in many ways, the Corps and other federal and state agencies learned about these needs as they went throughout the 1950s.

Regardless of the inadequacies, the Corps began construction, and the FCD commenced its responsibilities. One of the first tasks the FCD faced was the acquisition of lands to be used as water conservation areas. As a preliminary step, the district made a restudy of how large the areas should be, using the "knowledge and experience of engineers familiar with the hydrology of the Everglades."[11] It recommended reductions in the three conservation areas proposed by the Corps in House Document 643 in order to keep valuable agricultural land and tracts held in trust for the Seminole Indians free from flooding. Smaller areas would also curb seepage rates, a problem because of the permeability of the limestone underlying the land. The FCD suggested that Water Conservation Area No. 1, originally proposed as 175,315 acres in the vicinity of Loxahatchee Marsh in Palm Beach County and supplied with water from the West Palm Beach and Hillsboro canals, be trimmed by 21,299 acres, while Conservation Area No. 2 in Broward County (containing water from the Hillsboro and North New River canals) be reduced from 142,259 acres to 135,187 acres. The largest decrease would occur in Conservation Area No. 3 in Dade and Broward counties (supplied by the

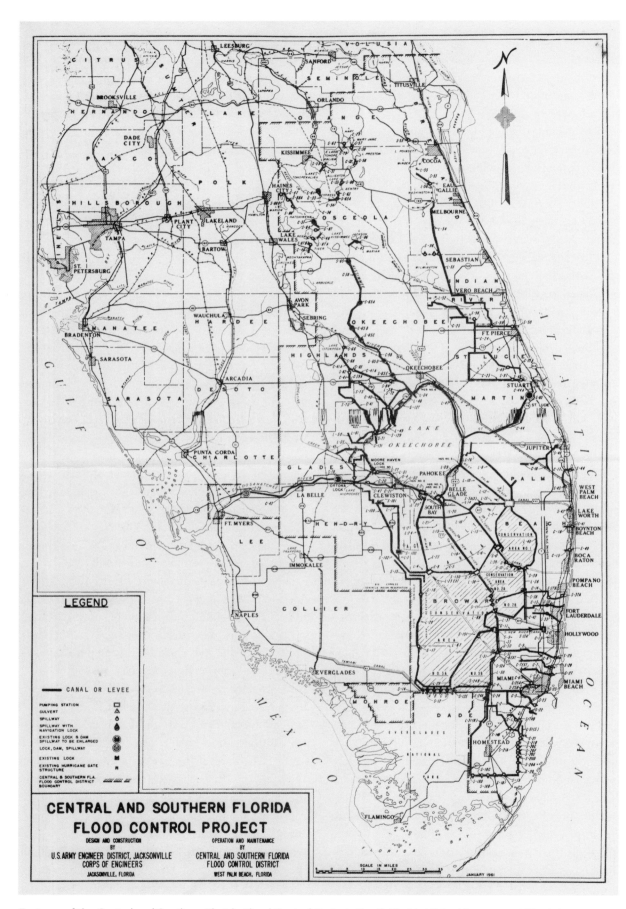

Features of the Central and Southern Florida Flood Control Project. (South Florida Water Management District)

North New River and Miami canals), which would be reduced from 671,411 acres to 563,724 acres. Over 130,000 total acres would be cut from the three areas, a 13.8 percent reduction.[12]

Despite the large acreage involved, the Corps agreed to the FCD's suggestions, and in the early 1950s, the FCD purchased land for the water conservation areas. According to Lamar Johnson, who had been appointed assistant engineer of the FCD, "the landowners' generally did not accept the appraised value of the lands," meaning that "most of the lands were acquired by condemnation."[13] However, some landowners insisted that they be allowed to retain their possessions because the possibility existed that they contained oil and gas. To appease these owners, the FCD acquired only flowage rights to the private land that it could not condemn, amounting to approximately 10 percent of the conservation areas. Although the FCD did not have full possession of this land, the flowage rights still allowed it "to flood the surface of the lands at any time and to any degree necessary."[14] The land acquisition program for the conservation areas continued until its completion in 1954, upon which the FCD had purchased approximately 860,000 acres.

Yet in its land acquisition efforts, the FCD ran into some trouble with the Seminole Indians. As explained earlier, the state of Florida had moved the Seminole reservation out of the proposed boundaries of Everglades National Park in 1935. The new location of the reservation, however, infringed on the area where the Corps and the FCD wanted to build Conservation Area No. 3. In 1950, the Corps proposed to construct L-28, a north-and-south levee that would help impound water in Conservation Area No. 3, three miles east of the Hendry-Broward county line. The Seminole objected to this plan because the levee would bisect their reservation and cause more than half of their grazing and hunting lands to be used for water impoundment, making them virtually worthless. After Corps and Bureau of Indian Affairs officials convinced the Seminole that alignment would not harm them, alleging that land to the west of the alignment could not be used for agriculture anyway, the Indians retracted their objections, allowing the levee's construction. Confirming Seminole fears, however, 16,000 acres east of the levee became part of Conservation Area No. 3, although the Indians could still use 12,000 acres to the west for grazing.[15]

As the FCD acquired land for the water conservation areas, it negotiated with both the FWS and the Florida Game and Fresh Water Fish Commission for the management of the areas. As early as 1946, the EDD had proposed that the FWS assume control over the conservation area in the vicinity of the Loxahatchee Marsh in order to provide a migratory bird refuge on the Atlantic and Mississippi flyways. The FWS agreed to the program, and when the area was finally created as Conservation Area No. 1 in 1950, the Service purchased a 50-year lease from the FCD. After some consultations, the Corps approved the lease as long as the FWS's management did not "interfere with the regulation and operation of conservation area 1

by the Corps of Engineers."[16] Thereafter, the FWS operated Conservation Area No. 1 as the Loxahatchee National Wildlife Refuge.

Yet tensions sometimes existed between the Corps and the FWS over Loxahatchee management. In 1952, for example, Roy Wood, the Service's regional supervisor complained that the Corps had organized an inspection trip of the C&SF Project for the House Public Works Committee, but had not included any FWS representatives in the planning or on the tour even though the FWS managed Conservation Area No. 1. This snub, Wood claimed, "clearly reveals the Corps of Engineers' mode of operation in the promotion of its program and perhaps the attitude which generally prevails in the Corps

The Loxahatchee National Wildlife Refuge. (South Florida Water Management District)

relative to active participation of other agencies in their affairs."[17] The Corps' oversight was probably more unintentional than deliberate, but Wood's complaint resonated with those who believed that the Corps did not regard fish and wildlife concerns as important as other parts of the C&SF Project.

In January 1952, the Florida Game and Fresh Water Fish Commission accepted responsibility over the other two water conservation areas, which were then designated as the Everglades Wildlife Management Area. According to the terms of the license agreement between Florida Game and the FCD, the commission would operate the areas "to attain the basic objectives of preservation, protection and propagation of wildlife and fish," as well as for recreational benefits. Measures would include developing wildlife environments and habitat, planting crops and plants "to increase the carrying capacity of the area for wildlife," and allowing controlled public hunting and fishing. However, the agreement clearly stated that the operation of the conservation areas for wildlife and fish objectives could not conflict with flood control and water retention.[18]

In addition to establishing the water conservation areas, the FCD and the Corps also investigated what other measures needed prioritization. One of the initial examinations was of the necessity of flood control work in the Kissimmee River Valley, located north of Lake Okeechobee. The Corps had performed survey work on the Kissimmee River, which flowed from Lake Tohopekaliga just south of Orlando into Lake Okeechobee, as early as 1901, receiving authorization under the River and Harbor Act of 13 June 1902 to maintain a channel in the river from 30 to 60 feet wide and three feet deep at ordinary low water from the town of Kissimmee to Fort Bassenger, a distance of about 100 miles. In the 1920s and 1930s, congressmen requested that the Corps investigate further improvements on the Kissimmee, including flood control, in order to make the land more suitable for ranching, but no action was taken.[19] When the Corps proposed Kissimmee River flood control as part of the C&SF plan, many Kissimmee Valley residents believed that they would finally receive the protection they desired. However, the Kissimmee plans were pushed aside in order to provide flood relief for the coastal communities and for the agricultural region south of Lake Okeechobee.

To alleviate the growing concerns of local citizens, the FCD held one of its first meetings in the town of Kissimmee.[20] At this gathering, Oscar Rawls of the Jacksonville District related that levees, improved channels, and impounding reservoirs were the three main ways to control floods in a valley. In the Kissimmee Basin, improved channels would be the most effective way, providing 90 percent of the flood relief. But since Kissimmee work was not part of the C&SF Project's first phase, the Corps could not act until Congress appropriated the necessary funds. According to U.S. Senator Claude Pepper, who also attended the meeting, "when the money will be available is a political problem rather than an engineering one." He promised the people that the Kissimmee region would be "taken care of in the course of the program," and counseled patience.[21] Kissimmee residents continued to clamor for flood control work, especially after more flooding in the latter part of 1949, but Chief of Engineers Major General Lewis A. Pick reported again that, although "the flood situation in the Kissimmee Valley is even more serious than

Opening construction blast for S-5A, 1952, (left to right) Col. H. W. Schull, Jr., Rep. Dwight L. Rogers, Sen. Spessard Holland, Gov. Fuller Warren, Col. W. K. Wilson, Jr. (South Florida Water Management District)

that revealed by the flood of 1947," the Corps could do nothing until Congress appropriated the necessary money.[22]

As concerns with the Kissimmee River Basin grew, the Corps investigated the feasibility of authorizing other phases of the C&SF Project. In November 1952, the Corps, the FCD, and the state of Florida held a conference to discuss the project's progress. In this meeting, the parties determined that the first phase of the program should be modified in order to complete an outer perimeter levee around the EAA and to begin work in the Kissimmee River and Upper St. Johns basins.[23] Before the Corps could get congressional authorization for this work, monetary problems developed. In the summer of 1953, Florida's two U.S. senators, Spessard Holland and George A. Smathers, criticized the Corps for delays in its construction schedule for the C&SF Project. Holland reported that the Corps had an unexpended balance for the 1953 fiscal year of over $6.5 million. Holland had been able to get additional amounts appropriated for the 1954 fiscal year, but he claimed that his job was more difficult because of "the slow handling of the program by the U.S. Engineers."[24] Smathers agreed, stating that "whatever victory we achieve in the legislative halls will be of little value unless the Corps of Engineers gets on the ball, and performs in a more satisfactory manner than has been the case in the past few years."[25]

Colonel H. W. Schull, Jr., District Engineer for the Jacksonville District, defended the Corps, explaining that the problems derived from "the system of appropriation and justification used on this project." Because the Corps could construct only works "approved by the Bureau of the Budget and defended before the Appropriations Committee," Schull said, it sometimes had to let funds lie until such approval had been obtained. The District Engineer explained further that the Corps was developing a system with the Bureau of the Budget "which will allow the construction agency more flexibility and which will still be acceptable to appropriations committees."[26] Instead of condemning the Corps, Congress should be proud of the effort the Jacksonville District had made to ensure that appropriations were judiciously and efficiently used. At the same time, however, Chief of Engineers Major General S. D. Sturgis, Jr., told Holland and Smathers that a lack of planning in the early stages of the project caused the delays because the Corps faced "many new problems" as construction continued. He pledged that more expert hydraulic engineers would be assigned to the project in order to "develop a backlog of plans."[27]

The problems with construction delays and the desire to modify the first phase of the C&SF Project led Smathers to ask Charles D. Curran, a senior specialist in engineering and public works, to make a study of the entire project and how it was progressing. Curran explained that, since 1947, the Corps had made additional examinations of the project area and determined that "the original plans were not completely adequate." It had thus made some "major design changes." Because of these alterations, Curran reported, the estimat-

ed cost of the first phase had risen from $70 million as originally authorized to $116 million. Addressing the delays in project expenditures, Curran stated that "the fault does not seem to lie in any one place or be the result of any one situation." He did admit that the bureaucracy surrounding appropriations caused problems, but he saw no solution. "It appears that the Central and Southern Florida Flood Control Project must progress somewhat slowly for reasons reflecting no discredit on the merits of the project itself," he concluded.[28]

Curran's report, which Smathers disseminated to interested parties, coupled with talks that the Corps was already holding with the Bureau of the Budget, convinced Congress in 1954 to authorize the entire C&SF Project, rather than continuing to allow the Corps to work in only approved phases. The Flood Control Act of 1954 provided the necessary permission. According to the legislation, Congress would determine how much local interests would pay for aspects of the project beyond the first phase "based on

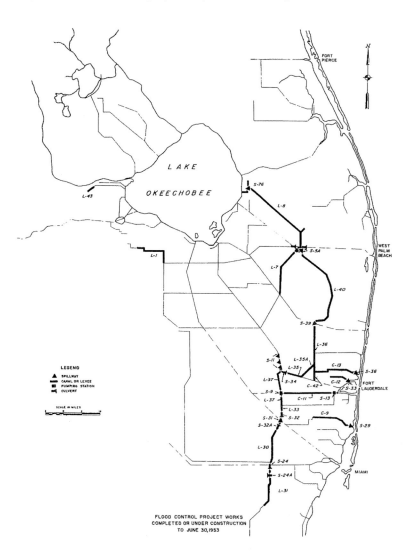

C&SF Project status, 1953. (U.S. Army Corps of Engineers, Jacksonville District)

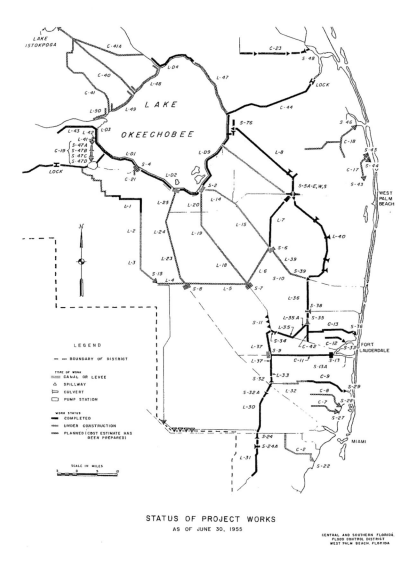

STATUS OF PROJECT WORKS
AS OF JUNE 30, 1955

CENTRAL AND SOUTHERN FLORIDA,
FLOOD CONTROL DISTRICT
WEST PALM BEACH, FLORIDA

C&SF Project status, 1955. (U.S. Army Corps of Engineers, Jacksonville District)

recommendations to be submitted at the earliest practicable date by the Chief of Engineers, through the Bureau of the Budget."[29] When those studies were completed in 1956, they determined that local interests would be responsible for 39.8 percent of the total cost of the entire project.[30]

The passage of the 1954 Flood Control Act meant that the Corps could now proceed with all aspects of construction. Some delays continued—Conservation Areas No. 2 and 3, for example, were not completed until the mid-1960s—but, for the most part, the Corps moved construction along expeditiously. In addition, new areas were gradually added to the C&SF Project as studies indicated the necessity of their inclusion. Thus, in 1958, Congress authorized work on 64 square miles in Hendry County west of the EAA and the water conservation areas, and in 1960, the Nicodemus Slough in Glades County was added to the project. Areas in south and southwest Dade County were included in the 1960s, as was Martin County in 1968.[31]

But as the work progressed, criticism and complaints about the C&SF Project began to develop. One of the key points was the effect of flood and water control on plants and wildlife within Everglades National Park. In 1949, Congress had authorized the secretary of the interior to obtain the rest of the acreage established as a minimum boundary for the park in 1944, thereby increasing the amount of park land to approximately 1,220,000 acres.[32] To manage this area, Park Superintendent Daniel Beard had a permanent staff of 20 people, seven of whom were in the field. This meant that each ranger had to patrol around 180,000 acres, which, according to chief ranger Earl M. Semingsen, was "too much to supervise and protect the way you'd like to see it done."[33] In addition to the problems of safeguarding the flora and fauna, personnel also had the task of figuring out just how the C&SF Project would impact the park, although officials held that the Corps should bear the responsibility of making these studies. Based on its own observations and on studies made by the U.S. Geological Survey, the NPS was convinced that a large water supply was critical, especially after more than 32 grass and forest fires exploded in the area in 1950.[34]

In order to maintain contact with the Corps about park needs, the NPS executed an agreement with the Jacksonville District "to discuss the project on the field level."[35] C. Raymond Vinten, coordinating superintendent of the Southeastern National Monuments in St. Augustine, Florida, was designated as the NPS representative. But in August 1949, NPS Region One Director Thomas J. Allen complained that the Corps had produced "no information whatsoever." Hearing about Corps proposals to improve the Caloosahatchee River, to construct a levee on the western side of the water conservation areas, and to build a levee south of Tamiami Trail, Allen worried that such construction would block necessary water from entering Everglades National Park. He emphasized to the Corps that the park "can be even more seriously affected by lack of water than it can be by an excess of water." Although the Corps had made general statements in House Document 643 about supplying water to the Everglades, Allen believed that this was not enough. "The water we need for dry periods," he stated, "involves the very life of the park through the maintenance of bird, animal, plant, and reptile life without interruption."[36] Park officials desired something more than general statements to convince them that the park would receive adequate water from the north and from the east.

District Engineer Colonel R. W. Pearson responded to Allen's complaints by insisting that the Corps had no new information to share. "This office is fully aware of the importance of proper supply and control of water for Everglades National Park," Pearson explained, agreeing to arrange conferences and "every possible degree of liaison and cooperation" with park officials once the Jacksonville District began developing detailed plans. He also attempted to alleviate Allen's fears by explaining that water storage in the water conservation areas would allow the Corps to release the resource "when

Everglades National Park in the 1950s. (The Florida Memory Project, State Library and Archives of Florida)

needed most," thereby creating "a regimen of flow . . . which in effect would tend to reduce the peaks and increase the valleys of the present natural flood hydrograph." Such conditions would be "far more desirable for the park area than the present experiences of too much or too little water." Finally, Pearson explained that the levees that concerned Allen were not designed to keep water out of the park, but to retain water in the Everglades. "It is regretted that your office has felt that it has not been properly informed," he wrote, but it was merely a misunderstanding. It was the Corps' "earnest desire . . . to work in close cooperation with your organization in all matters of mutual concern."[37] Allen thanked Pearson for his letter, explaining that it "clarifies the point that you are aware of the needs" of Everglades National Park.[38]

Less than a year later, however, such conciliatory attitudes had changed. After the NPS requested that the Corps make detailed hydrological studies to determine the water needs of Everglades Na-

tional Park, Pearson issued a rather stilted reply. Referring to the park as a "local interest," he stated that it had the responsibility of informing the Corps what its water needs were, and not the other way around. "Special investigations and studies related to the detailed determinations of requirements of local interest for water supply or other purposes . . . are not considered to be within the responsibilities or authorized functions of the Corps of Engineers," he declared. Pearson further explained that even though language in House Document 643 referred to restoring park water supplies to "natural conditions," that was not the purpose of the project. "Under natural conditions, the area was subjected to droughts, fires, and floods," he asserted, "none of which would tend to make the area attractive as a park area." Instead, the Corps would operate the project to provide "a regulated water supply," thereby promoting "optimum, or at least improved, conditions for growth of native vegetation." In addition, Pearson said, it was entirely possible that in some drought years, not enough water would be available from the conservation areas and Lake Okeechobee to serve all water needs. "In such cases," he continued, "Everglades National Park will compete with agricultural areas and urban centers for water supply" according to "an orderly plan and a recognized authority."[39]

Allen was uneasy with Pearson's letter, believing that the colonel's comments were "somewhat at variance with former official statements in the matter." Especially troubling was Pearson's reference to the park as a "local interest." The park was "a national project authorized by the United States Congress," he protested, "and cannot be disregarded in the planning by your organization of the flood control works." Allen also considered it well within Corps authority to ensure that the park received a proper supply of water since "any damage which will occur to Everglades National Park originates within, and only within, the limits of your project." Allen did not specifically address Pearson's claim about park competition with agricultural and municipal interests for water, but he did express hope that the C&SF Project could "guarantee the park an amount of water comparable to the 'normal' run-off and still attain its many conservation objectives." Based on measurements conducted at 23 discharge points along the Tamiami Trail, and following the recommendations of an FCD study, Allen insisted that 300,000 acre feet of water annually was "a very reasonable minimum annual flow for the park to expect the flood control project to provide under managed conditions."[40] Thus, by the summer of 1950, the NPS and the Corps had already drawn their lines in terms of water supply to Everglades National Park.

Although the Corps did not agree to perform a hydrological study of the needs of the park,[41] Lamar Johnson, the FCD engineer, assumed that function, having a "smoldering urge" to "analyze the park's water problem."[42] In 1950, the FCD published Johnson's report, which detailed the water resources of the park both in the pre-drainage and drainage eras. According to the report, a lack of records made it "impossible" to reconstruct accurately water flow into the Everglades before drainage, but Johnson still made an attempt, using rainfall and evaporation data and descriptions of the area before extensive drainage efforts began. He estimated that before drainage, the discharge into the region past the Tamiami Trail was "2,315,000 acre-feet in an average year; 10,744,000 acre-feet in a wet year; and negligible runoff into the Park during a dry year." In order to determine the amount of flow during the drainage era, Johnson used data obtained by the U.S. Geological Survey for the years 1940 to 1947, which contained "approximately normal years, a period of successive dry years and the wettest year of record." He concluded that "during successive average years a runoff of approximately 300,000 acre-feet could be expected for supply to the Park under existing conditions." Clearly, "water in primeval quantities cannot be made available," but 300,000 acre-feet as an annual minimum could "restore the former ecological balance of the Park—at least to a reasonable degree."[43]

Johnson also disagreed with Pearson's contention that "natural conditions" were not desirable for the park. "There is little doubt that the decision to approach primeval conditions, as nearly as possible, is the proper objective," he stated. Individuals in South Florida wanted the park "because they liked the flora and fauna as it is, or has been," Johnson continued, and "they will not be pleased by some brackish, bastard offspring sired by a fresh water deficiency."[44] To restore the balance between salt and fresh water in the park, Johnson proposed that some structures, such as knee-high overflow dikes, be placed within the park. He later recollected that the NPS "reacted with horror" to this suggestion because it did not want to

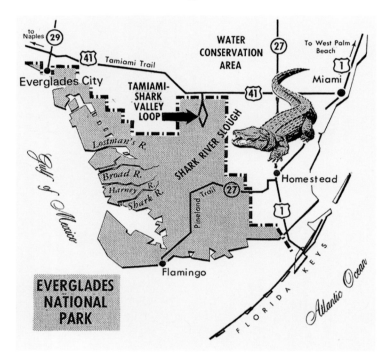

Map of Everglades National Park showing Shark River Slough. (The Florida Memory Project, State Library and Archives of Florida)

"interfere with nature by doing something artificial."[45] But Johnson could see no other solution, especially because "the wish and purpose of the majority of the people" was to use water for agriculture and municipal water supplies, not to maintain Everglades National Park. "The aesthetic appeal of the Park can never be as strong in the people as the demands of home and livelihood," Johnson claimed. "The manatee and the orchid mean something to most people in an abstract way, but the former cannot line their purse nor the latter fill their empty bellies." Regardless, Johnson recommended that "complete hydrological data" be gathered within the park since little information existed about "the influence of water on the gross ecology." The ultimate goal, he insisted, was to ensure that "one drop of water . . . preserve what two drops of water created."[46]

For the rest of the 1950s, the issue over water supply to Everglades National Park simmered on the NPS's backburner. One of the problems was that although NPS authorities believed that the park needed a certain amount of water, they were unsure how much this was, Johnson's conclusions notwithstanding. The superintendent of the park informed his superiors in 1957 that the Corps continued to request that park officials determine how much water they wanted, but park leaders knew only that they wanted "more water, but not too much."[47] Developing a definite figure was crucial in order to ensure that the C&SF Project supplied enough water to the park.

To obtain more specific figures, the NPS hired Johnson, who by now had left the FCD and was a private consultant, to conduct another study of park water needs in 1958. In many ways, Johnson's conclusions were no different from his 1950 determinations. He again estimated that a normal average flow into the Everglades before drainage was around 2.5 million acre-feet, although he did not believe that it was possible to provide water in that amount to the park. Instead, he stressed the importance of restoring the balance between salt and fresh water through control structures within park boundaries. Because there was more information in 1958 about how the water conservation areas would be operated, Johnson determined that the C&SF Project could provide "more water to the park in an average rainfall year than the old Everglades channel system had," although any supply from the conservation areas would have to be supplemented from other sources.[48] Therefore, he recommended that the NPS contact the Corps about diverting the runoff from a 745 square mile area in Collier and Hendry counties to the Shark River Slough within the park. Based on Johnson's conclusions, the NPS informed the Corps that the "optimum Park requirements" were "two or more million acre feet," including at least 150,000 acre-feet entering Shark River Slough each month in the spring.[49] More studies were necessary, however, to determine the minimum amount that the park needed. Yet the NPS did not heed many of Johnson's other suggestions; instead, Johnson recalled, park officials merely sat "like a fledgling egret on its nest, mouth open and squawking, waiting to be fed."[50]

While the NPS attempted to understand how much water it would receive from the C&SF Project and how this would affect plant and wildlife within Everglades National Park, the FWS and the Corps wrangled about how much water the water conservation areas could store. The Corps originally planned to maintain a constant level of 17 feet in Conservation Area No. 1 and 15.9 feet in Conservation Area No. 2. Engineering studies conducted in the 1950s, however, indicated that such stable levels were not "engineeringly feasible."[51] For one thing, a level of 15.9 feet in Conservation Area No. 2 would lead to seepage at rates that would prevent the maintenance of necessary levels for fish and wildlife. For another, engineers held that water as high as 15.9 feet would destroy vegetation and be susceptible to hurricane wind tides that could breach the levees and flood east coast communities. Therefore, the Corps proposed in 1956 to maintain seasonal levels between 12.5 and 15 feet in Conser-

vation Area No. 1 and between 10.1 and 13.0 feet in Conservation Area No. 2.[52]

When the FWS studied the problem, it decided that the proposed water levels would adversely affect fish and wildlife in the water conservation areas to the point of making any benefits negligible. The FWS therefore recommended a seasonal water level of between 14 and 17 feet for Conservation Area No. 1, which would "provide adequate water depths for waterfowl, frogs and other wildlife and

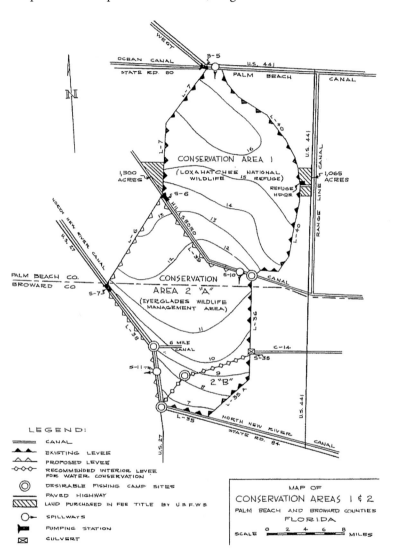

Conservation Area Nos. 1 and 2, 1958. (U.S. Army Corps of Engineers, Jacksonville District)

greatly increase fishing and other recreational use."[53] It also suggested that Conservation Area No. 2 be split into two pools (2A and 2B) by an interior levee in order to eliminate seepage loss, and that the level in Area 2A (the northwest portion) be maintained between 12 and 14.5 feet. Because of high seepage in Area 2B (which consisted of highly permeable soils over the Biscayne Aquifer), the FWS recommended that no high stage be maintained in 2B. No suggestions were made at that time for Conservation Area No. 3, which had yet to be completed, but it too was eventually partitioned into two sections (3A and 3B) to control seepage.[54]

Even though the Corps agreed to these changes, some of its leadership counseled the FWS to remember that it was only one of the interests involved in the overall water control program. Project works had to consist of "the most feasible plan of improvement, in accordance with the desires of all local interests" in order to be constructed.[55] The Corps would willingly work to minimize fish and wildlife damages, but could only do so in ways that would not affect primary project purposes. Likewise, B. F. Hyde, Jr., executive director of the FCD, insisted that the FCD's policy was "to preserve or enhance natural resources values wherever such is possible consistent with accomplishment of it's [sic] prime responsibility," namely "water control in the interest of all public needs and values." According to Hyde, the FCD tried to preserve fish and wildlife "to the maximum possible degree consistent with full consideration of all resources involved and recognition of limitations inherent to the Federal Flood Control Project."[56]

Such statements only confirmed a growing belief that the Corps and the FCD placed agricultural and urban interests above those of fish and wildlife.[57] One of the reasons for this perception was that agriculture and urban growth expanded considerably throughout the 1950s, increasing demands on water. Agricultural production escalated as the Corps built levees, canals, and pumping stations around the EAA in the 1950s, thereby walling it off from floodwaters and allowing needed irrigation in times of drought. More ranching occurred as well, in part because the Everglades Experiment Station indicated that St. Augustine grass, previously used only for lawns, was a nutritious forage well-suited for the Everglades. Sugar cane also maintained its place in the Everglades, although its largest boom would occur in the early 1960s. In addition, vegetable production continued in the EAA, mainly for winter markets.[58]

Meanwhile, urban populations, especially in Dade County, expanded considerably in the 1950s, as did the number of tourists to the region. Even though Americans had regarded Florida as a sun-drenched, desirable area since the 1920s, it was not until the post-World War II era that people began moving to the state in great numbers. Senior citizens migrated to St. Petersburg, Lake Worth, and Miami Beach in the 1940s, while Miami became noted in the 1940s and 1950s as "a winter playground for New Yorkers and a summer escape for Cubans."[59] By 1950, Dade County was the

Miami Beach, 1955. (The Florida Memory Project, State Library and Archives of Florida)

host of several interesting attractions, including college football's Orange Bowl, the Latin Quarter and Hialeah Race Track, Key Biscayne, and Brickell Avenue. In 1950, Miami had a population of 250,000 (the largest city in the state), and it only increased as the decade continued.

But as the population of Dade County skyrocketed, and as more and more tourists frequented the region, Dade County officials claimed that the Corps placed agricultural interests above urban needs. Therefore, Dade County officials asked W. Turner Wallis, a consulting engineer in Tallahassee formerly with the FCD, to prepare a report on water control in the area. Upon completing his examination, Wallis criticized the C&SF Project and the Corps for not heeding concerns voiced by representatives of Dade County. The county accounted for almost half of the population included in the project area and paid around two-thirds of the FCD's ad valorem tax, Wallis claimed, yet it had trouble getting the Corps to revise its plans as included in House Document 643. "Well over 50 percent of the total benefits claimed for the Central and Southern Florida Flood Control Project are based on land to be reclaimed for agricultural purposes," Wallis complained.[60] But the urban character of Dade County precluded it from obtaining any of these benefits; instead, county residents wanted more efforts to limit saltwater intrusion, an increased water supply for urban areas in the county, and recreation. Unfortunately, Wallis asserted, "the original project did not offer adequate measures in any of these three areas."[61] He called for the uniting of all interested parties in Dade County to pressure the Corps to address these concerns, thereby justifying the county's investment of millions of dollars in the C&SF Project. He also recommended that a better plan be devised for Dade County to address its ever-increasing water needs and that the county work more closely with the FCD to ensure that its needs were being met.

Wallis's report seemed to work; in 1960, Chief of Engineers Lieutenant General E. C. Itschner made a tour of Dade County and concluded that the Corps needed to build outlet structures through the Tamiami Trail and construct a diagonal levee northeastward from the Tamiami Trail through Conservation Area No. 3. Itschner also recommended the relocation of L-31N, a north-south levee south of the Tamiami Trail, farther west to the border of Everglades National Park in order to facilitate agriculture in that area.[62]

Despite Itschner's proposals, it was increasingly apparent that the county's needs for water would conflict with the requirements of other interested parties, including Everglades National Park. At a conference between the NPS, the FWS, the Florida Game and Freshwater Fish Commission, the FCD, and the Jacksonville District, representatives from the Corps noted that "sufficient water is not available to supply all demands, and methods to conserve water will have to be developed."[63] As growth continued in South Florida in the 1960s, the question of how water should be distributed would be hotly contested—especially by the NPS.

By the end of the 1950s, the Corps had made great strides in the construction of the C&SF Project. The FCD noted in 1960 that "128 miles of channels and canals have been dug, or improved, 300 miles of levees have been constructed and six pumping stations are serving the multiple purposes of flood control and water conservation."[64] The construction had occurred mainly along the east coast and Lake Okeechobee, creating both the EAA south of the lake and the water conservation areas between the EAA and the east coast. The FCD estimated that 60 percent of the levees surrounding the conservation areas were complete, 75 percent of the east coast levees were finished, and almost all of the levees surrounding the EAA were done.

But as this construction occurred, discontent emerged. Everglades National Park officials grew increasingly wary about the Corps' seeming lack of concern for water supply to the park, especially as Corps and FCD representatives insisted that fish and wildlife benefits were secondary to flood control and water supply. The growth of agricultural and urban interests in South Florida worsened the situation by elevating demands on water, and urban interests themselves complained about the Corps' operation of the project. By the end of the 1950s, various entities had drawn clear lines as to how they believed water should be managed in South Florida, and the purposes for which it should be used. Conflicts between these different interests seemed unavoidable as the 1960s dawned.

Endnotes

[1] Johnson, *Beyond the Fourth Generation*, 160-161; Lamar Johnson, Engineer, to Hon. Spessard L. Holland of Florida, July 17, 1948, File Flood Control Permanent (July-December 1948), Box 287, Holland Papers.

[2] Johnson, *Beyond the Fourth Generation*, 161; Jno R. Beacham, Chairman, Committee on Drainage and Water Conservation and Chairman, Committee on Drainage and Water Control, to Colonel Willis E. Teale, Corps of Engineers, April 8, 1949, File 1110-2-1150a (C&SF) Proj Genl—Flood Control (Jan 49-April 49), Box 8, Accession No. 077-01-0023, RG 77, FRC.

[3] Minutes, 14 July 1949, Volume 1 of Governing Board of Central and Southern Florida Flood Control District Minutes, Box 27, South Florida Water Management District administrative records, West Palm Beach, Florida [hereafter referred to as SFWMDAR].

[4] Quotation in Central and Southern Florida Flood Control District, *Facts About F.C.D.* (West Palm Beach, Fla.: Central and Southern Florida Flood Control District, 1955), 3, 7, 9; see also Minutes, 14 July 1949.

[5] John C. Stephens, "The Cooperative Water Control Program for Central and Southern Florida," paper presented at the Annual Winter Meeting of the American Society of Agricultural Engineers, 17 December 1958, in Library, Jacksonville District, U.S. Army Corps of Engineers, Jacksonville, Florida.

[6] Stephens, "The Cooperative Water Control Program for Central and Southern Florida."

[7] Johnson, *Beyond the Fourth Generation*, 161-163.

[8] Central and Southern Florida Flood Control District, *Ten Years of Progress: 1949-1959* (West Palm Beach, Fla.: Central and Southern Florida Flood Control District, 1959), n.p.

[9] "Round Table discussion on the drainage and flood control program for Central and Southern Florida, held at the annual convention of the Florida Wildlife Federation, November 3rd, at Daytona Beach, Florida," n.d., Folder COOP Drainage Wetlands—Florida, Box 1, Entry 57A0179, Atlanta Regional Office, Office of River Basin Studies, Wetlands, 1944-1956, RG 22, Records of the U.S. Fish and Wildlife Service, National Archives and Records Administration Southeast Region, Atlanta, Georgia [hereafter referred to as NARA-SE].

[10] W. Turner Wallis, "The Interests of Dade County in relation to the Cooperative Water Control Program for Central and Southern Florida," copy in South Florida Water Management District Reference Center, West Palm Beach, Florida.

[11] The Engineering Department of Central and Southern Florida Flood Control District, "Review of the Plan of Flood Control for Central and Southern Florida in connection with the proposed development of the Everglades area and the operation of the conservation areas," November 1949, 1, copy in South Florida Water Management District Reference Center, West Palm Beach, Florida.

[12] The Engineering Department of Central and Southern Florida Flood Control District, "Review of the Plan of Flood Control," 5-12.

[13] Johnson, *Beyond the Fourth Generation*, 181-182.

[14] Quotation in Stanley J. Niego, Attorney, to Annette Star Lustgarten, Assistant General Counsel, April 7, 1981, File Conservation Areas 1, 2, 3, 1970-1986, Box 02193, SFWMDAR; see also Game and Fresh Water Fish Commission to Hon. Richard W. Erwin, Attorney General, 30 March 1960, File E.C.A. High Water and Deer Herds, 1959-1974, Box 1, Series [S] 1719, Game and Fresh Water Fish Commission Everglades Conservation Files, 1958-1982, Florida State Archives, Tallahassee, Florida [hereafter referred to as FSA].

[15] Harry A. Kersey, Jr., "The East Big Cypress Case, 1948-1987: Environmental Politics, Law, and Florida Seminole Tribal Sovereignty," *The Florida Historical Quarterly* 69 (April 1991): 459-466.

[16] Quotation in R. W. Pearson, Colonel, Corps of Engineers, District Engineer, to The Division Engineer, 30 January 1951, File 1110-2-1150a (C&SF) Proj Genl—Flood Control (Jan 51-June 51), Box 7, Accession No. 077-01-0023, RG 77, FRC; see also Frank Pace, Jr., Secretary of the Army, to The Honorable The Secretary of the Interior, 29 May 1951, ibid.; Johnson, *Beyond the Fourth Generation*, 182-183.

[17] Roy Wood, Regional Supervisor, to Regional Director, 28 November 1952, File RB Coop US Corps Engineers (General), Box 3, Entry 57A0179, RG 22, NARA-SE.

[18] "Cooperative and License Agreement Between the Central and Southern Florida Flood Control District and the Game and Fresh Water Fish Commission," 18 January 1952, File GOV 2-16-03 WV 91 6160 DOT/Miccosukee Mediation, Box 19707, SFWMDAR.

[19] See Office, District Engineer, Jacksonville, Fla., to the Chief of Engineers, U.S. Army, 1 October 1929, File 1110-2-1150a (Kissimmee River) Project General—Flood Control 1924, Box 2, Accession No. 077-02-0048, RG 77, FRC; B. C. Dunn, Lt. Col., Corps of Engineers, District Engineer, to Hon. J. Mark Wilcox, House of Representatives, 10 January 1935, ibid.

[20] The FCD's governing board had decided soon after its formation to hold meetings "in different sections of the District in order to permit the people of the various areas involved to become better acquainted with the Board and its work." W. Turner Wallis, Secretary, to District Engineer, Jacksonville District, 17 September 1949, File 1110-2-1150a (C&SF) Proj—Genl—Flood Control (Aug 49-Oct 49), Box 8, Accession No. 077-01-0023, RG 77, FRC.

[21] Minutes, 3 September 1949, Volume 1 of Governing Board of Central and Southern Florida Flood Control District Minutes, Box 27, SFWMDAR.

[22] Lewis A. Pick, Major General, Chief of Engineers, to Honorable Spessard L. Holland, 25 October 1949, File 1110-2-1150a (C&SF) Proj—Genl—Flood Control (Aug 49-Oct 49), Box 8, Accession No. 077-01-0023, RG 77, FRC. The Chief of Engineers usually carries the rank of Lieutenant General (which Pick eventually assumed), but because Pick was only appointed Chief of Engineers in March 1949, his promotion was probably either not yet approved or delayed.

[23] C. H. Chorpening, Brigadier General, USA, Assistant Chief of Engineers for Civil Works, to The Division Engineer, South Atlantic Division, 10 December 1952, File 1110-2-1150a (C&SF) Proj Genl—Flood Control (Aug 52-Dec 52), Box 8, Accession No. 077-01-0023, RG 77, FRC.

[24] As cited in "Florida Senators Charge Flood Control Job Delayed," *St. Petersburg Times*, 24 June 1953.

[25] As cited in "Florida Senators Charge Flood Control Job Delayed," *St. Petersburg Times*, 24 June 1953.

[26] Quotations in H. W. Schull, Jr., Colonel, Corps of Engineers, District Engineer, to Honorable George A. Smathers, United States Senate, 5 July 1953, File 1110-2-1150a (C&SF) Proj Genl—Flood Control (July 1953-Nov 1953), Box 8, Accession No. 077-01-0023, RG 77, FRC; see also Schull to Honorable Spessard L. Holland, United States Senate, 5 July 1953, ibid.

[27] "Speed Assured on S. Florida Flood Project," *Miami Daily News*, 24 June 1953, clipping in File 1110-2-1150a (C&SF) Proj Genl—Flood Control (July 1953-Nov 1953), Box 8, Accession No. 077-01-0023, RG 77, FRC. As with Pick, Sturgis eventually was promoted to Lieutenant General, but at this time—only three months from the time he became Chief of Engineers—he was still listed as Major General.

[28] As quoted in *A Study of the Central and Southern Florida Flood Control Project* (Washington, D.C.: The Library of Congress, 1953), 6-7, 15-18, 42-46, copy in Library, Jacksonville District, U.S. Army Corps of Engineers, Jacksonville, Florida.

[29] Act of 3 September 1954 (68 Stat. 1248).

[30] Corps of Engineers, U.S. Army, Office of the District Engineer, Jacksonville, Fla., *Central and Southern Florida Project: Special Report on Local Cooperation in the Part of the Project Authorized by the Flood Control Act of 1954* (Jacksonville, Fla.: Corps of Engineers, U.S. Army, 1956), ii.

[31] Buker, *Sun, Sand and Water*, 107-108.

[32] Act of 10 October 1949 (63 Stat. 733).

[33] As quoted in Carl L. Biemiller, "The Water Wilderness—The Everglades," *Holiday* 10 (November 1951): 117.

[34] Thomas J. Allen, Regional Director, to District Engineer, 19 July 1950, File 1110-2-1150a (C&SF) Proj Genl—Flood Control (May 50-Aug 50), Box 9, Accession No. 077-01-0023, RG 77, FRC. See also Keith Wheeler, "Florida's Never-Never Land," *Science Digest* 29 (May 1951): 29-30.

[35] Thomas J. Allen, Regional Director, to District Engineer, Jacksonville District, Corps of Engineers, 16 August 1949, File 1110-2-1150a (C&SF) Proj—Genl—Flood Control (Aug 49-Oct 49), Box 8, Accession No. 077-01-0023, RG 77, FRC.

[36] Allen to District Engineer, 16 August 1949.

[37] R. W. Pearson, Colonel, Corps of Engineers, to Director, Region One, 26 August 1949, File 1110-2-1150a (C&SF) Proj—Genl—Flood Control (Aug 49-Oct 49), Box 8, Accession No. 077-01-0023, RG 77, FRC

[38] Allen to District Engineer, Jacksonville District, 1 September 1949, File 1110-2-1150a (C&SF) Proj—Genl—Flood Control (Aug 49-Oct 49), Box 8, Accession No. 077-01-0023, RG 77, FRC.

[39] R. W. Pearson, Colonel, Corps of Engineers, District Engineer, to Regional Director, U.S. Department of the Interior, National Park Service, 30 June 1950, File 1110-2-1150a (C&SF) Proj Genl—Flood Control (May 50-Aug 50), Box 9, Accession No. 077-01-0023, RG 77, FRC.

[40] Allen to District Engineer, 19 July 1950.

[41] Interestingly, in September 1951, Harold A. Scott, Chief of the Planning and Reports Branch of the Jacksonville District, presented a paper about water distribution from the C&SF Project. Using hydrologic studies completed by the Corps, he extensively discussed how much water would be available for agricultural and municipal areas and how it would be distributed. However, even though Everglades National Park was listed as a "major water-demand area," there was no discussion of how much water it would need or how it would get it. See Scott, "Distribution of Water in the Central and Southern Florida Project," 26 September 1951, 2, South Florida Water Management District Reference Center, West Palm Beach, Florida.

[42] Johnson, *Beyond the Fourth Generation*, 208-209.

[43] All quotations in Engineering Department, FCD, "A Report on Water Resources of Everglades National Park, Florida," 22 May 1950, 5-7, 10-11, South Florida Water Management District Reference Center, West Palm Beach, Florida.

[44] Engineering Department, FCD, "A Report on Water Resources of Everglades National Park, Florida," 11.

[45] Johnson, *Beyond the Fourth Generation*, 209; see also Engineering Department, FCD, "A Report on Water Resources of Everglades National Park, Florida," 12.

[46] All quotations in Engineering Department, FCD, "A Report on Water Resources of Everglades National Park, Florida," 13, 16.

[47] Superintendent, Everglades National Park, to Regional Director, Region One, 26 November 1957, EVER 22965, Central Records, Everglades National Park Archives, Everglades National Park, Homestead, Florida [hereafter referred to as CR-ENPA].

[48] Johnson, *Beyond the Fourth Generation*, 209-211. A copy of Johnson's report was not available, but Johnson summarized his conclusions in *Beyond the Fourth Generation*. For NPS comments on the report, see Acting Supervisory Engineer to Superintendent, Everglades National Park, 19 September 1958, EVER 22965, CR-ENPA; Chief, Water Resources Section, to Regional Director, Region One, 9 September 1958, ibid.; Superintendent, Everglades National Park, to Regional Director, Region One, 21 October 1958, ibid.

[49] Warren F. Hamilton, Superintendent, to Col. Paul D. Troxler, Chief, U.S. Corps of Engineers, Jacksonville, Florida, 29 December 1958, EVER 22965, CR-ENPA.

[50] Johnson, *Beyond the Fourth Generation*, 211.

[51] United States Department of the Interior, Bureau of Sport Fisheries and Wildlife, Region 4, *A Fish and Wildlife Report for Inclusion in the Corps of Engineers' General Design Memorandum, Part 1: Agricultural and Conservation Areas, Supplement 27—Plan of Regulation for Conservation Area #2, Central and Southern Florida Flood Control Project* (Vero Beach, Fla.: Branch of River Basins, 1958), 8-9.

[52] Bureau of Sport Fisheries and Wildlife, Region 4, *A Fish and Wildlife Report for Inclusion in the Corps of Engineers' General Design Memorandum, Part 1*, 8-9.

[53] Bureau of Sport Fisheries and Wildlife, Region 4, *A Fish and Wildlife Report for Inclusion in the Corps of Engineers' General Design Memorandum, Part 1*, 13.

[54] Bureau of Sport Fisheries and Wildlife, Region 4, *A Fish and Wildlife Report for Inclusion in the Corps of Engineers' General Design Memorandum, Part 1*, 17. The partitions of both Conservation Area No. 2 and No. 3 were largely completed in the 1960s. See Light and Dineen, "Water Control in the Everglades," 63-65.

[55] See Paul D. Troxler, District Engineer, to Regional Director, U.S. Fish and Wildlife Service, 3 October 1957, File 1110-2-1150a (C&SF) Kissimmee River Valley Study (Study for Navigation) Jan 1957-Dec 1957, Box 11, Accession No. 077-01-0023, RG 77, FRC.

[56] B. F. Hyde, Jr., Executive Director, to Mr. Walter A. Gresh, Regional Director, 7 October 1957, File 1110-2-1150a (C&SF) Kissimmee River Valley Study (Study for Navigation) Jan 1057-Dec 1957, Box 11, Accession No. 077-01-0023, RG 77, FRC.

[57] See, for example, undated newspaper clipping in File 1110-2-1150a (C&SF) Proj Genl—Flood Control (June 56-Dec 56), Box 7, Accession No. 077-01-0023, RG 77, FRC.

[58] G. H. Snyder and J. M. Davidson, "Everglades Agriculture: Past, Present, and Future," in *Everglades: The Ecosystem and Its Restoration*, 100-103.

[59] Gary R. Mormino, "Sunbelt Dreams and Altered States: A Social and Cultural History of Florida, 1950-2000," *The Florida Historical Quarterly* 81 (Summer 2002): 4, 6, 14.

[60] W. Turner Wallis, "The Interests of Dade County in relation to the Cooperative Water Control Program for Central and Southern Florida," copy in South Florida Water Management District Reference Center, West Palm Beach, Florida.

[61] Quotation in Wallis, "The Interests of Dade County in relation to the Cooperative Water Control Program for Central and Southern Florida"; see also "Dade County Water-Control Report," 35-36, attachment to Marion E. Boriss, Director of Public Works, to Honorable O. W. Campbell, County Manager, 28 November 1958, copy in South Florida Water Management District Reference Center, West Palm Beach, Florida.

[62] E. C. Itschner, Chief of Engineers, to Honorable A. S. Herlong, Jr., 26 April 1960, File 1110-2-1150a (C&SF) Dade County 1955-April 1960, Box 4, Accession No. 077-96-0038, RG 77, FRC.

[63] "Conference on Conservation Area No. 3, Jacksonville, Florida, 14 April 1960," File 1110-2-1150a (C&SF) Conservation Areas Jan 60-June 60, Box 15, Accession No. 077-96-0038, RG 77, FRC.

[64] Central and Southern Florida Flood Control District, *Ten Years of Progress*, n.p.; see also Light and Dineen, "Water Control in the Everglades," 60-63.

CHAPTER 4   **Conflicting Priorities:** Everglades National Park and
Water Supply in the 1960s

WHEN THE CORPS OF ENGINEERS first proposed the C&SF Project, the NPS and the U.S. Department of the Interior were both concerned about a lack of specifics in the plan about water supply to Everglades National Park. The Corps made general references to the necessity of providing adequate water to the park, but did not discuss explicit measures. These anxieties were heightened in the 1950s as project construction commenced, especially as the Corps and the FCD insisted that fish and wildlife preservation were secondary to flood control and urban and agricultural water supply. As population increased along Florida's southeast coast, and as sugar production exploded in the EAA, demands for water became more pressing. After the construction of C&SF Project works and consecutive drought years constricted the amount of water flowing into Everglades National Park, cries for a guaranteed supply of water became more pronounced, leading to discussions on water supply and ownership in South Florida. These pleas, as well as the efforts of a growing environmental movement in South Florida, led to the passage of a congressional mandate in 1970 that the C&SF Project deliver a certain amount of water to the park each year.

At the advent of the 1960s, NPS officials had been wrangling with the Corps over the issue of water supply to Everglades National Park for years. No one seemed to know exactly how much water the park required, but park authorities believed that the area needed the traditional overflows from Lake Okeechobee to course through its veins, especially between the months of October and May when rainfall was scarce. Unfortunately, the construction of drainage and flood control works constricted that southward flow, reducing the hydroperiod of the park, or the time when water enveloped the landscape. This left Everglades National Park parched and dusty when

rainfall ceased. The situation did not seem too severe in the 1950s, mainly because the construction of the East Coast Protective Levee allowed water flowing to the ocean to be diverted south through the Everglades.[1] As the Corps completed construction of L-29—the southern boundary of Water Conservation Area 3—these diversions were eliminated, causing clashes between the NPS and the Corps.

One of the primary agricultural industries that expanded considerably in the 1960s was sugar. Cane had been an important crop in the EAA since the 1920s, but because of the United States' sugar quota system, established in the 1934 Sugar Act, the sugar industry in Florida remained relatively small, confined mainly to the operations of Charles Mott's United States Sugar Corporation. In the early 1960s, however, the industry expanded greatly in Florida due to several factors. For one, Fidel Castro overthrew the Cuban government in 1959, leading the United States government to sever all ties with Cuba, one of the main suppliers of sugar to the United States. For another, some vegetable growers in the EAA, facing unstable markets, wanted to diversify their crops and saw sugar as a safe and profitable venture. In addition, Puerto Rican growers could not meet their production quotas, creating a void in the market.[2]

Because of these conditions, sugar production increased dramatically in Florida in the 1960s. Numerous new companies began operations, including the Osceola Farms Company, formed by a Cuban family, the Fanjuls, who would eventually become the second largest sugar producer in Florida, and the Sugar Growers Cooperative of Florida, established in 1962 by George Wedgworth, the son of South Florida farming pioneers. The Glades County Sugar Growers Cooperative Association, the Talisman Sugar Corporation, and the Atlantic Sugar Association were other fledgling organizations. This influx of companies expanded the amount of acreage under sugar production in Florida from 38,600 in 1954 to nearly 220,000 acres in 1964, mostly in the EAA.[3] As sugar became the dominant EAA crop, its growers and representatives became increasingly interested in how water was distributed throughout South Florida.

Sugar cane plants in South Florida. (South Florida Water Management District)

Castro's revolution also contributed to South Florida's growing population, as numerous Cubans moved to Miami and Dade County to escape communism. Because many Cubans located elsewhere after landing in Miami, and because others did not register upon their entry into Florida, it is difficult to estimate the number of Cubans that relocated to Dade County during this period. However, by 1970, over 300,000 Cubans lived in the county, accounting for approximately 22 percent of its total population of 1,267,792. Although immigrants from other countries in the Caribbean, Latin America, and Asia would enter Florida in large numbers in later decades, Cubans, according to historian Charlton W. Tebeau, "were by far the most significant addition to Florida's population in the sixties."[4] By 1970, the combined population of Dade, Broward, and Palm Beach counties almost reached two million. As the urban region became more populated, settlement extended southwest toward Homestead and closer to the boundaries of Everglades National Park, and the larger populations made increasing demands on water.[5]

Miami was the center for much of this urban growth. Construction of hotels along Miami Beach facilitated the tourist industry, as did the broadcasting of television shows on the beach, which showed millions of Americans the leisure opportunities that Miami offered. More permanent residents were attracted by burgeoning economic opportunities, such as the growth of the Miami International Airport, more jobs generated by the increasing popularity of the fast-food chain Burger King (headquartered in the area), and the booming real estate market. By the late 1960s, South Florida had

a developed area approaching 600 square miles, almost quadruple what it had been around 1955.[6]

This growth increased the demand for water, a situation that alarmed Everglades National Park officials, especially after the Corps began developing a South Dade County Project in the late 1950s. This plan had several components, including a proposal to use water from Conservation Area No. 3, which was supposed to store water for national park usage, to enlarge the county's water supply. The Corps also proposed to build a series of canals to drain land east and south of the park. Concerned that such waterways would divert water that normally drained into the park, NPS authorities protested.[7]

To address these concerns, the Corps held a conference with NPS, FWS, FCD, and Florida Game and Fresh Water Fish Commission representatives in April 1960. At this meeting, NPS representatives emphasized the park's need for a steady supply of water, especially in its southern and western sections and below the Tamiami Trail. The Corps understood these needs, but also reiterated its responsibilities to provide water for salinity control, sewage dilution, agriculture, and municipal purposes. "Methods to conserve water will have to be developed," Jacksonville District officials stated.[8] They also explained that although water from Conservation Area No. 3 would be used for Dade County, such utilization would not "greatly affect" flood discharges into the park from the north, "the principal source of outside water supply to the Everglades National Park."[9] The Corps worked for the next few years to build conveyance canals to route water from Southwest Dade County into the park, but this too generated criticism because it had the potential of bringing insecticides, pesticides, and fertilizers into the park.[10]

Yet it was clear that as Dade County continued to grow—and projections estimated that the county would reach two million by 1970 and four million by 1980—its population would need more and more water. This led Secretary of the Interior Stewart L. Udall to wonder about how the C&SF Project would affect Everglades National Park in regard to the amount, place, and time of water releases. Fearing that Dade County would encroach on park water, Udall asked that the Corps grant the park a guaranteed annual supply that municipal or agricultural demands could not reduce.[11]

Secretary of the Army Elvis J. Stahr, who would later become president of the National Audubon Society, explained that the Corps could not make such an assurance because it had no authority to grant water rights to any entity. "The Department of the Army does not acquire water rights for the construction and operation of Civil Works projects," Stahr claimed, "except as they may be connected with lands being acquired for a dam or a reservoir."[12] If the NPS officials wanted a guarantee, they would have to coordinate with the FCD or the state of Florida, but the FCD believed that no such assurance was possible because of the difficulty of predicting how much water each interest would need in a given year.

Miami Beach, 1963. (The Florida Memory Project, State Library and Archives of Florida)

The situation became more pronounced as drought ravaged the park. In 1961, much of Everglades National Park received only half of its normal rainfall, and, by March 1962, the park was littered with "remnants and carrion—but no life," according to *National Parks Magazine* contributor Gale Koschmann Zimmer.[13] The lack of water destroyed fish and shellfish populations, and, faced with the decimation of these food sources, birds either died or fled. At the same time, fire danger became high, and saltwater concentrations along coastal areas of the park became pronounced. "The whole effect of the drought upon the ecology of the Everglades cannot now be foretold," the park's chief naturalist Ernst Christensen explained, but "the impact upon park life is already serious."[14]

Park officials believed that C&SF Project features only exacerbated the drought because they eliminated traditional sheet flows into the area. They therefore demanded that the Corps give Everglades National Park as much water as it received before C&SF Project construction began. In addition, they asked the Corps to enlarge the water conservation areas to provide sufficient storage for the park's needs. Acting South Atlantic Division Engineer Colonel H. J. Kelly responded that the C&SF Project actually delivered more water than the park had received during Florida's drainage era, and that the conservation area solution was unrealistic because increased seepage and evaporation would offset any raises in water levels. But, Kelly continued, although the Corps could not fully satisfy the NPS's demands, it would search for "a middle ground of reasonable compromise" that would help the park receive more water.[15]

The NPS was especially concerned with the construction of Levee 29 and Control Structure 12, which would form the south boundary of Conservation Area No. 3. According to park officials, these devices would completely eliminate water entering the park

S-12C. (The Florida Memory Project, State Library and Archives of Florida)

from the north. The Corps proposed placing four major outlet spillways in the levee to discharge water into the park, as well as building transitions within the park so that the water could be effectively distributed. But NPS officials refused to allow the Corps to build any structures within the park, forcing the Jacksonville District to work outside park boundaries. Corps officials did not believe that this demand was too unreasonable, but at the same time, according to Colonel Kelly, it evinced an uncooperative, insular attitude that hindered discussion and negotiation.[16]

The positions of both sides hardened at an October 1961 conference between the NPS, the Corps, and the FCD in the Interior De-

partment offices in Washington, D.C. As reported by FCD engineer William V. Storch, the NPS reiterated the necessity of a guaranteed water supply to Everglades National Park, and declared that if the Corps would not grant one, the NPS would petition Congress to restrict C&SF Project funds until an agreement was reached. Yet Corps representatives insisted that a guarantee had to be arranged between the FCD and the NPS. FCD officials agreed with the Corps' position, but, they stated, no agreement could be made "until more accurate knowledge was available both as to Park minimum requirements and the east coastal demands."[17] Not all was lost for the park, however. According to Storch, the Corps did admit that House Document 643 contained "an apparent obligation . . . to provide positive water supply benefits to the Park," and it pledged that it would make "a thorough review of the overall water needs of the area" to determine how this could be accomplished.[18]

In 1962, tensions continued to simmer. When the Corps proposed to enlarge the lower 17 miles of the West Palm Beach Canal to facilitate floodwater discharge to the Atlantic Ocean, the NPS objected, stating that the Corps should expand storage facilities and divert the floodwater into the park. The Corps responded that such a proposal was not feasible because of the expense.[19] Moreover, the NPS made good on its threat to turn to Congress, and in the summer of 1962, the Senate Committee on Public Works passed a resolution asking the Board of Engineers for Rivers and Harbors to make a comprehensive survey of existing water supplies to the park and to recommend how it could receive more water.[20]

Before anything could be accomplished, trouble developed over Levee 29. Even though the Corps had placed four spillways within the structure to ensure that water reached the park, no water flowed through the levee between January and May 1962, causing, in the words of the USGS, "near-record low water levels" and saltwater encroachment in the southern portion of the park.[21] The Corps claimed that the situation resulted because it had to shut off water to complete additional construction in the area, but many questioned that position. Verne O. Williams, a reporter for the *Miami Daily News*, wrote that the only reason why Everglades National Park did not have enough water was because of a "man-made drouth," and he placed all of the blame on the Corps and its "costly drainage works," calling Levee 29 "a plug in the throat of a funnel."[22]

The FCD did not help matters by refusing to open Levee 29's gates once the Corps had finished construction. From 1963 to 1965, the gates remained shut, even though drought continued to ravage the Everglades. Although the FCD had legitimate reasons for closing the gates, such as the necessity of filling the finally completed Conservation Area No. 3 and of maintaining it at the desired level, many believed that the FCD was trying only to preserve more water for agricultural and urban interests.[23] Paul Tilden, a contributor to *National Parks Magazine*, claimed that even though the park received more than 500,000 visitors annually, the FCD and the Corps regarded it as an "afterthought" and an "appendage" that could get water only "after the Florida east coast cities, industries, and agricultural areas have been served."[24] This disregard, Tilden believed, mobilized individuals concerned with Florida's environment, and they increasingly called for a halt to C&SF Project construction until Everglades National Park received a minimum guarantee of water.

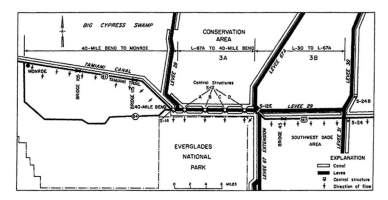

L-29 and its four spillway structures (S-12A, B, C, D). (U.S. Army Corps of Engineers, Jacksonville District)

Meanwhile, the Corps moved forward on its study of park water requirements. Yet its proposed plan of study focused on how engineering structures could bring more water to the area, rather than investigating how much water the park needed to survive.[25] Therefore, the NPS called on different government and private agencies to examine the park's water needs. Responding to these demands, the USGS, after correlating average monthly water stage data in the park with flows from the Tamiami Trail from 1953 to 1962, determined that water flows for that period averaged around 260,000 acre-feet at Shark River Slough and 55,000 acre-feet at Taylor Slough. This was a landmark finding even though park officials had no ecological data to show that this amount was necessary or sufficient to keep plant and wildlife alive. After receiving this information from the USGS, NPS officials agitated for an annual delivery of 315,000 acre-feet to the park, a figure they would continue to cite throughout the 1960s and 1970s.[26] This figure, of course, was drastically different from the more than two million acre-feet that Superintendent Warren Hamilton said was the park's optimum requirement in 1958. However, an Interior Department position paper published in 1964 clarified that the 315,000 acre-feet was merely what the park desired for an interim supply; it was not based on what was needed to maintain the park ecologically and should not be construed as such. Further long-term studies were necessary to determine the ecological needs of the park and its estuaries.[27]

Indeed, inquiries into the requirements of the Shark River and Taylor sloughs were ongoing. These sloughs were deep, wide water channels that conveyed water across the Everglades. Shark River Slough, the larger channel, was located south of Conservation Area

No. 3 and the Tamiami Trail, and flowed southwest into the Gulf of Mexico. Taylor Slough ran southwest from the park's eastern boundary, moving through the Royal Palm area into Florida Bay. If these sloughs did not receive enough water, the whole park suffered. In addition, a lack of water in Taylor Slough affected life in Florida Bay, an estuary that was a prime nursery for shrimp and coastal fishes. Shrimpers annually harvested $15 million worth of shrimp from Dry Tortugas, a cluster of seven islands located southwest from the bay, meaning that changes in water flow not only harmed the ecology of the bay, but a thriving South Florida industry as well.[28]

Aware of this situation, the Institute of Marine Science at the University of Miami conducted a study from 1963 to 1966 about the ecology of Everglades National Park's estuarine regions and the effects of water—or the lack of it—on these areas. Institute scientists especially wanted to see how salinity and temperature changes affected plant and animal communities between the upper Florida Keys and the Chatham River of the Ten Thousand Islands. They could then use these data to construct the freshwater requirements of the estuaries, allowing park officials to make a more informed recommendation as to how much water the park needed annually to protect not only the land-based ecology but the estuarine regions as well. The study concluded that variations in salinity had the greatest impacts on plant and animal life, and that ground water elevation in the Homestead well—designated as S-196A—had a direct relation to Florida Bay's salinity. Therefore, Everglades National Park had to have at least enough water to prevent high saline conditions in the bay.[29]

Meanwhile, the NPS received information that even though the Corps had not yet completed its restudy of water demands, Corps officials were planning works to supply water to Martin County. The NPS objected to such a program "until the project can and does supply the water needs of Everglades National Park."[30] In fact, between 1962 and 1965, the NPS consistently denounced Corps plans for any new construction on the C&SF Project because the Corps would not guarantee water for Everglades National Park. But the Corps insisted that it was giving every consideration to park needs and that it was trying to solve the problem within project parameters. It would support releases to the park as long as they did not, in the words of the Secretary of the Army, "override the basic purposes of the project or the higher priority needs of water supply based on the rapidly expanding population of Florida."[31] Indeed, primary project purposes, as defined by House Document 643, were flood control and water supply for agricultural and municipal uses; fish and wildlife preservation was only a secondary purpose. But the Interior Department had insisted from the beginning (and even in House Document 643 itself) that the Corps operate the project to benefit Everglades National Park, and the Corps had seemingly agreed to that arrangement.[32] Now, NPS officials charged, the Corps had reneged on those promises to the detriment of the park's ecology.

By 1965, the water situation in Everglades National Park had become critical. The Interior Department related that pools and marshes had evaporated, while saltwater intrusion along coastal areas had shrunk fish and wildlife habitat. At the same time, alligator holes dried up, forcing park officials to dynamite holes out of the limestone bedrock to provide adequate habitat for the animals. To alleviate the situation, the FCD worked on an emergency water release schedule for the park, whereby it would receive water from Conservation Area No. 3. This plan went into effect in 1965, but the NPS complained that it only provided at best one-tenth of the park's monthly requirements. Meanwhile, because Lake Okeechobee was experiencing high water levels in the spring of 1965, the Corps allowed 70,000 acre-feet of water to flow to the Atlantic Ocean and the Gulf of Mexico between April 7 and April 22.[33] The NPS loudly decried these releases because of the parched state of the Everglades, wondering why the Jacksonville District could not have sent the water directly to the park. The media picked up on these complaints, prominently displaying the park's dry condition and excoriating the Corps for the discharges. Based on these reports, outraged citizens began writing letters to the Corps demanding that water from the water conservation areas be released into the park. At the same time, Secretary of the Interior Stewart Udall demanded that the Corps study the park's water problem and develop a solution, and the Senate and House Committee on Public Works passed resolutions requesting such a study. The Corps began work thereafter, but it would take three years before its report was finished.[34]

In the meantime, the Corps and the FCD explained that the discharge was necessary to relieve the high water situation quickly and that canals were not designed to divert large volumes of water southward to the park.[35] In addition, William Storch, the director of the FCD's engineering division, emphasized that the FCD had made "a reasonable effort" to provide more water for Everglades National Park in accordance with "the water needs of the area contributing taxes to the support of the District," namely the EAA and east coast urban areas. Storch cautioned people to remember that water supply questions had difficult "social, economic and political considerations," and he admonished participants to leave emotion out of the decision-making process.[36]

The situation became less severe in September 1965 when Hurricane Betsy flooded the Everglades with six to ten inches of rain, but the overall problem of water supply to the park remained.[37] Therefore, after receiving recommendations from the NPS based on past water flows to the Everglades, the Corps and the FCD established an interim regulation schedule to supply water to the park until the Corps had completed its water study and constructed whatever works were necessary. According to the agreement, the FCD would pump water from Lake Okeechobee "in addition to or in conjunction with pumping for lake regulation as scheduled" and the Corps would reimburse the district's expenses for such

Deer in Everglades National Park. (The Florida Memory Project, State Library and Archives of Florida)

pumping based on the amounts that actually flowed to the park at S-12. The pumping would occur "whenever it is necessary to lower the lake level for flood control and at such other times when water is available in the lake," and the water thus pumped would be supplied "to the lower East Coast Area and to the Park."[38] In order to allow for such conveyance, the Corps would enlarge and extend the North New River Canal, the Miami Canal, and the L-67 Borrow Canal. In times of imminent emergency, the Corps would still have to send floodwater to the Atlantic Ocean and the Gulf of Mexico through the St. Lucie Canal and the Caloosahatchee River, but on other occasions the FCD could pump water from Lake Okeechobee to the water conservation areas for park use.[39] After much discussion with the Corps, the NPS approved the interim plan, and it went into effect in March 1966.[40]

But a comprehensive water plan was still necessary; as Michael Straight wrote in *National Parks Magazine*, "little can be gained by viewing the needs of the park only in emergency and in isolation."[41] Besides, the drought's effects on wildlife in the park had been startling; NPS officials estimated that only 5 percent of the alligator population had survived, and bird numbers were drastically lower as well. In the words of Park Superintendent Roger W. Allin, the drought years had "caused extensive changes in habitat which may have far-reaching influence on biotic balances."[42]

Regardless of the damage that the drought had caused, Everglades National Park received more than 1.2 million acre-feet of water in 1966.[43] Yet the impoundment of water in Conservation Area No. 3, coupled with heavy rainfall in the spring and summer of 1966, caused severe problems for deer herds in the region and placed both the Corps and the FCD under fire for allowing too *much* water. But Florida Game and Fresh Water Fish Commission Director O. E. Frye, Jr., claimed that several factors caused the high levels in Conservation Area No. 3. Because Everglades National Park demanded "a guaranteed amount of water introduced into the park on a daily basis," and because Lake Okeechobee's water levels exceeded its regulation stage, the Corps and the FCD had released "an unusual amount of water" from the lake and "conveyed [it] southward through various canals" to the water conservation areas. Frye continued that stands of sawgrass in the northern part of Conservation Area No. 3, coupled with the flat topography of the region, prevented water from flowing quickly to the park, making it "stack up in those parts of the conservation areas adjacent to the pumping stations."[44] Unfortunately, the region was the home of a large deer population which was fawning, and the high water had a devastating impact on those animals. As water levels increased, newspapers began publishing accounts of helpless and starving deer stranded in the area; environmentalists such as John "Johnny" Jones of the Florida Wildlife Federation characterized the situation as "a wildlife version of Auschwitz."[45]

To alleviate the problems, the state's cabinet issued an order to the FCD and the Board of Conservation on 12 April 1966 to halt pumping temporarily at pump station S-8, located in the northwestern corner of Conservation Area No. 3, so that water levels could decrease. When levels remained high, Florida Governor W. Haydon Burns ordered the pumping moratorium extended "until favorable conditions returned."[46]

Even though large-scale pumping ceased, the situation became grave in June when Hurricane Alma dumped large amounts of rain on South Florida, causing levels in Conservation Area No. 3 to rise another six inches and placing already-stressed deer in an emergency situation. In response, sportsmen organizations and other concerned citizens called on Governor Burns to take decisive action. Robert F. McDonald, a delegate of the Palm Beach County Airboat and Half Track Club, asked Burns to end "this senseless and shameful disregard of our precious remaining wildlife" by forcing the FCD to stop pumping, but both the FCD and the Corps insisted that it had to pump during heavy rainfall in order to prevent flooding in the EAA.[47]

With the deer herd facing catastrophe, Florida's cabinet created an interagency committee in July consisting of representatives from the Board of Conservation, the FCD, and the Game and Fresh Water Fish Commission "to develop a program to safeguard the Everglades deer herd and other wildlife from intermittent high waters."[48] The committee, known as the Everglades Natural Resources Coordinating Committee, consulted with state and federal management agencies to develop plans as to how the deer could be saved. These consisted of several temporary arrangements, including:

- Obtaining NPS approval to cut channels 200 feet wide and ½ mile long "immediately south of S-12," thereby increasing outflow to Everglades National Park (the NPS had previously refused to allow the construction of such structures);
- Increasing the flow of canals by sending water to coastal areas;
- Ceasing pumping at stations S-6, S-7, and S-8 and moving water from the EAA into Lake Okeechobee; and
- Moving some deer to higher ground.

Under the circumstances, Committee Chairman Randolph Hodges related, these were "the best solution[s] which could be evolved."[49]

Meanwhile, the Corps developed both immediate and long-term solutions to the problems. In the summer of 1966, the agency supplied mowers to cut sawgrass in the northwestern portion of Conservation Area No. 3; it also prohibited vehicles from traversing levee roadways so that deer would not experience "needless fright-induced activity," and it removed a plug at the intersection of the Tamiami Canal and Levee 67 Extension Canal so that more water could flow southward.[50] At the same time, the Corps proposed more long-standing answers, such as completing the construction of a canal running south from the Tamiami Canal on Everglades National Park's eastern boundary to increase water flow from the water conservation areas, and conducting studies into the feasibility of building another conveyance canal on the park's western border. The Corps also provided the Game and Fresh Water Fish Commission with a cost estimate for developing small islands in the conservation areas "above reasonable flood levels" so that deer could have "high-water grazing and refuge."[51] In addition, it proposed to build a conveyance canal through Conservation Area No. 3 so that water could more easily flow southward from the northern parts of that area. "All agencies concerned are cooperating fully and doing all possible to relieve the problem," the Corps concluded, insisting that it could not possibly be blamed for not foreseeing the "extremely wet season" that affected "an area which is primarily intended for water impoundment."[52]

But in the summer of 1966, the media continued to report that the C&SF Project was in large measure responsible for the deer situation, forcing the Corps to take a defensive stance. "The area in which these deer are located is a natural swamp," Acting Chief of Engineers Major General R. G. MacDonnell told one concerned citizen. If the Corps had not constructed the C&SF Project, MacDonnell stated, the water in Conservation Area No. 3 would have flooded cities on the east coast and "the major agricultural lands south of Lake Okeechobee."[53] Likewise, Joe J. Koperski, chief of the Jacksonville District's Engineering Division, informed a journalist that the C&SF Project had actually prevented $15 million in damages from the June rains. "If the large volumes of excess floodwater had not been pumped to the lake and conservation areas," he continued, "the deer situation would have been far overshadowed by headlines citing a disastrous flood in both urban and agricultural areas of south Florida." Koperski claimed that "conservation of natural resources" was a "primary function" of the C&SF Project, and he emphasized that using the water conservation areas for flood control did not necessarily make them incompatible with fish and wildlife propagation.[54] Ronald Wise, a commissioner with the Florida Game and Fresh Water Fish Commission, agreed, although he characterized the conservation areas' "primary purpose" as flood control and the storage of water to "guarantee" that Everglades National Park had a sufficient supply. Yet if the commission could construct "small islands at intervals throughout the conservation area," he concluded, wildlife did not have to suffer during times of high water.[55]

Accordingly, in 1967, the Game and Fresh Water Fish Commission began developing islands in Conservation Area No. 3, ensuring that they contained open sloughs on their sides so that water could continue to flow southward. In addition, the Corps started construction on the different canals and extensions that would facilitate water flow from and within the water conservation areas, including an extension of L-67 along the eastern boundary of Everglades National Park and a conveyance canal from L-67 to the park. It provided the spoil from these projects for the island development. According to Randolph Hodges, these measures were "the maximum compr[o]mise of flood control facilities possible at this time for wildlife preservation without endangering the primary purpose of the flood control project."[56]

In the meantime, another controversy arose in 1966 over the opening of the Aerojet Canal, or Canal 111, in Southwest Dade County. As part of the Dade County Project explained above, the Corps constructed the canal in the 1960s, running from just below Homestead to Barnes Sound. The initial purpose of the canal was to drain lands east and south of Everglades National Park, but after Aerojet General, a space technology company, built a rocket engine testing center in the region, critics saw the canal as creating a barge-accessible waterway for Aerojet's testing facility. In addition, the drainage aspect threatened to allow saltwater to creep up the canal and contaminate fresh water in the park in times of drought. To prevent water from interfering with bridge construction, the Corps had placed an earthen plug in the canal where it intersected U.S Highway 1 (about two miles inland from Biscayne Bay), and this prevented the flow of seawater. Yet upon completion of the bridge, the Corps would remove the plug, allowing saltwater to mingle with freshwater

during unusually high tides and strong winds. The NPS and environmental organizations petitioned the Corps to keep the plug in place, but Corps leaders proposed that it remove the plug and observe whether saltwater intrusion really occurred. Objecting to this plan, the National Audubon Society and other groups applied for a court injunction to maintain the plug. The Corps then informed the NPS that the plug would remain "indefinitely" while a plan was formulated to protect Everglades National Park, and by 1969, the Corps had constructed an earthen barrier with gated culverts downstream from the original plug.[57]

While the controversy raged over C-111, drought returned to South Florida in 1967, renewing cries for more water to the park. The battle was becoming more polarized as the 1960s progressed; essentially, it was a question of whether enough water existed for both Everglades National Park and agricultural and municipal purposes, or whether the FCD and the Corps had to choose among the three. As this polarization occurred, environmental organizations began to wade into the fray with increasing frequency. The National Parks Association asked Americans to contemplate whether sugar and cattle industries should be developed in Florida at the expense of the Everglades, and whether urban centers in South Florida should continue to grow if it endangered park tourism and the shrimp industry in Florida Bay.[58]

C-111. (U.S. Army Corps of Engineers, Jacksonville District)

But not all proponents of fish and wildlife viewed the supply of water to Everglades National Park in positive ways. O. E. Frye, Jr., director of the Florida Game and Fresh Water Fish Commission, for example, noted in 1968 that continual supplies of water to the park were creating critical situations for fish in the water conservation areas, and he requested, with the support of the governor's cabinet, that "if it became apparent that a fish kill was in the offing, releases to the Park . . . be discontinued."[59] Clearly, many factors were involved in water supply issues for the park, and as views became more hardened, the emotionalism decried by Storch became a larger component of water management.

Into this charged setting came the Corps' report on its restudy of water needs in South Florida as requested by Secretary Udall and as required by Congress. Although the Corps originally planned on releasing the report in the summer of 1967, delays, including efforts to address concerns expressed by the NPS, extended the completion date. In the fall of 1967, the Jacksonville District held public hearings in Belle Glade and Coral Gables on its preliminary findings. According to a notice of the hearing, the Corps recommended that in order to provide enough water for the needs of South Florida through 2000, it needed to modify the C&SF Project in the following ways:

- Raise Lake Okeechobee by four feet to a seasonal regulation range between 19.5 and 21.5 feet above mean sea level to provide for more water;
- Pump excess floodwater first to the water conservation areas before discharging it to the Atlantic Ocean and Gulf of Mexico;
- Backpump excess water from Martin and St. Lucie counties to Lake Okeechobee to increase available water;
- Allow several canals draining to the coast to backpump excess runoff to the conservation areas;
- Deliver 315,000 acre-feet to Everglades National Park annually; and
- Build conveyance canals to South Dade County and the Taylor Slough.[60]

The NPS offered its cautious approval to this plan, now believing that, according to available information, a minimum of 315,000 acre-feet a year would allow the park to "survive."[61] Others were not so sure; the Florida Game and Fresh Water Fish Commission, for example, supported the basic principles of the plan, but objected to several specific provisions, including the raising of Lake Okeechobee (which it claimed would have harmful effects on both vegetation and fish and wildlife) and the fact that the commission could find no evidence that the Corps had considered the ecological maintenance of the water conservation areas in its plan. Instead, it appeared to the commission, "the Conservation Areas will be drawn down and sacrificed for the benefit of the water demand areas."[62]

Still others were more concerned with the amount of water going to Everglades National Park. For instance, the National Parks Association objected strongly to the proposal, holding that Everglades National Park needed at least 400,000 acre feet of water and that this amount needed to be explicitly guaranteed. Representatives of the National Wildlife Federation agreed, claiming that the annual delivery needed to be "adjusted to account for the [park's] biological needs."[63] Therefore, the National Parks Association called on Congress to eliminate funding for more C&SF Project work in Florida "until the Nation as a whole has firm legal assurances, binding on the State of Florida and binding even on the Central Florida Flood Control District, guaranteeing the necessary water deliveries into Everglades National Park permanently."[64]

At the same time, agricultural and municipal interests were not pleased with the Corps' recommendations, believing that the Corps was providing too much water to Everglades National Park. Dade County Manager Porter Homer, for example, criticized the restudy, saying that "the 315,000 acre-feet per year used by the corps is not based on adequate research."[65] In the weeks following the public hearings, Corps officials seemed to pay more attention to agricultural and municipal complaints than to environmental criticisms. For one thing, the Corps rethought its proposal to deliver 315,000 acre-feet to the park. Even though NPS leaders insisted that this was a minimum amount that the park needed, South Atlantic Division Engineer Major General T. J. Hayes echoed Homer's complaints that no study existed showing that this was "the required amount to sustain the Everglades effectively" since the USGS had merely averaged the flow into the park from 1952 to 1961.[66] The Corps also refused to guarantee water to the park for several reasons, including its lack of jurisdiction and the fact that "parks do <u>not</u> have an established priority over other authorized project purposes."[67] In addition, members of the Jacksonville District did not want to upset Florida state officials who believed that an annual guarantee would completely halt any urban or agricultural development in South Florida. Finally, Corps representatives believed that they could provide "the basic water demands of the park" without making a guarantee.[68]

When the Corps issued its final report—prepared under the direction of Colonel Robert Tabb of the Jacksonville District—in May 1968, it recognized that "preservation of Everglades National Park is a project purpose and that available water should be provided on an equitable basis with other users." The Corps stated that the prolonged drought had shown that additional measures were necessary "to meet the growing urban and agricultural water-supply needs of South Florida while still providing flows sufficient to maintain the environment of Everglades National Park." To do so, the Corps proposed raising Lake Okeechobee levees so that the lake's level could be maintained at between 19.5 and 21.5 feet above mean sea level; backpumping excess water into the lake and the water conservation areas; improving the conveyance and distribution of water to the park through a system of canals, levees, pumping stations, and control structures; increasing the capacity of the North New River and Miami canals; and providing for recreational development. If such improvements were made, the Corps concluded, it could, on average, offer Everglades National Park 315,000 acre-feet of water while still maintaining South Florida's water needs. Although no specific guarantee was provided, the report seemed to make some concessions to the park and the environmental community. But to environmental supporters, the document's lack of a guarantee was proof that the Corps was either unable or unwilling to correct the damage that its project caused.[69]

Yet given the outcry that agricultural and municipal interests had raised, the Corps' avoidance of an explicit assurance seemed a logical and middle ground position to take, although not one popular with environmental interests. But in the minds of Corps leaders, there was little else the agency could do. Because flood control and water supply were higher priorities under the C&SF Project, the Corps could not specifically guarantee water to the park without congressional direction, especially if the state of Florida, for whom the project was built, was unwilling to compromise on the issue.

At the same time, however, the Corps' response was one that infuriated observers who noted that the Corps was not a passive agency, unable to do anything without congressional approval. Instead, critics charged, the Corps was a highly adaptable, fairly aggressive promoter of its own interests. It was especially difficult in the case of Everglades National Park to understand why the Corps could not merely direct the FCD to supply necessary water to the park, especially since benefiting fish and wildlife *was* a purpose of the C&SF Project, secondary or not. In the minds of many critics, the claim that the Corps just followed congressional instruction was disingenuous at best and historically inaccurate at worse. Arthur R. Morgan, a leading critic of the Corps who had formerly worked as Chief Engineer of the Miami Conservancy District and chairman of the Tennessee Valley Authority, for example, claimed that the real reason why the Corps did not guarantee sufficient water for the park was because it had not conducted "adequate engineering analysis" that focused on South Florida as "an environmental unit." "There is no reticence in the Corps about interfering with and changing legislation of public policy," Morgan argued. "It is only where the Corps wishes to prevent carrying through a program that it pleads its lack of power."[70]

Upset by the lack of an unambiguous guarantee, NPS Director George Hartzog, Jr., informed the Board of Engineers for Rivers and Harbors that unless the report stated "clearly and unequivocally" that Everglades National Park would receive a certain amount of water, the NPS would not concur with the report.[71] Unwilling to act on "national policy questions outside of the purview of the Board," the board emphasized to the Chief of Engineers the need for water in the park, suggesting that the chief should "clearly define the ecological objectives and the amounts of supplemental water needed to meet those objectives." But the board required no definite promise of water in the Corps' report.[72]

Receiving no help from the Board of Engineers, Assistant Secretary of the Interior Stanley A. Cain reiterated that the Interior Department could not approve the proposed project modifications unless it received "written assurance by the Secretary of the Army that he will provide the water supplies as set forth in the report, undiminished by new incursions."[73] Perhaps fearing that Congress would not approve the modifications unless the NPS gave its concurrence, or perhaps in agreement with the NPS's position, Major General F. J. Clarke, Acting Chief of Engineers, informed Secretary of the Interior Stewart L. Udall that "the Chief of Engineers will insure the project is regulated to deliver the water requirements of the Everglades National Park as so set forth in the report."[74] At a subsequent meeting between the Interior Department, the Department of the Army, and the Bureau of the Budget, the Corps assured Interior and the bureau verbally that it would provide 315,000 acre-feet of water to the park and that future demands would not reduce that figure, but it still would not place a specific guarantee in writing. Congress then published the Corps' report as House Document 369, and authorized the modifications, estimated to cost about $70 million, in the Flood Control Act of 1968.[75]

The state of Florida continued to resist any kind of water guarantee to the park. Accordingly, in the summer of 1968, the Corps tried to mediate between the state and the NPS to develop a memorandum of agreement that would assure 315,000 acre-feet of water to the park except in times of drought when it would share shortages with other users on a pro rata basis. The Florida Board of Conservation refused to approve the memorandum, believing that the agreement would forfeit its water rights and insisting that no water user in Florida should have priority over another.[76]

Faced with these problems, the secretary of the interior requested that the department's solicitor issue an opinion as to whether or not the Corps could require the FCD to deliver a certain amount of water to Everglades National Park each year. The solicitor argued that because Congress approved modifications to the C&SF Project upon the recommendation of the Bureau of the Budget, and because the Corps assured the bureau and the Interior Department in its July meeting that the park could receive 315,000 acre-feet, the law required the Secretary of the Army to manage the project "for the purpose of meeting the water requirements of the Everglades National Park." The solicitor continued that the Secretary of the Army "not only has the statutory authority but also a Congressional mandate to issue, unilaterally, regulations for the delivery of project water to the park."[77]

Nevertheless, the Corps began to renege on its verbal assurances, as Robert Jordan, Special Assistant to the Secretary for Civil Functions, insisted that the modification authorized the Corps to provide the 315,000 acre-feet as an *objective*, not as a guarantee.[78] In an attempt to resolve the problem, the U.S. Senate Committee on Interior and Insular Affairs held hearings in June 1969 on water supply to Everglades National Park. At these hearings, Nathaniel Reed, special assistant to Florida Governor Claude R. Kirk, Jr., expressed the state's concern for the park, but stated that it was impossible to guarantee a certain amount of water each year because of Florida's erratic rainfall. Drought might decimate water supplies to the point where the FCD could not supply a required amount. Reed also told the committee that certain priorities existed in Florida regarding water: man—meaning municipal supplies—was first, agriculture was second, and "somewhere along the line" was Everglades National

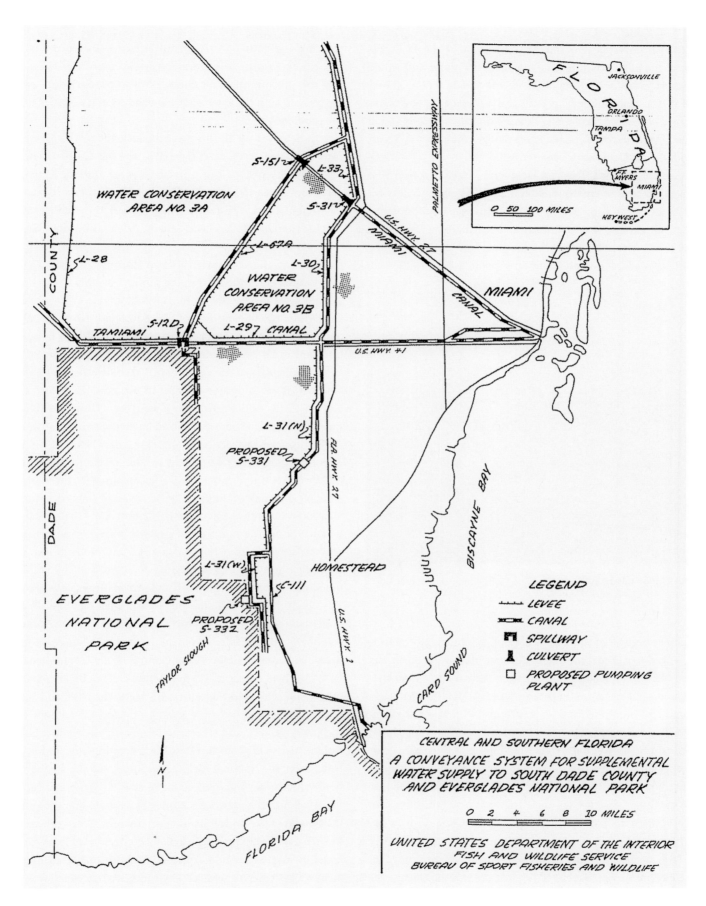

South Dade County Project map. (U.S. Fish and Wildlife Service, Vero Beach, Florida, administrative files)

Park. However, under a new interim schedule that the FCD was developing, the park would receive the necessary water and would only be short in times of drought when "everybody will be short." Robert Padrick, chairman of the FCD and a member of the Sierra Club, agreed with Reed, explaining that the interim schedule would deliver 260,000 acre-feet to Shark River Slough annually "in accordance with the park's monthly requirements."[79]

But Senator Gaylord Nelson, a Democrat from Wisconsin who had a strong interest in environmental matters, as evidenced by his support in this same time period for the National Environmental Policy Act, signed into law by President Richard Nixon on 1 January 1970, could not understand why the state would not agree to a guarantee. The federal government had expended $170 million on

Everglades National Park in the 1960s. (The Florida Memory Project, State Library and Archives of Florida)

the project, he argued, so the state could not claim that the resulting water belonged to it. The intransigence of state officials on the matter infuriated Nelson, who called the situation "ridiculous," "preposterous," and a "disgrace."[80] Acceding to the wishes of the National Parks Association and other environmental groups (who also testified at the hearings), he threatened to halt a proposed $9 million appropriation for the C&SF Project if the state would not give the park a guarantee of 315,000 acre-feet regardless of future demands on water.

Only days after the conclusion of the hearings, Nelson executed his threat, asking the Senate Public Works Subcommittee of the Committee on Appropriations to halt the C&SF Project's $9 million appropriation for fiscal year 1970 until the state and park officials had reached a water supply agreement. Accordingly, the committee's appropriations report directed the state of Florida, the Interior Department, and the Department of the Army to develop an operating agreement to ensure water deliveries to Everglades

National Park. But by 1970, the three parties had held no meetings to formulate a plan. Therefore, Senator Spessard Holland requested that the Subcommittee on Public Works Appropriations convene a conference with the interested state and federal agencies to discuss the problem.[81]

In February 1970, this meeting occurred, with representatives from the state of Florida, the NPS, and the Corps attending. To begin the discussion, NPS Director George Hartzog stated that he could not agree to any plan whereby the park had to share water in drought years with future users. Despite these declarations, the parties, aided by Holland and Senator Allen J. Ellender, chairman of the committee, made some progress and eventually agreed to several things. First, they concurred that an interim water supply delivery plan developed by the FCD in the summer of 1969 to simulate more accurately historical flow patterns would go into effect immediately, supplying 260,000 acre-feet of water to Shark River Slough (canal enlargement had to occur before the Taylor Slough and the eastern panhandle could receive 55,000 acre-feet). Second, when the Corps had enlarged the capacity of Lake Okeechobee to 17.5 acre-feet (which was supposed to occur in two years), the state, the NPS, and the Corps would review the plan to see if the park could receive more than 260,000 acre-feet. Third, once the Corps had completed the necessary construction to increase Lake Okeechobee's levels to 21.5 feet, the interim agreement would cease and the FCD would deliver 315,000 acre-feet annually. Fourth, in 1980, the Corps would conduct a restudy of the C&SF Project and of water demands to see what other action was necessary. The only issue that remained was whether or not the Corps could establish a priority of use that would protect the park from future demands, and Holland and Ellender strongly suggested that a meeting to solve that difference occur quickly so that appropriations for the C&SF Project could resume.[82]

On 16 March 1970, the three parties held another conference to discuss the water supply problem, but although some conciliation was offered, no suitable agreement resulted.[83] Therefore, in April, the Senate Subcommittee on Flood Control of the Committee on Public Works held a hearing on the matter. During this meeting, Senator Nelson reiterated that unless the state, the NPS, and the Corps reached an accord, he would again try to stop any appropriation for the C&SF Project, and representatives from environmental organizations such as the National Wildlife Federation, the Florida Wildlife Federation, the National Parks Association, and the National Audubon Association concurred with this stance. Harkening back to the July 1968 meeting between the Interior Department, the Corps, and the Bureau of the Budget, Nelson accused the Corps of reneging on its verbal pledge to provide 315,000 acre-feet to the park unencumbered by future uses, and expressed his hope that "escalating public concern in America over all environmental matters" would force the Corps and the state to guarantee a water supply.[84] Upon Nelson's conclusion, Senator Edmund S. Muskie, a Democrat from Maine

Everglades National Park. (The Florida Memory Project, State Library and Archives of Florida)

who was known for his support of environmental causes, proposed that the hearing investigate what protections Congress could provide to the park. Although no firm conclusions were reached, it was clear that some members of Congress would fight until Everglades National Park had its guaranteed water.

And, indeed, Nelson and Muskie did. Tired of the constant bickering between the state, the Corps, and the NPS, and resigned to the fact that no agreement was forthcoming, the two pushed a bill through Congress providing money for the conveyance canals and pumping stations proposed in the Corps' 1968 report. But the bill also contained a stipulation added by the Committee on Public Works, of which Muskie was a member, that as soon as practicable, and no later than when the Corps had completed the necessary works, Everglades National Park would receive either 315,000 acre-feet annually, prorated monthly according to an NPS schedule, or 16.5 percent of the total water deliveries from the project, whichever was less.[85] The committee's report explained that the proviso was added "to secure as promptly and regularly as possible delivery of water to the Everglades National Park" and to extinguish all questions of how much water the C&SF Project had to deliver to the park. Because the federal government had supplied so much money for the C&SF Project, and because the park was "a national asset to be preserved for our own and future generations," the committee believed it had the authority to make this stipulation.[86]

Although the NPS now seemed to have the guarantee of water that it desired, problems resulted almost immediately. Since language in the act required the stipulation to become effective as soon as practicable, the Corps and the FCD began implementing it in 1971, a year when little rain fell. Therefore, even though the park would have received more water under the FCD's interim plan, the FCD provided water throughout 1971 following Congress's requirement. This meant that the park received 20 to 25 percent less water than what it would have procured, while agricultural and urban interests continued to receive normal amounts, a situation that struck FCD Executive Director G. E. Dail, Jr., as unreasonable. "Since there is agreement that this formula is an extremely poor one," Dail told Jacksonville District Engineer Colonel A. S. Fullerton, "we do not believe that it should continue to be applied under current conditions," especially since projections showed that normal rainfall would allow "all essential demands" to be met "without the need to impose a curtailment of water use." Fullerton promised to investigate whether Congress intended the formula to apply immediately, but in the meantime, Everglades National Park faced a depleted water supply.[87]

Nevertheless, at least some strides had been made in providing necessary water to the park from the C&SF Project. Throughout the 1960s, the Corps, the FCD, and the NPS all had different viewpoints as to the water priority of Everglades National Park, and these disparities became glaringly apparent as drought ravaged the Everglades. When little water from the C&SF Project was forthcom-

ing, NPS officials demanded that the Corps guarantee to the park a certain amount of water untouchable by future demands. In the words of NPS Director George Hartzog, it was time for the Corps to stop paying mere "lip service to the preservation of the Everglades."[88] Corps leaders, however, claimed that they could not provide such a promise, insisting that only the state of Florida could make that assurance. Because of the phenomenal growth of South Florida, and because supplying water to the park could have adverse effects on fish and wildlife in the water conservation areas (as evidenced by the problems with deer herds in 1966), state officials refused to provide a guarantee. Despite the opposition of the state and the reluctance of the Corps to provide a specific written guarantee, the Corps, in the 1968 restudy report, did, for one of the first times since the authorization of the C&SF Project, admit that the project needed to supply sufficient water to Everglades National Park. This reiteration of the promise in the C&SF Project plan, although somewhat vague, showed that the drought of the 1960s and the work of park proponents was having some effect on the Corps' perception of how the project should be operated. It was a small step, but it set the stage for congressional leaders, such as Senators Gaylord Nelson and Edmund Muskie, to resolve the situation.

Despite the accomplishments, problems of water quality loomed on South Florida's horizon. The 1968 report's proposal to supplement Everglades National Park and Lake Okeechobee water by backpumping from east coast lands and agricultural areas, for example, produced new concerns about water quality, both in the lake and in the park, because the recycled water often contained pesticides, fertilizers, and other harmful chemicals. Even as the NPS fought for a guarantee of water, another danger threatened park ecology: a proposal to build a jetport in the Everglades region. In the 1970s, environmental forces first mobilized in the fight for a guaranteed water supply would need all of their resources to contend with these concerns.

Endnotes

[1] See S. D. Leach, Howard Klein, and E. R. Hampton, *Hydrologic Effects of Water Control and Management of Southeastern Florida*, Report of Investigations No. 60 (Tallahassee, Fla.: State of Florida, Bureau of Geology, 1972), 97.

[2] Wedgworth and Miedema interview. Wedgworth, one of the founders of the Sugar Cane Growers Cooperative of Florida in the 1960s, vehemently denies that Castro's revolution allowed Florida sugar production to increase. "Every pound of the quota that Cuba had," he claims, "was reallocated to 40-some odd foreign countries, and the domestic sugar producers . . . didn't get one pound of the Cuban quota."

[3] "Florida Acres in Sugarcane (1,000 acres)," document provided by George Wedgworth and Barbara Miedema, Sugar Cane Growers Cooperative of Florida, Belle Glade, Florida.

[4] Tebeau, *A History of Florida*, 434.

[5] "How the Florida Boom is Changing," *U.S. News & World Report* 46 (5 May 1969): 89; John G. Mitchell, "The Bitter Struggle for a National Park," *American Heritage* 21, No. 3 (1970): 99; Tabeau, *A History of Florida*, 431-434.

[6] Grunwald, *The Swamp*, 231-232.

[7] See Major General William F. Cassidy, Assistant Chief of Engineers for Civil Works, to Honorable Spessard L. Holland, United States Senate, 27 May 1960, File 1110-2-1150a (C&SF) Conservation Areas Jan 60-June 60, Box 15, Accession No. 077-96-0038, RG 77, FRC; Paul M. Tilden, "The Water Problem in Everglades National Park, Part II," *National Parks Magazine* 38 (March 1964): 9; Frank Nix, Hydraulic Engineer, Everglades, to Director, 21 June 1967, File L54 Levee L-31W, EVER 22965, CR-ENPA.

[8] "Conference on Conservation Area No. 3, Jacksonville, Florida, 14 April 1960," File 1110-2-1150a (C&SF) Conservation Areas Jan 60-June 60, Box 15, Accession No. 077-96-0038, RG 77, FRC.

[9] Cassidy to Holland, 27 May 1960.

[10] See Nix to Director, 21 June 1967; Stanley C. Joseph, Superintendent, to Mr. G. E. Dail, Executive Director, Central and Southern Florida Flood Control District, 13 December 1965, File L54 Levee L-31W, EVER 22965, CR-ENPA.

[11] Stewart L. Udall, Secretary of the Interior, to Mr. Secretary, 18 July 1961, File 1110-2-1150a (C&SF) Project Gen—Flood Control May 1961-Apr 62, Box 7, Accession No. 077-01-0023, RG 77, FRC; Tilden, "The Water Problem in Everglades National Park, Part II," 9.

[12] Elvis J. Stahr, Secretary of the Army, to The Honorable Steward L. Udall, The Secretary of the Interior, 7 September 1961, File 1110-2-1150a (C&SF) Project Gen—Flood Control May 1961-Apr 62, Box 7, Accession No. 077-01-0023, RG 77, FRC; see also Colonel J. V. Sollohub, District Engineer, to Chief of Engineers, 2 August 1961, ibid.

[13] Gale Koschmann Zimmer, "Unless the Rains Come Soon . . . " *National Parks Magazine* 36 (June 1962): 4-7.

[14] As quoted in Zimmer, "Unless the Rains Come Soon . . . " 4-7.

[15] Colonel H. J. Kelly, Acting Division Engineer, to Chief of Engineers, 4 August 1961, File 1110-2-1150a (C&SF) Project Gen—Flood Control May 1961-Apr 62, Box 7, Accession No. 077-01-0023, RG 77, FRC.

[16] Kelly to Chief of Engineers, 4 August 1961; Colonel J. V. Sollohub, District Engineer, to Chief of Engineers, 2 August 1961, File 1110-2-1150a (C&SF) Project Gen—Flood Control May 1961-Apr 62, Box 7, Accession No. 077-01-0023, RG 77, FRC.

[17] W. V. Storch to Files, 13 October 1961, File Conservation Area 1, 2, 3 1950-69 Deeds/General/Regulation/Petition for Change of Zoning, Box 02193, SFWMDAR.

[18] Storch to Files, 13 October 1961. The phrase "positive water supply benefits" was left undefined.

[19] See Assistant Secretary of the Interior to General Wilson, 11 May 1962, in U.S. Army Engineer District, Jacksonville, Corps of Engineers, *Central and Southern Florida Project, Plan of Survey: Everglades National Park Water Requirements* (Jacksonville, Fla.: U.S. Army Engineer District, Jacksonville, Corps of Engineers, 1964), A-1—A-3 [hereafter referred to as *Plan of Survey*]; Lieutenant General W. K. Wilson, Jr., Chief of Engineers, to The Honorable Stewart L. Udall, The Secretary of the Interior, 19 June 1962, File CE SE Central and South Florida FCP Everglades NP Basic Data, FWS-VBAR. The NPS had first proposed this idea in 1958 when Park Superintendent Warren F. Hamilton told District Engineer Colonel Paul D. Troxler that the NPS would prefer to have floodwater provided to the park through a floodway flowing south rather than going to the Atlantic Ocean through the Caloosahatchee River and St. Lucie Canal. See "Chronological Documentation of National Park Service Efforts and Corps of Engineers Responsibility to Assure Everglades National Park of Fresh Water Supply from Central and

Southern Florida Flood Control Project," 5, File CE-SE Central and Southern Florida FCP Everglades National Park Basic Data, FWSVBAR.

[20] Lieutenant General W. K. Wilson, Jr., Chief of Engineers, to The Honorable Stewart L. Udall, The Secretary of the Interior, File 1517-08 (C&SF) So. Dade Co. Mult. Pur. Svy Reso. 11/15/54, Box 5, Accession No. 077-96-0017, RG 77, FRC.

[21] J. H. Hartwell, H. Klein, and B. F. Joyner, *Preliminary Evaluation of Hydrologic Situation in Everglades National Park, Florida* (Miami, Fla.: United States Department of the Interior, Geological Survey, Water Resources Division, 1963), 5, 8.

[22] Verne O. Williams, "Man-Made Drouth Threatens Everglades National Park," *Audubon Magazine* 65 (September-October 1963): 290-291.

[23] U.S. Geological Survey, "The Road to Flamingo: An Evaluation of Flow Pattern Alterations and Salinity Intrusion in the Lower Glades, Everglades National Park," available at <http://sofia.usgs.gov/publications/ofr-02-59/culverts.html > (10 January 2005); "Summary of Everglades National Park and Its Water Problems," 1 June 1965, 4, File Everglades Park Area—Review of Central and Southern Florida Vol. III, FWSVBAR.

[24] Tilden, "The Water Problem in Everglades National Park, Part II," 10.

[25] *Plan of Survey*, 2-3, 8-9, B-1.

[26] Hartwell, Klein, and Joyner, *Preliminary Evaluation of Hydrologic Situation in Everglades National Park, Florida*; "Supplemental Statement of Harthon L. Bill," in Senate Committee on Public Works, Subcommittee on Flood Control—Rivers and Harbors, *Central and Southern Florida Flood Control Project: Hearing Before the Subcommittee on Flood Control—Rivers and Harbors of the Committee on Public Works, United States Senate*, 91st Cong., 2d sess., 1970, 178; E. W. Reed, Chief, Branch of Water Resources, to Mr. Wallis, 7 July 1964, File L54 Water Resources USGS FY 65, EVER 22965, CR-ENPA.

[27] United States Department of the Interior, "Position Paper: Water Problem, Everglades National Park," 2-3, EVER 22965, CR-ENPA.

[28] See "Research Plan," 12 December 1963, File CE SE Central and South Florida FCP Everglades NP Conveyance Canals (Taylor Slough), FWSVBAR.

[29] D. C. Tabb and T. M. Thomas, "Prediction of Freshwater Requirements of Everglades National Park," 3, 4, 12, copy in South Florida Water Management District Reference Center, West Palm Beach, Florida.

[30] George B. Hartzog, Jr., Director, to Major General Jackson Graham, Director of Civil Works, Office of the Chief of Engineers, ca. 22 September 1965, File 1517-08 (C&SF—Martin County) Multiple Purpose Survey—SR 7/22/50 August 1965-April 1966, Box 4, Accession No. 077-96-0017, RG 77, FRC. South Atlantic Division Engineer Major General George H. Walker recommended against parts of the Martin County plan, stating that "our relations at this time with various groups interested in the Everglades National Park do not shine bright with mutual trust and respect." To propose to divert water from Lake Okeechobee to Martin County before the water budget study was complete, Walker argued, was "prejudging our own findings." Walker to Chief of Engineers, 3 December 1965, ibid.

[31] See "Summary of Corps of Engineers' Reports on Central and Southern Florida Flood-Control Project," 7-8, File CE-SE Central and Southern Florida FCP Everglades National Park Basic Data, FWSVBAR; "Chronological Documentation of National Park Service Efforts and Corps of Engineers Responsibility to Assure Everglades National Park of Fresh Water Supply from Central and Southern Florida Flood Control Project," 13-14.

[32] See House, *Comprehensive Report on Central and Southern Florida for Flood Control and Other Purposes*.

[33] Senate Committee on Interior and Insular Affairs, *Everglades National Park: Hearings Before the Committee on Interior and Insular Affairs*,

United States Senate, Ninety-First Congress, First Session, on the Water Supply, the Environmental, and Jet Airport Problems of Everglades National Park, 91st Cong., 1st sess., 1969, 24; "Summary of Everglades National Park and Its Water Problems," 5; U.S. Army Corps of Engineers, "Fact Sheet: The Water Situation in Southern Florida and the Everglades," 30 June 1965, 4, File Everglades Park Area—Review of Central & Southern Florida Vol. III, FWSVBAR; Luther J. Carter, *The Florida Experience: Land and Water Policy in A Growth State* (Baltimore, Md.: Johns Hopkins University Press, 1974), 120.

[34] See Marian Sorenson, "The Everglades 'Drought,'" undated *Christian Science Monitor* clipping, File CE SE Central and South Florida FCP Everglades NP Basic Data, FWSVBAR; Lloyd R. Wilson to Lt. Gen. W. S. Cassidy, Chief of Engineers, 18 October 1965, File Everglades Park Area—Review of Central & Southern Florida Vol. III, FWSVBAR; Davis, *An Everglades Providence*, 406; Lieutenant General William F. Cassidy, Chief of Engineers, to The Secretary of the Army, 3 June 1968, in U.S. House, *Water Resources for Central and Southern Florida*, 90th Cong., 2d sess., 1968, H. Doc. 369, 1 [hereafter referred to as House, *Water Resources for Central and Southern Florida*].

[35] See U.S. Army Corps of Engineers, "Fact Sheet: The Water Situation in Southern Florida and the Everglades" (including marginalia), 8-9; Ed Buckow, "Unraveling the Everglades Furor," *Field & Stream* 71 (October 1966): 15. Critics asserted that pumping water southward through the canals would have lowered the ground water on cropland.

[36] William V. Storch, "South Florida Section, A.S.C.E., Fort Lauderdale, October 7, 1965," 5, 9, File CE SE Central and South Florida FCP Everglades NP Basic Data, FWSVBAR.

[37] For more information on Hurricane Betsy's effects on the Everglades, see Taylor R. Alexander, "Effect of Hurricane Betsy on the Southeastern Everglades," *Quarterly Journal of the Florida Academy of Science* 30 (1967): 10-24.

[38] "Agreement Between the Corps of Engineers and the Central and Southern Florida Flood Control District Establishing Basis of Payment for Pumping Water for Release to Everglades National Park," 26 August 1966 (emphasis in the original), File 1501-07 DACW17-72-A-0004 South Florida Water Management District (Lake Okeechobee Discharge into Everglades National Park), Box 19, JDAR.

[39] Department of the Interior—National Park Service, Department of the Army—Corps of Engineers, "Joint Fact Sheet on: Water Situation at Everglades National Park," 16 February 1966, File CE SE Central and South Florida FCP Everglades NP Basic Data, FWSVBAR; Wallace Stegner, "Last Chance for the Everglades," *Saturday Review* (6 May 1967): 72.

[40] Senate, Committee on Interior and Insular Affairs, *Everglades National Park*, 24.

[41] Michael Straight, "The Water Picture in Everglades National Park," *National Parks Magazine* 39 (August 1965): 7.

[42] Quotation in "Rains Fail to Wash Out Florida Worries," *The Evening Star* (Washington, D.C.), 24 June 1966; see also Stegner, "Last Chance for the Everglades," 72.

[43] U.S. Army Corps of Engineers, "Water Levels Fall in Conservation Area of Flood Control Project," File C&SF Flood Control Dist, Box 10, S1160, Florida State Board of Conservation Water Resources Subject Files, 1961-1968, FSA.

[44] O. E. Frye, Jr., Director, to Honorable Paul G. Rogers, Member, United States Congress, 31 May 1966, File Everglades High Water Correspondence: 1966, Box 1, S1719, Game & Fresh Water Fish Commission Everglades Conservation Files, 1958-1982, FSA.

[45] John C. Jones interview by Brian Gridley, 23 May 2001, 42, Everglades Interview No. 9, Samuel Proctor Oral History Program, University of Florida, Gainesville, Florida [hereafter referred to as Jones interview].

[46] Frye to Rogers, 31 May 1966.

[47] Quotation in Robert F. McDonald, Delegate of Palm Beach County Airboat and Half Track Club, to Hon. Haydon Burns, Governor of Florida, 11 July 1966, File Everglades Conservation Area: General Information Corr. 1964-1967, Box 1, S1719, Game & Fresh Water Fish Commission Everglades Conservation Files, 1958-1982, FSA; see also "Fact Sheet on the Deer Situation in Conservation Area 3," 8 August 1966, ibid.

[48] As quoted in Florida Cabinet Press Release, 12 July 1966, File Everglades Conservation Area: General Information Corr. 1964-1967, Box 1, S1719, Game & Fresh Water Fish Commission Everglades Conservation Files, 1958-1982, FSA.

[49] Randolph Hodges, Chairman, Everglades Natural Resources Coordinating Committee, to All Interested Persons, August 2, 1966, File Everglades Conservation Area: General Information Corr. 1964-1967, Box 1, S1719, Game & Fresh Water Fish Commission Everglades Conservation Files, 1958-1982, FSA.

[50] "Fact Sheet on the Deer Situation in Conservation Area 3," 8 August 1966, File Everglades Conservation Area: General Information Corr. 1964-1967, Box 1, S1719, Game & Fresh Water Fish Commission Conservation Files, 1958-1982, FSA.

[51] "Fact Sheet on the Deer Situation in Conservation Area 3," 8 August 1966.

[52] "Fact Sheet on the Deer Situation in Conservation Area 3," 8 August 1966.

[53] R. G. MacDonnell, Major General, USA, Acting Chief of Engineers, to Miss Linda K. Effler, 25 August 1966, File 1110-2-1150a (C&SF) Conservation Area, January 1966-August 1966, Box 15, Accession No. 077-96-0038, RG 77, FRC.

[54] Joe J. Koperski, Chief, Engineering Division, to Miss Vickie Smith, St. Paul Dispatch, St. Paul Pioneer Press, 11 August 1966, File 1110-2-1150a (C&SF) Conservation Area, January 1966-August 1966, Box 15, Accession No. 077-96-0038, RG 77, FRC.

[55] Ronald Wise, Commissioner, to Honorable Don Fuqua, Member of Congress, House of Representatives, 12 August 1966, File Everglades High Water Correspondence: 1966, Box 1, S1719, Game & Fresh Water Fish Commission Everglades Conservation Files, 1958-1982, FSA.

[56] Quotation in Randolph Hodges, Director, Board of Conservation, to Dr. O. E. Frye, Jr., Director, Game and Fresh Water Fish Commission, 9 February 1967, File Everglades Conservation Area: General Information Corr. 1964-1967, Box 1, S1719, Game & Fresh Water Fish Commission Everglades Conservation Files, 1958-1982, FSA; see also "Deer Island Project," File E.C.A. High Water & Deer Herds, 1959-1974, ibid.

[57] See "NPA Urges Protection from Everglades 'Salting,'" *National Parks Magazine* 40 (July 1966): 19; "Opening of Canal 111 Is Delayed for Study," *National Parks Magazine* 40 (August 1966): 24; Charles H. Callison, "National Outlook," *Audubon Magazine* 69 (May/June 1967): 56; Wallace Stegner, "Last Chance for the Everglades," *Saturday Review* (6 May 1967): 72; Blake, *Land Into Water*, 216; Light and Dineen, "Water Control in the Everglades," 70; "Aerojet Canal: No Barge Ever Came; Bridge Never Opened," *The South Dade News Leader*, 13 July 1971. Other canals proposed as part of the Dade County Project aroused similar controversy; in 1971, state officials halted digging of C-109 and C-110, located south of C-111, because agricultural interests north of the canals feared that the waterways would drain water that they needed. Park officials also worried that the canals would convey polluted water into the Everglades. See "Plug

to Remain in 2 SD Canals," *The South Dade News Leader*, 1 September 1971; Joe Brown, Superintendent, to Mr. Don Albright, State Planning and Development Clearinghouse, 6 July 1971, File L54 Canals 109-110 (106, 107, 108), EVER 22965, CR-ENPA.

[58] See "'Glades Vs. Florida' Issue Seen Brewing," *St. Petersburg Times*, 1 June 1967; "The Defense of the Everglades," *National Parks Magazine* 41 (August 1967): 2; "A Legal Ruling Needed on Everglades Water Rights," *Audubon* 69 (July/August 1967): 5.

[59] O. E. Frye, Jr., Director, to Mr. Randolph Hodges, Director, State Board of Conservation, 2 May 1968, File E.C.A. High Water & Deer Herds, 1959-1974, Box 1, S1719, Game & Fresh Water Fish Commission Everglades Conservation Files, 1958-1962, FSA.

[60] Department of the Army, Jacksonville District, Corps of Engineers, "Notice of Public Hearings on Improvements for Water Resources for Central and Southern Florida," 27 October 1967, 3, File Draft Reports: Water Resources 1967, 1968, Box II-13, Office of History, Headquarters, U.S. Army Corps of Engineers, Alexandria, Virginia [hereafter referred to as HQUSACE].

[61] Deputy Director to Brig. Gen. H. G. Woodbury, Jr., Director of Civil Works, Office of the Chief of Engineers, 20 October 1967, EVER 22965, CR-ENPA. Another factor in the NPS's willingness to accept the 315,000 acre-feet figure could have been the realization that, given the state of Florida's concerted opposition to even a 315,000 acre-feet guarantee, obtaining more than 315,000 acre-feet was unlikely.

[62] "Comments on the Survey Review Report on Water Resources for Central and Southern Florida Project by the Florida Game and Fresh Water Fish Commission," 5, File E.C.A. Control Water Study Plans & Reports, 1967-1970, Box 1, S1719, Game & Fresh Water Fish Commission Everglades Conservation Files, 1958-1982, FSA.

[63] Herbert L. Alley, Director Region 4, National Wildlife Federation, to Colonel R. B. Tabb, District Engineer, 20 October 1967, EVER 22965, CR-ENPA.

[64] "Water for Everglades National Park," *National Parks Magazine* 41 (December 1967): 2.

[65] As quoted in "Corps Water Proposal 'Failure,' Says Homer," *South Dade (Fla.) News Leader*, 16 November 1967.

[66] Bill to Brig. Gen. H. G. Woodbury, Jr., 20 October 1967, File Central Florida Water Supply, Central and Southern Florida 1967, Box II-13, Office of History, HQUSACE; Hayes to Brigadier General H. G. Woodbury, Jr., 10 January 1968, File Draft Reports: Water Resources 1967, 1968, Box II-13, Office of History, HQUSACE.

[67] "Draft, General Position on Everglades," 7 November 1967, File Central Florida Water Supply Central and Southern Florida 1967, Box II-13, Office of History, HQUSACE (emphasis in the original).

[68] Quotations in Chief, Engineering Division, to District Engineer, 4 January 1968, File Draft Reports: Water Resources 1967, 1968, Box II-13, Office of History HQUSACE; see also Hayes to Woodbury, 10 January 1968.

[69] Quotations in House, *Water Resources for Central and Southern Florida*, 1, 6-7; see also Department of the Army, Jacksonville District, Corps of Engineers, *Survey-Review Report on Central and Southern Florida Project: Water Resources for Central and Southern Florida, Main Report* (Jacksonville, Fla.: Department of the Army, Jacksonville District, Corps of Engineers, 1968), 29, 46; Davis, *An Everglades Providence*, 407-408.

[70] Arthur E. Morgan, *Dams and Other Disasters: A Century of the Army Corps of Engineers in Civil Works* (Boston: Porter Sargent Publisher, 1971), 386.

[71] George B. Hartzog, Jr., Director, to Chairman, Board of Engineers for Rivers and Harbors, 11 April 1968, File BERH Public Notices 1968, Box II-13, Office of History, HQUSACE. For more information about Hartzog's role in the fight for Everglades water supply, see George B. Hartzog, Jr., *Battling for the National Parks* (Mt. Kisco, N.Y.: Moyer Bell Limited, 1988), 225-231.

[72] See Major General R. G. MacDonnell, Chairman, to Chief of Engineers, Department of the Army, 7 May 1968, File BERH Public Notices 1968, Box II-13, Office of History, HQUSACE.

[73] Stanley A. Cain, Assistant Secretary of the Interior, to General Cassidy, 12 June 1968, File 1517-08 (C&SF Martain [*sic*] County) Svy Multiple Purposed—SR 7/22/50 July 1967, Box 4, Accession No. 077-96-0017, RG 77, FRC.

[74] Major General F. J. Clarke, Acting Chief of Engineers, to The Honorable Stewart L. Udall, The Secretary of the Interior, 14 June 1968, File 1517-08 (C&SF Martain [*sic*] County) Svy Multiple Purposed—SR 7/22/50 July 1967, Box 4, Accession No. 077-96-0017, RG 77, FRC.

[75] Senate Committee on Appropriations Subcommittee on Public Works, *Water Supply for Central and Southern Florida and Everglades National Park: Meeting Arranged by Subcommittee of the Committee on Appropriations, United States Senate*, 91st Cong., 2d sess., 1970, 21-22, 25-26; Act of 13 August 1968 (82 Stat. 731).

[76] "Transcription of Information Given by Brigadier General Charles C. Noble, OCE, in Telephone Conversation with FBC for a Proposed Memorandum of Agreement Between OCE, BOB, and Dept. of Interior, to be Contained in a Letter from the National Park Service to the Chief of Engineers, 19 July 1968, File Everglades, Box 6, S949, Governor's Office, Jay Landers, Subject Files, FSA; Randolph Hodges, Director, to Brigadier General Charles C. Noble, 23 July 1968, ibid.

[77] Quotation in Solicitor to Secretary of the Interior, 8 October 1968; see also Senate Committee on Appropriations Subcommittee on Public Works, *Water Supply for Central and Southern Florida and Everglades National Park*, 21-22, 25-26.

[78] Senate Committee on Appropriations Subcommittee on Public Works, *Water Supply for Central and Southern Florida and Everglades National Park*, 26.

[79] Reed and Padrick quotations are both in Senate Committee on Interior and Insular Affairs, *Everglades National Park*, 50-59, 65.

[80] Senate Committee on Interior and Insular Affairs, *Everglades National Park*, 32-33, 44-46; see also "Nelson Asks Water for Everglades," *The Miami Herald*, 5 August 1969.

[81] Spessard L. Holland to Hon. Allen J. Ellender, Chairman, Public Works Subcommittee, 22 January 1970, in Senate Committee on Appropriations Subcommittee on Public Works, *Water Supply for Central and Southern Florida and Everglades National Park*, 1-2; Senate, *Public Works for Water, Pollution Control, and Power Development and Atomic Energy Commission Appropriation Bill, 1970*, 91st Cong., 1st sess., 1969, S. Rept. 91-528, Serial 12834-4, 24-25; "Battle Rages Over Everglades Park," *The Christian Science Monitor*, 14 June 1969.

[82] Senate Committee on Appropriations Subcommittee on Public Works, *Water Supply for Central and Southern Florida and Everglades National Park*, 20, 39-40.

[83] "Report of Meeting with Representatives of the Departments of Army and Interior and State of Florida on Water Supply to Everglades National Park, Miami, Fla., March 12, 1970," in Senate Committee on Public Works Subcommittee on Flood Control—Rivers and Harbors, *Central and Southern Florida Flood Control Project*, 98-100.

[84] Senate Committee on Public Works Subcommittee on Flood Control—Rivers and Harbors, *Central and Southern Florida Flood Control Project*, 106, 111, 151, 228-232, 236-240.

[85] Act of 19 June 1970 (84 Stat. 310); see also Blake, *Land Into Water*, 194; Carter, *The Florida Experience*, 124.

[86] Senate, *River Basin Monetary Authorizations and Miscellaneous Civil Works Amendments*, 91st Cong., 2d sess., 1970, S. Rept. 91-895, Serial 12881-3, 16-17. The report further explained how the committee reached the 16.5 percent formula. The Corps had estimated in its 1968 report that the C&SF Project could deliver 1,905,000 acre-feet of water. Three hundred fifteen thousand acre-feet was approximately 16.5 percent of that figure. Therefore, whenever the project supplied water at its normal capacity, the park would receive at least 315,000 acre-feet. In times of drought, "the park guarantee of 315,000 acre-feet will be proportionately reduced." This formula eliminated "priorities of use between present and future water users" and did not "rest on the reliability of Corps projections of future demand and water supply—concepts which have been the subject of continuing dispute and misunderstanding" (pp. 18-19).

[87] See Colonel A. S. Fullerton, District Engineer, to Division Engineer, South Atlantic, 25 June 1971, File 1110-2-1150a (C&SF) Water Resources—Proj. Gen 1968 Authn Jan 1971-Dec 1971, Box 16, Accession No. 077-02-0048, RG 77, FRC; G. E. Dail, Jr., Executive Director, to District Engineer, Jacksonville District, 18 October 1971, ibid.; James L. Garland, Chief, Engineering Division, to Superintendent, Everglades National Park, 28 October 1971, ibid.

[88] Hartzog, *Battling for the National Parks*, 228.

CHAPTER 5

Flexing the Environmental Muscle: The Cross-Florida Barge Canal, the Everglades Jetport, and Big Cypress Swamp

IN THE LATE 1960S, environmentalists came to the defense of Everglades National Park and its water needs. Two other controversies in the late 1960s and early 1970s—the proposal to build a jetport in Big Cypress Swamp and the construction of the Cross-Florida Barge Canal—would mobilize and crystallize the environmental movement in Florida to an even greater degree. On their faces, these two skirmishes, which have already been widely discussed by environmentalists, journalists, historians, and political scientists, seem to have little place in a history of the C&SF Project. The jetport, for example, generated little Corps involvement, in part because it had no direct impact on the C&SF Project.[1] The barge canal—although planned and constructed by the Corps—was located in northern Florida, outside of the scope of the C&SF Project. But for several reasons, both of these stories must be told in order to comprehend the full history of water management in South Florida. Both highlighted growing concerns with water quality in Florida in the late 1960s and early 1970s, concerns that would eventually reach an apex with the debate over the Kissimmee River and Lake Okeechobee in the 1970s. Both dealt with how industrial and engineering structures could harm a unique ecosystem, be it Big Cypress Swamp or the Oklawaha River Valley. Perhaps most importantly, both showed the increasing influence of the environmental movement in Florida in water management matters. In both instances, environmentalists were able to focus national attention on the controversies, forcing both the legislative and executive branches of the federal government to become involved. The jetport controversy and the debate over the Cross-Florida Barge Canal thus foreshadowed how environmentalists would handle water management issues in South Florida in the 1980s and 1990s.

In the 1960s, environmentalism became an established force in the United States. The conservation movement of the late 1800s and early 1900s provided a greater awareness of the environment, but it was not until the 1960s that an actual movement—"concerted, populous, vocal, influential, active"—coalesced.[2] Several factors contributed to this, including the expansion of the nation's economy in the 1940s and 1950s, which created a more affluent society and increased the number of college educated, middle-class Americans who had time to think about and work for a better quality of life.[3] This was significant, as it focused citizens more on a holistic view of the environment and the importance of environmental quality, rather than just wise use, efficiency, and the use of technology to help humans get the most from natural resources.

Likewise, the acceptance of environmental causes as a legitimate aspect of the liberal agenda, the grass roots activism of middle-class women and men, and an infusion of energy by the United States' counterculture played a large role in heightening concern for the environment. Democratic politicians, for example, saw environmental preservation as a worthwhile cause. President John F. Kennedy sponsored a White House Conference on Conservation in 1962 and appointed environmental enthusiast Stewart L. Udall as his secretary of the interior, while President Lyndon B. Johnson pushed the environmental agenda even further as part of his "Great Society" plan, in part because of the influence of his wife, Lady Bird Johnson. Indeed, women were an essential part of the expanding environmental movement, just as they had been an important component of the conservation movement. Many women protested environmental degradation in the 1960s as part of their domestic sphere responsibilities: poor water quality or contaminated milk could affect the health of their children. Other women found the environmental cause liberating and a way to become more involved in politics and economics. Finally, many young activists in America embraced environmentalism as a part of their war against authority, consumerism, and large corporations, especially in the late 1960s. "Hippies" founded communes based on becoming one with the earth, while student radicals equated the use of chemical defoliants in the Vietnam War with oil spills and other environmental destruction in the United States. The vigor of these activists infused the environmental movement with necessary energy.[4]

As evidence of environmental destruction, environmentalists turned to ecologists for support. Ecology (a term first used by German zoologist Ernst Haeckel in 1866) had slowly evolved in the nineteenth and early twentieth centuries into a stand-alone scientific discipline focused on the study of how animals relate to their inorganic and organic environments. The Ecological Society of America was formed in 1915, and the first ecology departments at universities were established in the 1950s. By that same decade, the examination of all elements in a bounded environment, or ecosystem, and the effects that individual actions had on other aspects of the system, had become an essential part of ecology, influenced by the work of E. P. Odum.[5]

As the environmental movement gained in momentum, it used the ecosystem concept to show the consequences of human actions on the environment, and ecologists, in turn, became caught up in the environmental movement; scientists began to write books and articles for a more general audience, as well as giving public lectures, in order to obtain public support for funding and "to educate the public about the history of science as well as the significance of current research."[6] Rachel Carson, a marine biologist, for example, published *Silent Spring* in 1962, a book that, in the eyes of many, ushered in the environmental movement. Other scientists followed, including biologist Barry Commoner, who published *The Closing Circle*, and Paul Ehrlich, an entomologist whose book *The Population Bomb* warned about the dangers of overpopulation. Spurred on by these publications, environmentalism became more prominent in American society in the 1960s; the number of articles on environmental topics in national magazines increased by more than 300 percent from the late 1950s to the late 1960s. Membership in the Sierra Club grew from 15,000 in 1960 to 113,000 in 1970, while the National Audubon Society expanded from 32,000 constituents in 1960 to 148,000 in 1970.[7]

By the end of the 1960s, environmentalism had become a hot political topic, and senators such as Wisconsin's Gaylord Nelson, Maine's Edmund Muskie, and Washington's Henry Jackson made environmental protection one of their primary focuses in Congress. Due to their influence, Congress passed a law in December 1969 declaring the federal government's responsibility towards the environment—the National Environmental Policy Act (NEPA). It stipulated

that the government would cooperate with state and local entities to ensure the coexistence of man and nature "in productive harmony." The law established a Council on Environmental Quality in the Executive Office of the President "to appraise programs and activities of the Federal Government," and it also required federal agencies to prepare environmental impact statements (EISs) whenever they conducted activities "significantly affecting the quality of the human environment."[8] In accordance with the policy established by NEPA, Congress and the White House created the Environmental Protection Agency (EPA) soon after NEPA's passage to regulate actions affecting the nation's environment.[9]

With the aid of NEPA, environmental groups began to attack the Corps with more frequency and with more concerted approaches. Because the law required federal agencies to produce EISs for their projects, it opened federal construction proposals to more public scrutiny than ever before. The law therefore forced the Corps and other federal agencies to consider environmentalist concerns in their endeavors, heightening the already-burgeoning power of the movement.[10] Nowhere is this more apparent than in the issues surrounding the Cross-Florida Barge Canal, the Everglades Jetport, and Big Cypress Swamp in the late 1960s and early 1970s.

As environmental organizations increased the visibility of construction projects in Florida, several Floridians increased their prominence in the national eye. Joseph B. Browder, for example, a former television producer who quit his job to focus on environmental issues, served as the southeastern regional representative of the National Audubon Society and was instrumental in forming the Everglades Coalition to defeat the jetport. During the debate over the jetport, he testified before numerous congressional committees about the airport's potential effects on Everglades National Park. Browder also convinced Marjory Stoneman Douglas, the author of *The Everglades: River of Grass*, to found Friends of the Everglades in 1969 to fight the jetport proposal. Arthur R. Marshall, a marine biologist who worked at the Vero Beach office of the U.S. Fish and Wildlife Service until 1970 (when he took a position at the University of Miami), spent countless hours educating people on the South Florida water system, believing that the Everglades needed its natural flow restored in order to prevent the region from dying. Marshall also criticized Florida's grow-at-all-costs approach to land use and water planning, believing that some restrictions were necessary to preserve the state's water supply.[11]

In both federal and state offices, Browder, Douglas, and Marshall had some receptive audiences; the importance of ecological issues in Florida transcended political parties. Although President Richard Nixon, a Republican, did not agree with much of the environmental movement, he understood politics well enough to support some key issues, such as NEPA and the Clean Air Act of 1970, in order to deflect the political influence of rivals such as Edmund Muskie and Henry Jackson. Nixon also appointed some crucial

Governor Claude Kirk (left) presenting an award to Nathaniel Reed (right). (The Florida Memory Project, State Library and Archives of Florida)

gulf coast at Yankeetown (approximately 70 miles north of Tampa), where the Withlacoochee River flowed. The canal would follow the Withlacoochee east to Dunnellon, and then northeast (but south of Ocala) to the Oklawaha River. Following the Oklawaha, it would connect to the St. Johns River at Palatka, eventually emptying into the Atlantic Ocean at Jacksonville. President Franklin D. Roosevelt authorized using emergency funds for the construction of this route in order to provide jobs in Florida, leading the Corps to begin construction on the waterway. After spending $5 million and clearing nearly 5,000 acres of land in the late 1930s, however, the project was abandoned, largely because of opposition from railroads and other entities, which claimed that poor water quality and aquifer contamination would result. Therefore, the Corps developed a new plan in 1943, proposing that the canal be a lock, rather than a sea-level structure, that would serve barges instead of ships. The 12-foot deep waterway would contain five locks and two dams, including the Rodman Dam and Eureka Dam across the Oklawaha River. However, due to the United States' participation in the Second World War, the canal received little federal support.[14]

environmental officers, including Deputy Assistant to the President for Domestic Affairs John Whitaker, who held a deep concern for the environment; Russell Train, the undersecretary of the interior who became chairman of the Council on Environmental Quality in 1970; and Nathaniel Reed, special environmental assistant to Florida Governor Claude Kirk, who became assistant secretary of the interior for fish, wildlife, and parks. Because of Reed's familiarity with Florida issues, he was instrumental in achieving national concern for problems affecting the Everglades and South Florida. On the state level, Governor Claude R. Kirk, Jr. (Republican, 1967-1971) understood the political benefits of supporting environmental causes, while his successor, Reubin Askew (Democrat, 1971-1979) was more committed personally to environmental action, as was Jay Landers, his environmental adviser. Because of the efforts of these officials, environmentalists were able to achieve some worthy goals in Florida in the late 1960s and early 1970s, especially revolving around Big Cypress Swamp and a proposed jetport in the area.[12]

In order to defeat the jetport, environmentalists used tactics pioneered in the fight against a Florida construction project, albeit in northern Florida, planned by the Corps of Engineers: the Cross-Florida Barge Canal. The canal had deep historical roots. The idea for a waterway connecting one side of Florida to the other had existed since the initial Spanish occupation of Florida in the 1500s, and Floridians had made several proposals of a trans-Florida canal in the 1800s and early 1900s.[13] With General Charles P. Summerall, a retired four-star general who had served as Chief of Staff from 1926 to 1930 heading the efforts, support for the canal gained momentum in the 1930s, in part because it promised jobs for a depression-ridden state, and in part because the Corps determined that a feasible route existed. The Corps concluded that the best path for the canal, which would be a sea-level ship canal, would begin on Florida's western

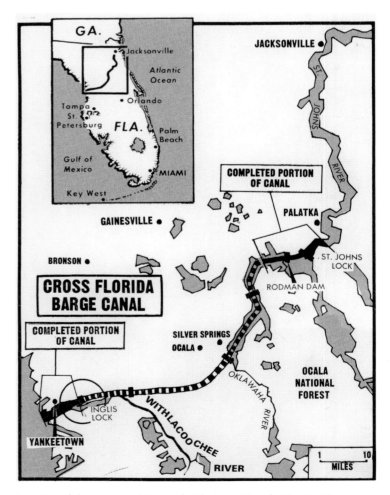

Location of the proposed Cross-Florida Barge Canal. (The Florida Memory Project, State Library and Archives of Florida)

The major push for construction of the barge canal came in the 1960s after John F. Kennedy won the presidency, partly on a platform guaranteeing the waterway's construction. His support, coupled with state backing engineered by Governor Farris Bryan, pushed Congress to appropriate funds for the canal's construction in 1962. On 27 February 1964, President Lyndon B. Johnson presided over a groundbreaking ceremony in Palatka that commenced canal construction once again.[15]

However, opposition to the canal gradually coalesced, largely because of its potential environmental harm. In 1962, after seeing a presentation by the Alachua Audubon Society, Marjorie Carr, a resident of Gainesville and wife of University of Florida zoologist Archie Carr, became convinced that the canal would destroy much of the lower stretches of the Oklawaha River. This river meandered for 60 miles through northern Florida, east of Ocala, the largest city near the river, as an outlet for the Oklawaha chain of lakes, including Lake Apopka. Beginning at Lake Griffin, the river ran through a subtropical hardwood forest on its way to the St. Johns River, providing habitat for limpkin, otter, and alligator, as well as numerous game

fish such as bass. Although farmers had diked the upper portion of the river in the 1800s, and although the timber industry extracted numerous trees from the forest in the 1880s, the Oklawaha still had, in the words of journalist Luther Carter, a "wild and junglelike character."[16] Realizing the beauty and importance of the Oklawaha ecosystem, Marjorie Carr, together with biochemist David S. Anthony of the University of Florida, began a society-sponsored study of the barge canal's potential environmental effects.[17] After deciding that the canal and the construction of Rodman Dam and Reservoir would largely destroy 40 out of the 50 miles of the Oklawaha that still flowed freely, Carr, the Alachua Audubon Society, and the Florida Audubon Society asked Congress to investigate alternate routes for the waterway, bypassing the river. Stating that the Corps claimed that environmental damage would be minimal, Congress refused.[18]

Yet Carr influenced others, and they began to agitate for the preservation of the Oklawaha. In 1966, over 350 people attended a state-sponsored public hearing on the canal, which, according to William N. Partington of the Florida Audubon Society, was "the largest of its kind to be held on a Florida conservation issue."[19] Critics,

Governor Claude Kirk (left) presenting an award to Marjorie Carr (center), the driving force behind environmental opposition to the Cross-Florida Barge Canal. Carr's husband Archie (right) looks on. (The Florida Memory Project, State Library and Archives of Florida)

including a group called Citizens for the Conservation of Florida's Natural and Economic Resources, told state leaders that the canal and Rodman Reservoir would kill the Oklawaha's natural beauty. According to Partington, Florida Secretary of State Thomas Adams and other officials, using arguments that jetport proponents would also make, dismissed these concerns as "birdwatchers let[ting] off steam" and counseled environmentalists to move out of the way "so that orderly progress could be made."[20] Despite the unproductive nature of the meeting, Partington believed it to be a turning point in the history of Florida's environmental movement because it was the first time that individuals and disparate groups united behind a common ecological cause.

In March 1966, state officials formally endorsed the project, and for the next few years, the Corps worked on channel construction and building other works, including Rodman Reservoir. But when the Corps filled the reservoir in 1968 and 1969, water hyacinth began to flourish, validating a 1967 report by the Federal Water Pollution Control Administration indicating that algal blooms were likely in the reservoir. The Corps continued its work on the Eureka Lock and Dam on the Oklawaha, but the condition of Rodman Reservoir led Carr and others, who originally wanted the Corps to change only the course of the canal, to call for a complete halt to construction.[21]

In order to effectuate a work stoppage, Florida environmentalists formed the Florida Defenders of the Environment in July 1969 to coordinate legal work with the Environmental Defense Fund, Inc., an organization established in 1967 to litigate against ecological despoilers, specifically against the use of the pesticide DDT (one of its founders, Victor J. Yannacone, Jr., lived by the motto "Sue the Bastards").[22] Using environmental litigation to stop potentially destructive projects was a relatively new tactic, having been pioneered in 1965 by the Sierra Club and other environmental groups to stop the construction of a hydroelectric project above the Hudson River. Yet it had proved enormously effective, paving the way for the establishment of Florida Defenders, with Partington as chairman and Carr as assistant general chairman. Having commissioned a study of the canal's ecological effects, Florida Defenders, assisted by the Environmental Defense Fund, filed a suit against the Corps on 15 September 1969, charging it with violating the constitutional rights of American citizens by destroying the natural resources of the Oklawaha River Valley. The litigation asked that the U.S. District Court in Washington, D.C., enjoin the Corps from further work on the canal until a study on social costs and benefits could be performed.[23]

Meanwhile, the Florida Game and Fresh Water Fish Commission and the FWS both determined that the canal would result in drastic changes in the Oklawaha ecosystem and that the Rodman Reservoir would degrade quickly into a stagnant nutrient trap. Both agencies recommended a detailed ecological study of the canal's impacts.[24] At the same time, Florida Defenders of the Environment

completed its ecosystem study in March 1970. It foresaw only ecological disaster for the Oklawaha Valley—"the only large wild area remaining that supports the full spectrum of plant and animal life native to north-central Florida"—if the canal was completed. The organization, therefore, recommended the halt of further appropriations for the waterway, the draining of Rodman Reservoir, and the return of the Oklawaha to its "natural free-flowing condition."[25]

In 1970, an article in *Reader's Digest*, which had 18 million subscribers, attacked the project and the Corps further, influencing hundreds of people to write letters to the secretary of the interior about the project. In this essay, entitled "Rape of the Oklawaha," James Nathan Miller, an environmentalist, characterized the Corps as "the most damaging single force at work on the U.S. countryside" and the canal as merely one more pork-barrel boondoggle. He accused Corps leaders of deliberately massaging the canal's benefit-cost ratio in order to justify it economically. Miller asked for not only a stoppage of construction, but also recommended that the federal government either eliminate benefit-cost analyses altogether (because no economic price could be placed on environmental values) or provide ways to "inject *human judgment* into a formula that now accepts only dollar signs."[26]

The Corps also began facing battles on the economic front, as Congress cut congressional appropriations for the canal. This led to a slowdown in construction, and the delay allowed canal enemies to increase their efforts. Blazing a path that jetport opponents would follow, environmentalists decided to petition the federal government for help. In January 1970, 162 prominent scientists, including environmental leaders throughout the United States, sent a letter to President Richard M. Nixon, asking him to dismiss the project to prevent "further degenerative manipulation of one of the most valuable natural ecosystems of Florida" and to preserve "the quality of the subsurface water supply of Central Florida."[27] In June, Secretary of the Interior Walter Hickel asked the Secretary of the Army to implement a moratorium on construction until new ecological and economic studies could be completed. After some resistance, Corps leaders agreed to a six- to twelve-month moratorium.[28]

Meanwhile, the President's Council on Environmental Quality (CEQ) investigated the canal situation. After perusing several ecological studies, the CEQ concluded that the canal would destroy the unique characteristics of the Oklawaha River Valley, causing water weed infestation in the area, polluting surface and subsurface water, and changing the river from "a cool, highly enriched, densely shaded, flowing" waterway to "a warm water, highly enriched, unshaded, shallow watercourse, with little or no flow."[29] Because of this potential damage, Russell Train, chairman of the CEQ, recommended to John C. Whitaker, Deputy Assistant to the President, that project construction halt.

Whitaker forwarded Train's recommendation, as well as a separate decision paper Whitaker had composed, to John Ehrlichman,

An editorial cartoon from *The Palm Beach Post* depicts the "crazy" reasoning that barge canal proponents used to justify the canal.

President Richard Nixon's aide over domestic affairs. After reviewing these documents, Ehrlichman decided that the CEQ had valid reasons for wanting the project halted, so he told the Chief of Engineers to end construction. "It's doing terrific damage," Ehrlichman recalled saying to the general, and "the cost-benefit basis doesn't prove out to me."[30] The Corps did not necessarily disagree, and its Environmental Advisory Board (first established in April 1970 to provide advice to Corps leaders on ecological concerns) recommended a thorough review of the project in December 1970.[31]

Before the Corps could make a comprehensive examination, U.S. District Court Judge Barrington Parker issued a preliminary injunction barring the Corps from further work on the canal. Only four days later, on 19 January 1971, Nixon released a written directive that the Corps cease work on the canal to preserve the Oklawaha environment. Not only would the canal significantly harm "a uniquely beautiful, semi-tropical stream," Nixon stated, but it was also economically unjustified. "The step I have taken today," the President explained, "will prevent a past mistake from causing permanent damage."[32]

But Nixon's order had repercussions, as both state officials and canal proponents believed that he had exceeded his authority.[33] Accordingly, the Authority filed a suit in the Jacksonville Federal District Court against the United States, stating that the President did not have the power to halt construction.[34] The litigation continued for the next three years, and on 31 January 1974, U.S. Circuit Court Judge Harvey M. Johnsen ruled that Nixon did not have the proper authority to halt the canal, stating that such power rested only with Congress. Canal proponents celebrated this victory, but it seemed hollow, primarily because Johnsen also issued a permanent injunction on further construction until the Corps completed

a comprehensive environmental impact statement (EIS) with a revised benefit-cost ratio. Johnsen's ruling eroded state support of the project, as Florida Governor Reubin Askew stated that he and his cabinet would not ask for any additional canal appropriations until the Corps had completed the EIS, and the Florida Department of Natural Resources rescinded its previous support of the canal until it had examined the EIS and the economic report.[35]

The state's position on the canal was clarified in a two-day public hearing held in December 1976. Three hundred fifty people attended, some of them wearing green signs proclaiming "Stop the Canal" or "Save the Oklawaha," while others had red and blue buttons declaring "I Support the Canal." After hearing testimony from both sides, the cabinet voted six to one to withdraw state support for the canal, and on 17 January 1977, it passed a resolution recommending against further construction and asking Congress to deauthorize the project.[36]

With no further state backing, and realizing that the issue had become so politicized that the Corps could not win, the Jacksonville District's EIS, published in 1977, recommended against further construction. Jacksonville District Engineer Colonel Donald Wisdom still believed that the canal was both economically and ecologically viable, but only if both sides were willing to compromise. Unfortunately, according to Wisdom, canal opponents "no longer could look at anything but total stoppage of the canal"; there was no chance of conciliation.[37] Chief of Engineers Lieutenant General J. W. Morris concurred in the Jacksonville District's decision to abandon the canal, declaring that environmental concerns precluded the Corps from continuing the project. The only things left to accomplish were the Oklawaha River's restoration and the project's deauthorization, things that took several years to accomplish. Ultimately, however, the state designated canal route lands that it owned as the Cross Florida Greenway State Recreation and Conservation Area, renamed the Marjorie Harris Carr Cross Florida Greenway in 1998.[38]

The Everglades Jetport
The battle over the Cross-Florida Barge Canal was not an isolated incident; instead, there were several examples in the 1960s and 1970s of environmental interests halting or rejecting Corps projects. A proposed dam and reservoir on the Meramec River in eastern Missouri, first planned in the 1930s, met its ultimate demise in August 1978 when voters voted against the project's continuation for both economic and environmental reasons. Likewise, in southwestern Wisconsin, environmentalists banded with congressional leaders such as U.S. Senator Gaylord Nelson to prevent the Corps from building a dam and reservoir for flood protection on the La Farge River.[39]

Another example came in Florida itself, where environmentalists used many of the same tactics that they employed in the Cross-Florida Barge Canal fight to defeat a proposed jetport in Big Cypress Swamp in the late 1960s and early 1970s. The jetport proposal stemmed from the increasing growth and rapidly expanding population of South Florida in the 1960s. Dade County, for example, saw its population climb from approximately 500,000 people in 1950 to nearly 1.5 million in 1970. The larger number of residents created a real estate boom that showed no signs of stopping; Florida was projected in the 1960s to become the third largest state in the United States by 2000 (a prediction slightly off-the-mark, as the state had the fourth largest population in 2005). But the state itself did not have adequate planning measures for controlling the effects of this expansion on vital natural resources, including water. As Paul Brooks explained in an article in *Audubon* in 1969, "pressures on the land and water are at a maximum; zoning for their protection at a minimum."[40] As Florida continued to grow, the stress on its natural resources increased as well.

These problems were clearly seen when Dade County proposed the building of a jetport in Big Cypress Swamp on the northwest boundary of Everglades National Park. The swamp itself was a mix of marsh and lowland forest, containing sloughs, tree islands, bay and cypress trees, orchids, ferns, and bromeliads on limestone and sand formations. The area was almost completely flat, and it was estimated that 50 percent of the surface water running into Everglades National Park (or 9 percent of the park's total water) came from the swamp's extremely slow-moving sheet flow. The area also housed 17 endangered species, including the Florida panther, the American alligator, and the roseate spoonbill.[41]

Despite its ecological importance, many believed that the swamp was the ideal place for a new jetport in South Florida. An airport was necessary, proponents claimed, because the increasing number of tourists going to Miami and South Florida's east coast brought an ever-growing number of flights and travelers to Miami International Airport. The fact that Miami was a good departure point for transoceanic travel and that domestic carriers conducted many training flights in the area only compounded the problem. In the mid-1960s, transportation experts estimated that Miami International Airport, which saw 10 million passengers and 500 million pounds of air cargo a year, would reach its air traffic saturation point by 1973. Therefore, the Federal Aviation Administration (FAA) and the Department of Housing and Urban Development began searching for sites where a new training facility could be located, thereby relieving some of the airport's pressure.[42]

Working jointly with the Dade County Port Authority, which the Florida state legislature had created in 1945, and desiring that the site be somewhere remote from human habitation, the FAA determined in April 1966 that the best site was north of the Tamiami Trail in Water Conservation Area No. 3, close by the boundary of Everglades National Park. But no one consulted with either the NPS or the FCD about this location until February 1967. At that time, the FCD announced its opposition to the site because it believed an airport was incompatible with the objectives of the water conservation

areas, and the Dade County Port Authority and the FAA decided to search for a new location.[43]

Initially, the agencies investigated areas south of Tamiami Trail and next to Everglades National Park, but park officials complained that aircraft noise would disrupt wildlife in those locations. The Port Authority therefore turned its attention to southwestern Florida, and in November 1967, leaders of Dade and Collier counties announced that they had agreed to the construction of a jetport on a 39-square mile tract within Big Cypress Swamp, six miles north of the park's Forty-Mile Bend Ranger Station, with an eastern boundary common with Conservation Area No. 3's western border. Two runways would be completed within five years to begin pilot training, but the Port Authority envisioned that the jetport would eventually have another two to four runways and that it would begin conducting domestic and international commercial flights when Miami International Airport reached its saturation point. Preliminary construction plans commenced almost immediately.[44]

The proposal failed to produce any opposition in its first few months. The Florida Game and Fresh Water Fish Commission reviewed the plans and offered no objection; Director O. E. Frye, Jr., told the Dade County Port Authority that he was concerned with possible jet fuel contamination of Conservation Area No. 3, but he dropped the matter after a Port Authority representative assured him that no problems would occur. Instead, Frye complimented the planners, envisioning "the creation of extensive waterways resulting from the construction of elevated runways which could afford virtually unlimited fishing possibilities."[45] According to journalist Luther Carter, the FCD, the State Board of Conservation, and the trustees of the Internal Improvement Fund also reviewed the plans and made no objections.[46] NPS officials did voice some concern about the location, fearing a jetport would contaminate water flowing into the park, but these protests were only made to Florida Game officials.[47]

With only limited opposition, the Dade County Port Authority held a groundbreaking ceremony on 18 September 1968. Governor Kirk and U.S. Secretary of Transportation Alan Boyd did not attend the festivities, but Kirk sent a statement praising the jetport while Boyd participated by telephone. This spirit of cooperation ended in October during a meeting between the FCD and the State Road Department when Robert Padrick, chairman of the FCD and a member of the Sierra Club, discovered that the alignment of proposed Interstate 75 had been changed to cross through the middle of Conservation Area No. 3 in order to facilitate travel from Miami to the jetport. Because such a placement would have bisected the conservation area, potentially destroying its ecological values, Padrick, in the words of John Maloy, an engineer with the FCD, "sounded the clarion call," writing to more than 100 Florida environmentalists, including Nathaniel Reed in the governor's office, to mobilize opposition to the plan.[48]

Padrick also called a meeting in December 1968 with representatives from the U.S. Army Corps of Engineers, the Florida Game and Fresh Water Fish Commission, the NPS, Everglades National Park, the FWS, the USGS, the Sierra Club, and the National Audubon Society to discuss how to proceed. Park leaders again raised concerns that the jetport would pollute water coming into the park, while others worried about the impacts of industrial and housing developments that would certainly follow the airport's construction. Joseph Browder of the National Audubon Society and Gary Soucie of the Sierra Club indicated that the group should focus on relocating the facility, but others seemed unwilling to pursue that option. Instead, the gathering decided to submit questions and concerns to the Dade County Port Authority for its consideration.[49]

In the meantime, the jetport proposal began receiving national attention. The *New York Times* covered the issue extensively, in part because the New York Port Authority and the Metropolitan Trans-portation Authority believed that the completion of the jetport would divert international travelers to Miami. Some even speculated that the jetport would be bigger than the New York, Los Angeles, and Washington airports combined.[50] Anthony Wayne Smith, president of the National Parks Association, published an editorial against the facility in *National Parks Magazine*. Calling the jetport the latest of numerous environmental follies in Florida, Smith wondered why the United States in general and Florida in particular had such difficulty with "economic, social, and governmental planning." Could people not see that the jetport "greatly imperiled" a national park on which the public had "invested vast efforts and millions of dollars?" Could not effective land or water planning be implemented to prevent such travesties? Not only would the park suffer, Smith claimed, but the Miccosukee Indians, who were related to the Seminole and who had a state reservation in the area, would as well since the facility covered their traditional hunting

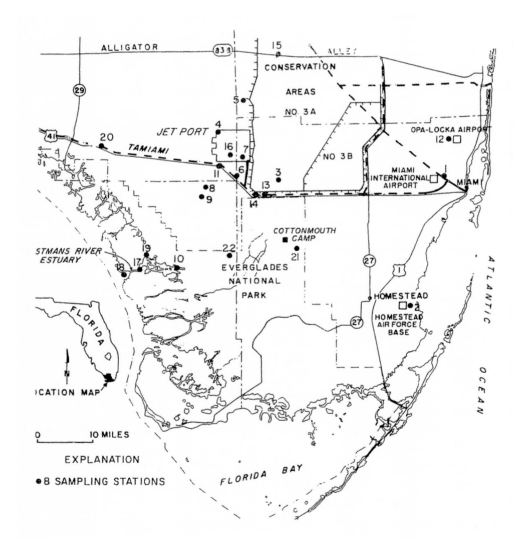

Map showing the location of the proposed Everglades Jetport. (U.S. Geological Survey, "Preliminary Determinations of Hydrobiological Conditions in the Vicinity of the Proposed Jetport and Other Airports in South Florida," 1969)

An editorial cartoon from *Audubon Magazine* shows the conflicts between airplanes and wildlife that the jetport would produce.

grounds. He called for concerned citizens to write to President-elect Richard Nixon and ask for his help.[51]

As the criticism mounted, the Dade County Port Authority decided to hold a public hearing on 28 February 1969 to answer growing concerns. At this meeting, Port Authority officials addressed the questions submitted to it by federal and state leaders, including what other locations had been considered, how the jetport would be operated, what steps the Port Authority would take to guard against water pollution, and what overall regional development planning had been made. Yet the Port Authority did not resolve any of these issues, answering most of them with a perfunctory "this question is presently under study."[52] Despite this unaccommodating attitude, federal agency representatives persisted in their opposition. John C. Raftery, superintendent of Everglades National Park, discussed the "enormously complex problems" that the jetport would cause, "including disruption of the Park's remaining natural water supply, introduction of pollutants and destruction of Park's wilderness values."[53] According to one observer, Arthur Marshall, representing the Interior Department, stated that the main problem was the environmental damage the jetport could wreak on South Florida, especially Big Cypress Swamp, Conservation Area No. 3, Everglades National Park, South Florida estuaries, and the land of the Miccosukee Indians. Air, noise, and water pollution were all potential effects, Marshall continued, as was the possibility of a reduction in water reaching the park. Because of this, Marshall proposed that an interagency working committee be appointed consisting of representatives from state and federal agencies, as well as the Miccosukee, to provide solutions to these issues.[54]

Marshall's suggestion fell on deaf ears, leading Browder to declare that unless the Port Authority could provide assurances that the jetport would not harm the Everglades ecosystem, he would wage a national campaign to stop its construction.[55] State officials, however, seemed largely pacified after the hearing. Reed informed Governor Kirk that the building of the jetport in Big Cypress Swamp was inevitable, meaning that the state should work to ensure that it

became a "great" facility with "minimal disturbance of natural values and historic water sloughes [*sic*]." By using "careful planning, zoning, and enforcement," Reed continued, these goals could be reached; he also argued that development of the area by the Port Authority—"a well financed agency"—was preferable to actions by individual landowners.[56]

Facing the intransigence of the Port Authority and the passivity of the state, environmental organizations took another approach. In April 1969, Smith and Browder formed the Everglades Coalition as a way for different national associations to work together for the stoppage of the jetport proposal. Smith and Elvis Stahr, former Secretary of the Army who served as president of the National Audubon Society, co-chaired the organization, while Browder served as its coordinator. This group contained representatives from most of the major environmental organizations in the United States, including the National Parks Association, National Audubon Society, Wilderness Society, Sierra Club, Nature Conservancy, National Wildlife Federation, and the Friends of the Earth—in the words of Smith, "practically the entire conservation movement."[57] Other organizations, such as the United Automobile Workers of America and the United Steelworkers of America also joined. The major objectives of the Everglades Coalition were to stop jetport construction, to preserve Big Cypress Swamp, and to protect Everglades National Park.

Meanwhile, Marjory Stoneman Douglas formed another organization whose initial purpose was jetport opposition. One night in a Miami grocery store, Douglas encountered Susan Wilson, one of Browder's assistants, and told her how impressed she was with Browder's work on the jetport problem. Instead of accepting the compliment, Wilson asked Douglas what she was doing to help the Everglades. "Oh me?" Douglas answered. "I wrote the book." Wilson, quick to seize the opportunity, rejoined "That's not enough," informing Douglas that they needed more help. A bit taken aback, Douglas mumbled that she was willing to do whatever she could. The next day, Browder called her and asked her to write a "ringing denunciation of the jetport" in the press.[58] When Douglas demurred, explaining that such statements were better coming from organizations, Browder told her to form one, explaining that she could unite some of the local individuals and organizations interested in preserving the Everglades in the same way that he had brought national interests together. Accepting Browder's challenge, Douglas created the Friends of the Everglades and opened it to all interested parties, requiring only a membership fee of $1. It grew steadily over the next few years as Douglas and other members traveled throughout South Florida, informing citizens about the jetport and the destruction it could cause.[59]

The Everglades Coalition and the Friends of the Everglades heightened public awareness about the jetport, and they also implemented a new strategy to stop the development. In the Department of Transportation Act of 15 October 1966, Congress had inserted a

Marjory Stoneman Douglas, founder of Friends of the Everglades. (The Florida Memory Project, State Library and Archives of Florida)

proviso that the secretary of transportation could not approve any undertaking using land from "a public park, recreation area, wildlife and waterfowl refuge, or historic site" unless he or she had first determined that no other feasible alternative existed and that the program had implemented sufficient mitigations to "minimize harm" to such areas.[60] Because the jetport required the rerouting of Interstate 75 through Conservation Area No. 3, representatives of the Everglades Coalition, Sierra Club, National Audubon Society, and other organizations argued that the airport's construction fell under the authority of the Transportation Act. Secretary of Transportation John Volpe had not made any studies of feasible alternatives or of environmental effects, they claimed, meaning that he had not complied with the law. In April 1969, Everglades Coalition members sent a letter to Volpe, urging him to conform to the act by stopping construction and relocating the airport. "We would hope that the burden of resolving this conflict would not have to fall upon the shoulders of the President of the United States," they concluded.[61]

But environmentalists were well aware that the involvement of high-level federal officials, and perhaps even President Nixon, might be necessary to prevent the jetport's construction.[62] Fortunately for them, they had an ally in Secretary of the Interior Walter Hickel. In March 1969, Hickel had toured South Florida to attract attention to alligator poaching in Everglades National Park. While there, he flew over the proposed jetport site, observing the completed runway and contemplating the "long-term damage" that the facility could cause.[63] Hickel and other Interior officials were especially worried about water pollution, stemming both from the jetport itself and from the construction of industrial and residential areas around the facility. Such development, Hickel believed, would dump fertilizer, insecticides, and sewage into water flowing into the park. After his return, Hickel contacted Volpe to express his concerns.[64]

Due to Hickel's pressure, as well as the constant criticism of environmental organizations, Volpe agreed in June 1969 to the creation of a joint committee of Interior and Transportation representatives to conduct studies on the jetport. The Interior Department took the lead on the examination of environmental effects, designating Dr. Luna Leopold, a USGS research hydrologist who was one of the most prominent geomorphologists of the twentieth century, as well as former head of the USGS's water resources division and the son of famed wildlife conservationist Aldo Leopold, as the coordinator of the study, with Arthur Marshall serving as the Florida liaison and Manuel Morris of the NPS as the federal contact. Governor Kirk, together with Reed, applauded the idea. Apparently, public discontent with the proposed jetport had convinced Kirk and Reed to cooperate with the environmental study.[65]

As the study commenced, the U.S. Senate Committee on Interior and Insular Affairs, under the leadership of Senator Henry M. Jackson of Washington, held hearings on Everglades National Park water issues, including the jetport. All interested parties were represented, such as the Interior Department, the FCD, the Corps of Engineers, the EC, the National Audubon Society, the Sierra Club, and the Dade County Port Authority. Critics of the jetport explained that they wanted the Port Authority to find another location for the facility; they were not asking for its complete elimination. The Port Authority, however, represented by William W. Gibbs and C. H. Peterson, doubted that another feasible site existed. Besides, they testified, the Port Authority had only plans to construct a training facility; it would not build a full-fledged jetport "until it can be clearly proven that such development will not have an adverse effect" on the park.[66] Senator Gaylord Nelson found that difficult to believe, especially because the Port Authority's 1968 annual report had delineated plans to convert the training operation into a commercial jetport by 1980 at the latest. In addition, Gibbs and Peterson angered Nelson by telling him that the Port Authority had no responsibility for any kind of development that occurred outside the 39-square-mile area, even though Port Authority Director Allen Stewart and Deputy Director Richard Judy kept boasting about the huge growth that would follow the jetport's construction. Who would take responsibility for ensuring that development did not harm the park, Nelson wondered. Peterson answered that it was a county duty, but that did not appease Nelson who decried the lack of land and water planning in Florida.[67]

The flippant attitude of the Dade County Port Authority regarding Big Cypress development upset environmentalists, as did several inflammatory quotations attributed to jetport supporters in the press. Michael O'Neil, Florida's secretary of transportation, for example, told reporters that he did not care if the jetport harmed alligators because the animals "make nice shoes and pocketbooks." Meanwhile, Judy proclaimed that "Big Cypress Swamp is just typical South Florida real estate" that would eventually be "one of the great population centers of America," while Stewart announced that "a new city is going to rise up in the middle of Florida . . . whether you like it or not."[68]

As jetport proponents made such bold pronouncements, and as the September opening of the first runway neared, a spate of critical articles appeared in national publications. In July, *National Parks Magazine* published a piece, complete with photographs of bulldozers and downed trees, calling attention to the "serious new threat" that the jetport posed to Everglades National Park.[69] That same month, an article in *Audubon* by environmentalist Paul Brooks condemned the jetport, quoting Park Superintendent John C. Raftery as stating that the park faced "slow death" if the facility became a reality. "As now located," Brooks declared, "the Everglades jetport is an abortive offspring of the unholy wedlock of the booster and the engineer."[70] Only by ensuring its removal could environmentalists protect the park from ultimate destruction.

General news magazines also provided publicity. *Time* called the battle over the jetport a "testing ground for U.S. environmental policies," stating that environmentalists feared the impacts of

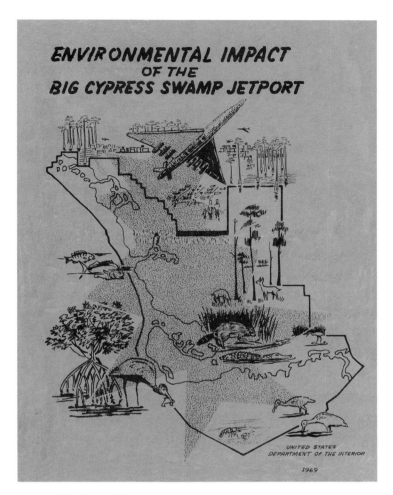

Cover of the Leopold Report.

"jet noise, exhaust fallout, fuel and oil spills" on Everglades National Park, as well as "the prospect of helter-skelter development around the airport."[71] *Look* issued a photo essay depicting "the assault on the Everglades,"[72] while *Life* published an article by Florida mystery writer John D. MacDonald, arguing that the jetport would eliminate the westward flow of water from Big Cypress Swamp, the last "reasonably natural" water supply to the park.[73] These articles all mentioned that the joint Department of Transportation/Department of Interior study of ecological effects was in process, but, as *Look* pessimistically related, "there is no assurance that the county will be willing to abandon years of ambitious planning" even if the examination proved that such an action was necessary to save the park.[74]

As the news media continued its discussion of the jetport, representatives of the Department of Transportation and the Department of the Interior completed their environmental examination, issuing it in September 1969.[75] The first sentence of the document, usually referred to as the Leopold Report after Luna Leopold, pulled no punches. It proclaimed that the

> development of the proposed jetport and its attendant facilities will lead to land drainage and development for

agriculture, industry, housing, transportation, and services in the Big Cypress Swamp which will inexorably destroy the south Florida ecosystem and thus the Everglades National Park.[76]

The major problems, the report continued, would result from the air, noise, and water pollution produced by the jetport and any commercial development, affecting plant and wildlife, the Miccosukee Indians, and tourists visiting the park. The report then outlined three alternatives that Florida officials could take: first, they could allow construction of the training facility, the subsequent jetport, and the commercial development to occur, thereby resulting in complete ecological devastation; second, they could allow the existing runway to be used as a training facility with no other expansion, which would give the state enough time to develop proper planning and land use regulations; or third, they could convince the Port Authority to remove the runway and abandon the site, which would "inhibit greatly the forces tending toward development in Big Cypress Swamp."[77] The report made no recommendation as to the appropriate course to take, content only to describe the environmental effects of each measure.

But to any careful reader, three conclusions were clear. First, jetport development should be abandoned and the runway should be removed in order to preserve Everglades National Park. Second, Big Cypress Swamp, as a watershed important to the park, needed additional forms of protection. And, third, the state of Florida needed to implement land use planning laws to safeguard its natural resources. Two subsequent reports from other sources bolstered these conclusions. The first, sponsored by the National Academy of Science, declared that the jetport would considerably damage Big Cypress and recommended instead that regional planning and Big Cypress preservation be implemented.[78] The second, conducted by a task force called Overview, which was chaired by former Secretary of the Interior Stewart Udall and commissioned by the Dade County Port Authority, outlined ways in which the jetport and the park could coexist, but ultimately called for the acquisition of Big Cypress Swamp by state or federal officials.[79]

With the growing amount of hard evidence that the jetport and commercial development in Big Cypress would have deleterious effects, state officials finally acted. Convinced that "poorly planned development" of the Big Cypress Swamp had harmed both Everglades National Park and South Florida's "ecological balance," Reed began agitating for regional planning and "enforceable land use programs that protect the environment while allowing the private owner use of his land." He asked a commission composed of representatives from Dade, Collier, and Monroe counties to develop "a regional land use program to protect the Big Cypress Water Shed," and he recommended the establishment of a state task force to aid Transportation and Interior in the selection of a new site.[80] At the same time, Kirk informed Hickel and Undersecretary of the Interior Russell Train that

the state no longer supported the jetport, and he requested abandonment of the Big Cypress site. The Everglades Coalition, meanwhile, filed a petition in October requesting that Volpe disapprove the jetport under the authority of the Department of Transportation Act of 1966, while Hickel told John Ehrlichman, Nixon's aide over domestic affairs, that the FAA had the power to delay and hinder the project, thereby making it too expensive for the Port Authority's liking.[81]

Hickel and Train also gave a copy of the Leopold Report to Ehrlichman and to John C. Whitaker, Deputy Assistant to President Nixon, asking that the White House back the jetport's relocation. Ehrlichman then prepared a summary of the issues and gave it to Nixon. After reading the brief, Nixon informed Ehrlichman that the South Florida airport must not be developed in Big Cypress, and that as soon as another location became viable, the training runway should be eliminated. He directed Ehrlichman to have Interior and Transportation officials begin negotiations with the Dade County Port Authority and the state of Florida to implement these actions.[82]

Nixon's efforts to prevent jetport construction came at a time when the President was first beginning to embrace a strategy of addressing environmental concerns proactively, resulting in part from favorable publicity that Nixon received for his support and signing of NEPA. Nixon's State of the Union address in January 1970, for example, would discuss the importance of the environment, and the President was also preparing an environmental message for Congress. Although Nixon would sour on environmental issues late in his presidency, his early administration sought to mine ecological concerns for political gold. Halting jetport construction early in 1970 fit into this scheme; the concerns of Hickel, Whitaker, and Ehrlichman also played into the decision.[83]

Regardless, for the next several weeks discussions occurred between the Interior Department, the Department of Transportation, the Florida governor's office, and the Dade County Port Authority about what to do with the runway and the ultimate development of the jetport. Finally, on 16 January 1970, all sides signed "The Everglades Jetport Pact."[84] This agreement recognized that South Florida needed another airport to relieve congestion at Miami International Airport, and it also acknowledged the Port Authority's efforts at finding a reasonable site. However, because studies had concluded that the jetport "would not be compatible with the preservation and protection" of Everglades National Park and that unregulated operation of the training facility would "produce serious environmental and ecological effects," all sides agreed to certain stipulations. The Port Authority assented to operate the training facility as a single runway, and it agreed to "immediately" institute measures to find another jetport site, submitting quarterly reports of its progress to the United States. If the federal government deemed that the Port Authority was not diligently pursuing another site, it could terminate the pact. Otherwise, when an appropriate location was found, the United States would purchase it for the Port Authority. The state of Florida would "diligently assist" the Port Authority in its search and would convey any state lands free of charge to it. Once the Port Authority had constructed a suitable airport, it would then abandon the runway in the Big Cypress Swamp. It also consented to a list of measures to prevent fuel or oil contamination of land or water by the existing facility, and it agreed not to drain the land or use herbicides, insecticides, or fertilizers. The United States would monitor these operations to ensure that no harm came to the park. In addition, the United States would conduct an ecological study of the Big Cypress Swamp in order to develop planning that would preserve and protect the park and its water supply.[85]

Yet the fight was not over. Soon after the execution of the pact, the secretary of the interior, the secretary of transportation, the FAA administrator, Florida's governor, and the Dade County board of county commissioners established an interdisciplinary team that began searching for a new jetport location. Over the next few years, the group evaluated 36 sites, and eventually decided that Site 14 in northwest Dade County by the Broward County line was the best location. This site was approximately 15 miles northeast of Everglades National Park along the transition between the Everglades and the Atlantic Coastal Ridge, covering approximately 48 square miles, two-thirds of which was in Conservation Area 3B. Several objections were raised to this location; the U.S. Army Corps of Engineers protested that it would affect C&SF Project works, while environmentalists worried about its impacts on the Everglades Kite, an endangered bird. An environmental impact statement was prepared, but by the late 1970s, use of the training facility in Big Cypress Swamp had drastically declined from 100,000 flights to 20,000, leading some to wonder whether a new site was really necessary. The debate over this issue eventually led to a temporary disbanding of the Everglades Coalition due to internal conflicts, as some members wanted no new jetport while others believed one was necessary. By the 1980s, no final jetport resolution had been reached, although Site 14 was still the desired location.[86]

Preservation of Big Cypress Swamp

Meanwhile, federal and state authorities wrestled with the problem of what to do with Big Cypress Swamp; in the press conference announcing the signing of the Everglades Jetport Pact, Secretary of the Interior Hickel had, in the words of historian J. Brooks Flippen,

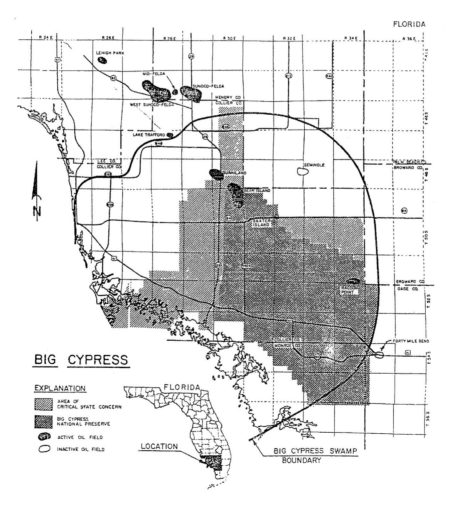

Map of the Big Cypress area. (Big Cypress Area Management Task Force, "Report to Governor and Members of the Cabinet," 1983)

"promised further administration action to protect the entire Big Cypress Swamp."[87] Indeed, many did not want to leave protection up to Collier County officials because of their action (or non-action) regarding Golden Gate Estates. In the 1960s, the Gulf American Land Corporation, led by brothers Leonard and Julius Rosen, had marketed 113,000 acres of land in the Big Cypress Swamp as Golden Gate Estates, a huge housing subdivision only a few miles from Naples and the Gulf of Mexico. To prepare for the development, the company built 171 miles of canals and 807 miles of roads, effectively draining much of the area and altering the ecosystem. But by the 1970s, only a few dozen families lived in the development, mainly because of the legal problems of Gulf American and its successor, GAC Properties. Collier County authorities could have prevented the road and canal construction, but instead encouraged it, even though Gulf American had filed no firm plans for the development, because several Golden Gate promoters sat on the board of county commissioners. Therefore, state and federal officials had little faith in Collier County developing any meaningful protective measures, especially since Florida had outlawed county zoning.[88]

In April 1971, the Everglades-Jetport Advisory Board, a commission consisting of the heads of the seven agencies composing the Interior Department as well as the department's solicitor, issued a study of how Big Cypress could be preserved. It concluded that outright purchase of the land would cost $155.6 million, so it recommended that the federal government acquire approximately 38,000 acres adjoining the Tamiami Trail and operate it as the Tamiami Trail National Parkway. The rest of the Big Cypress would be subject to compensable land use restrictions, meaning that no drainage or construction would be allowed, but landowners could file claims with the U.S. Court of Claims for compensation, which would have a limit of $10 million for all awards.[89]

Many environmentalists, including Browder and Marshall, disagreed with this recommendation, claiming that the only way to save Big Cypress and to protect the Everglades' water supply was through purchasing the entire area. They convinced Florida Governor Reubin Askew of this necessity, and in July 1971, he told Secretary of the Interior Rogers Morton (who had replaced Hickel in 1970) that "acquisition is the only sure method to protect the heart of this natural ecosystem," a stand supported by the entire cabinet sitting as the trustees of the Internal Improvement Fund.[90] With Askew's backing, Florida's two senators, Democrat Lawton Chiles and Republican Edward Gurney, introduced a bill (drafted by Browder) into Congress in August, stipulating that the federal government purchase 547,000 acres in Big Cypress Swamp and designate it as the Everglades-Big Cypress National Recreation Area. As this bill made its way through Congress, Reed, who had become assistant secretary of the interior for fish, wildlife, and parks, began pushing for the Nixon administration to support the acquisition, as did other prominent environmentalists such as Elvis Stahr of the National Audubon Society,

Anthony Wayne Smith of the National Parks Association, Browder, and Leopold. The Environmental Coalition for North America, an organization working for national environmental causes, pledged its backing as well. These individuals and groups had a ready ally in the White House in John Whitaker, Deputy Assistant to President Nixon. Because of Whitaker's and Reed's influence, and realizing the importance of obtaining Florida votes in the 1972 presidential election, Nixon issued a statement in November declaring that it was "essential for the federal government to acquire this unique and vital Watershed."[91]

Only a day after Nixon's proclamation of support, Senator Henry Jackson, chairman of the Senate Committee on Interior and Insular Affairs who had his own presidential aspirations, held a hearing on Chiles' legislation in Miami. State officials and environmentalists made a united stand on the purchase, but landowners in Collier County complained about the measure, stating that 35,000 landowners would be ruined by the acquisition. Some even likened the proposal to oppression by the Soviet Union. Former Florida Governor Fuller Warren, representing the landowners, stated that the government would severely cripple Collier County by removing so much land from the tax rolls, for "next to the air we breathe, the most essential ingredient of life is revenue."[92] Yet state officials and the Nixon administration continued to support acquisition; Nixon even sent his daughter, Julie Nixon Eisenhower, to tour the area with Secretary Morton in January 1972, while the administration introduced its own Big Cypress purchasing bill into Congress (S. 3139).[93]

In April 1972, the Senate Subcommittee on Parks and Recreation of the Committee on Interior and Insular Affairs held a hearing on the two bills, which were essentially similar except for three major points: S. 3139 created a national freshwater reserve rather than a recreation area, eliminated acquisition by legislative taking, and provided for joint state-federal management of the Big Cypress area. This time, however, Senator Alan Bible from Nevada presided, and he was not as favorably inclined toward the acquisition as Jackson. Bible found S. 3139 more palatable than Chiles' bill, but he still found problems with the legislation, including the cost of acquisition and the provision in S. 3139 that landowners be compensated over 10 years rather than immediately. He made his views known throughout the hearing, giving a sympathetic ear to Collier County landowners.[94]

In the late spring of 1972, Bible's opposition solidified when Robert O. Vernon, Florida's state geologist, claimed that Big Cypress Swamp runoff was not essential for the park's water supply because surface runoff accounted for only 11 percent of the park's total water.[95] Hearing this, Bible announced that he would not allow the Subcommittee on Parks and Recreation to release S. 3139 to the Senate "until the people of Florida resolve their differences on the Big Cypress question."[96] Environmentalists and other scientists vehemently disagreed that water from Big Cypress was unessential, but the damage had been done.

Meanwhile, the Seminole and Miccosukee Indians objected to the Big Cypress plan, fearing its effects on their land. In 1957, the Seminole Tribe of Florida had organized itself under the authority of the Indian Reorganization Act of 1934. In 1962, Indians living along the Tamiami Trail, who considered themselves as distinct from the Seminole (even though non-Indians generally referred to them as Seminole), had organized into a separate entity known as the Miccosukee Tribe of Indians. In order to ensure that the Miccosukee had a land base, the state had divided the Big Cypress Reservation in 1965, giving the lower 76,000 acres to the Miccosukee and allowing the Seminole to retain the upper 28,000 acres adjoining the federal tract.[97] Both tribes worried that this land would be included in the Big Cypress boundaries. Howard Tommie, chairman of the Seminole Tribal Council, and Fred Smith, president of the Seminole Tribe, thus counseled legislators to forestall such an action, claiming that the Seminole already managed 62 percent of their land for natural resources. "We don't want to be told what to do on our land," Tommie explained, while Smith insisted that the Seminole were

"more ecology-minded than some of the professional ecologists."[98] The major concern of both the Seminole and the Miccosukee was that the federal government would not allow them to maintain their traditional ways of life, including hunting and fishing, on any land included in the preserve. These fears evaporated, however, after Congress included specific language in the Big Cypress legislation allowing Indians the "usual and customary use and occupancy" of their lands, including hunting, fishing, trapping, and the conducting of tribal ceremonies.[99]

With the Seminole and Miccosukee on board, Governor Askew and the state legislature took some significant action in the spring of 1973, spurred on by Florida Wildlife Federation President John "Johnny" Jones. Jones, one of the most effective lobbyists in Florida history, was strongly in favor of Big Cypress preservation and obtained a meeting with Bible where he asked him why he opposed the bill. According to Jones, Bible told him that the federal government had already spent enough money to acquire Everglades National Park; he asked Jones, "When is Florida going to put the money into this damn thing?" Jones asked him how much he wanted Florida to contribute, and Bible said $40 million. Jones then informed Askew

Big Cypress National Preserve. (South Florida Water Management District)

Miccosukee and Seminole representatives before the Florida legislature.
(The Florida Memory Project, State Library and Archives of Florida)

"exceptional values or qualities illustrating the natural heritage of the Nation," including "ecological communities, . . . natural phenomena, or climax communities." Under this bill, the NPS would manage the preserve to maintain "the natural and scientific values of the area."[103]

After passing the House, the legislation went to the Senate where it sat for several months due to Bible's opposition. Finally, Thomas Kimball of the National Wildlife Federation approached Senator Henry Jackson and, according to Jones, "told him [Jackson] what a dirty deal Bible had pulled."[104] Jackson then pushed the bill, forcing Bible's subcommittee to consider it. The Subcommittee on Parks and Recreation made several changes and recommended its passage to the Senate. The alterations included replacing the legislative taking aspect of acquisition to "normal acquisition procedures" (meaning that landowners would receive compensation over a six-year period) and allowing "all improved residential and commercial property, including mineral estate" to remain in the Big Cypress area as long as it was not "detrimental to the Preserve."[105] The House agreed to these changes, and the bill went to President Gerald Ford for his signature. It became law on 11 October 1974, allowing $116 million for the purchase of 574,000 acres in the Big Cypress Swamp (the state of Florida would still contribute its $40 million).[106] Although some details still had to be finalized with the acquisition, environmentalists, the state of Florida, and the federal government had effectively ensured the preservation of part of the Big Cypress Swamp, an area important not only for its water supply to Everglades National Park but for its own ecology as well.

The battles over the jetport and the barge canal, coupled with the passage of NEPA in 1970 and the growing use of environmental law, ushered in a new way of doing business for the Corps. Both of these controversies indicated that environmentalists now had the power to halt projects that they considered to be ecologically damaging. In the aftermath of these fights, the Corps acknowledged that it had to consider environmental concerns more closely, something which it had vocalized since the late 1960s. The Corps would frequently encounter bumps and setbacks as it began to change its mission-oriented focus to one that accepted the necessity of considering environmental concerns, but by the mid-1970s, the Corps was clearly on its way to making such changes permanent. As historian George E. Buker has indicated, the Cross-Florida Barge Canal was "the last major engineering project" in Florida that "ignored the protests of the environmentalists."[107] Part of the reason for this was that Corps leaders, such as Colonel Wisdom, were willing to consider carefully environmental concerns. Wisdom himself denied Section 404 permits on Marco Island to the Deltona Corporation in 1975, for example, inaugurating "the most important single event during the post-NEPA period" that "improve[ed] the Corps' environmental image."[108]

But another reason was merely the increasing influence of environmentalists. Victories in halting both jetport and canal construction and in obtaining protection for Big Cypress Swamp gave

of Bible's request, and Askew had State Senator Daniel Robert "Bob" Graham, the future governor of Florida, propose a bill in the Florida legislature that the state contribute $40 million for land acquisition.[100] That same law, known as the Big Cypress Conservation Act of 1973, also designated approximately 574,000 acres of Big Cypress Swamp, as well as an additional 285,000 acre buffer zone (including Okaloacoochee Slough, the Fakahatchee Strand, and the northern Ten Thousand Islands) as an "area of critical state concern."[101] This designation was created in the Florida Environmental Land and Water Management Act of 1972 to allow the state to prevent development in and provide other protection to environmentally important regions.

Despite the passage of this legislation, Bible continued to oppose the bill, "double-crossing" Florida, in the words of Jones.[102] But in the fall of 1973, the House of Representatives passed H. R. 10088, a bill introduced by Representative James Haley of Florida and sponsored by the rest of Florida's congressional delegation. Similar to Chiles' bill, it had one major difference: instead of establishing a national recreation area, it would create the Big Cypress Water Preserve, a new unit of the national park system. The House Committee on Interior and Insular Affairs envisioned preserves as areas with

the environmental movement increasing confidence and more unity and cohesion, and highlighted its growing strength within Florida and the nation as a whole. Environmentalists were now major players in water management issues in Florida, and they had developed the organizational ability and the tactics to attack projects that could potentially degrade ecological values. The work of organizations such as the Everglades Coalition and the Friends of the Everglades, as well as individuals such as Joseph Browder, Arthur Marshall, and Marjorie Carr, pushed the administrations of Claude Kirk, Reubin Askew, and Richard Nixon to look more closely at environmental issues in Florida no matter what their political party. At the same time, the jetport and barge canal battles forced federal, state, and local officials to realize two things: first, that the state of Florida, in the face of continued growth, had inadequate measures to protect natural resources within its borders, and second, that the state's water resources—especially in terms of quality—needed to be addressed. For the rest of the 1970s, all water management players would have the opportunity to apply the lessons learned from the jetport and the barge canal as they tackled a problem that threatened the entire South Florida ecosystem—the degradation of the Kissimmee River and Lake Okeechobee.

Endnotes

[1] At a 1969 hearing before Congress, Chief of Engineers Lieutenant General William F. Cassidy explained that Big Cypress Swamp was outside of the boundaries of the C&SF Project, meaning that the Corps had no jurisdiction over the jetport. If the structure was actually built, the Dade County Port Authority would have to receive Corps approval before discharging water into any of the water conservation areas, but until that time, the Corps had no authority. Therefore, Corps participation in the jetport controversy was relegated to occasional attendance at meetings and hearings. Senate Committee on Interior and Insular Affairs, *Everglades National Park: Hearings Before the Committee on Interior and Insular Affairs, United States Senate, Ninety-First Congress, First Session, on the Water Supply, the Environmental, and Jet Airport Problems of Everglades National Park*, 91st Cong., 1st sess., 1969, 27 [hereafter referred to as Jetport Hearing]. As evidence of this, note that Cassidy's testimony runs for over 20 pages, but only about six short paragraphs have any information about the jetport; the rest deals with water supply to Everglades National Park.

[2] Kirkpatrick Sale, *The Green Revolution: The American Environmental Movement, 1962-1992* (New York: Hill and Wang, 1993), 6.

[3] Samuel P. Hays, *Beauty, Health, and Permanence: Environmental Politics in the United States, 1955-1985* (Cambridge: Cambridge University Press, 1987), 2-5, 34-35; Samuel P. Hays, *A History of Environmental Politics Since 1945* (Pittsburgh, Penn.: University of Pittsburgh Press, 2000), 15-19, 22-23; Adam Rome, "'Give Earth a Chance': The Environmental Movement and the Sixties," *Journal of American History* 90 (September 2003): 527-530; Lynton Keith Caldwell, *The National Environmental Policy Act: An Agenda for the Future* (Bloomington: Indiana University Press, 1998), 26-28.

[4] Rome, "'Give Earth a Chance,'" 525-554; J. Brooks Flippen, *Nixon and the Environment* (Albuquerque: University of New Mexico Press, 2000), 3-5.

[5] Robert P. McIntosh, *The Background of Ecology: Concept and Theory* (Cambridge: Cambridge University Press, 1985), 2, 194-196; Frank Benjamin Golley, *A History of the Ecosystem Concept in Ecology: More Than the Sum of the Parts* (New Haven, Conn.: Yale University Press, 1993), 1-3; Dorothy Nelkin, "Scientists and Professional Responsibility: The Experience of American Ecologists," *Social Studies of Science* 7 (1977): 79-80.

[6] Quotation in Michael A. Bryson, *Visions of the Land: Science, Literature, and the American Environment from the Era of Exploration to the Age of Ecology* (Charlottesville: University Press of Virginia, 2002), 134-135; see also Nelkin, "Scientists and Professional Responsibility," 80.

[7] Rome, "'Give Earth a Chance,'" 527; Flippen, *Nixon and the Environment*, 5; Michael E. Kraft, "U.S. Environmental Policy and Politics: From the 1960s to the 1990s," *Journal of Policy History* 12, no. 1 (2000): 23; Sale, *The Green Revolution*, 1-3, 20-22.

[8] The National Environmental Policy Act of 1969 (83 Stat. 852).

[9] Caldwell, *The National Environmental Policy Act*, 28-30, 37.

[10] See Jeffrey K. Stine, "Environmental Politics and Water Resources Development: The Case of the Army Corps of Engineers during the 1970s," Ph.D. dissertation, University of California at Santa Barbara, 1984, 25, 35.

[11] Arthur R. Marshall, "Repairing the Florida Everglades Basin," 11 June 1971, File Everglades National Park 1958-86 General/Resolutions/Agreements, Box 02161, SFWMDAR; Blake, *Land Into Water*, 196-197.

[12] Flippen, *Nixon and the Environment*, 8-11; Blake, *Land Into Water*, 197-198; Carter, *The Florida Experience*, 50-53; Hays, *Beauty, Health, and Permanence*, 57-58.

[13] Blake, *Land Into Water*, 150-151. For a history of the canal up to the 1920s, see Charles E. Bennett, "Early History of the Cross-Florida Barge Canal," *The Florida Historical Quarterly* 45, no. 2 (1966): 132-144. This is not an entirely objective piece, as Bennett, a U.S. representative from Florida, was a major proponent of the canal.

[14] See C. P. Summerall, Chairman, The Ship Canal Authority of the State of Florida, to Governor, 16 January 1943, File Canal 1940-1949, Box 2, Robert N. "Bert" Dosh Papers, Manuscript Series 25, Special and Area Studies Collections, George A. Smathers Library (East), University of Florida, Gainesville, Florida [hereafter referred to as Dosh Papers]; Walter F. Coachman, Jr., to The Members of the Executive Committee of Canal Counties, 24 January 1942, ibid.; U.S. Engineer Office, *Definite Project Report on Cross-Florida Barge Canal* (Jacksonville, Fla.: U.S. Engineer Office, 1943).

[15] William N. Partington, "History of the Cross-Florida Canal," in Florida Defenders of the Environment, *Environmental Impact of the Cross-Florida Barge Canal* (Gainesville, Fla.: Florida Defenders of the Environment, 1970), 55; Blake, *Land Into Water*, 152-162, 164-165, 201-203; Carter, *The Florida Experience*, 271-273, 276-278; J. Richard Sewell, "Cross-Florida Barge Canal, 1927-1968," *The Florida Historical Quarterly* 46 (April 1968): 371, 374-375, 379-381; Flippen, *Nixon and the Environment*, 58. Kennedy and Johnson's support was not necessarily out of a belief in the benefits of the canal; at least some came from political expedience. In July 1963, Major Holbrook Scott, an Ocala resident, claimed that the "Barge Canal issue can make or break the Federal Administration in Florida in 1964." The situation was probably not that dramatic, but it is clear that both Kennedy and Johnson realized the political value of supporting the canal, at least in these early stages. See Scott to Hon. Clarence Cannon, M. C. from Missouri, 29 July 1963, Folder Canal 1960-1965, Box 2, Dosh Papers.

[16] Carter, *The Florida Experience*, 267.

[17] "Canal Fight Started At Audubon Meeting," *The Tampa Tribune-Times*, 24 January 1971.

[18] Quotation in "Canal Fight Started at Audubon Meeting"; see also Stephen Trumbull, "The River Spoilers," *Audubon* 68 (March-April 1966): 109-110.

[19] Partington, "History of the Cross-Florida Canal," 55.

[20] Quotation in William M. Partington, "Oklawaha—The Fight Is On Again!" *Living Wilderness* 33 (Autumn 1969): 19; see also Carter, *The Florida Experience*, 281-282; "Canal Fight Started At Audubon Meeting"; Partington, "History of the Cross-Florida Canal," 55; "Cross Florida Barge Canal Chronological Development," 12 February 1971, 4, File Cross Fla Barge Canal—General, Box 2, Davis Papers. For an example of the recreational benefits of Rodman Reservoir, see Department of the Army, U.S. Army Engineer District, Jacksonville, Florida, News Release, "Cross-Florida Barge Canal Project," n.d., File Barge Canal 1969-70, Box 2, S949, FSA.

[21] Partington, "History of the Cross-Florida Canal," 55-56; "Canal Fight Started At Audubon Meeting"; Carter, *The Florida Experience*, 283.

[22] For more information on the formation of the Environmental Defense Fund and its role in the fight against DDT, see Thomas R. Dunlap, *DDT: Scientists, Citizens, and Public Policy* (Princeton: Princeton University Press, 1981).

[23] Partington, "Oklawaha—The Fight Is On Again!," 19-22; "Canal Fight Started At Audubon Meeting"; Partington, "History of the Cross-Florida Canal," 56; Carter, *The Florida Experience*, 284-287; Stine, "Environmental Politics and Water Resources Development," 51-52, 59.

[24] See Florida Game and Fresh Water Fish Commission, "A Brief Assessment of the Ecological Impact of the Cross Florida Barge Canal," November 1969, File Fla. Game and Fresh Water Fish Comm., Box 1, John Henry Davis Papers, Manuscript Series 23, Special and Area Studies Collections, George A. Smathers Library (East), University of Florida, Gainesville, Florida [hereafter referred to as Davis Papers]; United States Department of the Interior, Fish and Wildlife Service, Bureau of Sport Fisheries and Wildlife, "Review and Appraisal of the Cross Florida Barge Canal," 30 March 1970, File Florida Defenders of the Environment, ibid.

[25] Florida Defenders of the Environment, *Environmental Impact of the Cross-Florida Barge Canal*, 1-5.

[26] Quotations in James Nathan Miller, "Rape on the Oklawaha," *Reader's Digest* 96 (January 1970): 54-60 (emphasis in the original); see also Carter, *The Florida Experience*, 291-292.

[27] "Text of a Letter Addressed to President Nixon, Dated January 27, 1970, Mailed February 6, 1970 and Signed by 162 Environmental Scientists," File Clippings 1969-1974, Box 3, Davis Papers; Carter, *The Florida Experience*, 290.

[28] Flippen, *Nixon and the Environment*, 105; Robert E. Jordan III, Special Assistant to the Secretary of the Army (Civil Functions), to Honorable John C. Whitaker, Deputy Assistant to the President, 11 November 1970, File Barge Canal 2-71/8-71 (Governor Askew), Box 2, S949, FSA; "Canal Foe Coalition Seeks Drive Support," *The Tampa Tribune*, 9 September 1970; "Army to Reject Halting Barge Canal?" *Miami Herald*, 19 June 1970; "Environmental Veto Threatens Florida Canal," *Christian Science Monitor*, 29 June 1970; "Cross Florida Barge Canal Chronological Development."

[29] Quotation in Executive Office of the President, Council on Environmental Quality, "Summary of Environmental Considerations Involved in the Recommendation for Termination of Construction of the Cross Florida Barge Canal," File U.S. Council on Environmental Quality, Box 3, Davis Papers; see also Russell E. Train, Chairman, Memorandum for Mr. Whitaker, 1 December 1970, in Carter, *The Florida Experience*, 311-312.

[30] John D. Ehrlichman, "Presidential Assistant with a Bias for Parks," an oral history conducted in 1991 by William Duddleson, in *Saving Point Reyes National Seashore, 1969-1970: An Oral History of Citizen Action in Conservation*, Regional Oral History Office, The Bancroft Library, University of California, Berkeley, 1993, 368-369 [hereafter referred to as Ehrlichman interview]; see also Carter, *The Florida Experience*, 296-298.

[31] Stine, "Environmental Politics and Water Resources Development," 59-60.

[32] Quotation in Office of White House Press Secretary, Statement by the President, 19 January 1971, in Carter, *The Florida Experience*, 312; see also "President Blocks Canal in Florida," *New York Times*, 20 January 1971; Blake, *Land Into Water*, 209; Flippen, *Nixon and the Environment*, 130; Stine, "Environmental Politics and Water Resources Development," 60-61. Whitaker told journalist Luther Carter that Parker's court injunction did not influence Nixon to halt construction; the White House had received Whitaker's decision paper several days before the ruling was announced. Carter, *The Florida Experience*, 298-299.

[33] See Ehrlichman interview, 369; Martin Reuss, *Shaping Environmental Awareness: The United States Army Corps of Engineers Environmental Advisory Board, 1970-1980* (Alexandria, Va.: Historical Division, Office of Administrative Services, Office of the Chief of Engineers, 1983), 19.

[34] "Wanted Dead or Alive? Cross Florida Canal," *Pensacola (Fla) News-Journal*, 18 April 1971; Robert L. Shevin, Attorney General, to The Honorable Reubin O'D. Askew, 15 February 1971, File Barge Canal 2-71/8-71 (Governor Askew), Box 2, S949, FSA; L. C. Ringhaver, Chairman, to Richard M. Nixon, 12 February 1971 (with attachment), ibid.; Blake, *Land Into Water*, 212.

[35] "Nixon Can't Halt Canal, U.S. Judge Johnsen Rules"; "Resolution," File Cross-Florida Barge Canal, Box 9, S1160, Florida State Board of Conservation Water Resources Subject Files, 1961-1968, FSA; Blake, *Land Into Water*, 213. The Corps had prepared an EIS in 1970, but environmentalists believed that this seven-page paper was insufficient and the court agreed.

[36] "Corps Strives To Blunt Barge Canal Opposition," *The Florida Times-Union*, 17 December 1976; "Cabinet Vote Doesn't Bury the Barge Canal," *The Florida Times-Union*, 19 December 1976; "Resolution," 17 January 1977, in U.S. Army Corps of Engineers, Jacksonville District, *Cross Florida Barge Canal Restudy Report: Final Summary* (Jacksonville, Fla.: Department of the Army, Jacksonville District, Corps of Engineers, 1977), A-1—A-4; Blake, *Land Into Water*, 214.

[37] Colonel Donald A. Wisdom interview by George E. Buker, 22 and 23 December 1978, Jacksonville, Florida, 22, transcript in Library, Jacksonville District, U.S. Army Corps of Engineers, Jacksonville, Florida [hereafter referred to as Wisdom interview]; "Cross Florida Barge Canal Restudy Report, Appendix D, Major Issues," in Jacksonville District, *Cross Florida Barge Canal Restudy Report: Final Summary*, D-12.

[38] Jacksonville District, *Cross Florida Barge Canal Restudy Report*; Wisdom interview, 22; Florida Department of Environmental Protection, "Marjorie Harris Carr Cross Florida Greenway—History," <http://www.dep.state.fl.us/gwt/cfg/history.htm> (18 January 2005); Florida Defenders of the Environment, "Restoring the Ocklawaha River Ecosystem" <http://www.fladefenders.org/publications/restoring3.html> (18 January 2005). Ocklawaha is a variant spelling of the name of the river.

[39] For more information about the La Farge Project, see Theodore Catton and Matthew C. Godfrey, "Steward of Headwaters: U.S. Army Corps of Engineers, St. Paul District, 1975-2000," 75-82, manuscript prepared for the St. Paul District, U.S. Army Corps of Engineers, St. Paul, Minnesota. For more information about the Meramec Dam, see T. Michael Ruddy, "Damming the Meramec: The Elusive Public Interest, 1927-1949," *Gateway Heritage* 10 (Winter 1989-1990): 36-45.

[40] Paul Brooks, "Superjetport or Everglades Park?" *Audubon* 71 (July 1969): 5; see also Planning Research Section, Department of Planning and

Zoning, Miami-Dade County, "Demographic Profile, Miami-Dade County, Florida, 1960-2000," September 2003, 1 <http://www.co.miami-dade.fl.us/planzone/Library/Census/demographic_profile.pdf> (21 January 2005).

[41] Everglades-Jetport Advisory Board, "The Big Cypress Watershed: A Report to The Secretary of the Interior, April 19, 1971," in Senate Committee on Interior and Insular Affairs Subcommittee on Parks and Recreation, *Everglades-Big Cypress National Recreation Area: Hearing Before the Subcommittee on Parks and Recreation of the Committee on Interior and Insular Affairs, United States Senate*, 92d Cong., 1st sess., 1971, 137-139.

[42] Robert S. Gilmour and John A. McCauley, "Environmental Preservation and Politics: The Significance of 'Everglades Jetport,'" *Political Science Quarterly* 90 (Winter 1975-1976): 721-722; Carter, *The Florida Experience*, 188.

[43] Gilmour and McCauley, "Environmental Preservation and Politics," 723; "Miami Jetport and Interstate 75: Everglades National Park," File Jetport Correspondence 1971, Box 8, S949, Governor's Office Jay Landers Subject Files, FSA.

[44] Gilmour and McCauley, "Environmental Preservation and Politics," 723; Carter, *The Florida Experience*, 189-191; "Miami Jetport and Interstate 75." For a full discussion of Port Authority consultations with the NPS during this process, see J. D. Brama, Assistant Secretary for Urban Systems and the Environment, to Hon. Henry M. Jackson, Chairman, Interior and Insular Affairs Committee, 27 June 1969, in Jetport Hearing, 38-39.

[45] O. E. Frye, Jr., Director, to Mr. Alan C. Stewart, Dade County Port Authority, 22 February 1968, File Big Cypress: Jetport (Environmental Studies), Box 1, S1719, Game & Fresh Water Fish Commission Everglades Conservation Files, 1958-1982, FSA.

[46] Carter, *The Florida Experience*, 193.

[47] "Report on Ecology Conference Relative to the Proposed Collier-Dade Training Airport, June 20, 1968," File Big Cypress: Jetport (Environmental Studies), Box 1, S1719, FSA.

[48] John "Jack" Maloy interview by Matthew Godfrey, 14 July 2004, West Palm Beach, Florida [hereafter referred to as Maloy interview]; Jetport Hearing, 68, 87-88; Gilmour and McCauley, "Environmental Preservation and Politics," 725; Carter, *The Florida Experience*, 194-195; Blake, *Land Into Water*, 218.

[49] "Minutes of a Meeting of the Governing Board of the Central and Southern Florida Flood Control District and the Interested Parties Who Were Present to Discuss the Dade County Port Authority's Proposed Jet Port Site Held at the Riviera Country Club in Coral Gables, Florida, Friday, December 13, 1968," File Big Cypress: Jetport (Environmental Studies), Box 1, S1719, FSA; Carter, *The Florida Experience*, 195-196.

[50] Flippen, *Nixon and the Environment*, 31-32.

[51] "Folly in Florida," *National Parks Magazine* 43 (January 1969): 2.

[52] For the written answers to the questions, see "Answers to Questions Submitted by Central and South Florida Flood Control District, February 3, 1969," in Jetport Hearing, 74-79.

[53] "Statement of John C. Raftery, Superintendent, Everglades National Park at Jetport Meeting, Miami Springs, Florida, February 28, 1969," File Park Problems Big Cypress Jetport, EVER 22965, CR-ENPA.

[54] As quoted in C. Edward Carlson, Acting Regional Coordinator, to Mr. Allen Stewart, Director, Dade County Port Authority, 7 March 1969, File Big Cypress: Jetport (Environmental Studies), Box 1, S1719, FSA; see also Carter, *The Florida Experience*, 196.

[55] Jim Smith to Senator Adams, 7 March 1969, File Jetport Correspondence, Box 8, S949, FSA.

[56] Nat Reed to Governor Kirk, 5 March 1969, File Jetport—Everglades, Box 9, S949, FSA. Michael Grunwald, citing an interview with Reed, claims

that Reed, attending the meeting with Arthur Marshall, "jumped to his feet and berated Dade County mayor Chuck Hall for wasting everyone's time" with the dismissive answers. However, such an action is not in accord with the letter that Reed wrote to Kirk a week after the meeting, and no other accounts of the meeting mention an outburst by Reed. See Grunwald, *The Swamp*, 255-256.

[57] As quoted in Senate Committee on Public Works Subcommittee on Flood Control—Rivers and Harbors, *Central and Southern Florida Flood Control Project: Hearing Before the Subcommittee on Flood Control—Rivers and Harbors of the Committee on Public Works, United States Senate*, 91st Cong., 2d sess., 1970, 228-230; see also Joseph Browder interview by Theodore Catton, 17 November 2004, Washington, D.C. [hereafter referred to as Browder interview]; "Coalition Forms to Fight Florida Jetport," *National Parks Magazine* 43 (May 1969): 28; Carter, *The Florida Experience*, 196-197; Blake, *Land Into Water*, 218.

[58] All quotations in Marjory Stoneman Douglas with John Rothchild, *Voice of the River* (Sarasota, Fla.: Pineapple Press, 1987), 224-226; see also Martha Munzer, "The Everglades and a Few Friends," *South Florida History Magazine* 23 (Winter 1995): 12-13; Davis, *An Everglades Providence*, 473-474.

[59] Douglas, *Voice of the River*, 224-226; Munzer, "The Everglades and a Few Friends," 12-13; Grunwald, *The Swamp*, 257-258; Davis, *An Everglades Providence*, 477-480.

[60] Act of 15 October 1966 (80 Stat. 931).

[61] National Parks Association et al., to Hon. John A. Volpe, Secretary of Transportation, 17 April 1969, in Jetport Hearing, 126-128. For examples of the arguments dealing with the applicability of the Department of Transportation Act, see Jetport Hearing, 121-122.

[62] See Joe B. Browder, Southeastern Representative, to Mr. Charles H. Callison, Executive Vice President, National Audubon Society, 2 May 1969, File Jetport Correspondence 1971, Box 8, S949, FSA.

[63] Walter J. Hickel, *Who Owns America?* (Englewood Cliffs, N.J.: Prentice-Hall, 1971), 101-102.

[64] Hickel, *Who Owns America?*, 101-102; Walter J. Hickel, Secretary of the Interior, to Mr. Secretary, 30 April 1969, File Jetport Correspondence 1971, Box 8, S949, FSA; "Clash Seen Over Jetport Plan," *The Washington Post*, 15 May 1969; Carter, *The Florida Experience*, 197-198; "Conservationists Urge Halt on Jetport Work," unidentified newspaper clipping, File Jetport Correspondence 1971, Box 8, S949, FSA; Blake, *Land Into Water*, 218; Flippen, *Nixon and the Environment*, 39.

[65] Russell S. Train, The Under Secretary, to All Assistant Secretaries and Heads of Bureaus and Offices, 9 June 1969, File I.#2.A. Leopold Report, Box 2, Accession No. 412-91-0041, RG 412, Records of the Environmental Protection Agency, NARA-SE; Carter, *The Florida Experience*, 198-200.

[66] As quoted in Jetport Hearing, 101-102.

[67] Jetport Hearing, 104-106.

[68] O'Neil and Judy quotations in John G. Mitchell, "The Bitter Struggle for a National Park," *American Heritage* 22, no. 3 (1970): 100; Stewart quotation in Brooks, "Superjetport or Everglades Park?" 5.

[69] "'Progress' Menaces the Everglades," *National Parks Magazine* 43 (July 1969): 8-10.

[70] Brooks, "Superjetport or Everglades Park?" 5-11.

[71] "Conservation: Jets v. Everglades," *Time* (22 August 1969): 42-43.

[72] Anthony Wolff, "The Assault on the Everglades," *Look* (9 September 1969): 44-52.

[73] John D. MacDonald, "Threatened America—Last Chance to Save the Everglades." *Life* (5 September 1969): 58-66.

[74] Wolff, "The Assault on the Everglades," 44. For more information on the national coverage, see "Jetport Fight Stirs National Interest," *The Miami News*, 28 August 1969.

[75] Although the Department of Transportation was also supposed to have a voice in the report, it submitted its revisions too late to have them effectively implemented in the final version. Transportation officials had a different view of the problems than Leopold and his team, believing that any threat to the park would exist with or without the jetport, and that commercial development would not be a great problem. The differences between the two viewpoints were so great, and the deadline for publication so near, that Leopold merely submitted the report to Under Secretary Russell Train with Transportation's name eliminated from the cover. However, when Hickel transmitted the report to Governor Kirk, he stated that the conclusions were made "in consultation with Secretary of Transportation John A. Volpe." Carter, *The Florida Experience*, 204-205; Walter J. Hickel, Secretary of the Interior, to Governor Kirk, 7 October 1969, File Jetport Correspondence 1971, Box 8, S949, FSA.

[76] United States Department of the Interior and Luna B. Leopold, *Environmental Impact of the Big Cypress Swamp Jetport* (Washington, D.C.: United States Department of the Interior, 1969), 1-2 [hereafter referred to as Leopold Report].

[77] Leopold Report, 136-144; see also "Miccosukee See Airport as Final Destruction," *The Florida Times-Union*, 12 September 1969.

[78] Blake, *Land Into Water*, 220; Gilmore and McCauley, "Environmental Preservation and Politics," 729.

[79] For a reprinting of Overview's report, see "Overview: There's a Jetport in the Future If South Florida Plans Ahead," *The Miami Herald*, 11 December 1969.

[80] "Dictated by Nathaniel Reed—Thursday—September 11, 1969," File Big Cypress: Jetport (Environmental Studies), Box 1, S1719, FSA

[81] "Policy on Everglades Jetport," File Big Cypress: Jetport (Environmental Studies), Box 1, S1719, FSA; Carter, *The Florida Experience*, 205; Gilmour and McCauley, "Environmental Preservation and Politics," 730; Flippen, *Nixon and the Environment*, 41.

[82] Gilmour and McCauley, "Environmental Preservation and Politics," 731-732; Carter, *The Florida Experience*, 207-208.

[83] Flippen, *Nixon and the Environment*, 9, 55-56, 221.

[84] Gilmour and McCauley, "Environmental Preservation and Politics," 731-732; Carter, *The Florida Experience*, 207-208. Collier County was supposed to be a party to the agreement as well, but its officials never signed it.

[85] "The Everglades Jetport Pact: Articles of Agreement by and between The United States, State of Florida, Dade County Port Authority, Collier County," File Jetport Correspondence 1971, Box 8, S949, FSA.

[86] Peter L. Cook, Acting Director, Office of Federal Activites, U.S. Environmental Protection Agency, Memorandum for Barbara Blum, 21 October 1977, File Jetport Background Info 1976-1979 DEIS, Box 1, Accession No. 412-91-0041, RG 412, NARA-SE; Colonel Emmett C. Lee, Jr., District Engineer, to Division Engineer, South Atlantic, 7 November 1972, File 1110-2-1150a (C&SF) Jetport 1972, Box 9, Accession No. 077-01-0023, FRC; Browder interview; Blake, *Land Into Water*, 221-222.

[87] Flippen, *Nixon and the Environment*, 56.

[88] Carter, *The Florida Experience*, 232-242; Grunwald, *The Swamp*, 234-235, 244; "Drainage Halt Urged In Big Cypress Swamp," *The Miami Herald*, 30 August 1970.

[89] Everglades-Jetport Advisory Board, "The Big Cypress Watershed: A Report to The Secretary of the Interior, April 19, 1971," in Senate Committee on Interior and Insular Affairs Subcommittee on Parks and Recreation, *Everglades-Big Cypress National Recreation Area: Hearing Before the Subcom-*

mittee on Parks and Recreation of the Committee on Interior and Insular Affairs, United States Senate, 92d Cong., 1st sess., 1971, 131-181; Carter, *The Florida Experience*, 242-244; Department of the Interior, "Interior Secretary Morton Releases Land Use Plan for Big Cypress Swamp," n.d., File Big Cypress, Box 3, S949, FSA; Secretary of the Interior to Director, National Park Service, et al., 19 November 1969, File USDI Everglades Jetport Advisory Board—File #1 Advisory Board Meeting Notes & Progress Report, Box 1, ibid.

[90] Reubin Askew, Governor, to Honorable Rogers Morton, 20 July 1971, File Big Cypress, Box 3, S949, FSA; Resolution, 23 November 1971, ibid.

[91] Quotation in The White House, Statement By the President, Big Cypress National Fresh Water Reserve, November 1971, File Legislation Big Cypress—Area of Critical Concern—Background Material 1973-1974, Box 20, D. Robert "Bob" Graham Papers, Manuscript Series 148, Special and Area Studies Collections, George A. Smathers Library (East), University of Florida, Gainesville, Florida; see also Carter, *The Florida Experience*, 245; Grunwald, *The Swamp*, 258. Grunwald argues that Nixon made his announcement because he caught wind that Senator Henry Jackson would be holding a hearing on Big Cypress in Miami and he wanted to "knock Jackson out of the box in Florida!"

[92] Senate Subcommittee on Parks and Recreation, *Everglades-Big Cypress National Recreation Area*, 53-54, 84-86.

[93] "Julie and Rogers View Cypress Swamp Tract," *The Evening Star*, 5 January 1972; Nathaniel Reed, Assistant Secretary of the Interior, to Hon. Henry M. Jackson, Chairman, Committee on Interior and Insular Affairs, 19 April 1972, in Senate Committee on Interior and Insular Affairs Subcommittee on Parks and Recreation, *Everglades-Big Cypress National Recreation Area: Hearings Before the Subcommittee on Parks and Recreation of the Committee on Interior and Insular Affairs, United States Senate, Part 2*, 92d Cong., 2d sess., 1972, 15-17.

[94] Senate Subcommittee on Parks and Recreation, *Everglades-Big Cypress National Recreation Area Part 2*, 29, 42.

[95] Robert O. Vernon, Director, Division of Interior Resources, to The Honorable Reubin O'D. Askew, Governor, 15 June 1972, File Big Cypress, Box 3, S949, FSA.

[96] Quotation in "Second Look at Big Cypress Bill," *Fort Myers News-Press*, 6 June 1972.

[97] Harry A. Kersey, Jr., "The East Big Cypress Case, 1948-1987: Environmental Politics, Law, and Florida Seminole Tribal Sovereignty," *The Florida Historical Quarterly* 69 (April 1991): 458-459. In 1984, this state reservation became a federal Indian reservation held in trust by the United States. See Hobbs, Straus, Dean & Wilder to Miccosukee Tribe of Indians of Florida, 20 March 1989, File Indian Affairs Miccosukee Research 94, Box 22792, SFWMDAR.

[98] "Seminoles Ask Exemption of Lands in Swamp Deal," unidentified newspaper clipping in Folder Legislation Big Cypress, 1972-1973, Box 20, Graham Papers.

[99] Quotations in Senate, *Establishing the Big Cypress National Preserve, Florida*, 93d Cong., 2d sess., 1974, S. Rept. 93-1128, Serial 13057-7, 7; see also Act of 11 October 1974 (88 Stat. 1258).

[100] Jones interview, 20.

[101] Quotation in "Big Cypress Conservation Act," copy in File Legislation Big Cypress Conservation Act 1973, Box 21, Graham Papers; see also "Big Cypress Eminent Domain Bill Filed," *Orlando Sentinel-Star*, 13 April 1973; "Senator Proposes State Preserve Big Cypress Area," *Sarasota Herald-Tribune*, 6 March 1973; "Memorandum for Record," 13 March 1973, File Legislation Big Cypress 1972-1973, Box 20, Graham Papers; Senator D. Robert Graham Press Release, n.d., ibid.; "Big Cypress Act Defended,"

Tallahassee Democrat, 10 November 1973; Carter, *The Florida Experience*, 246-247.

[102] Jones interview, 20.

[103] Quotation in House, *Establishing the Big Cypress National Preserve in the State of Florida, and for Other Purposes*, 93d Cong., 1st sess., 1973, H. Rept. 93-502, Serial 13020-5, 6-7; see also "State and Federal Efforts to Preserve the Big Cypress Swamp," *Tropical Audubon Bulletin*, n.d., copy in File Legislation Big Cypress—Area of Critical Concern 1973-1974, Box 20, Graham Papers.

[104] Jones interview, 20.

[105] Senate, *Establishing the Big Cypress National Preserve, Florida*, 93d Cong., 2d sess., 1974, S. Rept. 93-1128, Serial 13057-7, 7.

[106] Act of 11 October 1974 (88 Stat. 1258); Blake, *Land Into Water*, 234.

[107] Quotation in George E. Buker, *The Third E: A History of the Jacksonville District, U.S. Army Corps of Engineers, 1975-1998* (Jacksonville, Fla.: U.S. Army Corps of Engineers, 1998); see also Stine, "Environmental Politics and Water Resources Development," 39-40, 51.

[108] Quotations in Jeffrey K. Stine, "Regulating Wetlands in the 1970s: U.S. Army Corps of Engineers and the Environmental Organizations," *Journal of Forest History* 27 (April 1983): 71; see also Stine, "Environmental Politics and Water Resources Development," 147-157.

CHAPTER 6   **The Liquid Heart of Florida:** Lake Okeechobee and the
Kissimmee River in the 1970s

BY THE END OF 1971, environmentalists had a few successes to celebrate in Florida. They had halted construction of the Cross-Florida Barge Canal and forced proponents of a proposed jetport to rethink their Big Cypress Swamp location. But other problems loomed on the horizon, ones that would not see such immediate resolutions. "There is a water crisis in South Florida today," a group of academic, government, and environmental scholars told Florida Governor Reubin Askew in September 1971, predicting a dire future for the region unless the state instituted land and water planning.

One of the major reasons for this pessimism was the condition of Lake Okeechobee. By the early 1970s, many scientists were forecasting the imminent demise of the lake because of a heavy influx of nutrients, especially from the Kissimmee River, which the U.S. Army Corps of Engineers had channelized in the 1960s for flood control. All of these problems led Florida state officials to take major action in the early 1970s. During the 1972 legislative session, the Florida legislature passed several land and water planning measures, including authorization of a major study on the eutrophication of Lake Okeechobee. At the same time, environmentalists called for the restoration of the Kissimmee River, believing that this was one of the best ways to heal the lake. Despite all of these measures, little firm action had been taken to resolve the river and the lake's problems by the end of the 1970s. The state had authorized additional studies and had formed a coordinating council to deal with Kissimmee restoration, but the river remained channelized and nutrient-rich water continued to pour into Lake Okeechobee, in part because of disagreement among environmentalists, state, and Corps officials as to the best remedy for the lake's sickness.

In 1970, environmentalists such as Arthur Marshall predicted that Florida would soon suffer from a water shortage if development continued at its current pace. "It became common, and indeed fashionable," scholar Robert Healy has argued, "to question the value of growth itself" in 1970s Florida because of the state's tremendous growth.[1] Florida as a whole had doubled in population every 22 years since 1920, while South Florida counties had doubled every 14 years in that same time span, adding at times an average of 110,000 people a year for an average annual growth rate of 3.08 percent.[2]

Compounding South Florida's expansion was the construction of Disney World in Orlando on the northern border of the C&SF Project. Walt Disney Productions had secretly purchased approximately 27,000 acres near Orlando in the 1960s for $5 million, desiring to build a park five times larger than southern California's Disneyland and to sequester it from the rest of the region. Construction began in the mid-1960s; the corporation hoped to have the amusement park in place by the end of 1971, as well as hotels, motels, and boating and golfing facilities. The company also planned on building an industrial park, an airport, 2,500 additional hotel and motel units, and a 50-acre shopping and recreation complex. In order to maintain water levels in the development and to prevent flooding, Disney also constructed a water conservation system of 40 miles of canals and 16 structures, using Major General William E. Potter, a former District and Division Engineer for the Corps, as its supervisor. As the construction occurred, the population of Orlando began climbing rapidly in the late 1960s, reaching 100,000 in 1970. Residents, as well as the new facilities, demanded not only water, but also dumped sewage and other wastes into existing waterways, creating water quality problems.[3]

Concerned by these conditions, scientists and environmentalists wondered what the future held for South Florida, both in terms of water quantity and quality. By the 1970s, ecologists had begun to focus research efforts on an aspect of the science known as systems ecology. An outgrowth of the study of ecosystems, systems ecology, in the words of ecologist George Van Dyne, was "the study of the development, dynamics, and disruption of ecosystems." Interdisciplinary in nature, systems ecology tried to integrate mathematics, engineering, and social science in its studies, which primarily focused on "large-scale biological communities or ecosystems of very great complexity." Because it examined "inanimate processes of the ecosystem," those involved in systems ecology had to have a "knowledge of physics, chemistry, geology, geochemistry, meteorology, and hydrology beyond that of traditional ecologists." Thus, systems ecology differed in five major ways from more general ecology:

- it examined "ecological phenomena at large spatial, temporal, or organizational scales;
- it used methodologies from other disciplines;
- it emphasized mathematical models;
- it used digital and analog computers in its modeling;
- it embraced "a willingness to formulate hypotheses about the nature of ecosystems."[4]

One of the areas that seemed well suited to the application of systems ecology was South Florida, an ecosystem of immense complexity. Therefore, scientists in the 1970s, such as Arthur Marshall, began to embrace the methods of systems ecology in their examinations of water issues in South Florida.

As Marshall continued his studies, he did not see bright prospects. An employee of the U.S. Fish and Wildlife Service for 20 years, Marshall joined the University of Miami in 1970 because, according to some observers, his position with the FWS did not allow him to voice publicly his true convictions about South Florida's ecosystem. Having removed those constraints, Marshall began speaking frequently to private groups and organizations, state officials, and the media about his concerns with man's destruction of the Everglades in South Florida.[5]

Marshall promulgated his views as a member of the Special Study Team on the Florida Everglades, a group formed by the FCD and the Florida Game and Fresh Water Fish Commission in 1970 to investigate wildlife issues within the region. The genesis of this report stemmed from issues involving deer and water levels in Conservation Area No. 3. As explained previously, heavy rains in the spring and summer of 1966 had imperiled the deer herd in that region, and the same situation occurred in 1968 and again in 1970 despite attempts to provide long-term solutions to the problem, such as the Corps' construction of 315 islands for refuge. Although the endangerment of deer did not seem to have any direct correlation to water quality in South Florida, the issue involved water management in the region in general, especially since the water conservation areas were

Arthur R. Marshall. (The Florida Memory Project, State Library and Archives of Florida)

storage areas for excess water. As the water rose in Conservation Area No. 3, the Corps again faced criticisms and attacks from the media and environmental organizations that it and the FCD placed agricultural and municipal interests over those of wildlife. This was especially prevalent in 1970 because high water in Conservation Area No. 3 that year was caused in part by the pumping of large amounts of excess water from the EAA, which experienced heavy rainfall in March. Although, in the words of a press release from the Florida Wildlife Federation, the preservation of the Everglades was a "more far reaching [problem] than saving the deer herd," environmentalists considered deer to be "an indicator animal" signifying the health of the region. The fact that deer had suffered in both 1966 and 1970 because of high water levels showed, according to the Federation, that "the ecology of the Everglades is being altered."[6]

At the same time, wildlife problems occurred in the Loxahatchee National Wildlife Refuge (Conservation Area No. 1), albeit from fluctuating levels of water, rather than from an excess of the resource. In May 1970, FWS officials investigated the effects of a rapid drop in the water level of interior and exterior canals (estimated as a decline of over three feet in 30 days) and discovered that the drawdown might have harmed fledgling populations of the Everglades Kite, an endangered bird. By July, it was clear that fluctuating water levels in the refuge—caused by the Corps' regulation of water—had adversely affected several other species as well, including the rare Florida Sandhill crane, the gallinule, and the alligator. John R. Eadie, manager of the refuge, emphasized that the problem was not that periods of high and low water existed, but that man had "artificially manipulate[d] the water levels in a short period," leaving nature to "react violently to try to adjust the animal population to the reduced carrying capacity of the land."[7] Clearly, environmentalists believed that water management of the water conservation areas was significantly harming Everglades wildlife.

These problems led the FCD and the Florida Game and Fresh Water Fish Commission to ask the Florida Chapter of The Wildlife Society to commission a special study team in March 1970 to investigate the problem and to propose solutions. This team consisted of George W. Cornwall, a professor of wildlife management at the University of Florida; Robert L. Downing, a wildlife research biologist with the FWS; James N. Layne, director of research at the American Museum of Natural History in Lake Placid, Florida; Charles M. Loveless, assistant director of the FWS's Denver Wildlife Research Center; and Arthur Marshall, director of the Laboratory for Estuarine Research at the University of Miami. Because the team's primary goal was to examine wildlife problems in the Everglades, the majority of its final report, issued in August 1970, dealt with wildlife matters, including the problems that fluctuating water levels in the water conservation areas had on deer. Along with specific recommendations about how to manage Conservation Area No. 3 to preserve deer life and about revisiting the regulation schedules for all of the conservation areas, the document related that concern for any individual species had to be "viewed in the context of the total problem." It therefore suggested that the natural hydroperiod of the Everglades be restored, or at least be approximated as closely as possible, and it recommended that an interagency coordinating committee be established to allow for "interaction and information exchange" between those "agencies and groups" responsible for natural resource management in the Everglades.[8]

The report also examined the effects of poor water quality on flora and fauna in the Everglades. As such, the group mirrored larger concerns in the United States about water quality. As urban areas expanded, especially in the eastern United States, Americans became more concerned about how urbanization affected the quality of water. Therefore, in 1965, Congress passed the Water Quality Act to increase the amount of federal funding available for sewage treatment plants and to charge states with developing water quality standards. Shortly thereafter, jurisdiction over the Federal Water Pollution Control Administration shifted from the Department of

Health, Education and Welfare to the Department of the Interior. By the early 1970s, some states, such as Maine, had already enacted significant measures to deal with water quality. Other programs were not as strong, perhaps in part because scientific technology had not yet advanced to the stage where it could accurately test and measure the "toxicity of chemicals to aquatic organisms." Instead, administrators focused more on biological observations to determine where water quality problems existed.[9]

In Florida, the upper Kissimmee River Basin—the headwaters of the entire Florida watershed—exhibited water quality problems in the early 1970s. The Kissimmee chain of lakes, especially Lake Tohopekaliga, faced pollution from the dumping of cattle excrement and fertilizers into the water by dairies, ranches, and farms. These pollutants subsequently flowed down the Kissimmee River into Lake Okeechobee. Backpumping from the Everglades Agricultural Area (EAA) also contributed nutrients and pesticides to the lake. "It is therefore imperative that the quality of the water in the Everglades ecosystem be continually monitored," the special study team's report declared, "and that steps be taken to maintain high water quality standards."[10]

Over the next few months, Marshall developed some of these ideas in his own speeches. At a state water resources conference in January 1971, he told Governor Askew and his cabinet that South Florida's ecology was under "stress" from a variety of factors. "I view environment—human ecology—as the number one problem of Florida," he declared.[11] In June 1971, he produced a paper entitled "Repairing the Florida Everglades Basin," claiming that drainage had wreaked devastation on Everglades ecology, not only because it had reduced the amount of water flowing through the area, but also because it had shortened the basin's hydroperiod. This caused saltwater intrusion, salinity concentration in estuaries, and soil subsidence. Almost more damaging, however, was that drainage allowed farming and settlement in vast areas of South Florida, creating a water shortage by increasing demand while reducing supply. Marshall also expressed concern for the quality of water in Florida, especially the overenrichment of Lake Okeechobee and the Kissimmee lakes.[12]

To deal with these concerns, Marshall proposed a series of measures for the state of Florida to take. These included improving the quality of water in the Kissimmee lakes, restoring the channelized Kissimmee River, slowing the Kissimmee's run-off into Lake Okeechobee, setting Lake Okeechobee's water levels at 15.5 to 17.5 feet (rather than the 17.5-21.5 feet schedule proposed by the Corps), restoring coastal bays such as the St. Lucie Estuary, preventing waste and nutrients from flowing to Lake Okeechobee, and establishing constraints on urban and agricultural settlement in South Florida. "We must change direction," Marshall pleaded. "Our exploitive and technological orientation must be re-directed in favor of more considerate uses of natural systems." Otherwise, South Florida would continue to face "accelerating impoverishment of its natural and human resources."[13]

While admitting that South Florida had serious water problems in need of resolution, some scientists believed that Marshall was unnecessarily foisting "doomsday predictions" on Floridians and that the situation was not as dire as he forecasted.[14] Others criticized him as not being realistic, of wanting to eliminate all human occupation of South Florida. William Storch, chief engineer of the FCD, called Marshall a polemicist and accused him of taking "immoral" actions to scare the public. "You seek to polarize rather than unite," Storch said.[15] But Garald G. Parker, a longtime hydrologist in South Florida, agreed with many of Marshall's conclusions and had an even more extreme solution. "The only way to save [the Everglades]," he asserted, "is to move man off them, keep them flooded, and let nature, in her implacable way, start all over again."[16]

Marshall's statements came during a year of severe drought in Florida, when rainfall amounts were 22 inches below normal, forcing the FCD to pump surface water into Miami's wells and causing fires in the Everglades that burned 500,000 acres. Marshall therefore caught Governor Reubin Askew's attention, and the governor de-

Water Management
BULLETIN

VOLUME 5 NUMBER 2   OCTOBER-NOVEMBER 1971

Five Major Problems
Listed By Governor

Governor Rubin Askew posed five major problems facing central and south Florida in his opening remarks at the recent Governor's Conference on the Everglades held in Miami Beach.

Governor Askew told the more than 100 persons on the panel that "We must build a 'peace' in South Florida, a peace between the people and their place, between the natural environment and the man made settlement, between the creek and the canal, between the works and

needs of men and women, and the life of mankind itself.

It was hoped that a full text of the panel's recommendations would be available for publication in this issue of the Water Management Bulletin. However, at press time the final text was not completed. We hope that the full findings and recommendations be printed in the next Bulletin.

Reed Praises FCD.,
Calls for Redirection

Recent actions by the Central and Southern Florida Flood Control District (FCD) received praise from a Washington source at the recently held Governor's Conference on the Everglades.

Nathaniel Reed, assistant Secretary of the Interior, speaking during the closing hours of the three day Miami Beach meeting told the more than 100 delegates: "In recent years there has been a considerable shift in the management philosophy of the FCD. The new policy to reflood portions of the Kissimmee floodplain is but one example among many of a proper direction in rethinking project goals."

He warned, however, that the FCD Governing Board must exercise caution to insure that any corrective action taken alleviates rather than compounds the situation in the Kissimmee Basin and Lake

Okeechobee. He urged the Board to let the biologist take another crack at the restoration proposal before moving head-long into a well meant effort to make sure that the specifics are correct and will accomplish the desired goals prior to tampering further with an already disturbed situation.

The Assistant Secretary then commented: "Lest these last comments be construed as damning with faint praise, let me say here that Bob Padrick (Chairman of the FCD Board) and his colleagues on the Board deserve the public's respect for having gone further than any previous Board in facing up to the tremendous problems of the project. I respect their courage and wisdom. I do not know of a more sensitive Board on which to sit. The decisions are all tough, there are no easy (Continued on Page 2)

(Continued on Page 2)

Cover of the FCD's *Water Management Bulletin* detailing Governor Askew's conference on Florida water issues.

cided to call a special conference on water management in South Florida in September 1971. He asked some of the top scholars in ecological and water issues to congregate in Miami for discussions of what the state could do to maintain water supply and quality as the region continued to grow. Participants included John M. DeGrove, dean of the College of Social Sciences at Florida Atlantic University who chaired the conference, Marshall, State Senator Daniel Robert ("Bob") Graham, Florida Wildlife Federation president John Jones, environmentalist William Partington, many scientists and engineers from Florida universities, and representatives from Everglades National Park, the Florida Game and Fresh Water Fish Commission, the U.S. Geological Survey, and the U.S. Sugar Corporation. In his opening remarks, Askew told the gathering that he wanted answers—"stated clearly, bluntly and forcefully"—to five questions: how muck fires and saltwater intrusion could be halted; how an impending shortage of high quality water could be prevented; how soil subsidence could be curbed; whether there should be a limit on South Florida's population growth; and who should manage South Florida's natural resources. "I realize that no study and no three-day conference on Miami Beach is going to solve our water management and pollution problems," Askew said, but—adopting a phrase first coined by landscape architect and regional planner Ian McHarg in 1969—he wanted the meeting to mark "the beginning of a new 'design with nature' for South Florida."[17]

After studying the issues, conference participants developed a statement of solutions for the governor; Marshall served as one of the prime authors. "There is a water crisis in South Florida today," the statement proclaimed, recommending that the state immediately institute "an enforceable comprehensive land and water use plan" to limit population in certain areas. To solve the water quality issues, the statement suggested that Kissimmee marshes be restored and that backpumping from the EAA into Lake Okeechobee be eliminated, or at least not continued until backpumped water could be treated. It also recommended that, in order to preserve the animal and plant life immediately around the lakeshore, the lake's level not exceed 17.5 feet, even though the U.S. Army Corps of Engineers had believed that the maintenance of a higher level, coupled with backpumping, could provide more water for South Florida. The statement asked that the state establish an agency or board comprised of nine gubernatorial appointees to manage Florida's land and water use plan, and that the board assume a wide range of responsibilities, including "managing water quality and quantity for the long term benefit of the environment of the region and the State" and "establishing policy and guidelines for such activities as drainage, water use, well drilling, land use, estuary protection, watershed management, flood control and soil conservation."[18]

After reading the statement, Askew established a Task Force on Resource Management to draft legislation implementing the recommendations. This committee had several key members, including

DeGrove, who had written his Ph.D. dissertation on the C&SF Project; Marshall; and Graham, whose background in Miami real estate, coupled with his desire for environmental preservation, allowed him to see issues from both sides. Fred P. Bosselman, an attorney from Chicago who had been instrumental in the preparation of the American Law Institute's Model Land Development Code, served as a consultant. Largely influenced by the institute's code, which outlined how states could designate environmentally unique regions as areas of critical concern, the task force developed several bills for introduction, including an environmental land and water management act, a comprehensive planning act, a water resource act, and an act asking for a $200 million bond issue to purchase environmentally endangered lands.[19]

These bills were not without controversy; many special interests and large-scale developers did not agree with the proposals and lobbied hard for their defeat. But the proposals had the backing of several prominent individuals, including Governor Askew, who sent summaries of the "highest priority" bills to interested parties, telling them to inform their senators and representatives of their "strong support" for the legislation.[20] Other important supporters included members of Conservation 70s, an organization formed in 1969 by Lyman Rogers, an environmental adviser to Governor Claude Kirk, to lobby environmental measures in the Florida legislature. Consisting of many state officials and legislators, the group had a great deal of influence in the early 1970s, and during the 1972 session, according to journalist Luther Carter, a well-known environmental reporter, it "was working the capitol corridors full time."[21] Senator Graham and Representative Richard Pettigrew, who sponsored the environmental land and water management legislation in their various chambers, also expended a great deal of effort, as did Representative Jack Shreve, who helped to usher the measures through Florida's House of Representatives. Due to these exertions, the Florida Environmental Land and Water Management Act passed in 1972, as did the Water Resources Act, the Land Conservation Act, and the Florida Comprehensive Planning Act.[22]

These laws implemented many of the measures desired by Askew and the task force. The Land Conservation Act created a $200 million fund for the purchase of environmentally endangered lands, while the Comprehensive Planning Act formed the Division of State Planning and authorized it to prepare a comprehensive land and water plan for Florida. The Florida Environmental Land and Water Management Act, according to Graham, provided "a strong state role in those land use decisions which transcend the jurisdiction of individual local governments."[23] It allowed the governor and the cabinet, upon recommendations from the Division of State Planning, to designate regions as areas of critical state concern if they met environmental or historical standards. In such cases, local governmental agencies would compose and administer land development regulations, subject to the approval of the governor and cabinet. The

Governor Reubin Askew. (The Florida Memory Project, State Library and Archives of Florida)

state also had the power under the law to declare certain land developments as developments of regional impact when they affected more than one county in terms of health, safety, or welfare. The local government would then have to ensure that any construction conformed to the state land development plan.[24]

The Water Resources Act of 1972, meanwhile, created five regional water management districts to make all water resource decisions—be they flood control, drainage, water supply, or whatever else—in the counties over which they had jurisdiction. As part of this, the FCD was reorganized as the South Florida Water Management District (SFWMD)—although this did not officially occur until 1977—and the Northwest Florida, Suwannee River Basin, St. Johns River Basin, and Southwest Florida water management districts were established. When the FCD became the SFWMD in 1977, several significant changes were made. For one thing, it fell under the supervision of the Florida Department of Environmental Regulation (formerly the Department of Pollution Control), although the governor and the cabinet still had the ability to rescind or modify district policies. For another, it received the responsibilities of maintaining

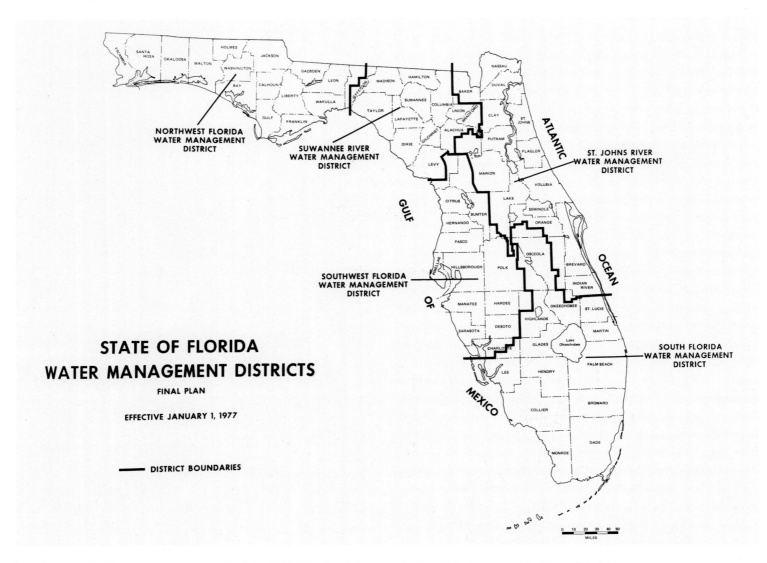

Boundaries of Florida water management districts, 1977. (The Florida Memory Project, State Library and Archives of Florida)

water supply and water quality as well as ensuring flood control. The importance of that change cannot be overemphasized, as it meant that the water management district would now be in a position to manage water in ways that did not harm the environment. Indeed, according to Executive Director John "Jack" Maloy, the district established an Environmental Sciences Division soon after the reorganization "in order to understand the effects of the (drainage) system."[25] All of these pieces of legislation greatly impacted water management in South Florida. According to Luther Carter, environmentalists were "jubilant" over the acts, but their effectiveness remained to be seen.[26]

As state legislators enacted measures to ensure better land and water planning in South Florida, the state also worked in cooperation with the USGS to prepare a report analyzing the effects that water control and management had had on South Florida since the establishment of the C&SF Project. Three USGS hydrologists—S. D. Leach, Howard Klein, and E. R. Hampton—studied the matter in cooperation with the FCD and with the financial backing of the Florida

Department of Natural Resources, the counties of Broward, Dade, and Palm Beach, the cities of Fort Lauderdale, Miami Beach, and West Palm Beach, the NPS, and the U.S. Navy. This study admitted that "the prime effect of the water-control works in South Florida" was the "changing [of] the spatial and temporal distribution of runoff from the Everglades," but it also pointed to the positive results of C&SF Project works, including a reduction in the amount of water discharged to the ocean from the Miami, North New River, Hillsboro, and West Palm Beach canals and the successful prevention of saltwater intrusion into the Biscayne aquifer. "Additional improvements in the hydrologic situation in places in southeast Florida can be achieved by applying existing hydrologic management practices to smaller, specific areas of need," the report concluded.[27]

Yet despite the general positive nature of the USGS's report on the C&SF Project, controversy swirled around the Kissimmee River and Lake Okeechobee. Marshall had noted in several of his speeches that concerns existed about the U.S. Army Corps of Engineers' channelization of the Kissimmee River as part of the C&SF Project. Before the 1960s, the river, which began near the town of Kissimmee, meandered along a 92-mile course through central Florida, eventually reaching Lake Okeechobee. A lyrical description of the river in a turn of the century edition of *The Kissimmee Valley Gazette* showed the appreciation that many observers had of the river's beauty:

> It is an extraordinary river in its narrowness, in the rampant growth of water plants along its low banks, in the unbroken flatness of the landscape, in the variety and quantity of its bird life, in the labyrinth of by-channels and cutoffs and dead rivers that best its sluggish course, and above all in the appalling, incredible, bewildering crookedness of its serpentine body. There are bends where it takes nearly an hour's steaming to reach a spot less than 100 yards ahead of the bow.[28]

But the river flooded often, causing consternation for ranchers who wanted to raise cattle on the floodplain. Hamilton Disston had initially proposed channelizing the Kissimmee in the 1880s, but he had not made much progress by the time of his death. Therefore, when the C&SF Project was authorized, the Corps included flood control for the Kissimmee River Valley in its plans. The 1954 Flood Control Act allowed the Corps to begin its efforts in that basin, including the construction of eight water control structures in the Kissimmee's upper headwater lakes, the straightening of the river itself, and the building of six water control structures within it. Essentially, the Corps removed the meanders and turns of the river and created Canal 38, a 52-mile waterway running to Lake Okeechobee with five different pools, each containing a water control structure and a lock.[29]

Some agencies objected to the channelization almost immediately. The U.S. Fish and Wildlife Service and the Florida Game and Fresh Water Fish Commission both claimed that the Corps' actions would destroy fish and wildlife in the Kissimmee Valley. They proposed that the Corps investigate other alternatives, but the Corps believed that channelization was the only effective means of dealing with the flooding problems.[30] Therefore, straightening proceeded, leading to other protests. The Florida Audubon Society passed a resolution in 1966 opposing the project, fearing "further destruction of the Kissimmee river and its wild tributaries," while several individuals contacted Florida's congressional delegation, requesting that construction be stopped.[31] "We are aware that a straight, wide, deep canal is not as esthetically pleasing as a winding natural stream," Jacksonville District Engineer Colonel R. P. Tabb responded, "but it does have distinct advantages where economics and water conveyance are concerned."[32]

The Kissimmee River before channelization. (South Florida Water Management District)

The Corps completed the channelization of the river, which cost approximately $30 million, in 1971, leaving it as a straight waterway interrupted by five shallow pools along the way. Not long after, the Corps manipulated Taylor Creek, Nubbin Slough, and other tributaries of the Kissimmee located north and northeast of Lake Okeechobee into one basin, totaling 116,000 acres, so that they would all drain into the lake. The Corps lauded these completions, believing that they prevented $12.1 million in flood damages between 1971 and 1978.[33] But environmentalists were outraged, both because of the destruction of fish and wildlife and because they believed that the Corps had created "a sewer that funneled pollutants and nutrients straight into [Lake Okeechobee,] choking it."[34]

Indeed, Lake Okeechobee experienced some problems at the beginning of the 1970s. Not only did C&SF Project canals bring EAA farmers lake water in times of drought, they also conveyed water from the farmlands back to the lake in times of excess rain—a process known as backpumping. Because such water contained fertilizers, pesticides, and other nutrients, environmentalists believed it contributed heavily to the eutrophication of the lake. Eutrophication essentially consisted of the contamination of surface waters by an influx of nutrients, usually nitrogen and phosphorus. It was the process of turning "clear, sandy-bottomed lakes filled with bass" into bodies of water "algae laden and swarming with gizzard shad."[35] Although all lakes experienced gradual natural eutrophication over an extended time span, cultural eutrophication, or the adding of nutrients by human land use, accelerated the process, killing lakes in a relatively short time. Florida's Lake Apopka, located some miles west of Orlando in Central Florida, had become hypereutrophic through human interference, for example, and in the late 1960s and early 1970s, algal blooms indicated that Lake Okeechobee was on a path to the same fate. Many thought that the channelization of the Kissimmee compounded Lake Okeechobee's problems, mainly because straightening the river had eliminated nutrient-filtering marshes in the Kissimmee Valley and had greatly shortened the basin's hydroperiod, meaning the amount of time that water actually stood on the land.[36]

Critics pointed to another problem that, they claimed, channelization of the Kissimmee had exacerbated: regulatory releases from Lake Okeechobee to the St. Lucie and Caloosahatchee estuaries. Under the C&SF Project, the Corps had enlarged the Caloosahatchee River and had constructed additional canals facilitating the flow of water to the St. Lucie estuary. The Corps then used these structures, as well as the existing St. Lucie Canal, to regulate the level of Lake Okeechobee, sending water down the waterways when the lake got too high. Although the Corps enlarged the Caloosahatchee River in 1970 to allow more water to flow down to the Gulf of Mexico, the St. Lucie estuary bore the brunt of the releases. Channelization of the Kissimmee River, some stated, worsened the situation by forcing the Corps to send even more "mud-laden" and polluted water to the

estuaries.[37] Residents of Martin County and FWS officials especially protested such releases, charging that they damaged estuary life, both because of the increased sedimentation that they caused and because the unnatural quantities disrupted the balance of fresh and salt water, driving fish from the estuaries. Because of these conditions, the Corps revisited the lake's regulation schedule in the 1970s and informed interested parties that it was "constantly" pursuing ways to "alleviate the situation."[38]

Although concerns about Lake Okeechobee estuary releases had existed since the 1950s, Arthur Marshall was one of the first to raise the alarm about the effects of Kissimmee channelization on Lake Okeechobee. Because the lake served as the "liquid heart" of South Florida's water system, he explained, any problems with its water quality affected the region as a whole. Marshall called channelization "an abuse of the public's water supply and wildlife resources,"[39] while claiming that "re[c]ent analyses of algal content in Okeechobee waters clearly indicate approach of eutrophication. There is no question as to whether this will occur," he continued, "it is a question of when."[40] To halt the process, the state of Florida could reflood Kissimmee marshes, thereby slowing the rate of runoff and allowing cleansing to occur. The final report of the Governor's Conference on Water Management in South Florida suggested the same thing, recommending as well that backpumped water either be treated before flowing into the lake or not allowed at all. Meanwhile, the results of a USGS study, published in 1971, explained that Lake Okeechobee was in an early state of eutrophication and that many tributaries draining into the water body contained excessive amounts of nutrients.[41]

But some remained skeptical about Lake Okeechobee's condition. William Storch, the FCD's chief engineer, for example, claimed that the USGS study actually showed that the lake was not in danger of dying and that its nutrient concentrations were not excessive. The study did highlight that poor quality water flowed into the lake, Storch explained, but that did not mean that the lake was in imminent danger. In addition, Storch and other FCD officials, as well as the Corps, disputed whether Kissimmee River channelization really had a detrimental effect on the lake.[42]

Despite these doubts, the FCD did agree that the restoration of some Kissimmee marshes was desirable, mainly for fish and wildlife purposes. Accordingly, in the spring of 1972, the FCD's board approved a restoration plan of approximately 9,000 acres, costing $400,000 in land acquisition costs. Because the FCD did not have any eminent domain authority, it needed the approval of the governor and cabinet in order to implement the proposal, and in the summer and fall of 1972, it prepared a presentation for the cabinet.[43]

In November, the FCD held a public hearing about the matter in West Palm Beach, obtaining testimony from those interested in Kissimmee restoration. Although admitting that marsh restoration was important, many environmentalists were disappointed at the small scale of efforts proposed by the FCD. FCD board member

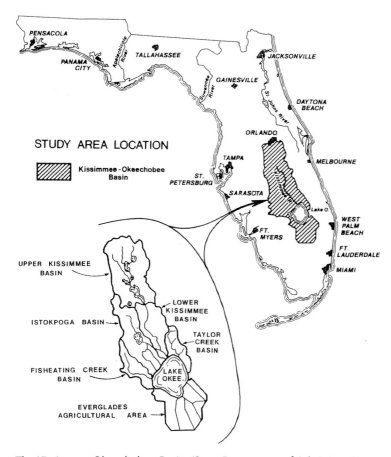

The Kissimmee-Okeechobee Basin. (State Department of Administration, "Findings and Recommendations from the Special Project to Prevent the Eutrophication of Lake Okeechobee" (1976)

Don Morgan responded that it was "the best we can do with a flood control program," but Marshall and other ecologists, including Dr. Robert Harris of the Florida State University Marine Laboratory, averred that more was necessary to prevent Lake Okeechobee's eutrophication. On the other hand, Andrew Lamonds of the USGS contended that Kissimmee River channelization was not the only thing causing problems in Lake Okeechobee; population increases in the 1960s would have resulted in nutrient addition regardless of the channelization. "The rate of flow is not the primary concern," he insisted. But others, including Marshall and O. Earle Frye, director of the Florida Game and Fresh Water Fish Commission, disagreed. "Channelization has worsened conditions for fish and wildlife, and has reduced the buffering effect of marshes," Frye stated. "We would like to see as much of the river put back in a natural state as possible." John Jones, representing the Florida Wildlife Federation, agreed, calling river restoration the first step in restoring water quality to Lake Okeechobee. Most participants realized, however, that financing and land acquisition were obstacles in any reflooding plan. Representatives of landowners in the Kissimmee Valley, who vehemently opposed marsh restoration, emphasized these issues, declaring that alternative measures for improving Lake Okeechobee's water quality should be studied.[44]

At the conclusion of the hearing, the FCD's governing board finalized their findings and recommendations for the governor and the cabinet. Lake Okeechobee water quality, the statement began, was a "serious and perplexing problem" that required more authority and responsibility than any existing agency had. "Total restoration of the Kissimmee River marshes," it continued, "may or may not be an effective solution by itself, in view of other possible grave consequences, especially flood control."[45] Because acquiring all of the lands in the river's floodplain would require $88 million, the board recommended that its limited program be implemented and that polluted water be treated before entering the lake. Then additional studies could be made to discover whether or not complete restoration was necessary or possible.

On 12 December 1972, the FCD presented these findings and suggestions to Governor Askew and the cabinet at a four-hour hearing where representatives from the FCD, the Corps, and environmental groups testified on the condition of Lake Okeechobee and the channelization of the Kissimmee. After the FCD made its presentation, Marshall reported on analyses that the Center for Urban and Regional Studies at the University of Miami had conducted, stating that these showed that the state needed to take immediate action, including restoration of the whole Kissimmee River, in order to prevent further water quality loss in Lake Okeechobee. "The water quality situation in Lake Okeechobee is tending rapidly toward irrevocable misfortune," he argued.[46] To curb the destruction, Marshall wanted the governor to appoint a water quality master for the Kissimmee-Okeechobee Basin to oversee nutrient-removal efforts. Marshall also presented the governor and cabinet with a copy of the center's report, entitled *The Kissimmee-Okeechobee Basin*. Colonel Emmett Lee, District Engineer of the Jacksonville District, however, opposed wholesale restoration, believing that it would return flooding problems to the Kissimmee Basin. After hearing these different viewpoints, the cabinet voted to implement a program to correct existing pollution in the Kissimmee Valley, to monitor water quality in the Kissimmee Basin and Lake Okeechobee, and to establish an interdisciplinary team of scientists to study whether or not restoration was necessary.[47]

The Florida legislature passed measures during its 1973 session to implement the governor's and cabinet's requests, including one creating a "Special Project to Prevent the Eutrophication of Lake Okeechobee" to conduct a study of the lake's water quality problems. The Division of State Planning received the responsibility of overseeing the effort, while the Florida Department of Pollution Control and the FCD were charged with water quality and quantity data collection and analysis. Federal, state, and local agencies, universities, and private consultants also contributed; Dale Walker, a critic of Kissimmee channelization who had worked for the Florida Game and Fresh Water Fish Commission, was appointed project leader. The study's main goal was to comprehend "the Lake Okeechobee ecosystem sufficiently to derive a land and water management plan which, when implemented, will prevent further cultural eutrophication of the lake."[48]

As the study commenced, Corps leaders decided that an objective examination of the Kissimmee River Basin and the effects of channelization on water quality was necessary. Fearing that Floridians did not have sufficient emotional detachment to make an objective analysis, the Corps hired Atlantis Scientific, an environmental auditing firm in Beverly Hills, California, to conduct the study. In April 1973, Atlantis filed its report with the Jacksonville District after concluding two months of fieldwork and consultations in Florida. It found little evidence that the channelization of the Kissimmee had adversely affected Lake Okeechobee's water quality and no definitive results as to how well marshlands would remove nutrients from water. Besides, the report argued, marshlands would serve only as a holding place for nutrients; any nutrients removed from the water would merely sit in the vegetation or soil until a future inundation released them back into water. This was a conclusion that seemed to fly in the face of assertions by environmentalists and scientists such as Marshall that the marshes acted as "scrubbers" to prevent nutrient-loaded water. The report also claimed that the most polluted water reaching Lake Okeechobee came from EAA backpumping. Finally, Atlantis stated that no clear evidence existed that Lake Okeechobee was in an accelerated eutrophic state. "Eutrophication is the natural aging process of bodies of water," the report related, and "every body of water is in some stage of eutrophication." Although "components or constituent matter contributing toward eutrophication" all existed in the lake, "there is no evidence of the synergism necessary to supposedly expedite the process."[49]

Supporters of Kissimmee restoration severely criticized Atlantis's report, claiming that it had no objectivity because it was performed at the Corps' bequest. Others saw the study as "a quiet effort [by the Corps] to discredit environmentalists' proposals for restoring the channelized Kissimmee River."[50] Atlantis, which referred to its work as an "environmental audit," insisted that it had no responsibility "to sustain the judgment of our client nor to refute the testimony of concerned citizens" and that its "first obligation" was "to our own professional integrity to provide an impartial and qualified assessment," but its actions in Florida belied that statement.[51]

David S. Anthony, a biochemist with the University of Florida, for example, claimed that the Atlantis team had employed "deliberately deceptive behavior" in order to deflect attention from its relationship with the Corps. "I, personally, was given an evasive answer which contained no mention of the Corps when I asked one of the team what his mission in Florida was," Anthony related. Two other scientists, he continued, "were given an answer that was a flat untruth when they asked the same question of one of the consultants."[52] It seems unlikely that the Corps pressured Atlantis to mirror the Corps' own conclusions, and, indeed, scientists eventually came to accept some of Atlantis's conclusions, including its contention that the Kissimmee River was not the major polluter of the lake. However, other findings, which directly contradicted conclusions reached by prominent Florida scientists (who had been studying the issues for years), indicate that the California firm may have been unqualified to analyze the pertinent subjects. Since the company based its conclusions on already-existing scientific literature, interviews with "a broad spectrum" of individuals, and an inspection of the area, rather than any scientific studies it conducted itself, this view seems justified.

Cattle grazing around Lake Okeechobee. (South Florida Water Management District)

Meanwhile, the FCD conducted additional studies of Lake Okeechobee and the Kissimmee River, including one examining how manipulating water levels in impoundment pools on the river might affect vegetation. After extended observations, the report concluded that raising water levels two feet above their normal control stage would help to reproduce natural marsh conditions and enhance survival rates of the fish and birds.[53] The FCD also studied how it could reduce nutrient loads in water flowing into Lake Okeechobee, especially from three sources: the Taylor Creek/Nubbin Slough drainage area (the location of numerous dairies and cattle ranches), the north-central part of the EAA, and areas in the lower Kissimmee Basin including and below pools S-65D and S-65E. The FCD recommended that Taylor Creek/Nubbin Slough and Lower Kissimmee Basin farmers use land and water management techniques to prevent large-scale concentrations of nutrients, and that EAA agriculturists store runoff water for reuse in land between the Miami and North New River canals known as the Holey Land tract. This was a 55-square mile area in southwestern Palm Beach County that served as a kind of wildlife buffer zone for Conservation Area No. 3, protecting wildlife in that area from development. "The time has come to begin to move out of the study phase and into the action phase," the FCD concluded, but "there must be assurance that action is not taken just for the sake of action," especially since another FCD study had determined that nitrogen and phosphorus levels in Lake Okeechobee had not significantly increased in the last five years.[54]

As different agencies performed their own analyses, some environmentalists, eager for concrete action, became angry. Lyman Rogers, environmental adviser to Governor Claude Kirk and a founder of Conservation 70s, complained that despite the clear recommendations of the Governor's Conference on Water Resources in South Florida, the state had implemented only "studies" and "studies to study the studies." In the meantime, he argued, "Lake Okeechobee is DYING" and would "continue to eutrophicate, until it becomes a giant sized Lake Apopka." Rogers called Askew an environmental phony, saying that he promised Florida "all kinds of cures, and has given us none." Askew needed to provide specific solutions to Lake Okeechobee's problems, Rogers declared, rather than just commission more studies.[55] Lieutenant Governor James H. Williams responded that Askew was "deeply committed" to finding a cure for Lake Okeechobee, but that "simple solutions do not solve complex problems." He counseled patience, explaining that the report by the Special Project to Prevent the Eutrophication of Lake Okeechobee would contain detailed management plans to reduce Lake Okeechobee's nutrient content.[56]

Before the publication of the Special Project's findings and recommendations, the Florida Sugar Cane League, which represented the major sugar producers in Florida, commissioned its own examination of Lake Okeechobee's problems, specifically focusing on backpumping. Black, Crow & Eidsness, Inc., a Gainesville firm, completed this study, which claimed that backpumping from the EAA supplied only 7.7 percent of the lake's phosphorus and 20.2 percent of its nitrogen. The Taylor Creek/Nubbin Slough watershed, on the other hand, contributed 33.7 percent of phosphorus and 7.7 percent of nitrogen, while the Kissimmee River supplied 30.5 percent of phosphorus and 36.4 percent of nitrogen. Citing studies of eutrophication in the Great Lakes, Black, Crow & Eidsness argued that phosphorus, and not nitrogen, was the limiting nutrient for algal and plant growth. Since the EAA was not a heavy supplier of phosphorus, the company recommended that backpumping continue while more investigations were performed to determine whether phosphorus or nitrogen served as the limiting nutrient in Lake Okeechobee. If phosphorus played the most important role, state officials should examine ways of reducing input from the Taylor Creek/Nubbin Slough. If nitrogen was the critical element, backpumping from the EAA should be reduced or eliminated.[57]

In December 1975, the FCD released its findings as to the engineering and environmental feasibility of storing backpumped water in the Holey Land area, stating that such a project could work. It therefore recommended that a reservoir be constructed on the Holey Land and the adjacent Rotenberger Tract and that it have a regulation schedule of 12 to 15 feet. In order to make the plan work, the state would have to acquire private lands on the Rotenberger Tract, and to do so, the FCD suggested that an exchange for state-owned lands be made. In addition, the Corps would have to enlarge the Miami and North New River canals in order to ensure that runoff went to the reservoir instead of the lake. The project would cost approximately $14 million, but if constructed, would divert 203,910 acre feet of runoff annually away from the lake.[58]

Some disagreed with the FCD's recommendations, in part because they wanted to maintain the Holey Land and Rotenberger tracts as buffer zones for wildlife. Others insisted that the areas were largely void of wildlife, using a report conducted by Ecoimpact, Inc., in 1974, to support their views. Ecoimpact's study, however, was widely panned by environmentalists, in large part because they viewed it as a hatchet job performed at the bequest of sugar interests wanting to use the tracts for cultivation.[59] Still others, such as the Florida Department of Environment Regulation, which had succeeded the Department of Pollution Control, rejected the FCD's suggestions because they wanted a complete cessation of backpumping. The department's *Report of Investigations in the Kissimmee River-Lake Okeechobee Watershed*, which summarized all of the studies the department had performed as part of the Special Project, claimed that EAA backpumping contributed more phosphorus to the lake than the Black, Crow & Eidsness report had indicated. C&SF Project pumping stations S-2 and S-3 alone contributed 10.9 percent of the lake's phosphorus, the department contended, and when one added backpumping from private interests and small drainage districts, the total approached 45 percent. The department also concluded that channelization of the Kissimmee did impact Lake Okeechobee eutrophication because of the elimination of marshes and the development of higher flow rates that caused larger nutrient releases, conclusions that clashed with those presented in the Corps-sponsored Atlantis study.[60]

The Department of Environmental Regulation's report, issued in March 1976, was the precursor to the Division of State Planning's final Special Project report, which was not officially published until November 1976. The Special Project's findings and recommendations, however, were provided to the state legislature in April. The major conclusion of the study was that Lake Okeechobee was "of such eutrophic condition that present nutrient loads must be substantially reduced." Nutrients came from various sources, but EAA backpumping was an especially egregious supplier. To correct this problem, the Special Project recommended that backpumping from the EAA "be eliminated or reduced to the maximum degree feasible," and it suggested that an impoundment reservoir be constructed on the Holey Land Tract in order to store water for reuse. The report did not recommend complete restoration of the Kissimmee River, but it did suggest that marshes be re-established in order to aid upland retention of water.[61]

Despite its moderate findings, or perhaps because of them, the Special Project's report met opposition from all sides. Colonel Donald A. Wisdom, District Engineer of the Jacksonville District, criticized the document for containing many "purely subjective" statements "designed to sell an idea by eliciting an emotional response in the reader." The report did not address what the C&SF Project had done for South Florida in terms of "agricultural and human productivity," Wisdom complained, although it delineated extensively "what has been lost in natural productivity."[62] Agricultural interests in the EAA, including sugar growers, did not like the backpumping recommendation, nor did they agree with the proposal to build a reservoir on the Holeyland, mainly because sugar producers wanted to expand into that area. Hunters did not like the Holeyland suggestion either, as it would eliminate an excellent deer hunting spot. Environmentalists, meanwhile, wanted the full restoration of the Kissimmee River, not just a reflooding of some of its marshes. All of these interests conveyed their displeasure to Florida senators and representatives. Especially vocal was Johnny Jones of the Florida Wildlife Federation, who was convinced that dechannelization of the Kissimmee River was the only way to save Lake Okeechobee. Jones wrote a bill mandating the restoration of the river and, with the approval of Marshall, sent it to the Florida legislature in its 1976 session.[63]

This bill, which was sponsored by Representative A. H. "Gus" Craig in the House and Senator Jon Thomas in the Senate, recognized the findings of the Special Project, but went further in its recommendations. It mandated the restoration of the Kissimmee River to its natural channel, and allowed for the reflooding of natural marshes in the Taylor Creek-Nubbin Slough basin. It also proposed that cattle and dairy farmers in the Taylor Creek-Nubbin Slough Basin build polishing ponds to remove nutrients before releasing water into Lake Okeechobee. These ponds would hold water for a short period of time, generally one to three days, in order to extract nutrients through biological processes. Moreover, the bill established a four-person advisory council to oversee restoration efforts—the Coordinating Council of the Restoration of the Kissimmee River Valley and the Taylor Creek-Nubbin Slough Basin—composed of the executive directors of the Florida Department of Natural Resources, the Florida Game and Fresh Water Fish Commission, the FCD, and the secretary of the Department of Environmental Regulation. Jones lobbied for the bill incessantly, proclaiming to the media that "the

View of the Kissimmee River. (South Florida Water Management District)

Florida Legislature can pass a Kissimmee River bill or we in South Florida can all move out."[64] His tactics worked; the Senate and the House approved the measure unanimously, and Askew signed it in June 1976.[65]

The passage of the bill meant that the state now fully supported the restoration of the Kissimmee River. However, there were still a few unresolved issues. For one, the legislation did not clearly define what restoration meant. Thomas believed that it denoted returning the Kissimmee to its natural channel and "recreating the natural marshes and flood plain" in order to "enhance the water storage capabilities" of the Kissimmee Valley, improve Lake Okeechobee's water quality, and increase wetland vegetation and wildlife.[66] Others were not so certain. Colonel Wisdom, for example, who would have charge of the restoration since Kissimmee channelization was an authorized component of the C&SF Project, was not convinced that dechannelization was either necessary or mandated. "There is a large communication gap between the environmental investigators and the hydrologists and water resources managers," he explained.[67]

Nowhere was that gap more clearly seen than in two accounts of the debate over Kissimmee River channelization and what the restoration bill actually meant. An article in *ENFO*, a periodical published by the Environmental Information Center of the Florida Conservation Foundation, depicted the initial channelization of the Kissimmee as the product of Corps leaders intent on steamrolling any opposition to straightening the river. "The project was promoted in the name of flood control," the article argued, "and its opponents never had a chance." The essay disputed that flood control really resulted from the channelization, claiming that it enabled settlement in the floodplain, an area obviously more prone to flooding, the Corps project notwithstanding. In addition, the article claimed that channelization had changed the Kissimmee Valley from an area with thriving fish populations, "hundreds of thousands of wading birds and waterfowl," and a "healthy ecosystem" to a place of "stagnant water," "noxious aquatic weeds," "foul-smelling gas," and "a biological desert." Because the channelization of the Kissimmee sent pollutants from the Upper Kissimmee Valley to Lake Okeechobee, it threatened to give the "liquid heart" a massive "heart attack." Therefore, the essay concluded, somewhat misleadingly, the Florida legislature had mandated complete restoration of the Kissimmee in the 1976 legislation; any alternative was out of compliance with the law.[68]

Patrick McCaffrey, staff director of the Coordinating Council of the Restoration of the Kissimmee River Valley and the Taylor Creek-Nubbin Slough Basin, had an entirely different perspective. He claimed that initial opponents of Kissimmee channelization, although unable to halt the process, forced the Corps to make major design modifications to accommodate fish and wildlife. The "fruits of their labors may not have been as sweet as expected," McCaffrey explained, "but in the context of the times they were major concessions by the Corps of Engineers." McCaffrey also expressed doubts that

The ENFO publication heavily criticizing the channelization of the Kissimmee River and calling for restoration.

restoring the river would improve water quality in Lake Okeechobee, a view, he asserted, that the Special Project report supported. Because data showed that routing flows through natural or man-made marshes and nonpoint source control had a greater impact on Lake Okeechobee water quality than Kissimmee River restoration, the Special Project had dismissed reinstating the river as a viable option. Although pro-restoration forces claimed that the 1976 law mandated complete restoration of the Kissimmee, McCaffrey and FCD leaders believed that it merely required the Coordinating Council to "develop measures . . . to restore water quality," and those measures could consist of marsh reflooding, partial restoration, or other solutions.[69] This, then, was the point of contention: environmentalists (as well as the bill's sponsors) believed that the law mandated dechannelization, but others, including FCD and Corps officials, interpreted it as requiring the restoration of water quality to the Kissimmee Basin in whatever ways the Coordinating Council deemed necessary, an opinion supported by Florida's attorney general.[70]

Despite the disagreements, the Coordinating Council began operations in the summer of 1976, believing that it had the responsibility for investigating different options for restoring good quality of water to the Kissimmee Valley. It quickly established an ad hoc advisory committee and an interagency technical committee to provide advice and assistance. The advisory committee contained representatives from environmental organizations such as the Florida Audubon Society and the Sierra Club, as well as members of agricultural groups such as the Florida Sugar Cane League and the Florida Cattleman's Association. The interagency committee had representatives from the FCD, the Corps, several state agencies, and the U.S. Department of the Interior. With the help of these groups, the Coordinating Council developed 11 actions that the state could take to improve water quality in the Kissimmee Basin. These included dechannelizing the river through plugging the pools in the canal, recreating marshland through pool manipulation and tributary marsh impoundments, and backfilling Canal 38 to restore the river to its natural course. After holding public hearings on the alternatives in February 1977, the Coordinating Council made its recommendations to the state legislature in March.[71]

The Coordinating Council explained that the best way to restore water quality to the Kissimmee River Valley was by treating agri-cultural pollution at its source in the Taylor Creek-Nubbin Slough basin. This included creating an upland detention/retention project and implementing on-farm monitoring programs. As far as restoring the river was concerned, the Coordinating Council decided to let the legislature decide. If the legislature wanted complete restoration, the council suggested that a partial backfilling method be used, whereby 60 percent of the canal would be refilled, restoring two-thirds of the marshland. The state would need to obtain congressional approval in this case since the Corps did not have authorization to undo a project unless Congress specifically mandated it. If the legislature did not intend for the Kissimmee to be dechannelized, the Coordinating Council recommended that pool stages be implemented in order to create impounded wetlands. The choice, however, solely rested with Florida's legislators.[72]

As the 1977 legislative session began, environmentalists moved into action, believing that the impounded wetlands idea was merely, in the words of McCaffrey, "an attempt to prevent dechannelization."[73] Jones again lobbied hard for the legislature to mandate complete restoration, and initially it looked as though he would succeed, as both houses passed resolutions requiring dechannelization. But

View of the Taylor Creek-Nubbin Slough area. (South Florida Water Management District)

when the actual legislation came forward, agricultural interests influenced legislators to kill the bill unless other measures were implemented. With almost no chance of passing an act requiring restoration, proponents had to compromise, and the measure that emerged merely requested that the state ask Congress to authorize a Corps restudy of the river. Several state agencies issued resolutions supporting this action, and Congress authorized the restudy in April 1978, appropriating money for the examination in September.[74] The Corps clearly saw the examination as a way of investigating a variety of options for the river; dechannelization would only be an "alternative" under study, not the main purpose of the analysis.[75]

Likewise, little firm action was forthcoming on other issues pertaining to Lake Okeechobee water quality. Despite the Special Project's recommendation that backpumping from the EAA cease, it continued. Because the pollutants resulting from backpumping exceeded state water quality standards, the Florida Department of Environmental Regulation was required to issue a permit to the South Florida Water Management District (as discussed above, the name and organization had changed in 1977) before backpumping could occur, but the state did not enforce that requirement until faced with litigation. In 1977, the state asked the SFWMD to apply for a permit, and after the district did so, the Department of Environmental Regulation issued a temporary operating permit with the understanding that the SFWMD would develop an Interim Action Plan to reduce nutrients flowing into the lake from the EAA. The plan did not diminish how much water was backpumped from the area, however; it merely redirected some of the backpumped water to the water conservation areas instead of the lake. Environmentalists were livid with the Department of Environmental Regulation for issuing the permit, believing that the state should require stricter measures to curtail backpumping, but agricultural interests, especially the sugar industry, protested that halting backpumping would have detrimental effects on farming activities.[76]

In many ways, the conflicts over Lake Okeechobee and the Kissimmee River in the 1970s represented a failure for the environmental community. Although it had successfully halted jetport construction and forced a halt to the Cross-Florida Barge Canal, the problems with water quality in Lake Okeechobee and the channelization of the Kissimmee River remained. Yet environmentalists had called attention to serious water quality issues in South Florida, and had forced state officials to take significant measures to ensure a clean and adequate water supply for the region in the future. The 1972 legislative session saw the passage of several land and water planning laws, while a plethora of scientific studies on Lake Okeechobee and the Kissimmee River were produced. At the least, environmentalists had set the necessary background for more stringent measures to occur at a later time.

But why was the environmental community not able to stop backpumping to Lake Okeechobee or to force the Corps to restore the Kissimmee River, especially in light of the jetport and barge canal successes? First, the problems surrounding Lake Okeechobee and the Kissimmee River did not receive significant national attention, and there were few in the federal government interested in these endeavors or willing to push legislation to resolve the issues. In addition, it was a different matter to get the Corps to halt a project under construction than it was to destroy a project already completed. Had the cry about the Kissimmee River been stronger during its actual construction (rather than just muted complaints from a few individuals in the U.S. Fish and Wildlife Service and the Florida Game and Fresh Water Fish Commission), it might have been easier to stop channelization. For example, environmentalists in southern California in the late 1960s and early 1970s had successfully prevented the Corps from channelizing the Sierra Madre Wash through numerous protests and through the active efforts of city council personnel opposed to the project.[77] Florida did not see the same scale of efforts when channelization of the Kissimmee River was proposed; instead, as former SFWMD executive director John Maloy related, the hot environmental issues in Florida in the 1960s were the barge canal and the condition of Lake Apopka—"the Kissimmee kind of slipped underneath the threshold and didn't gain a lot of attention."[78] Finally, state officials, including Governor Askew, waffled as to their commitment to full restoration of the Kissimmee or to the cessation of backpumping. Although Askew had strongly acted in his first term to preserve Florida's environment, several forces prevented the implementation of stringent measures regarding Lake Okeechobee and the Kissimmee River. For one, the sugar industry, which was increasing in political strength, vehemently opposed the stoppage of backpumping, as did agricultural interests in the Kissimmee Valley. For another, despite all of the studies that had been completed, an air of uncertainty still existed at the end of the 1970s as to whether complete restoration of the Kissimmee was really the best step to take, or whether water quality could be improved through other means. Also important was the issue of funding. Dechannelizing the Kissimmee would require a large amount of money, at least some of which would probably have to come from Florida. It would take several more years of studies, including water quality examinations conducted by the Lake Okeechobee Technical Advisory Committee and some strong gubernatorial support, before Kissimmee restoration and stringent measures to protect Lake Okeechobee's water quality would become a reality.

Endnotes

[1] Robert G. Healy, *Land Use and the States* (Baltimore, Md.: The Johns Hopkins University Press, 1976), 109.

[2] Senator D. Robert Graham, "A Quiet Revolution: Florida's Future on Trial," *The Florida Naturalist* 45 (October 1972): 146-147.

[3] See "How the Florida Boom is Changing," *U.S. News & World Report* 66 (5 May 1969): 88; "Florida's Land Boom Gets Down to Earth," *Business Week* (6 April 1968): 136; U.S. Army Corps of Engineers, Office of the Chief of Engineers, *Engineering Memoirs: Major General William E. Potter*, EP 870-1-12 (Washington, D.C.: U.S. Army Corps of Engineers, 1983), 189-194.

[4] All quotations (including Van Dyne's) in McIntosh, *The Background of Ecology*, 200, 203, 229.

[5] "Arthur Marshall, Director of the Division of Applied Ecology, The Center for Urban Studies, University of Miami," Folder 8, Box 4, Arthur R. Marshall Papers, Manuscript Series 73, Special and Area Studies Collections, George A. Smathers Library (East), University of Florida, Gainesville, Florida [hereafter referred to as Marshall Papers]. For more information on Marshall, see Grunwald, *The Swamp*, 246-248.

[6] Quotation in Florida Wildlife Federation press release, 8 April 1970, File Everglades High Water & Deer Situation: U.S. Senate Subcommittee Hearing, Box 1, S1719, Game & Fresh Water Fish Commission Everglades Conservation Files, 1958-1982, FSA. For more information about the deer problems in 1970, see "Information Sheet on Spring 1970 High Water Situation—Southern Florida," File 1110-2-1150a (C&SF) Conservation Area Jan 1970-Dec 1970, Box 15, Accession No. 077-96-0038, RG 77, FRC.

[7] Quotations in "Adequate Water Essential for Wildlife Survival," Bureau of Sport Fisheries and Wildlife, South Florida National Wildlife Refuges press release, File Conservation Areas 1, 2, 3, 1970-86, Box 02193, SFWMDAR; see also Joseph D. Carroll, Jr., Acting Field Supervisor, to Regional Director, 7 May 1970, no file name, FWSVBAR.

[8] "Report of the Special Study Team on the Florida Everglades, August 1970," 3, 5-8, 12, 17-27, copy in South Florida Water Management District Reference Center, West Palm Beach, Florida.

[9] Henry B. Sirgo, "Water Policy Decision-Making and Implementation in the Johnson Administration," *Journal of Political Science* 12, nos. 1-2 (1985): 53-58.

[10] "Report of the Special Study Team on the Florida Everglades, August 1970," 13-15. The report did not dwell on the EAA backpumping issue, only listing it as one of the factors contributing to poor water quality in the lake.

[11] Arthur R. Marshall, "Water Problems of South Florida," 26 January 1971, File Water Resources Development Comm. General/Meetings 1957-73, Box 02500, SFWMDAR.

[12] Arthur R. Marshall, "Repairing the Florida Everglades Basin," 11 June 1971, copy in South Florida Water Management District Reference Center, West Palm Beach, Florida.

[13] Marshall, "Repairing the Florida Everglades Basin."

[14] See G. W. Cornwell, "The Everglades—Nature's Yo-Yo," Folder 16, Box 4, Marshall Papers

[15] W. V. Storch to Professor Arthur R. Marshall, 6 July 1971, attachment to Marshall, "Repairing the Florida Everglades Basin."

[16] Garald G. Parker, Chief Hydrologist, to Dale Twachtmann, Executive Director, 3 August 1971, Folder 5, Box 7, Marshall Papers.

[17] Quotations in "Remarks of Reubin O'D. Askew, Governor of Florida, at the Governor's Conference on Water Management in South Florida, Miami Beach, Florida, September 22, 1971," Folder 4, Box 6, Marshall Papers; see also "Five Major Problems Listed By Governor," *Water Management Bulletin* 5 (October-November 1971): 2; Graham, "A Quiet Revolution," 146-147; Grunwald, *The Swamp*, 260; "Lake Okeechobee Health Center of Rising Dispute," *The Miami Herald*, 20 September 1971. McHarg's book *Design With Nature* (Garden City, N.Y.: Natural History Press, 1969) emphasized the importance of environmental planning in the protection of ecosystems.

[18] Quotations in "A Statement to Reubin O'D. Askew, Governor, State of Florida, from the Governor's Conference on Water Management in South Florida," September 1971, 1-6, File Wetlands Regulation Federal, State, Local Correspondence/General 1973-75, Box 02500, SFWMDAR; see also "Population, Drainage Limits Urged," *The Miami Herald*, 25 September 1971; Grunwald, *The Swamp*, 260.

[19] Reubin Askew, Governor, to Friend, 7 March 1972 (with attachment), File L54 Water Resources Jan 72-June 72, EVER 22965, CR-ENPA; Askew to Conference Participant, 30 December 1971, ibid.; Carter, *The Florida Experience*, 126-131; Healy, *Land Use and the States*, 110-111.

[20] Askew to Friend, 7 March 1972.

[21] Carter, *The Florida Experience*, 136.

[22] Carter, *The Florida Experience*, 132-136; Grunwald, *The Swamp*, 261; Jones interview, 20.

[23] Graham, "A Quiet Revolution," 148; see also The Florida Environmental Land and Water Management Act, copy in EVER 22965, CR-ENPA.

[24] "The Florida Environmental Land and Water Management Act of 1972," summary in File Legislation Environmental Land & Water Management Act of 1972—General, Box 25, Graham Papers; "Water Management in Florida," 2 November 1973, ibid.; Carter, *The Florida Experience*, 133; Blake, *Land Into Water*, 230.

[25] As quoted in Stephen Glass, "Rebirth of a River," *Restoration & Management Notes* 5 (Summer 1987): 7; see also Jack Shreve to Senator, 3 April 1972 (with attachment), File Legislative Water Resources Act 1973-1975, Box 38, Graham Papers; "Water Management in Florida," 2 November 1973, ibid.; "Water Management Act of 1972," File Legislative Water Management Districts 1974-1975, Box 38, Graham Papers; Forrest T. Izuno and A. B. Bottcher, "The History of Water Management in South Florida," in *Everglades Agricultural Area (EAA): Water, Soil, Crop, and Environmental Management*, A. B. Bottcher and F. T. Izuno, eds. (Gainesville: University Press of Florida, 1994), 21-22; Maloy interview; Carter, *The Florida Experience*, 133; Blake, *Land Into Water*, 230. Because the FCD did not become the SFWMD until 1977, we will continue to refer to it as the FCD until our discussion reaches that year.

[26] Carter, *The Florida Experience*, 137.

[27] Leach, Klein, and Hampton, *Hydrologic Effects of Water Control and Management in Southeastern Florida*, 1-3.

[28] As quoted in M. Timothy O'Keefe, "Cows, Crackers and Cades: Bloody, Muddy and 'Unbent,' The Kissimmee River Has Seen Some Weird Goings-On," *Florida Sportsman* (August 1978): 20.

[29] "Presentation by Colonel James W. R. Adams, District Engineer, Jacksonville District, to the Florida Wildlife Federation, Orlando, Florida, 23 September 1978: 'Two Rivers—Kissimmee-Oklawaha—Where We Are Today,'" 1-3, Kissimmee Binder, Box 4383, JDAR.

[30] Unsigned letter to District Engineer, 17 December 1958, copy in South Florida Water Management District Reference Center, West Palm Beach, Florida; Blake, *Land Into Water*, 260.

[31] C. Russell Mason to General W. S. Cassidy, Chief Army Engineer, 4 February 1966, File 1110-2-1150a (C&SF) Kissimmee River Valley Study Aug 1965-Feb. 1967, Box 10, Accession No. 077-01-0023, RG 77, FRC.

[32] Colonel R. P. Tabb, District Engineer, to Honorable George A. Smathers, 17 October 1967, File 1110-2-1150a (C&SF) Kissimmee River Valley Study March 1967-December 1968, Box 10, Accession No. 077-01-0023, RG 77, FRC.

[33] "Kissimmee River Survey Review, Town Hall Meetings, Information Packet," Binder Kissimmee, Box 4383, U.S. Army Corps of Engineers, JDAR; "Presentation by Colonel James W. R. Adams, District Engineer, Jacksonville District, to the Florida Wildlife Federation, Orlando, Florida, 23 September 1978," 3; James L. Garland, Chief Engineering Division, to Mr. Alexander

L. Crosby, 25 May 1973, File 1110-2-1150a (C&SF) Kissimmee River Valley Study Jan 1973-Dec 1973, Box 9, Accession No. 077-01-0023, RG 77, FRC.

[34] M. Timothy O'Keefe, "The Kissimmee Problems: Trying to Unmuddle Man's Meddling," *Florida Sportsman* (August 1978): 29.

[35] Carter, *The Florida Experience*, 26.

[36] Carter, *The Florida Experience*, 26, 123.

[37] Arthur R. Marshall, "Water Problems of South Florida," 26 January 1971, copy in File Water Resources Development Commission, General/Meetings, 1957-73, Box 02500, SFWMDAR.

[38] Quotation in Colonel Emmett G. Lee, Jr., District Engineer, to Honorable Lawton Chiles, United States Senate, 27 August 1974, File 1110-2-1150a (C&SF Martin County) Project—General—1968 Authn 1968-1979, Box 1, Accession No. 077-01-0023, RG 77, FRC; see also Donald B. Benedict, Martin County Conservation Alliance, to Mr. Carol White, Corps of Engineers, 3 June 1976, ibid.; Joseph D. Carroll, Jr., Field Supervisor, Fish and Wildlife Service, to District Engineer, 4 June 1976, File 1520-03 (C&SF) Water Control Jan. 1976-Dec. 1976, Box 3579, JDAR.

[39] Quotation in Marshall, "Water Problems of South Florida," 1; see also "Report of the Special Study Team on the Florida Everglades, August 1970," 14.

[40] Quotation in Arthur R. Marshall, "Are The Everglades Nearing Extinction?" *The Florida Naturalist* 44 (July 1971): 80-81 (this is a reprint of Marshall's statement); see also Marshall, "Repairing the Florida Everglades Basin," 4.

[41] "A Statement to Reubin O'D. Askew, Governor, State of Florida, from the Governor's Conference on Water Management in South Florida," 1; see also Lothian A. Ager, Fishery Biologist, to John W. Woods, Fisheries Division, 16 September 1971, File LO Historical Data/Text, Box 18060, SFWMDAR.

[42] "Lake Okeechobee Not Dying, Official Study Shows," *The Palm Beach Times*, 24 August 1971.

[43] J. Walter Dineen to Jack Maloy, 3 August 1971, Folder 5, Box 3, Marshall Papers; "FCD To Seek Restoration of Kissimmee Marshes," *The Palm Beach Times*, 1 December 1972; "Urges Support Be Given To Restoring Kissimmee," *Stuart News*, 19 November 1972. In August 1972, Marshall received an appointment to the FCD's board, but he resigned in 1973 because he believed that other board members did not take environmental matters seriously. See Governor to Mr. Art Marshall, 17 August 1972, Folder 9, Box 4, Marshall Papers; Askew to Marshall, 17 July 1973, ibid.

[44] All quotations in "Summary of Kissimmee River Hearing—At West Palm Beach—November 15, 1972," 3, 5-10, File Kissimmee River Restoration History, Box 17166, SFWMDAR.

[45] "Findings and Recommendations of the Governing Board, Central and Southern Florida Flood Control District, as the Result of Public Hearing Concerning Alleged Environmental Damage Resulting from Channelization of the Kissimmee River, November 15, 1972," copy in South Florida Water Management District Reference Center, West Palm Beach, Florida.

[46] *The Kissimmee-Okeechobee Basin: A Report to the Cabinet of Florida* (Miami, Fla.: Division of Applied Ecology, Center for Urban and Regional Studies, University of Miami, 1972), 2, 48-50.

[47] "December 12, 1972, Cabinet Meeting, 9:00 A.M.," 36-37, 53, 57, 67, File Lakes & Streams Kissimmee River Basin (FCD) 1969, Box 1, S1570, Game & Fresh Water Fish Commission, Lake & Stream Restoration Project Files, 1950-1986, FSA.

[48] Quotation in *Interim Report on the Special Project to Prevent the Eutrophication of Lake Okeechobee* (Tallahassee: Florida Department of Administration, Division of State Planning, 1975), 1; see also Blake, *Land Into Water*, 263.

[49] Atlantis Scientific, "An Assessment of Water Resource Management in the Central and Southern Florida Flood Control District: A Review and Evaluation of Environmental Reports on the Kissimmee River and Lake Okeechobee," 2 April 1973, 49-50, 52, 59-60, 69, 77, copy in Library, Jacksonville District, U.S. Army Corps of Engineers, Jacksonville, Florida.

[50] "Corps Is Quietly Attacking River Restoration Proposal," *The Miami Herald*, 27 May 1973.

[51] Atlantis Scientific, "An Assessment of Water Resource Management," 2.

[52] David S. Anthony, "Water Problems in the Kissimmee_Okeechobee Basin in Florida," 6, File Kissimmee River Restoration History, Box 17166, SFWMDAR.

[53] Robert L. Goodrick and James F. Milleson, *Studies of Floodplain Vegetation and Water Level Fluctuation in the Kissimmee River Valley*, Technical Publication No. 74-2 (West Palm Beach, Fla.: Central and Southern Florida Flood Control District, 1974); "Lake Okeechobee—Kissimmee Basin, Proposals for Management Actions, Central and Southern Florida Flood Control District, March 20, 1975," copy in South Florida Water Management District Reference Center, West Palm Beach, Florida.

[54] "Lake Okeechobee—Kissimmee Basin, Proposals for Management Actions, Central and Southern Florida Flood Control District, March 20, 1975"; Frederick E. Davis and Michael L. Marshall, *Chemical and Biological Investigations of Lake Okeechobee, January 1973-June 1974, Interim Report*, Technical Publication No. 75-1 (West Palm Beach, Fla.: Central and Southern Florida Flood Control District, 1975). The Division of State Planning published an interim report on the SPPELO in 1975 that noted the FCD's conclusions, but explained that other studies of vegetation and sediments showed that phosphorus and nitrogen levels had tremendously increased. It expected to have "hard data" about the lake's eutrophic condition by the end of the study. *Interim Report on the Special Project to Prevent the Eutrophication of Lake Okeechobee* (Tallahassee: Florida Department of Administration, Division of State Planning, 1975), 32.

[55] Lyman E. Rogers to Lt. Governor J. H. 'Jim' Williams, n.d., Folder 45, Box 1, Marshall Papers (emphasis in the original); see also Lyman to Art, 28 August 1975, ibid.

[56] Williams to Rogers, 11 September 1975, Folder 45, Box 1, Marshall Papers.

[57] Black, Crow & Eidsness, Inc., "Report to Florida Cane Sugar League on Eutrophication of Lake Okeechobee," December 1975, xii-xiii, xv, 1-2, 2-7, copy in South Florida Water Management District Reference Center, West Palm Beach, Florida.

[58] Central and Southern Florida Flood Control District, "Report on Investigation of Back-Pumping Reversal and Alternative Water Retention Sites, Miami Canal and North New River Canal Basins, Everglades Agricultural Area," December 1975, i-iv, copy in Folder 12, Box 3, Marshall Papers. The Rotenberger Tract was named for Ray Rotenberger, a man who built a camp and airfield in the area in the late 1950s and early 1960s. The state purchased 6,300 acres of land in the area in 1975. Florida Fish and Wildlife Conservation Commission, "Rotenberger Wildlife Management Area" <http://www.floridaconservation.org/recreation/rotenberger/history.asp> (March 22, 2005).

[59] See J. Walter Dineen, Director, Environmental Sciences Division, to Director, Resource Planning Department, 16 September 1974, Folder 35, Box 2, Marshall Papers; Dineen to Director, Resource Planning Department, 19 November 1974, ibid.; John C. Jones, Executive Director, Florida Wildlife Federation, to Honorable Reubin Askew, Governor of Florida, 4 December 1974, ibid.

[60] State of Florida Department of Environmental Regulation, *Report of Investigations in the Kissimmee River-Lake Okeechobee Watershed* (Tallahassee: State of Florida Department of Environmental Regulation, 1976), 2, 4, 13.

[61] Department of Administration, Division of State Planning, Bureau of Comprehensive Planning, "Findings and Recommendations from the Special Project to Prevent the Eutrophication of Lake Okeechobee," May 1976, copy in South Florida Water Management District Reference Center, West Palm Beach, Florida. For the full report, see Florida Department of Administration, Division of State Planning, *Final Report on the Special Project to Prevent Eutrophication of Lake Okeechobee* (Tallahassee: Florida Department of Administration, Division of State Planning, 1976).

[62] Colonel Donald A. Wisdom, District Engineer, to Honorable John H. Williams, Lieutenant Governor of Florida, 13 April 1976, File C&SF Special Project to Prevent Eutrofication [*sic*] of Lake Okeechobee 1975, Box 3693, JDAR.

[63] Jones interview, 13-14; "Conservationist Urges River Bill Passage," *The Palm Beach Times*, 24 May 1976; Patrick M. McCaffrey, "Politics of the Kissimmee River Survey Review," 51-56, Folder 7, Box 3, Marshall Papers.

[64] "Conservationist Urges River Bill Passage," *The Palm Beach Times*, 24 May 1976.

[65] "New Law May Help Stricken Kissimmee River," unidentified newspaper clipping, File Kissimmee River News Clippings, Magazine Articles, Box 17166, SFWMDAR; A. H. "Gus" Craig to Mr. Joseph W. Landers, Jr., 21 June 1976, Folder 6, Box 3, Marshall Papers. Most accounts say that the bill passed unanimously, but "New Law May Help Stricken Kissimmee River" reported that two members of the House voted against it.

[66] Jon C. Thomas to Mr. Jay Landers, Secretary, Department of Environmental Regulation, 22 June 1976, Folder 6, Box 3, Marshall Papers.

[67] Colonel Donald A. Wisdom, District Engineer, to Honorable J. H. "Jim" Williams, 4 February 1976, File C&SF Special Project to Prevent Eutrofication [*sic*] of Lake Okeechobee 1975, Box 3693, JDAR.

[68] "Restoring the Kissimmee River May Be Florida's Environmental Armageddon," *ENFO*, n.d., copy in Folder 7, Box 3, Marshall Papers.

[69] McCaffrey, "Politics of the Kissimmee River Survey Review," 51-56.

[70] Robert L. Shevin, Attorney General, to Honorable A. H. "Gus" Craig, Representative, 28th District, 15 April 1977, File 10-1-7a (Kissimmee River Restoration-State Coop Study) May 1977-August 1977, Box 3693, JDAR.

[71] Coordinating Council on the Restoration of the Kissimmee River Valley and Taylor Creek-Nubbin Slough Basin [Coordinating Council], *Legislative Summary: First Annual Report to the Florida Legislature* (Tallahassee, Fla.: Coordinating Council on the Restoration of the Kissimmee River Valley and Taylor Creek-Nubbin Slough Basin, 1977), 1-1—1-4.

[72] Coordinating Council, *Legislative Summary*, 1-5—1-8; McCaffrey, "Politics of the Kissimmee River Survey Review," 57.

[73] McCaffrey, "Politics of the Kissimmee River Survey Review," 58.

[74] Colonel Donald A. Wisdom, District Engineer, to Honorable J. Herbert Burke, 24 August 1977, File 10-1-7a (Kissimmee River Restoration-State Coop Study) May 1977-August 1977, Box 3693, JDAR; "Kissimmee River Battle Subsides," *The Palm Beach Times*, 21 September 1977.

[75] Colonel Donald A. Wisdom, District Engineer, to Arthur J. Fox, Jr., Editor, Engineering News Record, 7 July 1977, File 10-1-7a (Kissimmee River Restoration-State Coop Study) May 1977-August 1977, Box 3693, JDAR.

[76] Thomas G. Tomasello, Counsel, to Colonel James W. R. Adams, District Engineer, 28 June 1979, File C&SF Special Project to Prevent Eutrofication [*sic*] of Lake Okeechobee 1975, Box 3693, JDAR; 12; South Florida Water Management District, "Executive Summary: Water Quality Management Strategy for Lake Okeechobee," 11 December 1981, Folder 16, Box 3, Marshall Papers; Forrest T. Izuno and A. B. Bottcher, "Introduction," in *Everglades Agricultural Area (EAA): Water, Soil, Crop, and Environmental Management*, A. B. Bottcher and F. T. Izuno, eds. (Gainesville: University Press, of Florida, 1994), 8.

[77] See Jared Orsi, *Hazardous Metropolis: Flooding and Urban Ecology in Los Angeles* (Berkeley: University of California Press, 2004), 134-137.

[78] Maloy interview.

CHAPTER 7

"Save Our Everglades": Reagan's New Federalism and Governor Bob Graham in the 1980s

IN THE EARLY 1980S, two political leaders brought strikingly different agendas to bear on Florida's water management problems. Daniel Robert "Bob" Graham, a Florida Democrat with a strong record on the environment in the state senate, was elected governor of the state in 1978. Ronald Reagan, the former governor of California who blamed many of the nation's economic woes on "environmental extremists," was elected president of the United States two years later. The two politicians moved forward with their respective agendas at different speeds but with telling synchrony. Upon taking office in 1981, President Reagan began an immediate overhaul of environmental regulations that had developed over the past decade, and he reduced federal funding across a broad range of environmental agencies and programs. Reagan's environmental policy provoked strong reaction in Congress and among the general public, forcing the president (along with some other factors) to dismiss the two cabinet members who were the most identified with his environmental agenda, EPA Administrator Ann Gorsuch and Secretary of the Interior James G. Watt. Governor Graham treaded cautiously in the environmental arena in his first two years in the state house, but began wading into environmental issues shortly after Reagan became president. In 1983—the high water mark of public consternation with Reagan's environmental policy—Graham made "Save Our Everglades" a major element in his political program for Florida, continuing and strengthening the environmental concern first demonstrated by Reubin Askew's administration.

The thrust of Reagan's environmental agenda was to shift responsibility from the federal government to the states. Graham's environmental program revolved around central planning and public land acquisition. In the face of Reagan's electoral triumph and the contraction of federal leadership in environmental affairs, Graham marshaled the state's resources to undertake the mammoth task of restoring the Everglades ecosystem. Together, the policies of these two contrasting leaders—building on the foundation constructed by Askew's administration—rearranged the political landscape of South Florida water management.

In February 1981, *Sports Illustrated* published a hard-hitting article about environmental degradation in Florida. Authors Robert H. Boyle and Rose Mary Mechem described the state's rampant population growth and frenetic new construction and noted that Governor Bob Graham had declared his administration in full support of bringing in more industry. "The sad fact is that Florida is going down the tube," the authors wrote. "Indeed, in no state is the environment being wrecked faster and on a larger scale." The authors went on to cite dire warnings of ecological collapse by such prominent Florida environmentalists as Marjory Stoneman Douglas, Charles Lee, John "Johnny" Jones, Arthur R. Marshall, and Nathaniel P. Reed.[1] What set this article apart from a dozen other contemporary essays about Florida's ailing environment was its prominent placement in a magazine with a broad readership. This was *Sports Illustrated's* hot-selling annual swimsuit issue, and the 10-page article on Florida's environment dovetailed with a 14-page spread of bathing beauties on Florida beaches.

The instigator of the *Sports Illustrated* article was none other than the brash and effective political lobbyist Johnny Jones, executive director of the Florida Wildlife Federation and an instrumental player in Kissimmee and Okeechobee issues in the 1970s. Negative publicity was just what Jones was seeking. In 1980, he began calling newspaper editors and outdoors writers around the state, feeding them information for stories on the environment. After some success at the state level, he approached *Sports Illustrated*. Jones' idea at that point was to challenge the governor by making caustic remarks about Bob Graham's commitment to the environment in the national magazine.[2] "There never was a better environmental senator than Bob Graham," Jones was quoted as saying in the article, referring to Graham's legislative achievements in the state senate. "But as governor he has wandered away from us. I can't even get in to talk with him, and I run the biggest conservation organization in Florida. As a governor, he ain't got it." Jones went on to charge Graham with forsaking environmentalists, reaching out to sugarcane growers and agribusiness, and generally moving to the political center because he had presidential ambitions.[3] Jones' words were harsh but measured; privately he held the governor in high esteem. Yet he knew from experience, he told an interviewer many years later, "that if you want somebody to move in government . . . you have to pick up a two-by-four and hit him upside the head."[4]

Graham's commitment to the environment was both personal and complex. Born in the Miami suburb of Coral Gables in 1936, he

Governor Bob Graham. (The Florida Memory Project, State Library and Archives of Florida)

was no stranger to South Florida's growth issues. His father, Ernest Graham, had moved to Florida from Chicago and, in the words of Marjory Stoneman Douglas, "turned a Miami dairy farm into a real estate fortune."[5] As a young man, Bob Graham took a turn with cattle raising and home construction, before venturing into politics. He served in the Florida House of Representatives from 1967 to 1970, and in the Florida Senate from 1971 to 1978, where he got several important environmental laws enacted. In 1978, he was elected governor. One of his first acts was to create the Office of Planning and Budgeting, which was aimed at giving state planners more influence in shaping the state budget.[6] In this regard, Graham's approach to governance contrasted with that of President Ronald Reagan, who made masterly use of the federal budget as a tool for shaping federal policy. While both Graham and Reagan appreciated the nexus of budget and policy, Reagan approached it from the opposite direction, using the Office of Management and Budget (OMB) to degrade federal programs he did not like—including many environmental programs affecting South Florida.

When Graham took office, one of the key planners in Florida state government was Estus Whitfield. Whitfield was the principal author of the state's first land development plan. In 1979, Governor Graham asked Whitfield to join his staff, requesting that Whitfield attend some cabinet briefings. According to Whitfield, "the environmental stuff was always the most significant part of the Cabinet meetings. It did not necessarily take the most time, but it was always difficult, controversial, because you had the issues of the use of state-sovereign lands."[7] Consistent with the acts he had sponsored in the state legislature in the early 1970s, Governor Graham wanted to facilitate dialogue between advocates of growth and environmental protection. Not surprisingly, the criticism of the governor in the *Sports Illustrated* piece found its mark.

Whitfield remembers bringing the governor a copy of the magazine. "There is some good news and some bad news here," Whitfield said, "and the good news is on the front, Christie Brinkley in her swimsuit." The bad news was inside. To see such a popular national magazine lambasting Florida's mismanagement of the Everglades, and to read the criticism by Jones—a political friend and supporter who had a big following in the state—made a powerful impression on Graham. "I do not ever remember a time thereafter," Whitfield remarked, "that the environment was not on the top of the agenda."[8]

The environmental problems that the governor faced were tied inextricably to Florida's continuing rampant growth. By the 1980 census, the state had nearly 10 million people and had risen to the rank of seventh largest state in the union. Two cities, Miami and Orlando, were growing apace. Miami, long established as the nation's gateway to Latin America, had become, like New York and Los Angeles, one of the nation's great immigrant cities. Its population contained not only Cuban exiles, but also large numbers of non-Cuban

Hispanics, Caribbeans, and Asians. In the early 1980s, Miami captured national headlines as it coped with a floodtide of refugee "boat people" and rising ethnic violence. Orlando, meanwhile, continued to grow as a destination resort and service center for Disneyworld's Magic Kingdom. Sprawling across four counties, this metropolitan area was attracting some 150 new residents daily by the 1980s.[9] The phenomenal growth of Miami and Orlando, together with the development of other cities and innumerable retirement communities, placed increasing demands on the water supply of South Florida.

Agriculture, too, continued to grow, imposing its own set of water demands. Sugar cane, in particular, consumed a huge quantity of water. While dairy farms and citrus groves north of Lake Kissimmee disappeared into subdivisions on the expanding fringes of Orlando, sugar cane interests further south increased their stake in the lands served by the C&SF Project. In the EAA, the farm crop amounted to $700 million in 1981, of which $600 million was in sugar cane. That year, Florida surpassed Hawaii as the nation's top producer of sugar. Ornamental tree farms were another significant agricultural interest. In Dade County's agricultural areas, intensive fruit and vegetable farming yielded 75 percent of the winter vegetables and 95 percent of the limes consumed in the entire country.[10]

One concern for many people was the water supply for Dade County. Three million people in South Florida depended on the Biscayne aquifer for their sole source of drinking water. As demands on the aquifer mounted, so did longstanding concerns about seawater infiltrating into the groundwater. Another concern was contamination of the water supply by chemicals and sewage. There were known hazardous waste sites all over South Florida where no barrier existed to stop potentially hazardous wastes from leeching into the groundwater. People, too, worried about contamination of shore waters. Pollution in Tampa Bay was so bad that parts of the bay were off-limits to swimmers.[11]

But the most pressing issue was the decline of the Everglades. By the early 1980s, experts were agreeing that the health of the Everglades ecosystem was deteriorating at an accelerating rate. The signs of ecological imbalance were many: soil subsidence; water scarcity and pollution; alteration and elimination of vegetation, wildlife, and fisheries; and intrusion of exotic species. If there was one ray of hope,

Miami, 1985. (The Florida Memory Project, State Library and Archives of Florida)

it was in the growing recognition that water management was the key to saving the Everglades. It was not just a matter of protecting the quantity and quality of water in the natural system; it was imperative that resource managers discover how to distribute the water so that it closely paralleled the historic sheet flow and the region's annual rainfall cycle.[12]

These ideas crystallized in 1982 when another deer crisis occurred in Water Conservation Area No. 3. From 1980 to 1981, South Florida had experienced a drought, and the low water, coupled with fewer hunting opportunities, had caused the deer population to expand. Then, in the spring of 1982, heavy rains began falling, raising water levels in Conservation Area No. 3 to 11 feet, significantly above the regulation schedule of 9.5 to 10.5 feet. The heavy rains also forced the Corps to pump water from the EAA to the water conservation areas, exacerbating the condition. With deer unable to find forage due to the high water, the Florida Game and Fresh Water Fish Commission called for an out-of-season hunt to reduce the deer population and enable more of the animals to survive. On 18 and 19 July 1982, hunters killed 722 deer. But animal-rights activists protested strongly against the hunt, vilifying both the Corps and the Game and Fresh Water Fish Commission, and the plight of the deer became nightly fodder on the national evening news. This negative publicity prompted Graham to create the Everglades Wildlife Management Committee, composed of representatives from state and federal agencies and headed by Estus Whitfield. Graham asked the committee to develop a wildlife management plan in harmony with water management goals.[13]

The Everglades Wildlife Management Committee held public hearings to gauge what could be done about the deer. Numerous animal-rights activists came, as did many environmentalists. According to Whitfield, "the common theme, which was repeated over and over and over again," was that "poor water management" was the problem. Therefore, when the committee issued its report, it stated, in Whitfield's words, that the deer should not only be managed at lower levels, but that "the water-management system is flawed and . . . needs to be altered."[14]

One of the ways to change water management was by implementing steps that Arthur Marshall had been promoting for years—something that Johnny Jones referred to as the "Marshall Plan." Jones gave the program this designation both to honor his friend and for rhetorical effect, as it echoed the economic rebuilding program developed by Secretary of State George C. Marshall after the Second World War and the plan promulgated by U.S. Geological Survey employee Robert Marshall to solve the water woes of the San Joaquin Valley in the 1920s (which eventually became the Central Valley Reclamation Project). The Arthur Marshall Plan was essentially what Marshall had advocated since the early 1970s—a restoration of natural sheet flow to the entire ecosystem, from the headwaters of the Kissimmee River to Florida Bay. It was then a prototype for the comprehensive plan for ecosystem restoration that would emerge in the 1990s. Marshall's ideas, however, focused on the problems of the upper basin, particularly along the Kissimmee River, and referred to the plan as a way to "repair" the Everglades, not "restore" it.[15] Regardless, the program provided a blueprint for ecological restoration—albeit encompassing a limited area of the Kissimmee-Okeechobee-Everglades ecosystem—and it assumed a large commitment of funding by the state government.

Ironically, Florida's latter-day Marshall Plan did not have any federal funding behind it even though the enormity of the proposal and the fact that it would affect Everglades National Park seemed to warrant it. Some even believed that the U.S. Army Corps of Engineers would oppose the plan, given the extensive modifications of the C&SF Project it would require. Nor was it clear whether Florida would get much support for its environmental initiatives from agencies such as the EPA, the FWS, and the NPS, even though all of these entities had supported environmental initiatives in the 1960s and 1970s. The reason was simple: about the same time that environmentalists succeeded in reawakening Graham to environmental issues in Florida, they lost whatever influence they had had in the executive branch of the federal government under Ronald Reagan. The newly elected president was avowedly pro-business, strenuously opposed to "big government," and hostile to most of the environmentalists' agenda. His election in 1980 created a new political context for environmental restoration initiatives in South Florida.

Reagan campaigned for the presidency on a theme of cutting taxes and government "red tape" in order to revitalize the economy. His vision of economic growth contrasted with President Jimmy Carter's message of sacrifice. Reagan's conviction that the United States could achieve energy independence by unlocking its own domestic resources contradicted Carter's emphasis on conservation and limits to growth. Reagan also presented "simple" alternatives to Carter's complicated analyses. Although environmental issues were not at the center of debate in the presidential campaign, the two candidates presented a stark difference on environmental policy.[16]

When Reagan won the election, environmentalists feared that the new president would undo many of the movement's accomplishments of the previous decade. Reagan, for his part, claimed that his landslide victory at the polls (he had defeated Carter by 489 to 49 electoral votes and garnered a plurality of popular votes in a three-way race with Independent John Anderson) gave him a popular mandate to reform the nation's economic and regulatory policies, including those shaping the environment. Reagan's environmental agenda was deregulation, reduction of programs, and the opening of public lands for energy development and other uses. But in spite of Reagan's decisive victory, it was questionable whether his environmental policy was in sync with majority opinion. Because most people based their votes on multiple issues, no one could be sure that voters had given Reagan a mandate to reform environmental policy.

Indeed, one strong indication that the public remained committed to environmental reforms of the 1970s was the fact that environmental organizations dramatically increased in membership during the Reagan administration. Analysts came to believe that the public consensus that had formed behind the environmental movement in the 1970s held together in the 1980s. As the Reagan administration moved ahead on its agenda, environmentalists sought to protect the status quo through Congress and the courts.[17]

Reagan pursued his environmental policies through three principal strategies. First, he appointed administrators who shared his conservative ideology and who were willing to undertake an extensive rewrite of federal regulations. Whereas Carter had appointed many administrators from the environmental community, Reagan recruited largely from the business arena and his appointees to environmental agencies typically had cut their teeth opposing govern-ment regulators. While some of Reagan's top appointments provoked congressional opposition, his lower level designees were too numerous for Congress, environmental groups, or the media to monitor. The Reagan administration's political appointments reached farther down in the ranks of the executive branch than in any previous modern presidency, and the President accomplished his goals in part through a systematic weakening of federal environmental regulations and policies at the hands of these administrators.[18]

Reagan's second principal strategy was to use the budget process to implement policy. He began his first term by pushing a package of massive tax and budget cuts through Congress. "No previous administration ever came into power more determined or better prepared to achieve substantial domestic policy change through the budgetary process," political scientist Robert V. Bartlett argued. In each ensuing budgetary cycle following the 1981 tax cut, Reagan

Ronald Reagan at a press conference in Florida. (The Florida Memory Project, State Library and Archives of Florida)

directed the inevitable reductions in federal appropriations toward those environmental programs he favored least. "Few environmental programs," Bartlett wrote, "escaped the Reagan budgetary scalpel."[19] While Congress began to restore some environmental program budgets to former levels in Reagan's second term, many programs—such as funding for research and development—bore lasting scars from this budget slashing approach to policymaking.

Reagan's third principal strategy was to avoid battles with Congress. He sought to weaken environmental laws not through legislative amendment but through selective non-enforcement of the laws based on regulatory revision and budget cuts. The Reagan administration made few legislative proposals in the environmental arena. Although Congress opposed the Reagan environmental policy agenda in many particulars, it went along with the president's tax and budget cuts in 1981. As a result, Congress's opposition to the president on environmental issues during the 1980s was largely confined to budget battles.[20]

The poster child for Reagan's environmental agenda was Secretary of the Interior James G. Watt, the most flamboyant and controversial of Reagan's appointees. A Wyoming lawyer who had spent years lobbying for and battling against Interior Department policies, most recently as a member of the conservative Mountain States Legal Foundation, Watt was a self-proclaimed leader in the "Sagebrush Rebellion." Centered in the West, the Sagebrush Rebellion was fueled by frustration over declining energy prices and protective land policies that were allegedly harming western rural communities. Its principal goal was to privatize certain public lands administered by federal agencies. Although the Sagebrush Rebellion focused most intently on federal wilderness areas and minerals administered by the U.S. Forest Service and the Bureau of Land Management—western issues that were seemingly remote from Florida—its attack on public land ownership was significant. One of Secretary Watt's first initiatives was to freeze expenditures by the Land and Water Conservation Fund.[21] This fund was used primarily for purchasing land in authorized additions to the national park system, and from 1965 to 1981, it had enabled the NPS to acquire 1.4 million acres—including land in Florida's Big Cypress National Preserve and Biscayne National Monument. As proposals for ecosystem restoration in South Florida increasingly pointed to the need for more publicly owned conservation lands, the Reagan administration did nothing to encourage a vigorous federal or state land acquisition program.

The operating budget of Everglades National Park took some direct hits in Secretary Watt's drive to reduce expenditures by the NPS. In 1982, the park was denied funding for restoration of natural drainage to the Turner River watershed in Big Cypress National Preserve and the western edge of Everglades National Park, a project that the Izaak Walton League claimed "would be truly a precedent setting action."[22] In 1983, the park lost funding of its exotic plant control program.

Watt also had a penchant for making flip remarks (Reagan once gave Watt a sculpture of a cowboy boot with a bullet hole through the toe to symbolize this characteristic) and this tendency was evident in his response to the ecological endangerment of Everglades National Park. In June 1983, at a time when the national news media were carrying stories about manmade flood disaster in the park, Watt said in a prepared address to the American Petroleum Institute, "I'm told that Everglades National Park is being improved and is in better shape than it has ever been."[23] Park officials told the press the next day that the secretary of the interior was vastly uninformed.

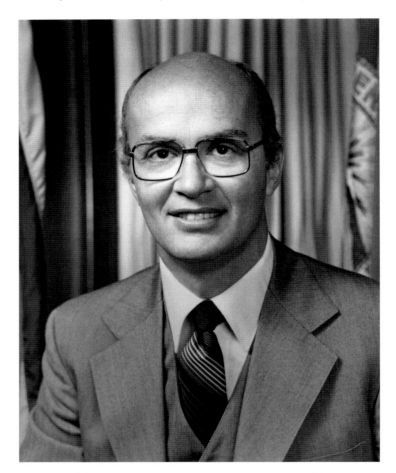

Secretary of the Interior James G. Watt. (U.S. Department of the Interior Library)

Watt's critics pounced on this flap as evidence that the secretary did not talk to his park superintendents and did not comprehend or care about the ecological integrity of national parks. Indeed, it seemed to be further proof that Watt was narrowly focused on infrastructure improvements in the national parks. Florida Audubon Society president Peter Mott saw it as evidence that the interior secretary was ignorant that the park had lost 90 percent of its wading bird population. "Or maybe he thinks visitor centers are more important so people can go there and see trees with no birds in them," Mott sarcastically commented.[24]

Another controversial Reagan appointee, Ann Gorsuch, presided over drastic reductions of budget, staffing, and regulatory enforcement at the EPA. Under her leadership, according to historian Edmund P. Russell, "EPA cut enforcement actions in half and morale plunged." Administration critics asserted that Gorsuch was unqualified and incompetent to lead EPA. Pressured by Congress, and with critics within the White House as well, she resigned in 1983, and Reagan brought back EPA's first administrator, William Ruckelshaus, to restore the agency's sense of mission. In one important respect EPA had been adrift since the Carter administration. Originally, the

Assistant Secretary of the Army (Civil Works) William R. Gianelli. (U.S. Army Corps of Engineers)

EPA had concerned itself with the total environment, but it had since moved toward a narrower focus on cancer-causing toxins.[25] In part because of this focus, EPA paid little attention to Everglades protection during the Reagan years.

Reagan's choice to oversee water resource development programs of the U.S. Army Corps of Engineers was William R. Gianelli, who became Assistant Secretary of the Army (Civil Works) in April 1981. Gianelli had served as head of California's Department of Water Resources when Reagan was governor of that state. Gianelli had two principal goals as the assistant secretary. First, as part of Rea-

gan's effort to deregulate, he wanted to curtail the Corps' Section 404 regulatory program. This program was named for Section 404 of the Clean Water Act of 1972, which assigned the Corps responsibility for issuing permits to dredge and fill wetlands. During the Carter years, the Section 404 program evolved so that the permitting process served to protect, or at least mitigate, wetlands loss. Gianelli believed this was wrong. In his view, the intent of Section 404 was to protect water quality, not wetlands. If Congress really wanted to prevent the destruction of wetlands, Gianelli maintained, then it needed to pass another law and assign responsibility to EPA or the FWS. So Gianelli embarked on a revision of the Section 404 regulations with the assistance of his deputy, Robert Dawson. The principal result of these efforts was to streamline the permitting process and allow more development in wetlands. Gianelli was particularly proud that the revised regulations reduced the influence of "single-purpose agencies" such as the FWS and the NPS.[26]

Gianelli's other main goal was to reform the system of funding for water development projects. Curiously, this was one environmental policy initiative the Reagan administration shared with the Carter administration. But Reagan would not repeat the mistake that Carter had made. In 1977, Carter had gone to Congress with a "hit list" of Corps of Engineers civil works projects that he considered unnecessary and worthy of deauthorization. Carter took aim at these projects because the Corps had a longstanding reputation as an agency that lent itself to "pork barrel" politics. Members of Congress had made an industry out of obtaining civil works projects for their local districts—projects that were often paid for by taxpayers at no additional cost to the local communities that were supposed to benefit from them. Moreover, the Corps' civil works program carried a backlog of projects that, in the context of the environmental movement, appeared misguided and detrimental to the environment. Carter, however, underestimated how jealously Congress guarded its prerogative to authorize these civil works projects, and he encountered a firestorm of congressional opposition. The resulting standoff killed any new water resource development acts—the semi-annual appropriation bills of the Corps' civil works program. Carter retreated from his hardline stance in the last two years of his administration, but the damage had been done.[27]

When President Reagan came into office in 1981, he was every bit as opposed to pork barrel politics as his predecessor. Yet Reagan strongly supported water resources development. Indeed, Carter's assault on water development projects was one of the wellsprings of the Sagebrush Rebellion that had helped elect Reagan president. Therefore, Reagan's prescription for the Corps' civil works program was to get it moving again by shifting a greater proportion of project costs to the states. Carter, too, had proposed new cost sharing arrangements, but Reagan took it further.

Gianelli, again with support from his deputy, Robert Dawson, proposed substantially increased "cost-sharing" between the federal government and non-federal sponsors for construction of new projects. Historically, a study by the Water Resources Council had shown, local interests had paid an average of 19 percent for federal flood control projects, but Gianelli, using figures developed by Robert Eiland, his special assistant, proposed that this figure be increased to 35 percent. Congress resisted this reform, and the stalemate over new project authorizations that had begun in the Carter administration continued through the first term of the Reagan administration. Finally, the Reagan administration succeeded in passing the Water Resources Development Act of 1986 (WRDA-86), which required local interests to enter into cost sharing agreements with the Corps for almost all new flood control projects. Local interests would bear at least 25 percent of the burden, instead of the federal government paying the full amount, and they would also supply 50 percent of the cost for feasibility studies. According to the Corps' chief counsel at the time, Les Edelman, it was Dawson who skillfully sold the cost-sharing idea to Congress, but congressional leaders, especially Senate Republicans such as Robert Dole, James Abdnor, and Mark Hatfield, were the ones who secured the legislation's passage.[28]

WRDA-86 was a significant achievement for an administration that put forth few legislative proposals in the environmental policy arena. The Reagan administration advanced its environmental agenda most effectively through the budget process, slashing appropriations for the Department of the Interior, EPA, and even the Corps, which lost nearly 3,000 positions in the civil works division from 1981 to 1983.[29] Therefore, as Florida positioned itself to undertake much more extensive ecosystem restoration in the early 1980s, it had little support from Washington. Reagan's new federalism would shape the direction of ecosystem restoration in South Florida for the next dozen years.

Reagan's new federalism was evident in the return of the jetport controversy. Although the first jetport proposal in the Big Cypress Swamp had been defeated, controversy swirled again over Site 14, a 26-square-mile area in northwest Dade County selected under the Everglades Jetport Pact in the 1970s. Construction had been delayed on the jetport, and by the late 1970s, environmentalists were bitterly divided over Site 14. Some environmental groups, notably the Florida Audubon Society, opposed Site 14, believing that the sacrifice of wetlands, the encroachment on critical habitat of several endangered species including the Everglades kite, and the disruption of sheet flow across Conservation Area No. 3B were unacceptable costs to accommodate further aviation needs.[30] But most of the Everglades Coalition supported the alternative site in the belief that, if it were not approved, Dade County would eventually build somewhere else in a location that would probably do more harm to Everglades National Park. Nathaniel Reed took this point of view, arguing that Conservation Area No. 3B, in which Site 14 was located, must be treated as multiple-use land in order to protect the national park and preserve land to the south and west. The dispute drove a wedge in the coalition.[31]

In October 1979, EPA raised the bar for Site 14 when it designated the Biscayne aquifer as a sole source aquifer under Section 1424(e) of the Safe Drinking Water Act. Now pollution discharged into the groundwater had to be considered in addition to the other environmental impacts of the proposed facility. EPA notified the FAA of its intent to review the project in light of its Section 1424(e) authority. By the spring of 1980, EPA officials in the Regional Administrator's office in Atlanta were warning the FAA that EPA had "several major concerns"—particularly since the Dade County Water and Sewer Authority had recently proposed to develop a major new well field for tapping the Biscayne aquifer near Site 14.[32]

EPA completed its review of the final EIS for Site 14 in February 1982, nearly a year into Reagan's first term. The agency reiterated its concern that the development of a commercial airport "within the cone-of-influence" of the newly developed well field represented a "potential threat to future drinking water quality in south Florida." Yet EPA would not use its Section 1424(e) authority to prevent the project, stating that it could not "conclusively demonstrate that an airport at Site 14 will lead to contamination of the Biscayne [aquifer]."[33]

Commenting on the final EIS in January 1982, the Jacksonville District of the Corps of Engineers expressed its own reservations about Site 14. The jetport facility would render Conservation Area No. 3B "virtually useless" for its intended purpose as a floodwater storage area. Moreover, the Corps objected to a statement in the final EIS that the Corps, together with other federal agencies, had determined that the jetport development was compatible with plans for the building of a conveyance canal from Conservation Area No. 3 to Everglades National Park. "The Corps of Engineers was not a party to this determination, and in fact disagrees with this determination," the Jacksonville District stated.[34] To the contrary, there was high potential for the jetport facility to contaminate water that was to be conveyed directly to Everglades National Park through the new canal. Therefore, the District requested that the FAA address these issues in its final EIS, noting that the FAA needed to follow the procedures outlined in Section 404(r) of the Clean Water Act as well.[35]

Faced with mounting criticisms of the project, Governor Graham formed a committee to study the issue and to recommend a final decision. He selected seven individuals: five from state government, one from private industry, and one from the University of Florida. The governor charged the group with making a thorough analysis of South Florida's future aviation needs. The group considered a range of other measures to improve commercial aviation service and determined that existing facilities at Miami International Airport could meet the region's demands through 2000. From a state perspective, further consideration of a proposed airport at Site 14 was unwarranted, the group decided. The governor accepted the group's findings.[36]

On 11 May 1983, Graham announced that he opposed the jetport plan and would not renew the state's participation in the Everglades Jetport Pact. "Our decision to withdraw the state's support for a Jetport in Dade County will mean enhanced protection of the Everglades, the 'river of grass' unique to the world," Graham stated. "This decision to withdraw state support of the study of a major new jetport in Dade County means that such a Jetport should not be built in this century."[37] Reminding people that the jetport proposal had been under consideration for a long time, the governor explained that public attitudes about land use had changed dramatically in the intervening years. Graham might have added, although he was too

Urban encroachment of Dade County. (South Florida Water Management District)

politically savvy to say it, that the federal perspective on land use had changed dramatically as well. If growth pressures in Dade County threatened Everglades National Park, it was now up to the state, not the federal government, to press for action.

At the same time that debate raged over the jetport proposal, the Corps struggled with another task that, in a different political climate, might have provided valuable help for addressing South Florida's water management problems. In 1980, the Jacksonville District initiated a study of South Florida's water supply mandated by Congress to resolve the questions of water supply to Everglades National Park. As explained previously, P.L. 91-282, passed in 1970, required that the Corps conduct a study in 1980 to "determine whether further modifications of the project [were] warranted, and [to] give further assurances of maintaining the essential water supply to insure the protection of the Park's ecosystem."[38] The study was funded as a component of the C&SF Project—another in a series of restudies of the huge project to gauge its progress and prospects. But instead of focusing on whether or not Everglades National Park was receiving enough water, the Jacksonville District decided to use the restudy to assess how the C&SF Project could increase water availability for the region's agricultural, municipal, and industrial needs. The restudy

of the early 1980s foreshadowed the extensive reexamination that would be undertaken in the following decade. Like the latter effort, its purpose was to identify means for expanding the water pie in South Florida so that all stakeholders in the project could get a larger piece. However, the water supply study of the early 1980s withered. It served to highlight needs and options, but did not result in any comprehensive plans or recommendations.

A drought beset South Florida at the same time that the study was commenced, heightening public interest in the water supply problem. Hardest hit by the drought was Lake Okeechobee. As the level of the lake dropped to its lowest point in history, the Corps was urged to reconsider old proposals to raise the lake level and increase its storage capacity. The problem was that the level of the lake directly affected its water quality and wildlife habitat. To resolve these conflicts, the Corps changed the lake-stage regulation schedule in 1978 in order to increase water storage, and the SFWMD curtailed backpumping of water from the EAA into Lake Okeechobee in order to improve water quality. However, due to the drought, the lake level dropped despite the new regulation schedule. In order to prevent levels from decreasing even more, the SFWMD allowed a resumption of backpumping through August 1982.[39]

As part of the water supply study, the Corps investigated options for raising the level of Lake Okeechobee. It also considered backpumping water from east coast canals into the lake. Other alternatives included the establishment of additional water conservation areas, storage of freshwater in deep aquifers so that it could be pumped

Lake Okeechobee. (South Florida Water Management District)

to the surface in times of shortage, desalinization of seawater, and water conservation. One concerned citizen wanted to import water to Lake Okeechobee from the St. Johns River basin to the north, a proposal that U.S. Senator Lawton Chiles (D-Florida) relayed to Colonel Alfred B. Devereaux, Jr., District Engineer of the Jacksonville District. Devereaux's deputy responded to Senator Chiles that interbasin diversion was not among the alternatives under consideration.[40]

Governor Graham agreed with the Corps on this matter, stating that interbasin transfers of water should be considered only as a last resort, and that the state and the Corps were investigating other means of improving South Florida's water supply. Graham noted that the array of water supply concerns included the protection of water quality, coordination of water use activities in each region, and "maintaining minimum flows for natural systems."[41] This last item was of crucial importance with respect to Everglades National Park.

Governor Graham was not the only one requesting that the Corps address the needs of Everglades National Park. John M. Morehead, superintendent of the park, was interested in the ultimate effects of the Corps' study. In a letter to Colonel Devereaux, he expressed hope that the study would "examine ways to rejoin the historical hydrological equilibrium between the east Everglades, the Water Conservation areas, and Everglades National Park."[42] Environmental groups, too, wanted assurance that the water supply study would adequately address park needs. It was not an unreasonable demand, since Congress originally authorized the study in 1970 as part of its efforts to protect water supply to the park.[43]

But in the view of the Corps, the multiple demands on South Florida's water supply by municipalities, industry, and agriculture required a broader approach. As the water supply study entered its third year, the Corps determined that the best way to complete it was to throw its effort behind an initiative of the SFWMD involving computer modeling. Thomas MacVicar, a state hydrologist in the Resource Planning Department of the SFWMD, had begun developing a computer model capable of simulating the hydrology of South Florida on a regional scale. As the model progressed, the Corps provided funding through contractual arrangements with the SFWMD. Gradually, the water supply study melded into the SFWMD's computer modeling strategy.[44] In its original published form (1984), the model was called the South Florida Water Management Model (version 1.1). It would be continually modified, upgraded, and populated with additional data over the next two decades, and in the early 1990s, it would form the basis for a Natural System Model that was crucial in developing a comprehensive plan for ecosystem restoration.[45]

MacVicar's original concept was to develop something that would simulate how water was distributed and flowed through the entire ecosystem so that managers could test how operational decisions in one locality would affect hydrologic conditions elsewhere. The hydrologic model simulated groundwater flow, surface water flow, and how hydrology would respond to hypothetical channel routings from changes in canals, levees, and other structures. Spatially, the model consisted of a grid-pattern overlay of South Florida composed of squares two miles on each side, with each point of intersection in the grid being a node in the computer model. For each node, the model was populated with data on topography, land use, and aquifer thickness and permeability. In terms of timing, the model used one-day intervals, and data were supplied for rainfall, well field withdrawal, and structure discharge for each day of simulation.[46] Because it explored the relationship of disparate regions within the entire ecosystem—showing how changes in one part of the area affected water distribution or other characteristics in another section—MacVicar's model foreshadowed the Corps' restudy of the C&SF Project in the 1990s. In many ways, it was one of the key factors allowing the concept of Everglades restoration to bloom.

In its beginning stages, MacVicar's program was used for developing "optimization" of the C&SF Project. MacVicar encoded the model so that it would compare actual water stages throughout the system with "optimum" stages. Actual amounts were computed by entering the previous day's hydrologic data into the program. Optimums for each canal were variable depending on time of year, hydrologic conditions, and other operational considerations. The model helped managers to minimize the "absolute deviation" between the actual and optimum stages in each canal or reservoir, thereby allowing agencies "to drive the system to operate as close as feasible to the optimum."[47]

Park Superintendent Morehead saw the possibility of adapting MacVicar's hydrologic program to predict the results of different C&SF Project modifications for water deliveries to Everglades National Park. Through the Corps, he supplied MacVicar with data on mean monthly delivery volumes for the years 1969-1975.[48] MacVicar then entered operational data—the spatial arrangement of levees, canals, and other structures—for the same period. As he explained in a meeting of Corps, SFWMD, and NPS hydrologists and engineers in May 1983, he could now run the computer model for "base" (current) or "historical" conditions. By operating the model on historical conditions, it was possible to rewind the clock on C&SF Project developments and simulate how the hydrology would respond. In this case, "historical conditions" referred only to the operational system in the years 1969-1975—not far in the past—but the idea was to model different scenarios for filling in or degrading existing canals and levees in order to achieve a measure of ecosystem restoration. Toney Lanier, the Corps' project manager for the water supply study, assisted the development of MacVicar's program by committing project monies for it.[49]

Yet the Corps never produced a final report with an analysis of alternatives, as was conceived at the outset of the study and in the examination's congressional authorization. Yet the agency never officially abandoned it either. Strapped for funds and personnel, and distracted by project cost-sharing issues, it merely utilized the SFWMD's hydrologic model of South Florida for information.[50] After Congress passed an emergency measure for Everglades National Park in November 1983, requiring the Corps and the NPS to implement a two-year experimental program of modified water delivery for the park, the Corps became firmly wedded to the SFWMD's computer program as a means of re-evaluating water supply options not only for the park but throughout the Kissimmee River-Lake Okeechobee-Everglades ecosystem. It had used the water supply study to assess how agricultural, industrial, and municipal needs could co-exist with the park's ecological needs, much to the chagrin of the park, and it subordinated the study to the state's own water management objectives.

In 1987, seven years after the water supply study was initiated, the Corps sought comment from the FWS on another iteration of the examination, offering, for unclear reasons, only a meager $5,000 transfer of project funds for the FWS review. Field Supervisor Joseph D. Carroll of the FWS's Vero Beach office responded indignantly to Jacksonville District Engineer Colonel Charles T. Myers, III. "What is impending is a request at the 11th hour by your staff to respond in two weeks or 30 days to a ream of computer data," Carroll wrote. "As indicated in the Scope of Work, your staff will want to know detailed biological effects on Lake Okeechobee, the water conservation areas, the Holeyland and Rotenberger tracts, and Everglades National Park (millions of acres). This just cannot be done!" Carroll accepted the $5,000 transfer but warned that such a trifling sum would merely pay for "the most superficial treatment of this huge project, largely based on past studies and experience."[51]

Meanwhile, Governor Graham advanced his environmental agenda through state legislation. Graham's first significant environmental law was his "Save Our Rivers" initiative, enacted in June 1981, which provided $320 million to the state's five water management districts over the next ten years for river cleanup. In 1983, the state legislature passed the Water Quality Assurance Act, creating a $100 million trust fund to help local governments upgrade sewage treatment plants. The law also established guidelines for protection of groundwater from industrial pollution and gave the Florida Department of Environmental Regulation $3 million for enforcement. In 1984, the legislature enacted the Wetlands Protection Act, which enlarged the department's jurisdiction over swamps, marshes, and floodplains by extending the list of plants that identified an area as wetland from 67 to 266 species. The law also gave the department permitting authority for development in wetlands—a responsibility that overlapped the Corps' regulatory program under Section 404 of the Clean Water Act. By 1985, officials in Tallahassee claimed that Florida was doing more than any other state to protect water supplies, with the possible exception of California. Despite these state efforts, Department of Environmental Regulation Secretary Victoria J. Tschinkel emphasized, Florida's water supply remained vulnerable to threats of pollution. More than nine out of ten Floridians depended on groundwater for their drinking water. "In south Florida," one environmentalist commented, "we live right on top of our water supply. It's like drinking out of the toilet." A noted expert on the state's water resources, John DeGrove, testified in hearings held by the department that "what we've seen here scares the daylights out of me."[52]

While President Reagan promised the nation regulatory relief, Governor Graham assured Floridians that they deserved more environmental safeguards—and as the governor's popularity rose, it increasingly appeared that that was what most Floridians wanted. Under the strong leadership of Secretary Tschinkel, the Department of Environmental Regulation moved ahead of both the Corps and EPA to address state water problems. Tschinkel announced a protective state water policy in 1981, followed by a coastal management plan soon after that. Impatient for EPA to mandate allowable limits for toxic chemicals in drinking water under the federal Clean Water Act, Tschinkel's agency established state standards, becoming the first state in the nation to do so.[53]

Governor Graham's enthusiasm for land-use planning to protect environmental quality culminated in Florida's Growth Management Act of 1985. Although the state had passed similar legislation a decade earlier, it was not effective in dealing with the

Mangroves. (National Park Service)

pressures of Florida's rapid population expansion. One of the most pressing issues for many Floridians was how to preserve the small-town character of communities that were becoming smothered by strip malls and homogenous residential subdivisions. The Growth Management Act of 1985 aimed to address numerous problems in conjunction with this development, including inadequate infrastructure to support growth, affordable housing, and urban renewal, as well as environmental degradation. In essence, the law required local governments to develop local comprehensive plans for land use.[54]

In the long run, this law failed in its objectives. The principal reason was that the legislation was not enforceable: it allowed for substantial local control, and when developers wanted to develop a particular parcel of land that was out of bounds, they lobbied the local government to have the comprehensive plan modified. The law made Florida's Department of Community Affairs responsible for enforcing the comprehensive plans, but it allowed the plans to be modified as often as twice a year. Another limitation of the Growth Management Act was that it did not attempt to coordinate land-use planning with conservation needs. Land use plans too often ran afoul of the Section 404 permitting process, for example, or of requirements under the Endangered Species Act.[55]

Growth management was fundamentally a problem for local and state governments, but clean water, protection of wetlands, and cleanup of hazardous waste sites were environmental issues that involved state and federal cooperation. While Graham was willing to put the resources of the state behind various initiatives, he was frustrated by the lack of federal support. "To date, we have sort of dragged the federal agencies with us," Graham remarked in 1986. "I'd like to see Washington move from a passive and reluctant partner to a full and enthusiastic one."[56]

Nowhere was the state in more need of federal assistance than in South Florida, where the Everglades continued to show signs of inexorable decline. Environmentalists had long insisted that the federal government was neglecting its stewardship responsibilities in South Florida. The national parks and other federal interests in South Florida ought to compel greater federal involvement in that region's ecological problems, environmentalists argued. Graham did not disagree, but he decided that the underlying problems were so broad and complex that environmental leadership had to come from the state. In particular, he was impatient with the Corps over its slow pace in studying how to repair environmental damage caused by the C-38 canal it had built down the Kissimmee River Valley. As Jones and Marshall reminded him following the exposé of Florida's environmental problems in *Sports Illustrated*, the Corps had resisted modifying the Kissimmee River project since it was first asked to reexamine it in 1976. "The governor is not going to wait forever for the Corps of Engineers to act," Graham's chief environmental aide, Estus Whitfield, informed the media.[57] Rather, the state would propose its own version of Kissimmee River restoration.

In late March 1983, Graham called a summit of his top environmental administrators. State agencies represented at the conference included the SFWMD, the DER, the DNR, the Florida Game and Fresh Water Fish Commission, and the Department of Community Affairs. He demanded interagency cooperation and gave meeting participants a 1 July deadline to develop a blueprint for saving the Everglades. Over the next three months, the governor held a series of other meetings on the Everglades with business leaders, environmentalists, and state officials. Although the governor conducted these conferences behind closed doors, word began to leak to the press that the plan would involve restoration of sheet flow through the entire ecosystem. It would likely mean changes of land use in two bitterly contested areas: the EAA (south of Lake Okeechobee) and the section of Dade County bordering Everglades National Park known as the East Everglades Area.[58]

On 9 August, Governor Graham unveiled his program of environmental initiatives for South Florida. Aimed at "rejuvenation" of the entire Kissimmee River-Lake Okeechobee-Everglades ecological system, the program took the name "Save Our Everglades" to highlight the significance of the effort for Everglades National Park. The program embraced six "Phase I" actions, presented roughly in north-south or downstream order. First, the state would seek federal cooperation in reestablishing the values of the Kissimmee River. Second, the Holey Land and Rotenberger tracts—lands within the EAA that were now mostly in state ownership—would be restored as wetlands for the benefit of water sheet flow and wildlife habitat. Third, the deer population in Conservation Area No. 3 would be managed so that high water levels would not cause massive die-offs. Fourth, the two highways traversing the Everglades east to west, Alligator Alley and the Tamiami Trail, would undergo extensive modification to reduce impoundment of sheet flow from north to south. Fifth, the state would acquire land, as well as encourage the federal government to acquire land, in the controversial East Everglades Area for the protection of Everglades National Park. In addition, the state would support the park's demands for a modified water delivery plan. Sixth and finally, state and federal land acquisition would be pushed ahead in Big Cypress National Preserve and Fakahatchee Strand for the protection of the Florida panther.[59]

Of the six actions, only the second and third were primarily achievable without federal participation. Therefore, Governor Graham informed President Reagan about the "Save Our Everglades" program in a personal letter delivered to the White House on 8 August, one day prior to the program's official disclosure. "Florida is undertaking an ambitious program to restore and preserve the Everglades, a national treasure and a key factor in the future prosperity of our State," the governor began. "We will need the assistance of federal agencies. I urge your cooperation in revitalizing the Everglades and the environment of South Florida." Graham then cited the actions that would depend heavily on federal support: restoration of fish and wildlife values in the Kissimmee River Valley, requiring the "expedited cooperation" of the Corps of Engineers; reconstruction of Alligator Alley as an interstate freeway, which would need the assistance of the U.S. Department of Transportation; mitigation of water management impacts on Everglades National Park, necessitating Corps help; and acquisition of lands adjacent to the park and within Big Cypress National Preserve, in cooperation with the Interior Department. The governor requested that President Reagan designate a federal coordinator "who would be charged with expediting the actions of federal agencies in concert with state and local governments."[60] Although this last request did not garner any response from the Reagan administration, it foreshadowed the establishment of a federal task force on South Florida ecosystem restoration a decade later.

"Save Our Everglades" was a program more than a plan: it put forth a public goal and six distinct "actions" that would be pursued more or less independently of each other. Nevertheless, it was a huge step forward in forging public support for an ambitious program of environmental action in South Florida. And unquestionably, the program hung together around the central concept of ecosystem restoration. When Governor Graham announced "Save Our Everglades" in a press conference in Tallahassee on 9 August 1983, he defined three public purposes for the project that were "fundamental priorities" of his administration: first, to avoid any further degradation of the Everglades and related natural systems from the headwaters of the Kissimmee River to Florida Bay; second, to reestablish the "natural ecological functions" of the ecosystem; and third, to improve the overall management of recreation, water, fish and wildlife for the Everglades and surrounding areas.[61] All three purposes emphasized the connectedness of the Everglades with all of South Florida. They could be summed up in three words: preservation, restoration, and use. The concept of ecosystem restoration presented in "Save Our Everglades," then, enlarged upon the core national park mandate of preservation and use; it related Everglades National Park to the rest of South Florida, and it posited that ecological functionality was vital to both. An 11-page issue paper that accompanied the

press release on "Save Our Everglades" made repeated references to "the Everglades and the environment of South Florida."[62] To save the Everglades was to save South Florida, home to six million people.

In the "Save Our Everglades" program, Graham offered a vision, or definition, for ecosystem restoration. The program was designed to provide "that the Everglades of the year 2000 looks and functions more like it did in 1900 than it does today."[63] The issue paper carried a "background statement" that sketched some history of human-induced changes to the environment from Hamilton Disston's drainage works in the 1880s to the initiation of the C&SF Project in the late 1940s. It then declared, "Although the system can never be the same as it was before Disston began his work, many of its natural functions and values can be restored while providing water supplies and flood protection to south Florida."[64]

Graham reiterated these themes on a barnstorming tour of South Florida on 10 August 1983, accompanied by state officials, members of the press, and Colonel Alfred Devereaux, District Engineer of the Jacksonville District. The governor emphasized the interdependence between the natural environment and the nearly six million people who lived in South Florida.[65] In pledging Florida's efforts to restore

The area south of the Tamiami Trail was a key component of Governor Graham's restoration plan. (Everglades National Park)

the Everglades to its turn of the century condition, Graham had in mind an idealized baseline when the environment of South Florida supported agriculture, small cities, and the natural wonders of the Everglades in more or less equilibrium. In implementing this plan, Graham was continuing a process begun by Governor Reubin Askew in the 1970s: that of state initiative in repairing and restoring the South Florida ecosystem. The irony of the situation did not escape *Audubon* magazine, however, which observed that it was the state of Florida that had requested the C&SF Project in the first place, and now that same state was providing the program to save South Florida from the environmental destruction of the "enormous surface plumbing system."[66] Yet even though the state faced a presidential administration hostile to environmental policies, officials knew that federal involvement in Save Our Everglades was crucial. Whether or not that help would be forthcoming remained to be seen.

Endnotes

[1] Robert H. Boyle and Rose Mary Mechem, "There's Trouble in Paradise," *Sports Illustrated* 54 (9 February 1981): 82-93 (quotation on p. 84).

[2] Jones interview, 12.

[3] Quotations in Boyle and Mechem, "There's Trouble in Paradise," 93; see also Grunwald, *The Swamp*, 273.

[4] Jones interview, 12.

[5] Douglas, *Voice of the River*, 245.

[6] Estus Whitfield interview by Brian Gridley, 15 May 2001, 4, Everglades Interview No. 8, Samuel Proctor Oral History Program, University of Florida, Gainesville, Florida [hereafter referred to as Whitfield interview]; Grunwald, *The Swamp*, 272.

[7] Whitfield interview, 12.

[8] Quotations in Whitfield interview, 14; see also Grunwald, *The Swamp*, 271-272.

[9] Gary R. Mormino, "Sunbelt Dreams and Altered States: A Social and Cultural History of Florida, 1950-2000," *The Florida Historical Quarterly* 81 (Summer 2002): 15-17.

[10] Kevin Hanson, "South Florida's Water Dilemma: A Trickle of Hope for the Everglades," *Environment* 26 (June 1984): 17.

[11] Boyle and Mechem, "There's Trouble in Paradise," 86.

[12] Hanson, "South Florida's Water Dilemma," 15; William J. Schneider and James H. Hartwell, "Troubled Waters of the Everglades," *Natural History* 93 (November 1984): 47.

[13] Untitled draft document, File 10-1-7a C&SF Wtr (1983), Central and Southern Florida Water Supply (January-September 1983), Box 3285, JDAR; "Everglades Deer Crisis," File Everglades Conservation Area: Correspondence & Studies, 1978-1982, Box 1, S1719, Game & Fresh Water Fish Commission Everglades Conservation Files, 1958-1982, FSA; "Everglades Hunt: The Deer Can't Win," *Newsweek* 100 (17 October 1982): 27; John Weiss, "Everglades Deer in Trouble," *Outdoor Life* 165 (April 1980): 32, 36.

[14] Whitfield interview, 14.

[15] Boyle and Mechem, "There's Trouble in Paradise," 94; George Reiger, "The River of Grass is Drying Up!" *National Wildlife* 12 (December/January 1974): 62.

[16] C. Brant Short, *Ronald Reagan and the Public Lands: America's Conservation Debate, 1979-1984* (College Station: Texas A&M University Press, 1989), passim.

[17] Robert Cameron Mitchell, "Public Opinion and Environmental Politics in the 1970s and 1980s," in *Environmental Policy in the 1980s: Reagan's New Agenda*, Norman J. Vig and Michael E. Kraft, eds. (Washington, D.C.: Congressional Quarterly, 1984), 51-74.

[18] J. Clarence Davies, "Environmental Institutions and the Reagan Administration," in *Environmental Policy in the 1980s: Reagan's New Agenda*, 145-148.

[19] Robert V. Bartlett, "The Budgetary Process and Environmental Policy," in *Environmental Policy in the 1980s: Reagan's New Agenda*, 121-122.

[20] Henry C. Kenski and Margaret Corgan Kenski, "Congress Against the President: The Struggle Over the Environment," in *Environmental Policy in the 1980s: Reagan's New Agenda*, 98-105.

[21] Paul J. Culhane, "Sagebrush Rebels in Office: Jim Watt's Land and Water Politics," in *Environmental Policy in the 1980s: Reagan's New Agenda*, 293, 308.

[22] Quotation in Franklin B. Adams, President, Florida Division of the Izaak Walton League of America to Robert Baker, Regional Director, 5 March 1982, and Baker to Adams, 8 April 1982, Folder 24, Box 3, Marshall Papers; Hansen, "South Florida's Water Dilemma: A Trickle of Hope," 20.

[23] As quoted in *Naples Daily News*, June 16, 1983.

[24] As quoted in *Naples Daily News*, June 16, 1983.

[25] Edmund P. Russell III, "Lost Among the Parts per Billion: Ecological Protection at the United States Environmental Protection Agency, 1970-1993," *Environmental History* 2 (January 1997): 36.

[26] U.S. Army Corps of Engineers, Office of the Chief of Engineers, *Water Resources, People and Issues: An Interview with William Gianelli*, EP 870-1-24 (Washington, D.C.: U.S. Army Corps of Engineers, 1985), 32-33, 36 [hereafter referred to as Gianelli interview].

[27] Martin Reuss, *Reshaping National Water Politics: The Emergence of the Water Resources Development Act of 1986*, IWR Policy Study 91-PS-1 (Fort Belvoir, Va.: U.S. Army Corps of Engineers, Institute for Water Resources, 1991), 48-64; Stine, "Environmental Politics and Water Resources Development," 179-182.

[28] Reuss, *Reshaping National Water Politics*, 82-83, 145-199; Gianelli interview, 16-17; Les Edelman interview by Theodore Catton, 17 November 2004, 3-4.

[29] J. Clarence Davies, "Environmental Institutions and the Reagan Administration," in *Environmental Policy in the 1980s: Reagan's New Agenda*, 147.

[30] Joseph Browder interview by Theodore Catton, 17 November 2004, 4 [hereafter cited as Browder interview].

[31] Browder interview, 3-4.

[32] Briefing Document, Site 14—Training Airport, 5 May 1980, File Jetport Important Correspondence 1979, Box 1, Accession 412-91-0041, RG 412, NARA-SE.

[33] Howard D. Zeller, Acting Assistant Regional Administrator for Policy and Management, to James E. Sheppard, Chief, Miami Airport District Office, 5 February 1982, Folder 18, Box 2, Marshall Papers.

[34] A. J. Salem, Acting Chief, Planning Division to James E. Sheppard, Chief, Miami Airport District Office, 22 January 1982, Folder 18, Box 2, Marshall Papers.

[35] Salem to Sheppard, 22 January 1982.

[36] Untitled memorandum, no date, File Jetport, Box 6, S172, Executive Office of the Governor Press Secretary Subject Files 1979-86, FSA.

[37] Graham Announces Opposition to Southeast Florida Jetport, 11 May 1983, File Jetport, Box 6, S172, FSA.

[38] Senate, *River Basin Monetary Authorizations and Miscellaneous Civil Works Amendments*, 91st Cong., 2d sess., S. Rept. 91-895, 1970, Serial 12881-3, 20.

[39] R. Thomas James, Val H. Smith, and Bradley L. Jones, "Historical Trends in the Lake Okeechobee Ecosystem, III. Water Quality," *Arch. Hydrobiology* 107 (January 1995): 52-53.

[40] Robert J. Waterston III, Deputy DE to Lawton M. Chiles, Jr., 19 August 1981, File 10-1-7a C&SF Wtr (80-81) Central and Southern Florida Water Supply (January 1980—September 1981), Box 3285, JDAR.

[41] Bob Graham to C. J. O'Brien, 7 July 1981, File 10-1-7a C&SF Wtr (80-81) Central and Southern Florida Water Supply (January 1980—September 1981), Box 3285, JDAR.

[42] John M. Morehead, Superintendent, to Colonel Alfred B. Devereaux, District Engineer, 17 September 1982, File Everglades National Park 1958-88, General/Resolutions and Agreements, Box 02161, SFWMDAR.

[43] Alice Wainwright, National Audubon Society, to John Seiberling, Chairman, House Subcommittee on Public Lands and National Parks, 25 February 1983, File 31, Box 2, Marshall Papers.

[44] Toney Lanier, Project Manager, Memorandum for the Record, 11 May 1983, File 10-1-7a C&SF Wtr (1983) Central & Southern Florida Water Supply Study (January—September 1983), Box 3285, JDAR.

[45] Robert J. Fennema et al., "A Computer Model to Simulate Natural Everglades Hydrology," in *Everglades: The Ecosystem and its Restoration*, Steven M. Davis and John C. Ogden, eds. (Delray Beach, Fla.: St. Lucie Press, 1994), 250-251.

[46] "Central and South Florida Water Supply Modeling," no date, File 10-1-7a C&SF Wtr (1983) Central & Southern Florida Water Supply Study (January—September 1983), Box 3285, JDAR.

[47] "Water Management Decision Model," no date, File 10-1-7a C&SF Wtr (1983) Central & Southern Florida Water Supply Study (January—September 1983), Box 3285, JDAR.

[48] John M. Morehead, Superintendent, to Colonel Alfred B. Devereaux, District Engineer, 17 September 1982, File Everglades National Park 1958-88, General/Resolutions and Agreements, Box 02161, SFWMDAR.

[49] Toney Lanier, Project Manager, Memorandum for the Record, 11 May 1983, File 10-1-7a C&SF Wtr (1983) Central & Southern Florida Water Supply Study (January—September 1983), Box 3285, JDAR.

[50] On cost-sharing see A. J. Salem, Acting Chief Planning Division, to Commander, South Atlantic Division, 3 February 1982, enclosing "Benefit Evaluation Central and South Florida Project," File 10-1-7a C&SF Wtr (1988) Central & Southern Florida Water Supply (1988), Box 3285, JDAR. On the use of computer modeling see Joseph D. Carroll, Jr., Field Supervisor, to Charles T. Myers III, District Engineer, 14 February 1985, File CE-SE Central and Southern Florida FCP Everglades National Park Water Requirements Study, FWSVBAR.

[51] Joseph D. Carroll, Jr., Field Supervisor, to Charles T. Myers III, District Engineer, 3 March 1987, File 10-1-7a C&SF Wtr (83-85) Central & Southern Florida Water Supply (October 1983—1985), Box 3285, JDAR.

[52] Both quotations in "Florida's Growth Straining Fragile Groundwater," *Engineering News-Record* 189 (3 January 1985): 26.

[53] Victoria J. Tschinkel to David H. Pingree, 17 July 1981, File Department of Environmental Regulation, Box 4, S172, FSA; "Suggested Comments for Governor Graham, National Wetlands News Conference," 12 January 1983, File Environment, ibid.

[54] Randall G. Holcombe, "Why Has Florida's Growth Management Act Been Ineffective?" *Journal of the James Madison Institute*, No. 28 (Spring/Summer 2004): 13-16.

[55] Holcombe, "Why Has Florida's Growth Management Act Been Ineffective?"; Jones interview, 31; Colonel Terry Rice interview by Theodore Catton, 6 December 2004, 9.

[56] As quoted in Ron Moreau, "Everglades Forever?" *Newsweek* 98 (7 April 1986): 72.

[57] As quoted in *The Post* (West Palm Beach), 8 April 1983.

[58] *Fort Lauderdale News*, 5 June 1983.

[59] "Save Our Everglades," 9 August 1983, SFWMDAR.

[60] Bob Graham, Governor, to Ronald Reagan, President, undated, Folder 16, Box 2, Marshall Papers.

[61] "Graham Announces Save Our Everglades Program," 9 August 1983, Folder 16, Box 2, Marshall Papers.

[62] "Save Our Everglades," 9 August 1983, 4, 11.

[63] "Graham Announces Save Our Everglades Program," 9 August 1983, Folder 16, Box 2, Marshall Papers.

[64] "Save Our Everglades," 9 August 1983, 4.

[65] Colonel Alfred B. Devereaux, Jr., District Engineer, Memorandum for the Record, 12 August 1983, File 1110-2-1150a (C&SF) Kissimmee River Valley Basin Jan 1983, Box 9, Accession 077-01-0023, RG 77, FRC.

[66] Steve Yates, "Saga of the Glades Continues," *Audubon* 87 (January 1985): 34.

CHAPTER 8

"That Damn Sewer Ditch": Kissimmee River Restoration Efforts, 1978–1988

FACING A FEDERAL GOVERNMENT largely at odds with environmental concerns in the 1980s, Governor D. Robert "Bob" Graham initiated what was essentially an ecosystem restoration plan for South Florida known as "Save Our Everglades." Developed through discussions with prominent environmentalists, including Marjory Stoneman Douglas, Johnny Jones, and Arthur Marshall, the plan acknowledged the interconnectedness of the Kissimmee-Okeechobee-Everglades ecosystems and outlined ways to restore the water health of the region. An integral part of that plan was dechannelizing the Kissimmee River in accordance with the 1976 state law requiring the restoration of water quality in the Kissimmee River Basin. Efforts in the 1980s to remove C-38—called "that damn sewer ditch" by some environmentalists[1]—were promoted most vigorously by Graham and the South Florida Water Management District. Although the Corps seemingly dragged its feet for most of the 1980s on Kissimmee restoration, either by design or because of a lack of authorization to do much more than study the issue, it received an appropriation from Congress under Section 1135 of the Water Resources Development Act of 1986 to begin restoration efforts. By 1988, then, several significant steps had been taken toward dechannelization, setting a foundation for actual restoration in the 1990s.

In September 1978, Congress, responding to the state of Florida's initiative, provided appropriations to the Corps for a restudy of the Kissimmee River, and the Corps began its work in 1979. According to a 1980 publication, the purpose of the study was "to determine the feasibility" of altering the Kissimmee River flood control system in order to enhance water quality and improve "environmental amenities" and "fish and wildlife resources," among other things.[2] In October 1979, the Corps completed its recon-

naissance report (Stage I) and began Stage II of the restudy, which would develop numerous alternatives that the Corps could take. Thereafter, Phase III would examine the feasibility of those plans and recommend one as the course to follow. Because the Corps had a large amount of data to analyze, it decided to use a data management system known as SAM (Spatial Analysis Methodology) for the study. SAM, which had been developed by the Corps' Hydrologic Engineering Center in Davis, California, could evaluate all study aspects, including economic, environmental, and hydraulic conditions. The Corps pledged to obtain as much public input as possible in its examinations by conducting public meetings and workshops, thereby allowing for comments from a broad constituency. Corps officials estimated that all three phases of the examination could be completed by August 1982, with a draft Stage III report issued by January 1982.[3]

For those who believed that the 1976 Florida law mandated dechannelization of the Kissimmee River, this timetable was too long. Likewise, Marshall and others felt that there was an urgency to the issue. "The effectiveness of all the elements" of the Marshall Plan, Marshall explained, were "totally dependent on filling the Kissimmee ditch." In fact, he continued, "dechannelization [was] the answer and the hope for repairing the Everglades system."[4] To pressure the Corps to expedite its study and to champion Kissimmee restoration, new environmental organizations appeared, including the Kissimmee Restoration Coalition and Marshall's Coalition to Repair the Everglades.[5] Meanwhile, the Friends of the Everglades, holding that "the opportunity for the State of Florida to dechannelize the lower Kissimmee will not remain long," prepared a petition requesting that the state disallow further floodplain development, that it purchase floodplain lands, and that Congress and the President of the United States order the Corps to restore the river.[6]

As the first years of the 1980s passed, the Corps increasingly fell behind schedule on its feasibility study, frustrating many state officials. Victoria Tschinkel, secretary of the Florida Department of Environmental Regulation, for example, told newspapers that "the Corps was very behind schedule and above budget on its plans to restore the Kissimmee River."[7] She and Governor Graham called on the Corps to accelerate its work, and Jacksonville District Engineer Colonel Alfred Devereaux responded by pledging to have a decision by the end of 1982 as to how restoration could occur.

Many critics claimed that the Corps was merely dragging its feet because it did not want to dechannelize the Kissimmee, an accusation that Devereaux denied. He blamed the delays on SAM, explaining that the program had never been used on such a large study as the Kissimmee River plan, and that, therefore, establishing parameters became a long, drawn-out process. It "took a lot longer to get working than expected," he said, estimating that the program "probably added a couple of years" to the study's completion time.[8] M.

Kent Loftin, an engineer in the Jacksonville District agreed, explaining that data compilation and the need to break the river into a grid of three and twelve-acre land cells caused the slowdown.[9]

But it was also clear that despite the growing power of environmental organizations in the 1970s and the Corps' own attempts to transform itself into a more environmentally friendly organization, the agency was experiencing some setbacks. For one thing, it was difficult to shift agency culture away from engineering and towards environmental restoration. Corps leaders who actually embraced the transformation, for example, found resistance from old-time engineers who, in the words of historian Jeffrey Stine, declared that "they did not join the Corps of Engineers to come up with non-structural solutions to flood control problems."[10] Environmental organizations had helped to make the Corps more accountable in the 1970s, but, as Kissimmee River restoration efforts in the 1980s demonstrated, a long journey still lay ahead.

Regardless of the reasons for the delay, environmentalists wanted the Corps to act quickly. This feeling was heightened in 1982 when several scientists, including Arthur Marshall, claimed that

The Kissimmee River. (South Florida Water Management District)

the channelization of the Kissimmee River had altered the region's normal rain cycle. Meteorologist Patrick Gannon first proposed this hypothesis in 1977 in a doctoral dissertation titled "On the Influence of Surface Thermal Properties and Clouds on the South Florida Sea Breeze,"[11] but the theory was not widely publicized until an article appeared in a March 1982 issue of *Sports Illustrated* titled "Anatomy of a Man-Made Drought." This essay, written by Robert H. Boyle and Rose Mary Mechem, cited Marshall's assertion that drought in the Kissimmee Valley—which had approximated a one-in-700 years event in 1981—was "a predictable consequence of the land development and the drainage of wetlands in the Everglades and the Kissimmee River basin." According to the article, Marshall explained that water that flowed from the Kissimmee River Basin to Lake Okeechobee to the Everglades was "the key to the region's abundant rainfall" because vast amounts of it evaporated quickly in the summer and descended in the form of afternoon rain. Marshall claimed that "almost all the water that had risen from the wetlands would come down again," replenishing water supplies. With Kissimmee River channelization and other developments, however, not enough water was available for evaporation, meaning that the "rain machine" could not function as in the past. Boyle and Mecham also quoted Gannon as saying that the "entire [weather] cycle has been altered, weakened and shifted," and "we're setting up a heat regime rather than a rainy regime in the summer period."[12]

After the publication of the *Sports Illustrated* article, the Florida Water Resources Research Center of the University of Florida sponsored a conference on 14 May 1982 to discuss drought, rain, and their causes in Florida. In the course of this meeting, several scientists raised doubts about Gannon and Marshall's theory, noting that the 1981 drought affected all of Florida, not just the Kissimmee River Basin, and that more studies were necessary before anyone could definitively say that channelization provoked drought. Garald Parker, a former hydrologist with the U.S. Geological Survey, who had been quoted by *Sports Illustrated* as supporting Marshall's position, distanced himself from the rain-machine theory, insisting that claims of channelization's effects on climate were "not supported by anything more than a superficial look at hydrology. . . . We know there's a whole lot more work to be done."[13] Gannon himself backed off slightly from his previous position, claiming that his research had focused only on urbanization's effects on Florida's coastal areas and that he had no expertise in Kissimmee River matters. However, discounting any human manipulation of nature, Gannon also noted that "if the entire 3,300 square mile basin was once shallow wetlands and is now no longer so," climate changes "had to have occurred."[14]

For the most part, the rainfall debate diminished after this May 1982 conference, but efforts to dechannelize the Kissimmee River did not. Johnny Jones of the Florida Wildlife Federation continued his lobbying efforts for restoration, telling Senator John Vogt,

chairman of the state senate's Natural Resources Committee that the Corps was deliberately delaying its studies. According to Boyle and Mechem's article, Jones then asked Vogt to propose a bill in the state legislature to use funds under Florida's Conservation and Recreation Lands Act and the Save Our Rivers Act to "start filling that ditch . . . if the feds don't get off their butts." Vogt agreed, concerned that Florida would "become a desert" if "unlimited development and drainage of wetlands" continued.[15]

Others had similar ideas. In February 1982, Nathaniel Reed, former assistant secretary of the interior for fish, wildlife, and parks, requested that state officials designate the Kissimmee River floodplain as an area of critical state concern under the Florida Environmental Land and Water Management Act of 1972. Likewise, Vince Williams, a fishery biologist with the Florida Game and Fresh Water Fish Commission, advocated the designation of the entire Upper Kissimmee River Basin (Lake Kissimmee northward) as an area of critical state concern, in part because the region suffered from significant fish and wildlife decline due to "deteriorating water quality and unregulated residential encroachment."[16]

In the meantime, the SFWMD decided to take matters into its own hands. In May 1982, its governing board approved a plan to install a two-foot-high metal extension on the lift gates of five water control structures separating the Kissimmee River into five pools. In the rainy season, the SFWMD would raise water levels in the pools by two feet, allowing drained marshes to reflood. During the dry season, the SFWMD would reduce each pool's level by one foot below its normal elevation so that the marshes could dry. In part, the SFWMD wanted to see the effects of such reflooding, but its scientists and engineers also believed that the program could "dramatically enhance fish and wildlife habitat."[17] The district noted that its plan, which it hoped to begin in the fall, would cost only $22,000, and it submitted an application for approval to the Coordinating Council for the Restoration of the Kissimmee River. According to John "Jack" Maloy, executive director of the SFWMD, the plan was "a way in which we can easily and inexpensively almost double the river's marshlands without jeopardizing flood control objectives."[18]

Although Graham and other state officials enthusiastically endorsed the SFWMD's plan, not all Florida residents were pleased. Kissimmee Valley ranchers appeared before the SFWMD's board in August and expressed concern with the reflooding. "The plan you are proposing is going to cripple every cattleman on this river-marsh," said Perry Smith, who owned a farm in Okeechobee County.[19] Others agreed; proprietors of McArthur Farms asked for a state administrative hearing because, they claimed, the SFWMD's plan would unconstitutionally prevent them from using their land. Because of these protests, the governing board voted to stop its reflooding plans until, according to one newspaper account, "staff members have the opportunity to further assess what the impact will be on the lands of ranchers on the river."[20]

Agriculturists continued to fight against Kissimmee River restoration in general. "The people we are facing are the environmentalists who want to increase the bird and fish population," said Mike Palmer, who owned a dairy in the Kissimmee Basin. "They want to help ducks and fish and forsake the land animals who have had 10 years to adapt to habitats created" by channelization.[21] Likewise, Paul Wilson, a rancher from Frostproof, insisted that he preferred the straightened river because "it handles the flow of water more efficiently," while Allen Whitston, director of the Upper Chain of Lakes Property Owners Association, claimed that the state's Kissimmee plans used too much "scientific theory" and ignored "historical documented fact."[22]

Despite ranchers' concerns, the move to do something on the Kissimmee River accelerated in 1983, a benchmark year in the push for dechannelization. For one thing, as we have already seen, Governor Graham instituted his "Save Our Everglades" program in August, in part from frustration with the lack of progress on the Kissimmee River. Indeed, one of the major components of the first phase of his program was to revitalize the river, and he called on both the state and the federal government to "recognize the problem and correct the wrong done to the Kissimmee and the people of Florida." Specifically, Graham asked the Coordinating Council on the Restoration of the Kissimmee River Valley to make a firm recommendation as to how "the natural values of the Kissimmee River" could be restored, and he called on President Ronald Reagan to facilitate federal cooperation with the state.[23] "The governor is not going to wait forever for a resolution to these problems," Estus Whitfield, environmental aide to Graham, said. "He wants to start doing something now."[24]

Under this pressure, the Coordinating Council asked the Corps, in the words of Colonel Devereaux, to interrupt its feasibility study and "pull together some options" about how restoration could proceed. According to Devereaux, the council then would study these choices and "decide where they wanted to go."[25] To fulfill the council's needs, the Corps presented it with three options: the do-nothing alternative, where the river would be left alone; the partial backfilling alternative, consisting of refilling a large part of the river with dredged spoil material to allow for marshland reformation; and the combined wetlands alternative, which would leave the channelized river in place, but would develop pockets of wetlands along the watercourse.[26]

Cattle wading through the Kissimmee River. (South Florida Water Management District)

Before making its final decision, the Coordinating Council held a series of public meetings in August.[27] Environmentalists championed the partial backfilling plan, asking the state to move forward with it even if the Corps refused to provide aid, and they disparaged the combined wetlands alternative as "highly structural" and "worse than what's out there now."[28] Driving these statements was an implicit distrust of the Corps' focus on structural solutions for water problems. As reported in an article in *Oceans*, Marshall and others believed that the Corps had "engineered" the state of Florida "nearly to death"; manipulating the system even more through the creation of artificial impoundments was not the answer.[29]

But ranchers and agriculturists in the Kissimmee River Basin expressed their opposition to backfilling, fearing that it would flood their lands. "We were told we would have flood control and our operation is based on that," cattle rancher Pat Wilson said. "With restoration, you want to bring that water right back to our fro[n]t door." Kent Bowen, manager of McArthur Farms, agreed. "We could lose up to 3,000 acres," he protested, and "that would make our ranching operation economically unviable."[30] At the very least, ranchers called on the state to do nothing until the Corps had

completed its feasibility study (now estimated to be finished in the spring of 1984).

The Coordinating Council did not take agriculturists' advice; instead, on 19 August, it declared that, "after careful consideration" of the Corps' preliminary findings, it supported the partial backfilling alternative. "As much of the original channel of the Kissimmee River should be restored as possible," the council stated, and "any alternative which continues the existence and function of the C-38 Canal" should be shelved. The council tempered its decision by saying that it wanted more information about whether or not backfilling would "materially affect existing levels of flood protection in the Upper Kissimmee Basin," but as long as flood control could continue, backfilling was the preferred option. The council also recommended that the state "assume primacy" in restoration efforts even though many state officials believed that the federal government had a "moral obligation" to participate since a federal project had caused the damage in the first place.[31] Unfortunately, "it seems unlikely that the Corps could participate in restoration under the current Administration's policies and guidelines," the council explained, "unless there are quantifiable economic benefits."[32]

Weirs placed in the Kissimmee as part of the Demonstration Project. (South Florida Water Management District)

Acting on the council's recommendations, the SFWMD took the lead in conducting state efforts. One reason for this, according to Stanley Hole, who was elected chairman of the SFWMD's governing board in 1985, was that Graham had replaced members of the board "who [did] not share his environmental commitment to broad restoration." This move, Hole continued, effectively "changed the character of the board," making the SFWMD a "natural resources" district interested in environmental quality. "The most recognizable change," Hole related, "is that we used to say, 'Just tell me where you want the water put,' and then we'd manage it. Now," Hole concluded, "we have to be concerned with the overall effects of everything we do."[33]

The SFWMD was not alone in making such an attitude adjustment. Other flood control districts in the United States, such as the Los Angeles County Flood Control District, also began exhibiting an increased awareness of environmental values in their water management efforts. The Los Angeles County Flood Control District, which had been formed primarily to operate Corps flood control operations on the Los Angeles River, stated as early as 1971 that it wanted "to make our engineers sensitive to possible social and environmental problems of each project."[34] It pledged to consider cultural values, recreation, aesthetics, and the environment in its operations, much like new SFWMD members promised to explore environmental quality measures in South Florida.

But even long-term SFWMD officials, such as Executive Director Jack Maloy, supported the restoration effort. Maloy initially proposed that the district fill in a ten-mile stretch of the river, at a cost of between $400,000 and $700,000, to observe whether positive ecological conditions would return.[35] On 9 September 1983, the governing board of the SFWMD met to discuss Maloy's plans, eventually adopting it as the best method to follow. Graham concurred on 11 September after meeting with Florida's congressional delegation and with authorities in the Reagan administration. Under Maloy's plan, the SFWMD would place a weir at the south end of Pool B of C-38, effectively "plugging" the pool, and then refill approximately four to eight miles of the river between S-65A and S-65B. The SFWMD proposed to begin constructing the weir on 1 January 1984, estimating a completion date of two years for the entire project. After that time, the district would "monitor and evaluate" the "environmental impacts and benefits," as well as how the reflooding affected area land use and whether it had detrimental consequences on flood control in the region north of Lake Kissimmee.[36]

Before any construction could begin, the SFWMD had to receive both state and federal permits for the project. After the SFWMD had submitted its applications to the State Department of Environmental Regulation and the Corps, the Latt Maxcy Corporation, a large cattle company in the region, filed a protest, arguing that the Corps had not finished its studies on restoration and that the SFWMD merely wanted to "dump 4,519,898 cubic yards of silty sands" in the river "without knowing the effect, method, and

cost." The corporation also contended that the demonstration project would forestall navigation of the Kissimmee, and that it would "destroy the biota and habitat" that had developed after the Corps straightened the river. Finally, the company argued, its operations "relied on the permanence of the canal" and any restoration efforts would "adversely" affect its land rights.[37]

It is unclear how much influence the protests of Latt Maxcy and other agriculturists had, but the Corps eventually rejected the SFWMD's application. According to Colonel Deveraux, the denial occurred for several reasons. First, Corps officials believed that the demonstration project was large enough to require an EIS, something that would take at least a year to produce. Second, Devereaux explained, backfilling any part of the river would alter the navigability and flood control intent of the Corps' original Kissimmee project, and that could not occur without congressional approval. Most importantly, Devereaux said, the project only "put dirt back in the ditch" and "did not generate any wetlands," meaning that it did not fulfill "many of the State objectives."[38]

Environmentalists, however, saw the action as more evidence that the Corps did not want to dechannelize the Kissimmee. According to Estus Whitfield, environmental aide to Graham, Corps officials were "quite reticent and not too thrilled with the state's and the South Florida Water Management District's exuberance to go out and fill in C-38." One problem, Whitfield stated, was that "some of the [engineers] who designed the Kissimmee channel were still there" and did not want to undo it.[39] Whitfield had a point; Devereaux himself characterized advocates for complete restoration as "starry-eyed folks" and claimed both privately and publicly that the combined wetlands alternative was the only feasible option.[40]

Regardless of the reasons for the permit's denial, Corps officials, including Deveraux, Assistant Secretary of the Army (Civil Works) William Gianelli, and Director of Civil Works Major General John Wall met with state and SFWMD authorities in January 1984 to develop an alternative plan. In February, the SFWMD proposed a new program. Under this plan, known as Kissimmee River Restoration Phase I, the SFWMD would place three metal sheet pile walls in Pool B in order to divert water into the river's natural channels. A navigation notch would be placed in the walls so that boats could continue to navigate the river, and the Corps would construct baffle blocks on structures S-65B, C, and D so that it could manipulate the river's water levels. If the state legislature approved the plan, the SFWMD proposed to begin work on the approximately $1.2 million project in the spring of 1984.[41]

Many SFWMD officials saw the new demonstration project as a way to determine "once and for all" whether it was "realistic" to restore the Kissimmee River to its "meandering, natural" state.[42] Jan Horvath, director of the SFWMD's Resource Coordination Department, also thought that the project could demonstrate the best way for the state to manage the floodplain, as well as show how

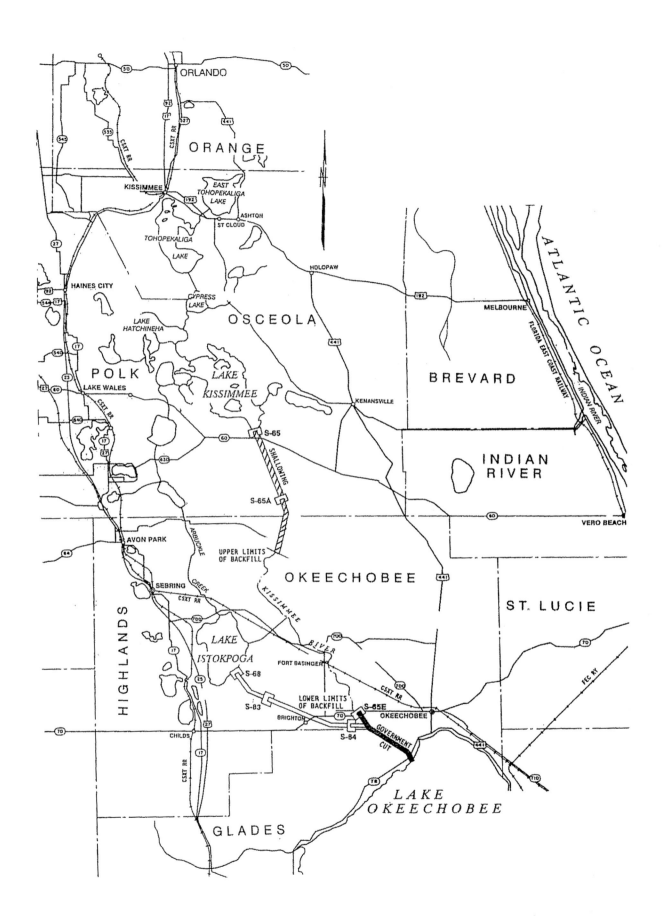

S-65 structures on C-38. (U.S. Army Corps of Engineers, Jacksonville District)

restoration would impact residents along the river. Yet others, including SFWMD Assistant Executive Director John Wodraska, worried that the SFWMD did not have enough information about the demonstration project's impacts, and he wondered if it would just create another "environmental disaster."[43]

A sign marks Phase 1 of the Demonstration Project. (South Florida Water Management District)

Meanwhile, the state took other measures to aid in Kissimmee River restoration. In November 1983, Graham issued an executive order creating the Kissimmee River-Lake Okeechobee-Everglades Coordinating Council to coordinate state and regional endeavors that would "restore and enhance the natural values and functions" of the Kissimmee-Okeechobee-Everglades ecosystem.[44] The council, which would basically have charge of Graham's "Save Our Everglades" program, would consist solely of state officials, including the secretaries of the departments of Environmental Regulation, Community Affairs, and Transportation; the commissioner of the Department of Agriculture and Consumer Services; and the executive directors of the Department of Natural Resources, the SFWMD, and the Game and Fresh Water Fish Commission. Likewise, these same entities concluded a memorandum of agreement on 1 November 1983, pledging to cooperate in dechannelization efforts and delineating specific responsibilities for each agency.[45]

One of the duties that the SFWMD assumed was purchasing floodplain lands in the Kissimmee Basin, both to expedite the demonstration project and to prevent further development in the area. In March 1984, the district's governing board decided to pursue a purchase plan for 40,000 acres in the Kissimmee Basin, even though the state would have to provide at least $40 million for that action.

According to board member Nathaniel Reed, price should not be a factor; "no blinking," he stated, "regardless of the financial crunch."[46] SFWMD officials proposed to fund the effort at least partly through the Water Management Lands Trust Fund, which was created by the state's Save Our Rivers Act of 1981 to allow the purchase of lands needed to conserve and protect water resources.

Yet an obstacle arose to both the purchasing plan and to the demonstration project. In March 1984, McArthur Farms Inc., Save Our Waterways Association, and Riley Miles (a Kissimmee resident and former SFWMD board member) opposed the SFWMD's new permit application to the Department of Environmental Regulation, charging, in the words of one newspaper account, that the demonstration project would "cause temporary and longterm pollution of the river" and would "drastically decrease the river's navigability."[47] Accordingly, the state held hearings on the application. In the course of these hearings, the SFWMD told concerned parties that the project would not adversely affect either navigation or flood control. The district also related the necessity of the demonstration project in order to determine exactly how restoring the river would impact the basin and whether or not changes in flora and fauna would occur. The project's purpose, the SFWMD reiterated, was to see whether "the historical ecological function of the river" could be restored through "the overall management of water, fish, and wildlife," in hopes that "further degradation of water quality" could be prevented and wildlife habitat restored.[48] The SFWMD's responses seemed to satisfy both the Corps and the Department of Environmental Regulation, and they issued permits for the Phase I work on 9 July and 29 July 1984.[49]

But even though the SFWMD took the lead on the demonstration project, many of its officials still had viewpoints that differed from the opinions of other state authorities. In May 1984, for example, Estus Whitfield composed a draft outlining the Kissimmee River restoration program and why it was necessary. John Wodraska, who had become executive director of the SFWMD upon Maloy's resignation, took issue with some of Whitfield's statements. Whereas Whitfield claimed that channelization of the Kissimmee caused much ecological destruction, Wodraska held that actual construction work caused some damage, but that the system had "healed" since that time and was now "a stabilized ecosystem."[50] Likewise, Whitfield insisted that channelization was not necessary for flood protection in the Upper Kissimmee Basin (since improving structures in the area could accomplish the same purpose), but Wodraska disagreed, stating that channelization provided "necessary 'getaway' for floodwaters from the upper basin." Finally, Whitfield asserted that a diminishment in the river's water quality resulted from channelization, while Wodraska claimed that degradation occurred because of "the development of intensive land use practices" rather than from a "reduction in wetlands."[51] Although both Whitfield and Wodraska agreed that some form of action was necessary to enhance ecological values in the Kissimmee Basin, they disputed how degraded the environment was and why that had occurred.

Such disconnect in views became more apparent in August 1984 when the Corps finally released its draft feasibility report on Kissimmee River restoration. This document noted that because of state requests, Congress had directed the Corps in 1978 to determine whether modifications to the congressionally authorized Kissimmee River project were "advisable." In discussing this question, the Corps noted that the construction of C-38 had "reduced flooding and enabled more intense land use," which some believed had led to "a number of adverse environmental effects." The Corps contended with claims that channelization had accelerated eutrophication of Lake Okeechobee, stating that the bigger problem was the Taylor Creek-Nubbin Slough area, which contributed most of the phosphorus to the lake, and the EAA, which supplied most of the nitrogen. "There is little evidence to suggest that water quality has been degraded in the Kissimmee basin as a result of channel modification," the report declared, "or that C&SF Project works . . . have accelerated the eutrophication of Lake Okeechobee."[52] These, of course, were arguments that the Corps had been making since the mid-1970s.

Aside from its conclusions regarding the condition of the Kissimmee River and its effects on Lake Okeechobee, the Corps discussed different alternatives that it could take to allow for at least a partial restoration of ecological conditions along the Kissimmee River. The six that it found feasible included partially backfilling C-38; constructing controlled wetlands; having agriculturists implement Best Management Practices (BMPs) to decrease the amount of nutrients in runoff; creating impounded wetlands at various points along the river; manipulating pool stages to increase wetlands; and restoring wetland conditions to Paradise Run, an eight-and-a-half mile stretch of the river in the southern part of the Kissimmee floodplain. However, the Corps had serious reservations about several of these, including the partial backfilling plan. Although environmentalists, state officials, and the SFWMD had all embraced either partial or full backfilling as the best program, the Corps disagreed. For one thing, it believed that backfilling would increase flooding risks in the Lower Kissimmee Basin. The Corps also claimed that partial backfilling could actually reduce the number of wetland acres, in part because not enough water would exist "to attain a natural hydroperiod." Indeed, the Corps asserted, it would produce only "a semi-natural riverine system."[53]

Kissimmee River restoration project area. (South Florida Water Management District)

Instead of partial backfilling, the Corps recommended the BMP, Paradise Run, and pool stage manipulation options because they would produce "the greatest benefit at the lowest cost." Yet the Corps claimed that it could not participate in these programs because "while generally beneficial for environmental concerns," they would not "contribute to the nation's economic development." Moreover, the Corps explained, if the state wanted to initiate the partial backfill plan or Phase II of the Demonstration Project, the Corps would have to obtain congressional authorization since these actions would "significantly alter the flood control and navigation purposes of the Kissimmee River project." The Corps believed that its report contained useful information that the state could use in developing its own restoration efforts, but from the Corps' perspective, "there is no basis for Federal implementation of modifications to the Kissimmee River Basin."[54]

Upon examining the draft feasibility report, state officials wondered about some of the Corps' conclusions. Governor Graham was especially concerned about the Corps' unfavorable partial backfilling assessment, fearing that it would "impede the State's restoration efforts," and he disagreed with the Corps' recommendation against federal participation.[55] In response, the Corps emphasized that it was required by law to recommend the plan that had "the greatest net economic benefit" and that it could find "no basis for Federal implementation of project modifications." According to the Corps, the Kissimmee River project was "functioning as designed," and altering it through partial backfilling would reduce "existing and future" economic project benefits, while producing only "limited environmental benefits." The Corps therefore had no "overriding reason" to suggest implementation of partial backfilling.[56] This view did not change; when the Corps issued its final feasibility report in September 1985, its conclusions were largely the same as in the draft report, although, based on comments from state agencies, it did revise upward its estimate of wetlands acreage produced by partial backfilling. Regardless, the report still stated that BMPs, pool stage manipulation, and restoring wetlands at Paradise Run provided the best economic benefits, and it maintained that no federal action was warranted.[57]

To many environmentalists and state authorities, this was just another example of the Corps stonewalling the issue because it was not interested in restoring the Kissimmee River. That charge, although technically true, did not take into account all of the nuances of the situation. Some Corps officials, such as Devereaux, were clearly against complete restoration. "The Kissimmee River is a man-changed system now, and it will always be one as far as I can see," he stated in 1984. "I don't see any reasonable way that it can ever go back to doing what people refer to as a pure kidney function."[58] Colonel Charles Myers III, who replaced Devereaux as District Engineer, agreed. "There are people in the Kissimmee Valley benefiting from the valley as it now exists," he said. "There's no way we can

back up to 1900."[59] Whether Devereaux and Myers took this position because they did not want to admit that channelization had been a mistake, or whether they truly believed that it was not possible to return the river to a natural state is unclear. On the one hand, the Corps' position was technically correct: it could not do anything to alter the original purpose of channelization without authorization from Congress, nor could it recommend a project if economic benefits did not justify it. The main problem, however, was that the Corps did not pursue restoration with enthusiasm, or support the idea in a meaningful way, effectively preventing the issuance of any congressional "authorization." In the eyes of many environmentalists and state officials, the agency was merely hiding behind its operating regulations to get what it truly wanted—the maintenance of C-38.[60]

With the Corps unwilling to participate in any restoration efforts, the state of Florida and environmentalists laid the groundwork for their own endeavors. In August 1984, Governor Graham oversaw the beginning of the SFWMD's demonstration project by symbolically planting a baby cypress tree on the banks of the Kissimmee River. He declared that the state's goal in the endeavor was that "by the year 2000, the water system will look and function more as it did in the year 1900 than it does today."[61] Graham also continued to call for federal participation in Kissimmee restoration, and environmental organizations sought to repair the breaches in the Everglades Coalition, banding together again in order to stimulate public involvement in South Florida ecological issues.[62]

At the same time, Graham adopted a seven-point plan for Kissimmee River restoration, beginning with Phase I of the demonstration project. Other steps included restoring wetlands in the Paradise Run area; expanding the Best Management Practice program to include not only Taylor Creek-Nubbin Slough, but also the lower Kissimmee River; developing modeling systems to measure hydraulic and sediment transport effects of restoration endeavors; and acquiring 50,000 total acres of the Kissimmee floodplain.[63] As part of this plan, Graham dismantled the Kissimmee-Okeechobee-Everglades Coordinating Council and gave responsibility for all restoration aspects, such as land acquisition, physical modeling, and the development of restoration alternatives, to the SFWMD.[64]

To help the SFWMD in its endeavors, Graham also established a 34-member Kissimmee River Resource Planning and Management Committee—composed of individuals from local, state, and federal agencies, including the Corps and the SFWMD—to review land and water problems in the Lower Kissimmee and Taylor Creek basins. In August 1984, Graham directed the committee to focus on land use management, land acquisition, water quality protection, and economic development in its examinations; by doing so, he hoped that the state could "guarantee the long-term health of the [Kissimmee] river system."[65]

One of the first tasks that the committee undertook was investigating land acquisition. This was important not only for the

Citrus groves near the Kissimmee River. (South Florida Water Management District)

demonstration project to occur, but also because of continuing agricultural encroachment into the Kissimmee floodplain, hastening drainage of the region. One account reported that between 1958 and 1972—the era when the Corps was straightening the river—agriculturists drained over half of the unimproved land in the region and planted it to Bahia grass for grazing. Then, in the early 1980s, citrus growers considerably increased their holdings in the Kissimmee area. Because of these endeavors, according to naturalist Ted Levin, "land that once spawned bobcats and sandhill cranes now [grew] cattle and oranges."[66]

To forestall further development, the Resource Planning and Management Committee proposed to develop "a workable land use strategy" to protect the river and allow for its restoration.[67] In January 1985, the committee sent to the state seven land management options that it considered viable. These ranged from doing nothing to acquiring land in fee simple to recommending that counties and the city of Okeechobee adopt a comprehensive land management and zoning plan.[68] After receiving these suggestions, the SFWMD decided to continue with its goal of purchasing 50,000 acres of the Kissimmee floodplain, and in January 1985, it bought 7,500 acres with funds provided under the Save Our Rivers Act. The district stated that public

management of half-mile strips of land on both sides of C-38 would be necessary for restoration to succeed, as well the acquisition of an additional 42,500 acres to protect the entire floodplain. According to Executive Director Wodraska, the purchase was "a giant stride" that would allow the SFWMD to see "if we can coax Nature to reestablish some of her lost beauties into the river's marshes."[69]

But as the end of 1985 approached, it was clear that, unless a change of attitude occurred, the state would have to generate any restoration effort without federal involvement. Colonel Charles T. Myers III, District Engineer of the Jacksonville District, for example, presented the Corps' final feasibility report to the Board of Engineers for Rivers and Harbors, recommending in person that no federal action be taken. According to Myers, a District Engineer usually did not present negative reports to the Board, but because the Jacksonville District's decision was "a very controversial" one "that necessitated lots of discussion," he believed it was necessary.[70]

After receiving the report, the Board deliberated on the recommendation, while Governor Graham lobbied for federal involvement. "As the State of Florida pursues its goal of restoring the Kissimmee River," he told the Board, "we will seek federal approval of and participation in this project." Graham claimed that the channelization of

Kissimmee River. (South Florida Water Management District)

the river had decreased the basin's original wetlands by "70 to 80 percent," and that this had led to degradation of water quality and loss of wetland habitat. Therefore, "just as the Corps has been a partner with the State in flood control, water supply, navigation, and other public works projects," Graham wanted it also to participate "in our new mission of environmental enhancement." Although the state could pledge "a great many dollars" towards Kissimmee River restoration, it still needed federal help in order to make a final restoration plan viable. Graham asked the Board to overturn the Jacksonville District's no federal participation recommendation, and he pledged to "work closely with the Corps" to develop "a specific restoration plan."[71] Despite Graham's efforts, the Board ultimately agreed with the District's decision, and in July 1987, Chief of Engineers Lieutenant General E. R. Heiberg III transmitted a report to Congress, stating that it was "not advisable" for the Corps to participate in project modifications "in the interest of water quality, flood control, recreation, navigation, loss of fish and wildlife resources, environmental problems, and loss of environmental amenities." Instead, Heiberg recommended that District Engineer Myers "continue to cooperate with the State of Florida under his existing authorities."[72]

In the meantime, Congress had passed the Water Resources Development Act of 1986 (WRDA-86), which authorized approximately $16 billion worth of water projects. Along with mandating cost sharing between local and federal interests on water projects,

the law also contained a section significant to the Kissimmee River controversy.[73] Riding the wave of environmental concerns with water resource development, Congress included Section 1135 in WRDA-86, authorizing the Corps to review existing projects and to "determine the need for modifications" in those projects in order to "improv[e] the quality of the environment in the public interest."[74] If the Corps made any modifications, the law directed, non-federal interests would pay 25 percent of the total cost.

Florida officials tried to get the Corps to undertake restoration of the Kissimmee River under the authority granted by Section 1135. In 1987, according to an Everglades status report issued by the governor's office, Governor Robert "Bob" Martinez, a Republican who had replaced Graham that same year (Graham had won an election bid for the U.S. Senate), informed Acting Assistant Secretary of the Army (Civil Works) John Doyle of "Florida's strong desire to restore the values of the Kissimmee River." Martinez asked Doyle to consider the Kissimmee "as it makes plans for implementing Section 1135."[75] Florida's congressional delegation, which now included Graham, requested the Corps to take the same action, but the politicians were not alone. Indeed, environmentalists, led by the Sierra Club and Theresa Woody, its Florida representative, made a push for Kissimmee River authorization under Section 1135. Their position was strengthened when the environmental community agreed that the only project it would request under Section 1135 was Kissimmee River restoration.[76] The Jacksonville District, led by Colonel Robert L. Herndon, District Engineer, nominated the project for Section 1135 consideration, but when it went to the Secretary of the Army (Civil Works) for approval, the Reagan administration determined that, according to Herndon, it was an "inappropriate use of federal funds to conduct such an environmental demonstration" and refused to transmit the request to Congress.[77] Regardless, Congress included $2 million in its 1988 fiscal year budget for a Corps Kissimmee River demonstration project. Unfortunately, the executive branch's Office of Management and Budget never allocated funds for that purpose, and Herndon was left to face environmentalist blame. "I would be more than willing to carry out environmental enhancement features of the Kissimmee River," Herndon related in 1989, but until he received authorization to use money for that purpose, "my hands are rather well-tied."[78]

In addition to this setback, some disagreements surfaced between environmentalists and state authorities as to what restoration meant. To people such as Richard Coleman, who spearheaded grassroots efforts supporting dechannelization, it meant "restoring [the Kissimmee] to what it was before, bend-for-bend, acre-for-acre."[79] State officials were not so sure. Louis Toth, who headed up the SFWMD Demonstration Project, defined restoration as "restoring a functioning ecosystem."[80] Stanley Hole, chairman of the SFWMD's governing board in the mid-1980s, agreed. "We can't just go in there and fill the [flood canal], no matter how the environmentalists cry for it," Hole stated. Instead, the SFWMD would try to "restor[e] the values the river offered in its pristine state without sacrificing the navigational and recreational benefits that channelization brought about."[81]

Despite these disagreements, the state and environmentalists had achieved some success on the Kissimmee front. Faced with a presidential administration largely uninterested in environmental quality, and with a Corps of Engineers that was, at best, unable to participate in restoration efforts and, at worst, dragging its feet because it did not want to dechannelize the Kissimmee, Governor Bob Graham and the SFWMD pushed Kissimmee restoration along. Because of the demonstration project (the construction of which the SFWMD had completed by 1986), the state now had a mechanism in place to observe how the environment would react if restoration occurred, and it had fully dedicated state resources to dechannelization. This commitment continued even when the Republican Martinez assumed the governorship from the Democrat Graham. With dechannelization, the state had taken its first steps along the road of ecosystem restoration, and it would move farther down that path in the 1990s.

Endnotes

[1] See "The 'Sewer Ditch' Undone," *Audubon* 89 (March 1987): 114.

[2] U.S. Army Corps of Engineers, Jacksonville District, *Kissimmee River Study Including Taylor Creek—Nubbin Slough Basins* (Jacksonville, Fla.: U.S. Army Corps of Engineers, 1980).

[3] Jacksonville District, *Kissimmee River Study Including Taylor Creek—Nubbin Slough Basins.*

[4] Arthur R. Marshall, "Repairing the Florida Everglades: A Brief Update to the members of the Coalition to Repair the Everglades," 2 April 1983, Folder 5, Box 1, Marshall Papers.

[5] "Report Backs Kissimmee River Restoration," *The News Tribune*, December 13, 1981; Steve Yates, "Marjory Stoneman Douglas and the Glades Crusade," *Audubon* 85 (March 1983): 118.

[6] "The Problem & The Plan," *For the Future of Florida: Repair the Everglades* 2 (1981): 2.

[7] As cited in "Kissimmee 'Disaster': Graham Asks Corps To Speed Up Repairs," *The Tampa Tribune*, 1 March 1982.

[8] Colonel Alfred B. Devereaux interview by George E. Buker, 23 April 1984, Jacksonville, Florida, 58-59, transcript in Library, Jacksonville District, U.S. Army Corps of Engineers, Jacksonville, Florida [hereafter referred to as Devereaux interview].

[9] "Political Fix For Wetland Woes," *ENR* 186 (2 September 1982): 33.

[10] Stine, "Environmental Politics and Water Resources Development," 65.

[11] Patrick Thomas Gannon, Sr., "On the Influence of Surface Thermal Properties and Clouds on the South Florida Sea Breeze" (Ph.D. diss., University of Miami, 1977).

[12] All quotations in Robert H. Boyle and Rose Mary Mechem, "Anatomy of a Man-Made Drought," *Sports Illustrated* 56 (15 March 1982): 46-48.

[13] As quoted in Margaret Yansura, Public Information Office, South Florida Water Management District, to Executive Director, Deputy Executive Director, and all Department Directors, n.d., File Kissimmee River, Box 17166, SFWMDAR.

[14] As quoted in Ron Mierau, "Trip Report—May 14, 1982, Regional Influence of Drainage on the Hydrologic Cycle in Florida," 19 May 1982, File Kissimmee River, Box 17166, SFWMDAR.

[15] As quoted in Boyle and Mechem, "Anatomy of a Man-Made Drought," 48.

[16] Quotation in Vince Williams, Fishery Biologist, to Mr. Richard Coleman, Polk County Sierra Group, 17 February 1982, Folder 8, Box 3, Marshall Papers; see also Nathaniel P. Reed to Victoria Tschinkel, et al., 2 February 1982, Folder 43, Box 1, ibid.

[17] South Florida Water Management District New Release, 24 May 1982, File Kissimmee River Restoration (File #1), Box 17166, SFWMDAR.

[18] South Florida Water Management District New Release, 24 May 1982.

[19] "Kissimmee Flood Plan Stalled After Protests," *Fort Lauderdale News and Sun-Sentinel*, 14 August 1982.

[20] "River Raising Halted," *Okeechobee News*, 20 August 1982.

[21] Quotation in "Ranchers: Kissimmee River Level Fine as Is," *The Palm Beach Post*, 5 September 1982; see also "Kissimmee Plans for Restoration Slowed by Corps?" *The Palm Beach Post*, 4 September 1982.

[22] Quotations in "Water-Management Fight Simmers Along Kissimmee," *The Miami Herald*, 6 March 1983.

23 Quotation in "Issue Paper: Save Our Everglades," 9 August 1983, 5, copy in Folder 16, Box 2, Marshall Papers; see also Bob Graham, Governor, to Honorable Ronald Reagan, 8 August 1983, ibid.

24 As quoted in "Graham Seeking Rescue Blueprint for Everglades," *The Palm Beach Post*, 8 April 1983.

25 Quotation in Devereaux interview, 59; see also Buker, *The Third E*, 85.

26 Coordinating Council on the Restoration of the Kissimmee River Valley and Taylor Creek—Nubbin Slough Basin, "Significant Findings Adopted by Majority Vote on 8/19/83," File Kissimmee River Restoration History (File #1), Box 17166, SFWMDAR; "Efforts to Restore Kissimmee Called Political," *The Miami Herald*, 10 August 1983; Devereaux interview, 59.

27 Coordinating Council on the Restoration of the Kissimmee River Valley and Taylor Creek—Nubbin Slough Basin, "Kissimmee River Restoration: Public Hearing Discussion Paper, Public Meetings, August 1983," File Kissimmee Valley Planning Area, Box 17166, SFWMDAR.

28 First quotation in "Battle Lines Flank the Kissimmee's Banks," *The Miami Herald*, 8 August 1983; second quotation in David S. Anthony, "Kissimmee River Survey Review: Reasons for Preferring Partial Backfilling Alternative," Folder 9, Box 3, Marshall Papers.

29 Sue Douglas, "Save the Everglades," *Oceans* 18 (March/April 1985): 4.

30 Both quotations in "Angry Ranchers Oppose River Project," *The Palm Beach Post*, 11 August 1983; see also "Water, Growth Hinge on Plan," *The Palm Beach Post*, 11 August 1983; "Governor Sees the Caged and Free Kissimmee," *The Miami Herald*, 11 August 1983; "Troubled Waters: Panel to Consider Fate, Direction of Kissimmee River," *Fort Lauderdale News and Sun-Sentinel*, 7 August 1983.

31 Quotations in Coordinating Council on the Restoration of the Kissimmee River Valley and Taylor Creek—Nubbin Slough Basin, "Council Recommendations on Kissimmee River Restoration, Adopted by Unanimous Vote on August 19, 1983," File Kissimmee River Restoration History (File #1), Box 17166, SFWMDAR; see also "Kissimmee Restoration Gets Boost," *The Miami Herald*, 20 August 1983.

32 Coordinating Council on the Restoration of the Kissimmee River Valley, "Kissimmee River Restoration: Public Hearing Discussion Paper."

33 As quoted in Douglas, "Save the Everglades," 4. Hole himself came from an interesting background, as he was previously an engineer for the Gulf American Corporation, the promoter of Golden Gate Estates. After leaving the corporation to become chairman of Collier County's water management advisory board, he continued to serve as a consultant to Gulf American. Carter, *The Florida Experience*, 241.

34 As quoted in Jared Orsi, *Hazardous Metropolis: Flooding and Urban Ecology in Los Angeles* (Berkeley: University of California Press, 2004), 130-131.

35 "Graham Applauds Decision on Kissimmee Restoration," *The Miami Herald*, 20 August 1983; "Kissimmee Council Decides to Undo Channelization," *The Stuart (Fla.) News*, 21 August 1983. Maloy later related that he came back from a Coordinating Council meeting, telephoned the SFWMD's director of operation and maintenance, and told him, "Bill, I want you to put a weir across the Kissimmee Channel." When the director expressed some disbelief, Maloy continued, "Let's see whether or not we can really impact the reestablishment of the oxbows and the way the river actually ran." See Maloy interview, 5-6.

36 Quotations in "Kissimmee River Demonstration Project," File Kissimmee River Demonstration Project, 1983 (File #1), Box 17166, SFWMDAR; see also Buker, *The Third E*, 86.

37 "Comments Submitted to the Department of Environmental Regulation by the Latt Maxcy Corporation," 31 October 1983, File Kissimmee River Demonstration Project, 1983 (File #1), Box 17166, SFWMDAR.

38 Devereaux interview, 61.

39 Whitfield interview, 20.

40 Devereaux interview, 58, 62.

41 Will Abberger and Estus Whitfield, *Save Our Everglades: The Kissimmee River* (Tallahassee, Fla.: State of Florida, Office of the Governor, Office of Planning Budgeting, Natural Resources, Unit, 1985), 27-28; see also "WMD Seeks Permits for River Restoration," *South Florida Water Management District Bulletin* 9 (Winter 1983): 1.

42 Jan Horvath, Director, Resource Coordination Department, to Mr. W. H. Morse, Chairman, Osceola Waterways Committee, Kissimmee/Osceola County Chamber of Commerce, 30 December 1983, File Kissimmee River Demonstration Project (File #1), 1983, Box 17166, SFWMDAR.

43 As quoted in "Saving the Everglades," *The Times-Union and Journal*, 8 January 1984.

44 State of Florida, Office of the Governor, Executive Order No. 83-178, 4 November 1983, copy in Abberger and Whitfield, *Save Our Everglades: The Kissimmee River*, Attachment 12, 100-106.

45 Memorandum of Agreement Between Florida Department of Environmental Regulation and the South Florida Water Management District and the Florida Game and Fresh Water Fish Commission, and the Florida Department of Agriculture and Consumer Services, and the Florida Department of Natural Resources, and the Florida Department of Community Affairs Concerning the Restoration of the Kissimmee River, 1 November 1983, copy in Abberger and Whitfield, *Save Our Everglades: The Kissimmee River*, Attachment 11, 89-99.

46 Quotations in "District Will Push Restoring of River," *Fort Lauderdale News and Sun-Sentinel*, 17 March 1984.

47 "SFWMD May Buy Kissimmee Flood Plain," *Fort Lauderdale News and Sun-Sentinel*, 16 March 1984.

48 Fred Schiller, Director, Community Relations Division, to John P. Clark, Public Issue Specialist, Community Relations Division, 28 June 1984, File Kissimmee River Demonstration Project (File #1), 1983, Box 17166, SFWMDAR.

49 Abberger and Whitfield, *Save Our Everglades: The Kissimmee River*, 28.

50 John R. Wodraska, Executive Director, to Mr. Estus Whitfield, Office of the Governor, 6 February 1985, File Kissimmee Restoration History (File #1), Box 17166, SFWMDAR.

51 Both quotations in Wodraska to Whitfield, 6 February 1985; see also "Draft—May 29, 1984, Kissimmee River Restoration," File Kissimmee Restoration History (File #1), Box 17166, SFWMDAR.

52 U.S. Army Corps of Engineers, Jacksonville District, *Central and Southern Florida, Kissimmee River: Executive Summary* (Jacksonville, Fla.: U.S. Army Corps of Engineers, 1984), 2-5, 17 [hereafter referred to as 1984 Executive Summary].

53 1984 Executive Summary, 6-10.

54 All quotations in 1984 Executive Summary, 22-23; see also Stuart Appelbaum interview by Brian Gridley, 22 February 2002, 8, Everglades Interview No. 11, Samuel Proctor Oral History Program, University of Florida, Gainesville, Florida.

55 As cited in untitled memorandum, 4 June 1985, File 1517-08 (Kissimmee River-Lake Okeechobee, FL-12222), Jan. 1985-Dec. 1985, Box 25, Accession No. 077-96-0033, RG 77, FRC.

56 Untitled memorandum, 4 June 1985.

[57] See U.S. Army Corps of Engineers, Jacksonville District, *Central and Southern Florida: Kissimmee River, Florida* (Jacksonville, Fla.: U.S. Army Corps of Engineers, 1985), i-ii; Kissimmee River-Lake Okeechobee-Everglades Coordinating Council, "Save Our Everglades: Annual Summary Report," January 1986, i.

[58] Devereaux interview, 67.

[59] As quoted in Ronald A. Taylor, "Saving a Fountain of Life," *U.S. News and World Report* 100 (24 February 1986): 64.

[60] See Buker, *The Third E*, 86-87.

[61] As quoted in Natalie Angier, "Now You See It, Now You Don't," *Time* 124 (6 August 1984): 56.

[62] See Robert Pierce, "South Florida's Land Puzzle: Federal, State, and Private Agencies Purchase Protection," *National Parks* 59 (July/August 1985): 17.

[63] Kissimmee River-Lake Okeechobee-Everglades Coordinating Council, "Save Our Everglades: Annual Summary Report," January 1986, 7.

[64] Attachment to Jon Glogau, Office of the Attorney General, Department of Legal Affairs, to Peter Antonacci, 17 September 1991, File PRO Background, Kissimmee River Restoration, Box 21213, SFWMDAR.

[65] Quotations in "Graham Panel to Study Growth Problems in Kissimmee River Region," State of Florida, Office of the Governor Press Release, 20 August 1984, File Kissimmee River Resource Planning and Management Committee, Box 17166, SFWMDAR; see also Bob Graham, Governor, to Colonel Charles T. Myers III, District Engineer, U.S. Army Corps of Engineers, 14 August 1984, File 1517-08 (Kissimmee River-Lake Okeechobee, Fla.) Study 12222, Box 25, Accession No. 077-96-0033, RG 77, FRC.

[66] Ted Levin, *Liquid Land: A Journey Through the Florida Everglades* (Athens: The University of Georgia Press, 2003), 238-239.

[67] "Issue Paper: Kissimmee River Resource Planning and Management Committee," August 1984, 4, File Kissimmee River Resource Planning and Management Committee, Box 17166, SFWMDAR.

[68] "Kissimmee River Resource Planning and Management Committee Land Acquisition Strategy Subcommittee Expanded Options A Through G," attachment to John P. Clark, Public Issue Specialist, Resource Coordination Department, to Mr. Dwaine T. Raynor, Bureau of State Land Planning, 10 January 1985, untitled file, Box 17166, SFWMDAR.

[69] South Florida Water Management District News Release, 23 January 1985, File Kissimmee River, Box 17166, SFWMDAR.

[70] Colonel Charles T. Myers III interview by George E. Buker, 30 December 1987, Jacksonville, Florida, 30, transcript in Library, Jacksonville District, U.S. Army Corps of Engineers, Jacksonville, Florida.

[71] Bob Graham, Governor, to Colonel John W. Devens, Resident Member, Board of Engineers for Rivers and Harbors, 28 January 1986, File 1517-08 (Kissimmee River-Lake Okeechobee, FL-12222), Jan. 1986-Dec. 1986, Box 25, Accession No. 077-96-0033, RG 77, FRC.

[72] Quotations in E. R. Heiberg III, Lieutenant General, USA, Chief of Engineers, to The Secretary of the Army, 6 July 1987, File 10-1-7a (Kissimmee River-Lake Okeechobee, FL) 12222, Box 25, Accession No. 077-96-0033, RG 77, FRC; see also Buker, *The Third E*, 87.

[73] For a thorough discussion of the Water Resources Development Act of 1986 and its significance, see Reuss, *Reshaping National Water Politics*.

[74] Act of 17 November 1986 (100 Stat. 4082, 4251).

[75] Quotations in "Everglades Status Report," 12 January 1988, File Everglades, Box 88-02, S1331, Executive Office of the Governor, Brian Ballard, Director of Operations, Subject Files, 1988, FSA; see also Dale Twachtmann, Secretary, Florida Department of Environmental Regulation, to Honorable Robert K. Dawson, Assistant Secretary of the Army (Civil Works), 11 February 1987, File LO Major Programs, Correspondence, Background, LOSAC, Box 18060, SFWMDAR.

[76] See Theresa Woody, "Grassroots in Action: The Sierra Club's Role in the Campaign to Restore the Kissimmee River," *Journal of North American Benthological Society* 12, No. 2 (1993): 203; Theresa Woody interview by Theodore Catton, 18 January 2005, Naples, Florida, 1; Theresa Woody, Southeast Associate Field Representative, Sierra Club, to John R. Wodraska, Executive Director, South Florida Water Management District, 5 June 1987, File LO Major Programs, Correspondence, Background, LOSAC, Box 18060, SFWMDAR.

[77] Colonel Robert L. Herndon interview by Joseph E. Taylor, 26 May 1989, Jacksonville, Florida, 19, transcript in Library, Jacksonville District, U.S. Army Corps of Engineers, Jacksonville, Florida [hereafter referred to as Herndon interview].

[78] Quotation in Herndon interview, 19; see also Buker, *The Third E*, 87-88.

[79] As cited in Glass, "Rebirth of a River," 13-14.

[80] As cited in Glass, "Rebirth of a River," 13-14.

[81] As quoted in Douglas, "Save the Everglades," 5.

CHAPTER 9 Lake Okeechobee II: Science in a Race with Politics and Nature

THROUGHOUT THE 1980s, two other problems, inextricably tied to the Kissimmee, simmered in Florida: how to improve the water quality of Lake Okeechobee and how to regulate the lake's stage in order to protect its littoral zone. Although studies were conducted and recommendations made about the littoral zone, most of the focus in the time period was on water quality. In December 1981, for example, the SFWMD adopted a management strategy for protecting water quality in Lake Okeechobee, a natural outgrowth of the SFWMD's institutional transformation nine years earlier when the flood control district became a water management district. Historically, the district had two primary goals in managing the waters of Lake Okeechobee: to control flooding in time of heavy rainfall, and to supply water for agriculture and the urban centers in South Florida in time of drought. Now the SFWMD recognized "a third major goal of equal importance . . . namely, to maintain and improve the quality of the water resources within the District."[1] But this bland pronouncement understated the extent that the ground was shifting under the SFWMD. If flood control and drainage were public works that abetted economic growth, water quality fundamentally involved the imposition of economic restraints. In the past, when the district was primarily concerned with water supply, it was able to treat water as a raw resource to be exploited for economic gain. Now that the district was responsible for water quality, it had to treat water as commons, not property. Neither the people within the institution nor the SFWMD's many partners in government and the private sector were prepared for such a fundamental shift in thinking.[2]

Yet, still, as discussed in previous chapters, it was really state and not federal initiatives in the 1980s that drove water management in South Florida, including the work on Lake Okeechobee. The Corps continued to operate the C&SF Project for flood protection and water supply (both to urban areas and to the Everglades), but these efforts were largely overshadowed by state promotion of the preservation of the South Florida ecosystem, in large part against the effects of the C&SF Project. Just as Kissimmee River restoration was one piece of the state plan to "Save Our Everglades," the SFWMD's concern with Lake Okeechobee water quality and regulation levels served as an essential part of the program.

In order to get a firm grip on the problems with water quality in Lake Okeechobee, scientists used a systems ecology approach, especially focusing on mathematical and computer models to determine nutrient loading in the lake. At the same time, advances in chemistry and other forms of scientific measurement that had been ongoing in the post–World War II era enabled scientists to measure smaller and smaller particles, allowing for better analyses of problems in waterbodies such as Lake Okeechobee. Scientists were thus able to use the techniques of physics, chemistry, geology, geochemistry, meteorology, and hydrology "to measure ecosystem parameters at increasingly sophisticated levels and to analyze large data bases."[3]

In using these methods to scrutinize Lake Okeechobee, scientists focused on two separate but related problems to the waterbody's eutrophication: how to get a better scientific understanding of what was causing the lake's eutrophication, and how to distribute the burden of economic restraints that would accompany control measures. It was evident that the SFWMD must find a way to reduce nutrient loading of the immense lake, but farmers and environmentalists disagreed strongly about where the control measures should fall and who should pay for them. Indeed, agricultural interests were divided among themselves. Opinion differed as to whether dairy farms north

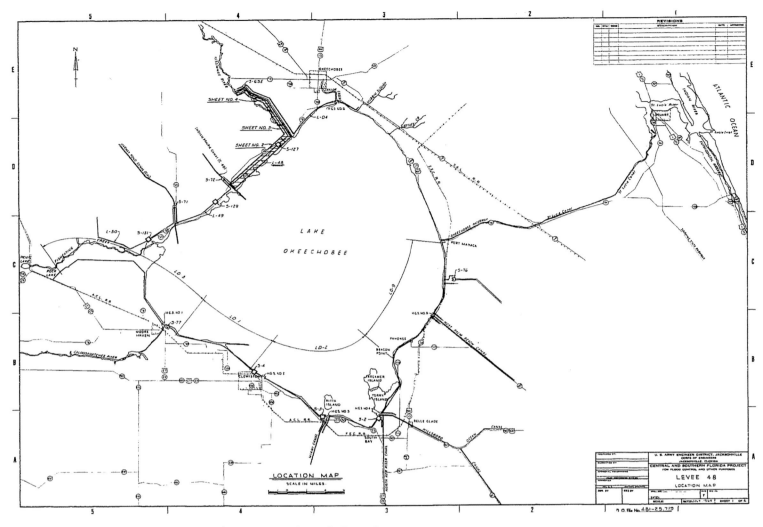

The Lake Okeechobee. (U.S. Army Corps of Engineers, Jacksonville District)

of the lake or sugar cane fields south of the lake were the main polluters. Opinion differed, too, on how fast eutrophication was occurring—an unknown that made it exceedingly difficult for the SFWMD to weigh management options on a cost-benefit scale.

One divisive issue was Lake Okeechobee's regulation schedule, which dictated lake "stages," or the quantity of water held in the lake from month to month. In 1978, the Corps of Engineers and the SFWMD implemented a new regulation schedule that raised the maximum lake stage level from 15.5 feet to 17.5 feet above sea level. While this increased the water supply, it had deleterious effects on water quality. High water affected marsh vegetation around the shoreline—the shallow lake's extensive "littoral zone," which accounted for more than one-fifth of the lake's surface. Continuous inundation of much of Lake Okeechobee's marsh area reduced the diversity of plant species, thereby affecting wading birds, waterfowl, reptiles, and fishes. In 1984, for example, the SFWMD published a study detailing the effects of high lake stages on wading bird utilization of Lake Okeechobee's littoral zone. The study concluded that "successful feeding conditions" for wading birds required a re-

ceding lake stage below 15.0 feet above mean sea level (msl), while successful nesting required "that the ground beneath the colony during the nesting period be flooded" from March to July.[4] Other examinations verified the damage that high lake levels caused to vegetation. Three years after the Corps had elevated the regulation schedule to 15.5-17.5 feet msl, scientists reported that "substantial changes" had occurred "in the composition and distribution of plant communities" in the littoral zone. These included the destruction of spikerush (*Eleocharis cellulose*), the proliferation of cattails, and the domination of torpedo grass (*Panicum repens*) in the mixed grass zone.[5]

The problem was that the Corps of Engineers and the SFWMD had to develop a regulation schedule for Lake Okeechobee that also took into account the water storage needs of urban and agricultural areas. Before 1974, the schedule had kept lake levels from exceeding 15.5 feet msl, while also allowing recessions down to 13.5 feet msl. This changed in the 1970s when the Corps elevated the schedule to between 14.5-16.0 feet msl in 1974 and then increased it to 15.5-17.5 feet msl in 1978 as explained above. The changes created less than optimal conditions for flora and fauna inhabiting the littoral zone, although scientists still needed more time to analyze how severe the effects really were.[6]

The same battle had been waged in the 1970s over the regulation schedules of the water conservation areas. Environmentalists deplored the Corps' drastic drawdowns of water levels in the areas, mainly because of the damage it caused to flora and fauna, and they also wanted better regulation in order to mitigate the effects of high water on deer populations. However, although the Corps reexamined its conservation area regulation schedules in the late 1970s, its final report—issued in October 1980—concluded that no reason existed for modifying the schedules.[7]

Environmentalists and sportsmen hoped for more success with Lake Okeechobee regulation. Yet sugar growers remained staunch advocates of the new regulation schedule since it protected them from drought. Thus, the schedule highlighted conflicts between the needs of water quality and water quantity, and the gulf in thinking about water as commodity or commons.[8]

At the same time, changes in marsh vegetation in turn affected the lake's ability to assimilate nitrogen and phosphorus inputs, inextricably tying protection of the littoral zone to Lake Okeechobee

water quality issues. One source of contention that aggravated the SFWMD's efforts to approach water quality evenhandedly and dispassionately was backpumping. Environmentalists focused on backpumping in part because of their animosity toward the sugar industry. But they had pragmatic reasons as well: it was easy to locate where the effluent was coming from (in contrast to "nonpoint source pollution"), and they could request public officials to stop it. Moreover, it was completely unnatural. By the flip of a switch, the SFWMD and the Corps of Engineers could activate the large S-2 and S-3 pumping stations on the south shore of Lake Okeechobee and reverse the flow of water through the three main canals braiding the 188,000-acre EAA, siphoning nutrient-laden water out of the sugar cane fields back into the lake. Environmentalists were appalled that the state would continue this practice in the face of mounting evidence that it was harming the lake. They were unmoved by arguments that backpumping was necessary during drought conditions to protect the water supply of South Florida. Although this activity was not the primary cause of nutrient loading of Lake Okeechobee, the S-2 and S-3 pump stations were obvious sources of agricultural pollution that the state controlled and could seemingly shut off at will. Therefore, environmentalists targeted backpumping as an evil,

Bank along Lake Okeechobee during low water level. (South Florida Water Management District)

S-135 pump station and lock; Lake Okeechobee during drought. (South Florida Water Management District)

pressured the SFWMD to stop it, and attacked the sugar growers by extension.[9]

The controversy over backpumping formed the immediate background to the SFWMD's formulation of a water quality management strategy. Environmentalists argued that the state should not allow the operation of the S-2 and S-3 pumps without a permit. In 1977, the Florida Department of Environmental Regulation issued a Temporary Operating Permit to the SFWMD to continue backpumping, pending the completion of the district's scientific investigation on the eutrophication of Lake Okeechobee. In 1979, the Florida Wildlife Federation and other environmental organizations brought suit against the Department of Environmental Regulation and the SFWMD, alleging that the backpumping of polluted water from the EAA into Lake Okeechobee violated state water quality standards. The Florida Sugar Cane League, Inc., and Dairy Farmers, Inc., intervened as interested parties in the lawsuit. The threat of litigation prompted the department to order the SFWMD to develop a water quality plan for Lake Okeechobee.[10] This was the political underpinning of the SFWMD's announcement in December 1981 of a new water quality management strategy. It was a necessary requirement to hold onto its Temporary Operating Permit.[11]

The SFWMD based the water quality management strategy on its newly completed scientific study of the lake. This study produced a conceptual model of Lake Okeechobee using extensive data on the lake's chemical and biological properties. The purpose of the model was to predict ecological change, test outcomes based on different inputs, and inform management guidelines—all with the goal of preventing catastrophic eutrophication of Lake Okeechobee.[12] Although overly simplistic by later standards, the model represented a sophisticated advance in water management and a first step on the path toward Everglades restoration.

As helpful as the conceptual model might be in selecting the appropriate management options for the protection of Lake Okeechobee, the model did not exist in a political vacuum, nor did it insulate management decisions from politics in the coming decade. Although the state successfully averted litigation over backpumping in the early 1980s, management of Lake Okeechobee continued to provoke considerable controversy. In 1985, Governor Graham established the Lake Okeechobee Technical Advisory Committee, made up of scientists from government, academia, and the private sector, to provide technical advice to the SFWMD in defining management options for Lake Okeechobee. Two years later, in 1987, the Florida legislature enacted the Surface Water Improvement and Management (SWIM) Act, requiring the SFWMD to develop a plan for Lake Okeechobee and other water bodies in South Florida.[13] Ultimately the same water pollution problems that were involved in the protection of Lake Okeechobee would form the basis for a federal lawsuit against the state of Florida in 1988, even though that suit would focus on waters entering Loxahatchee National Wildlife Refuge and Everglades National Park. Throughout the 1980s, the SFWMD put forward its conceptual model of Lake Okeechobee in part to keep the district on an even keel as it navigated these roiling political waters.

The use of a model to represent Lake Okeechobee's chemical and biological properties derived from a growing worldwide science on eutrophication of lakes and reservoirs. If the Kissimmee-Okeechobee-Everglades ecosystem was biologically unique, the accelerated eutrophication of Lake Okeechobee was not unusual at all. The same phenomenon had overtaken Lake Apopka in North Florida and was occurring in numerous water bodies all over the world. A by-product of human population growth, agricultural expansion, and increased use of fertilizers in crop production, "cultural eutrophication" was found to result when unnaturally large quantities of plant nutrients, mainly nitrogen and phosphorus, were loaded into lakes, thereby stimulating production of algae and other macrophytes and starting a train of other biological and chemical effects that could ultimately kill the lake. In the 1960s, scientists began to develop simple models using algorithms to approximate the real-world conditions of lakes that were undergoing accelerated eutrophication. The algorithms correlated such lake characteristics as water depth and surface area, water residence time (or flushing rate), and volume of nutrient loading. The use of models as a management tool for controlling eutrophication required the accumulation of empirical and statistical data over several years, and the selection of an appropriate model for the lake. By the mid-1970s, there were a handful of tried and tested models available to water managers, and the use of models had become an integral part of recommended management practice for controlling eutrophication.[14]

The SFWMD initiated a study of the biology and chemistry of Lake Okeechobee in 1973 aimed at developing the necessary data for modeling the lake. The study included four components: collection of lake water samples to obtain water chemistry data for trend analysis; development of a "material budget" (the measurement of the amount of water, phosphorus, nitrogen, and chloride coming into and leaving the lake at various points around the lakeshore); collection of data on the physical, biological, and chemical properties of the lake in spatial relationship; and finally, a trophic state assessment (to determine if the lake was oligotrophic, mesotrophic, or eutrophic). This major study continued through March 1980, yielding seven years of data for the SFWMD's initial modeling effort. In total, over 5,500 water samples were collected and analyzed for nitrogen species, phosphorus species, sodium, potassium, calcium, magnesium, chloride, alkalinity, color, turbidity, temperature, dissolved oxygen and specific conductivity, providing over 115,000 data points.[15]

Four scientists with the SFWMD—Anthony C. Federico, Kevin G. Dickson, Charles R. Kratzer, and Frederick E. Davis—analyzed

the data and produced a detailed report, "Lake Okeechobee Water Quality Studies and Eutrophication Assessment," in May 1981. One of the authors' findings was that the chemical and biological properties of Lake Okeechobee varied widely across its large expanse. The highest concentration of phosphorus occurred at the outlet of Taylor Creek/Nubbin Slough at S-191 downstream from the dairy farms. The highest nitrogen levels were found at the pump stations S-2 and S-3 at the head of the North New River and Hillsboro canals and the Miami Canal, respectively, where irrigation water from the EAA was backpumped into the lake. Thus, at the northern location there was an excess of phosphorus relative to what plants could absorb, while at the southern location there was more nitrogen than plants needed.[16] Since a major strategy in the control of eutrophication was to identify the limiting nutrient and reduce its input into the water body, this circumstance complicated the lake's management. Neither phosphorus nor nitrogen could be conclusively identified as the limiting nutrient, so both would have to be addressed.

Another result of the study was an improved understanding of the residence time of water in the lake. There were two elements to this: the average time that water took to move through the lake

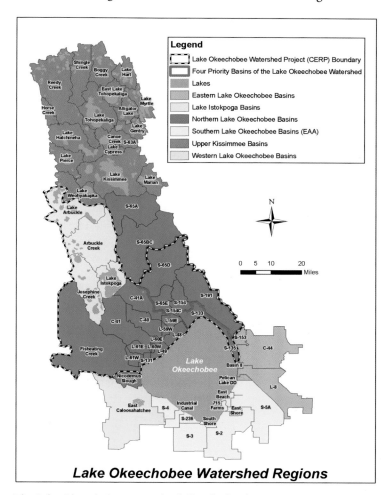

Lake Okeechobee Watershed Regions

The Lake Okeechobee watershed. (South Florida Water Management District)

(excluding evaporation) and the average elapsed time for water coming into the lake to replenish water going out of the lake (excluding rainfall). For this large, shallow lake, the average water residence time was a sluggish 3.47 years, the hydraulic loading rate a somewhat brisker 1.57 years. During the period 1973 to 1979, direct rainfall and the Kissimmee River amounted to approximately 70 percent of the water coming into the lake, while evaporation accounted for almost 66 percent of the water leaving the lake. These characteristics were important to understand in order to determine the impact that water from different sources had on Lake Okeechobee's phosphorus levels. The authors found that the longer it took water to move through the lake the more phosphorus and nitrogen was retained.[17]

Federico, Dickson, Kratzer, and Davis tested eight nutrient loading models for their applicability to Lake Okeechobee, and selected a modified Vollenweider model, published in 1976. R. A. Vollenweider was a prominent limnologist of Canada. As the authors noted, Vollenweider's first model, published in 1968, was based on a mathematical relationship between water depth and various measures of water quality.[18] A limitation of the model was that the measures of water quality were largely subjective. Subsequent models in the early 1970s refined Vollenweider's equation by factoring in trophic state indices, or quantitative indices used in categorizing lakes according to their place on a continuum from oligotrophic (nutrient-poor) to eutrophic (nutrient-rich). The EPA had developed one such index, as had a number of limnologists in the early to mid 1970s. Kratzer, one of the authors of the SFWMD report, had developed a trophic state indice specific to nitrogen levels in Florida lakes. Federico, Dickson, Kratzer, and Davis found that the indices were useful as far as they went, but the early indice-based models did not take into account water residence time and hydraulic loading rate—critical factors recognized by Vollenweider in his model of 1976.[19] The latter Vollenweider model discriminated among trophic states based on annual phosphorus loading and mean depth divided by hydraulic residence time.[20] It took form as the following mathematical expression:

$$TP = (L_p/q_s) (1 + \sqrt{t_w})$$

where:

TP = average annual in-lake total phosphorus concentration	
L_p = annual areal phosphorus loading	
q_s = annual areal water loading	
t_w = hydraulic residence time, and	
$\sqrt{}$ = mean depth[21]	

However, one problem remained with this Vollenweider model: it was oriented to lakes in northern temperate zones. Florida lakes, it appeared, could withstand higher total phosphorus concentrations before reaching the same level of production. Based on Kratzer's

trophic state indice for Florida lakes, the authors modified the Vollenweider model to allow higher loading rates. The final result was a plot curve showing "permissible" and "excessive" phosphorus inputs into Lake Okeechobee. As water residence time increased, so too did the permissible phosphorus load. The report included a similar plot curve for the permissible nitrogen load.[22]

Using the modified Vollenweider model, Federico, Dickson, Kratzer, and Davis concluded that Lake Okeechobee had a 78 percent probability of being eutrophic if phosphorus was the limiting nutrient, and a 79 percent probability if nitrogen was the limiting nutrient. The model further indicated that the phosphorus and nitrogen loads were 40 and 34 percent, respectively, above the excessive loading rate. Since the authors could not conclusively state that phosphorus or nitrogen was the limiting nutrient, they recommended that both phosphorus and nitrogen inputs be reduced to the permissible levels predicted by the modified Vollenweider model.[23]

The report by Federico, Dickson, Kratzer, and Davis provided the technical foundation for the water quality management strategy that the SFWMD adopted seven months later. In the latter document, which was approved by the district's governing board on 11 December 1981, SFWMD leadership accepted the finding that both nitrogen and phosphorus loading must be reduced, and it evaluated alternatives according to cost-effectiveness. Describing the control of eutrophication of Lake Okeechobee as "a very ambitious endeavor," it proposed a phased approach over a number of years. Phase I, of unspecified duration, would comprise five major activities. First, the SFWMD would continue the Interim Action Plan, with its provision for limited backpumping into the lake for five years. Second, the district would initiate development of a water retention facility within the EAA on the state-owned Holey Land tract. Third, it would accelerate use of Best Management Practices (BMPs) for dairy farms in the Taylor Creek/Nubbin Slough basin. Fourth, it would implement an expanded regulatory program with more stringent controls on any new construction of drainage systems within the lake's watershed. Finally, the district would coordinate with the Corps of Engineers in ensuring completion of the Kissimmee River Survey Review.[24]

The most controversial part of the strategy involved the Interim Action Plan. The SFWMD developed the plan as its initial response to the threat of litigation over its Temporary Operating Permit, issued to the district by the Department of Environmental Regulation so that the district could continue limited backpumping into the lake through pump stations S-2 and S-3. The plan promised to reduce the nutrient load on Lake Okeechobee by directing water from the sugar cane fields southward to the water conservation areas rather than northward back into the lake.[25] However, it also allowed for a resumption of backpumping in periods of drought in order to take advantage of Lake Okeechobee's water storage capacity. Just such drought conditions prevailed during the summer of 1981, and as the S-2 and S-3 pumps went into action, quantities of nitrogen far in excess of the permissible level were dumped into the lake. In a 12-month period during 1981-1982, water gauges recorded 14,250 tons of nitrogen pouring into the lake—nearly five times the recommended maximum.[26] To avoid a repetition of this event, district staff revised the Interim Action Plan to allow for backpumping in low volumes whenever the lake level fell below historical average.[27]

Reactions to the SFWMD's Lake Okeechobee water quality management strategy were mixed. The Department of Environmental Regulation commented favorably on the overall conceptual plan. However, it added a number of stipulations to the Temporary Operating Permit; for example, it required the SFWMD to report in detail on the implementation of BMPs in the Taylor Creek/Nubbin Slough basin.[28] The Florida Game and Fresh Water Fish Commission responded chiefly to the proposal to develop a water storage facility on the state-owned Holey Land tract. It had opposed this option in the past, partly because of the impact it would have on the local deer herd, but primarily because the facility would adversely affect Everglades wildlife habitat in Conservation Area No. 3 downstream from the site. It responded more favorably to this proposal because the plan incorporated the adjoining Rotenberger tract into the project, essentially designating one area for water storage and the other area for wildlife.[29]

The Rotenberger Tract. (South Florida Water Management District)

Environmental groups attacked the SFWMD's plan as both overdue and inadequate. Johnny Jones, executive director of the Florida Wildlife Federation, expressed outrage over the proposed use of the Holey Land area for water storage. "As we read the Management Strategy, it is nothing more than a smoke screen for the South Florida Water Management District's real agenda," he wrote. "That is a further state and federal subsidy for the agricultural inter-

ests of the EAA." The sugar growers, Jones insisted, should be forced to retain water on their own lands.[30] Paul C. Parks, commenting on behalf of the Florida Chapter of the Sierra Club, also objected to the use of state land "to take the pollution from farms," and declared, "this plan for the EAA is not going to be acceptable to the public." He was even more skeptical about the district's plan to implement BMPs among the dairy farmers in the Taylor Creek/Nubbin Slough basin. The district was naïve to expect significant results through volunteerism. "These dairies are not 'farms' in the usual sense," Parks wrote. "They are milk factories and their pollution ought to be regulated by the Department [of Environmental Regulation] like that of any other industry."[31]

Parks also voiced skepticism about the district's use of the modified Vollenweider model. He maintained that the model provided reasonable guidance for getting started on nutrient load reductions, but the Department of Environmental Regulation should require a continuing research program to assess whether the target amounts predicted by the model were adequate. The development of this model, he argued, was symptomatic of an unfortunate tendency to separate technical issues from policy issues. "The technical question is, can this be done; the policy question is, will this be done?"[32]

A month later, Parks wrote to the Department of Environmental Regulation again about the SFWMD's water quality management plan, arguing that it failed to justify an extension of the Temporary Operating Permit "because it is unlikely to result in nutrient loading rate reductions which are sufficient to meet the criteria established by the modified Vollenweider equation," that is, a 40 percent reduction of phosphorus loading and a 34 percent reduction of nitrogen loading. It was all very well for the SFWMD to establish targets for nutrient load reductions, Parks argued, but it also needed to demonstrate convincingly how it would achieve those levels.[33]

Initially, dairy farmers and sugar growers had little to say about the SFWMD's water quality management strategy. They gave it their tacit approval, a reasonable position considering some of the management options advocated by environmentalists that would have been more costly to them. Later, as drought conditions brought about a resumption of backpumping and public debate about Lake Okeechobee in the mid-1980s, the sugar growers would raise some objections. They criticized the SFWMD's strategy insofar as it did not pinpoint phosphorus as the chemical agent requiring the most stringent controls despite mounting evidence that phosphorus, not nitrogen, was the limiting nutrient in the eutrophication of Lake Okeechobee. Sugar growers also voiced skepticism about developing the Holey Land for water storage, as they eyed this area for future annexation to the EAA. Certainly they found this proposal preferable to taking land out of sugar cane production for water storage, however, so they muted their criticism.[34] As for the dairy farmers, their action in implementing BMPs spoke most directly to their attitudes about the SFWMD's strategy.

There were 24 separate dairy operations in the Taylor Creek/ Nubbin Slough basin in the mid-1980s, as well as an additional 12 dairy operations in the Lower Kissimmee Valley plus some beef cattle operations. In keeping with industry trends nationwide, these were large livestock operations that involved concentrated animal feeding stations as well as pasturing of cows. The average dairy farm in the region had about 1,000 cows. The cows were fed a phosphorus supplement to enhance milk production, which the cows were unable to absorb fully, excreting the unabsorbed portion in their manure. Treating and disposing of this animal waste was a challenge. Although some dairy farms employed state-of-the-art livestock waste management techniques, more stringent and systematic controls were needed.[35]

The SFWMD recommended a number of BMPs to the dairy farmers. These included better rotation of cows between pastures, feedlots, and barns to distribute manure more widely; fencing and other measures to keep cows away from watercourses; and various types of "biological nutrient removal," or use of aquatic plants to take up phosphorus that was already in the water. Such aquatic plants were called "scrubbers" or "polishers" since they had the effect of cleansing the water. In particular, there was a need to collect barn wash and direct this phosphorus-laden water through "oxidation/ polishing lagoons" for treatment before it was released as effluent to Lake Okeechobee. The SFWMD even proposed converting animal waste to methane gas for local energy use.[36]

Beginning in 1981, many dairy farmers in the Taylor Creek/ Nubbin Slough basin began upgrading their barns to improve treatment of barn wash (water used to rinse out dairy barns). They were aided by a federal grant under the federal Rural Clean Waters Project. For each barn, the federal government supplied 75 percent of the cost in a cost-share arrangement with the dairy farmer, up to a limit of $50,000 per barn. The SFWMD estimated the average cost at $100,000 per barn. Despite this funding shortfall, the dairy farmers were highly motivated to get their barns renovated. By 1987, all but three barns in the Taylor Creek/Nubbin Slough basin had undergone modification under the Rural Clean Waters program, with work in progress on the remaining three. In the Lower Kissimmee Valley, meanwhile, where the state contributed funds to a similar cost-share program initiated in 1986, all 19 livestock operators had signed up for renovation of their barns at a cost per barn of $170,000 by 1987.[37]

Nubbin Slough. (South Florida Water Management District)

Despite early skepticism, the state got surprisingly good co-operation from the dairy farmers in implementing the BMPs. State officials favored this approach as a low-cost management option that treated the problem at the source, and the farmers recognized the program's necessity. Although agriculturists benefited from federal and state funding supports, the program rested fundamentally on the farmers' voluntary efforts, which they made largely at their own expense.[38]

BMPs were only a starting point, however. The SFWMD's water quality management strategy called for additional control measures to reduce nutrient loading. District staff maintained that the results of BMPs had to be evaluated before pressing ahead with other, more expensive, engineering solutions such as the Holey Land reservoir. Such a systematic, fiscally conservative approach was standard practice in watershed management, but it carried the risk of doing too little too late. Indeed, the amount of nutrients pouring into Lake Okeechobee continued to exceed target levels in the mid-1980s and water quality monitoring showed that phosphorus concentrations were approaching the highest levels ever recorded. No one knew, of course, how long the excessive nutrient loading could persist before the ecological consequences became severe. Environmentalists argued that water managers, in the face of such uncertainty, must err on the side of caution, particularly since Lake Okeechobee was so large and central to South Florida's ecosystem.[39]

There were other signs that time was running out. Fishing guides and commercial fishermen reported extensive growths of filamentous blue-green algae on the lake surface. South shore residents complained that their drinking water had acquired a bad odor and taste.[40] Biologists studying the nesting success of wading birds in Lake Okeechobee's marshes found the birds' numbers declining because of damage to the littoral zone. When drought threatened in June 1985, causing the SFWMD to resume backpumping, the Florida Wildlife Federation again threatened to sue.[41] Amid heightening public concern, Governor Graham called for a comprehensive review. He asked the head of the Department of Environmental Regulation, Victoria Tschinkel, to put together a committee. Eager to bridge conflicts surrounding the lake's management, the governor wanted the review to "include consideration of the interests of federal, state and appropriate local government, agricultural and other users, environmentalists and sportsmen and other interests as may be appropriate."[42]

The Lake Okeechobee Technical Advisory Committee (LOTAC), as it became known, made a hurried study and issued preliminary findings in August 1986. LOTAC generally endorsed the SFWMD's approach, including the district's emphasis on BMPs, although it shortened the list of BMPs to just three in order to gain maximum compliance. These included a reiteration of the SFWMD's goal of fencing all animals away from watercourses, noting that

about 75 percent of the appropriate land area in the Taylor Creek/Nubbin Slough basin had been so fenced by 1987, much of it under the Rural Clean Waters Project; a prohibition of all direct discharge of barn wash into surface waters—a goal that was already practically attained in the Taylor Creek/Nubbin Slough basin and within reach for the dairy operators in the Lower Kissimmee Valley; and the implementation of measures to control storm water runoff from high intensity use areas. Ultimately, LOTAC declared, it might prove necessary for all dairy farms in the region to become "confinement dairies," where all runoff from the milking barns would be collected in a reservoir for treatment.[43]

An algae bloom on Lake Okeechobee. (South Florida Water Management District)

In addition to its discussion of BMPs, LOTAC accepted the modified Vollenweider model as the best available mathematical model for predicting permissible nutrient loading rates. It also affirmed the district's goal of reducing phosphorus loading by 40 percent. LOTAC went further than the SFWMD's water quality management plan, however, in identifying phosphorus as the limiting nutrient. Based on data accumulated since 1980, this conclusion was inescapable. While the amount of nitrogen had leveled off, the amount of phosphorus in the lake had doubled over the period 1973-1984. LOTAC theorized that the lake was losing its capacity to assimilate phosphorus because bottom sediments could no longer bind the mineral. Adjoining watersheds such as the Taylor Creek/Nubbin Slough basin were similarly unable to hold any more phosphorus. One scientist likened the situation to water dripping on a sponge: the sponge absorbed each drip until it became saturated, at which point the water passed right through.[44] It appeared that with background levels in the environment already high, phosphorus increases could soon accelerate. LOTAC recommended an intensified plan of research focusing on phosphorus loading, BMPs, effects of lake levels on biological communities, and downstream impacts of proposed diversions.[45]

Governor Graham turned LOTAC's recommendations into an executive order, promulgated on 23 August 1986. The order outlined more than a dozen action items—mostly research and monitoring—for the Department of Environmental Regulation, SFWMD, and four other state agencies, and it requested the Corps of Engineers and U.S. Department of Agriculture to participate in cost sharing and research efforts. The Department of Environmental Regulation was responsible for overall program coordination, and the agencies were to execute a memorandum of understanding and prepare a comprehensive plan by 1 November 1986. The governor's executive order contained one specific engineering requirement: it directed the SFWMD to coordinate with the Corps of Engineers on completion of a preliminary design for a diversion of waters from the Taylor Creek/Nubbin Slough basin.[46]

The Taylor Creek/Nubbin Slough diversion was not a new proposal; the Corps had recommended it to Congress in 1968, and the governing board of the SFWMD had requested that the Corps develop plans for it in 1979. The general plan was to divert waters from the basin to the St. Lucie Canal, which flowed east to the St. Lucie Estuary. Although the plan held some attraction to citrus growers in St. Lucie County because it would provide an alternative source of irrigation water during drought, it also raised concerns that the polluted water from the dairy farms would degrade the St. Lucie Estuary. After public review of seven alternatives, the Taylor Creek/Nubbin Slough diversion plan had been shelved in 1980. Now that the need to protect Lake Okeechobee appeared urgent, the SFWMD asked the Corps to give the plan further consideration. The project would be costly, primarily because it would involve the acquisition of a lot of private land, but it appeared to offer one of the fastest and most effective means of reducing phosphorus loading.[47]

At the same time, the SFWMD prepared cost estimates and fact sheets for the whole panoply of Lake Okeechobee protection options. The total cost, if all options were implemented, could run as high as $200 million, it suggested. But the SFWMD's current budget for Lake Okeechobee was a mere $4.4 million, and LOTAC-recommended projects for the current year would require an additional $5.2 million. Given the significance of Lake Okeechobee as a state resource, the district sought additional monies from the state's general fund. State legislators were sympathetic. In January 1987, the Senate Natural Resources Committee proposed a "Save Our Lakes" bill that would provide funding for protection and restoration through a documentary stamp tax. The House Natural Resources Committee considered a similar proposal. The SFWMD's Patricia A. Bidol, executive program director, presented the district's plans and cost estimates to the House committee at the end of January.[48]

Meanwhile, at the beginning of January 1987, there was a change of governors. Bob Graham went to the U.S. Senate and Robert "Bob" Martinez replaced him in the Florida statehouse. Governor Martinez, a Republican and former mayor of Tampa, promised

to continue his Democratic predecessor's popular environmental programs, including Save Our Everglades and the Lake Okeechobee protection plan. He indicated his support of the proposed legislation to protect and restore surface waters. However, soon after taking office the governor halted progress by the Department of Environmental Regulation and SFWMD in appointing a Lake Okeechobee Science Review Panel as recommended by LOTAC. Apparently, Martinez was responding to concerns by the sugar growers that the people making up the panel were all from out of state. Following the governor's lead, Dale Twachtmann, the newly appointed secretary of the Department of Environmental Regulation, placed a three-month hold on all LOTAC-directed activities by his department. On 11 April, Martinez and Twachtmann, accompanied by Estus Whitfield (who remained in his position as the governor's environmental advisor despite the change in governors) and three other staff, met with the sugar growers in Belle Glade.[49]

Governor Bob Martinez. (The Florida Memory Project, State Library and Archives of Florida)

Participating on behalf of sugar in this two-and-a-half hour meeting was the industry's Environmental Quality Committee, composed of the six most prominent men in the business—Alex Fanjul, Nelson Fairbanks, Jose Alvarez, John Hundley, George H. Wedgworth, and Joe Marlin Hilliard. Accompanying them were four staff and four consultants. The sugar growers told the governor that they had a history of involvement in environmental issues. Their Environmental Quality Committee, formed in 1963, had addressed the problem of air pollution in the 1960s and 1970s as the industry came under attack for its open-field burning of cane fields. The committee helped to develop regulations to limit open-field burning, and it oversaw technological improvements in the sugar mills so that they met Clean Air Act standards. In 1974, it initiated research on the eutrophication of Lake Okeechobee, and by 1987, it had completed more than a score of studies through contracts with environmental science and engineering firms. Many of the recommendations contained in these contracted studies, the agriculturists claimed, had appeared in the recommendations by LOTAC.[50]

The Environmental Quality Committee gave the LOTAC recommendations a strong endorsement. It agreed with LOTAC's selection of phosphorus as the limiting nutrient in the eutrophication of Lake Okeechobee. It supported the use of the modified Vollenweider model as a management tool. Finally, the committee approved of three project recommendations by LOTAC: the diversion of water from the S-4 basin on the west side of Lake Okeechobee to the Caloosahatchee River (similar to the Taylor Creek/Nubbin Slough diversion but much smaller), the Holey Land Reservoir project, and a pilot project to investigate the feasibility of Aquifer Storage Recovery. The agriculturalists favored these projects because they saw the necessity of increasing water supply for the EAA in place of back-pumping and water storage in Lake Okeechobee.[51]

The sugar growers tried to demonstrate their willingness to compromise, but they also shared some concerns about the SFWMD with the governor. "We are concerned that the District has utilized a fast track approach recently on the Lake Okeechobee matter," their written presentation stated. It seemed that the district's leadership had decided that throwing enough money at these problems would solve them. Perhaps this could be attributed to a desire to put programs into effect before the new governor changed the makeup of the governing board. Most disturbing to the agriculturists, John R. "Woody" Wodraska, who had become chairman of the governing board, continued to make divisive public statements concerning the need to control nitrogen, despite LOTAC's finding that phosphorus was the limiting nutrient.[52]

Although "Big Sugar" was greatly vilified in the public's eye, Governor Martinez decided to go to Belle Glade, a town largely populated by sugar growers and Haitian cane cutters, to consult on environmental policy for Lake Okeechobee. Perhaps because of this trip, Martinez was soon beset by charges that he would not support the pending legislation to protect Lake Okeechobee. On 4 May, the governor felt compelled to issue a statement aimed at correcting the "misunderstanding." Reiterating his support for the initiative to save Florida's imperiled lakes (now titled the Surface Water Improvement and Management bill, or SWIM), he explained that he merely opposed two features of the bill: the establishment of advisory councils to guide each lake's protection program—"new, unnecessary layers of bureaucracy"—and the use of state tax revenues to pay for the program.[53] Both of these features remained in the bill, however, when the state legislature passed the measure and Martinez signed it into law.

The Surface Water Improvement and Management Act marked a turning point after nearly two decades of plodding efforts to prevent the catastrophic eutrophication of Lake Okeechobee. The SWIM Act declared that "the declining quality of the state's surface waters has been detrimental to the public's right to enjoy these surface waters and it is the duty of the state, through the state's public agencies and subdivisions, to enhance the environmental and scenic value of surface waters."[54] The act mandated the establishment of a priority list of water bodies of regional and statewide significance, a list that began with Lake Okeechobee, Lake Apopka, Tampa Bay, Biscayne Bay, Indian River Lagoon, and Lower St. Johns River, and would grow to include 23 other water bodies by 1997.[55] For each listed water body, the law required the appropriate water management district to design and implement a surface water and improvement (SWIM) plan. It also created an advisory council (in the case of Lake Okeechobee, this was the second incarnation of LOTAC, known as LOTAC II), and it established a SWIM trust fund to provide financial support for planning and implementation efforts mandated under the law.

Scientific understanding of the eutrophication of Lake Okeechobee had progressed from the first study by the U.S. Geological Survey in 1969 to a myriad of studies by federal and state agencies, universities, and consultants in the 1970s and 1980s. The SFWMD had drawn water samples on a regular basis since 1973, and it had expanded its lake monitoring program in 1986 to encompass more than 50 sites in an effort to improve understanding of areal differences in water quality and the influence of the littoral zone and localized inflows. It had also begun to assess the effects of such lake reclamation activities as aquatic weed control and bottom sediment removal. Meanwhile, continuous gauging of phosphorus and nitrogen loading at all major surface water inflow structures around Lake Okeechobee enabled managers to determine the relative effects of different protection options such as the Taylor Creek/Nubbin Slough diversion. By 1987, managers had the requisite science to evaluate an array of potential engineering projects in various combinations, each project bearing an estimated cost, time of completion, and amount of phosphorus that would be subtracted from the total load going into Lake Okeechobee. The SWIM plan that emerged in 1989 included some phosphorus reductions from biological and chemical treatments, as well as implementation of further BMPs, but by and large it involved engineering solutions.[56]

Even before the development of the SWIM plan, scientific investigation had pointed the way to two critical decisions by water managers. The first was their acceptance of the modified Vollenweider model to establish maximum loadings of nitrogen and phosphorus that could be safely discharged into the lake. Modeling results called for reductions of nutrient loadings by 34 percent for nitrogen and 40 percent for phosphorus. When district managers were unable to achieve these reductions, it lent urgency to their request for state and federal support of the effort. Indeed, while nitrogen levels fell, concentrations of phosphorus more than doubled from the mid 1970s to the late 1980s.[57] The second important science-based decision by water managers was to adopt phosphorus control as the primary lake management strategy. Without that direction, water managers might have been faced with a standoff between the nitrogen-producing sugar growers on the south shore of the lake and the phosphorus-producing dairy farmers on the north shore.

But the SWIM plan did not only address water quality; it also recognized the need for action to protect Lake Okeechobee's littoral zone. According to the plan, "the most practical means" to ensure the propagation of littoral zone vegetation and wildlife was the "development of an appropriate regulation schedule." Yet the plan admitted that "the needs of natural systems in the Lake, especially the littoral zone plan communities have not yet been defined." It therefore called for the creation of a "special technical committee" to "define water level requirements of the littoral zone communities." Accordingly, in 1988, the Lake Okeechobee Littoral Zone Technical Group, composed of representatives from the SFWMD, the FWS, the Florida Game and Freshwater Fish Commission, Everglades National Park, the Florida Department of Natural Resources, the Florida Department of Environmental Regulation, and different universities, worked to develop a sense of how much water was needed to protect the littoral zone.[58]

Throughout the 1980s, then, the SFWMD wrestled with important water quality and littoral zone issues pertaining to Lake Okeechobee, using science as a major guide. Especially important in the time period was the realization that phosphorus was the limiting nutrient in Lake Okeechobee, and that concentration levels of the mineral were reaching dangerous proportions. In many ways, this reliance on science and the solutions that were ultimately proposed foreshadowed efforts in the 1990s to restore the South Florida ecosystem. At the same time, some actions taken to prevent the eutrophication of Lake Okeechobee had effects that proponents did not fully consider—another example of the law of unintended consequences. The curtailment of backpumping under the Interim Action Plan, for example, probably saved Lake Okeechobee from hypereutrophication, but it merely moved the problem to the water conservation areas. The gradual spread of cattails and an exotic plant called melaleuca soon began to tell a story of creeping eutrophication of the Everglades, just as algae blooms had alerted people to eutrophication of Lake Okeechobee a decade earlier. As water quality problems worsened in the Everglades, Florida officials proposed a new solution: the purchase of environmentally threatened lands.

Endnotes

[1] South Florida Water Management District, "Executive Summary, Water Quality Management Strategy for Lake Okeechobee," 11 December 1981, Folder 16, Box 3, Marshall Papers.

[2] Matthew Alan Cahn, *Environmental Deceptions: The Tension Between Liberalism and Environmental Policymaking in the United States* (Albany: State University of New York Press, 1995), 65-80.

[3] McIntosh, *The Background of Ecology*, 203.

[4] Quotation in Michael Zaffke, *Wading Bird Utilization of Lake Okeechobee Marshes, 1977-1981*, Technical Publication 84-9 (West Palm Beach, Fla.: South Florida Water Management District, Environmental Sciences Division, Resource Planning Department, 1984), 17; see also James F. Milleson, *Vegetation Changes in the Lake Okeechobee Littoral Zone, 1972 to 1982*, Technical Publication 87-3 (West Palm Beach, Fla.: South Florida Water Management District, Environmental Sciences Division, Resource Planning Department, 1987), i.

[5] Paul J. Trimble and Jorge A. Marban, "A Proposed Modification to Regulation of Lake Okeechobee," *Water Resources Bulletin* 25 (December 1989): 1250.

[6] Zaffke, *Wading Bird Utilization of Lake Okeechobee Marshes, 1977-1981*, 4.

[7] Department of the Army, Jacksonville District, Corps of Engineers, "Review of the Regulation Schedule for Water Conservation Area No. 3A, Central and Southern Florida, October 1980," File Conservation Areas 1, 2, 3, 1970-86, Box 02193, SFWMDAR; James W. R. Adams, Colonel, Corps of Engineers, District Engineer, to Honorable Bob Graham, Governor of Florida, 31 October 1980, ibid.

[8] "Summary of Florida Sugar Industry's Presentation to Governor Bob Martinez and FDER Secretary Dale Twachtmann, Belle Glade, Florida," 11 April 1987, 19, File LOTAC I/LOTAC II, Box 18060, SFWMDAR. *The Palm Beach Post*, 29 July 1985.

[9] Joseph D. Carroll, Jr., Director, Vero Beach Field Office, U.S. Fish and Wildlife Service to Field Supervisor, 23 June 1982, File NWR Fisheries Studies, Conservation Area 1 Loxahatchee, CE-SE Central and Southern Florida FCP, FWSVBAR; John C. Jones, Executive Director, Florida Wildlife Federation to Victoria Tschinkel, Secretary, FDER, 16 April 1982, Folder 16, Box 3, Marshall Papers.

[10] State of Florida, Division of Administrative Hearings, "Recommended Order" in *Florida Wildlife Federation et al. v. State of Florida et al.*, 8 November 1979; *The Palm Beach Post*, 29 July 1985.

[11] Paul C. Parks, Ph.D., Florida Chapter of the Sierra Club, to William Buzick, Deputy Director, Division of Permitting, Florida Department of Environmental Regulation, 1 June 1982, Folder 15, Box 3, Marshall Papers.

[12] Thomas James, telephone communication with Theodore Catton, 1 April 2005.

[13] South Florida Water Management District, "Draft Interim Surface Water Improvement and Management (SWIM) Plan for Lake Okeechobee," 10 October 1988, 1-2, File Okeechobee SWIM Plan (SFWMD) 1988, Box 1, S1497, Department of Agriculture and Consumer Services, Surface Water Improvement and Management Plan Files, FSA.

[14] Sven-Olof Ryding and Walter Rast, eds., *The Control of Eutrophication of Lakes and Reservoirs* (Park Ridge, N.J.: Parthenon Publishing Group, 1989), 85-94. The leading institution in the development of eutrophication control strategies was the Organization for Economic Cooperation and Development (OECD). Another source of support was UNESCO's Man and

the Biosphere Program. The United Nations designated the 1980s the "Water Decade." Within the United States, EPA and the Clean Water Act Amendments of 1972 provided guidance for water managers at the regional level.

[15] Anthony C. Federico, et al., *Lake Okeechobee Water Quality Studies and Eutrophication Assessment*, Technical Publication 81-2 (West Palm Beach, Fla.: South Florida Water Management District, Resource Planning Department, 1981), 3-8.

[16] Federico et al., *Lake Okeechobee Water Quality Studies and Eutrophication Assessment*, 21.

[17] Federico et al., *Lake Okeechobee Water Quality Studies and Eutrophication Assessment*, 22-23.

[18] Federico et al., *Lake Okeechobee Water Quality Studies and Eutrophication Assessment*, 215-216.

[19] Federico et al., *Lake Okeechobee Water Quality Studies and Eutrophication Assessment*, 202-205, 219.

[20] United Nations Environmental Programme, Division of Technology, Industry, and Economics, International Environmental Technology Centre, "Planning and Management of Lakes and Reservoirs: An Integrated Approach to Eutrophication" <http://www.unep.or.jp/publications/techpublications/techpub-11/1-4-2.asp> (5 April 2005).

[21] Ryding and Rast, eds., The Control of Eutroph*ication of Lakes and Reservoirs, 96; Federico et al., Lake Okeechobee Water Quality Studies and Eutrophication Assessment, 226.*

[22] Federico et al., *Lake Okeechobee Water Quality Studies and Eutrophication Assessment*, 223.

[23] Federico et al., *Lake Okeechobee Water Quality Studies and Eutrophication Assessment*, 24.

[24] South Florida Water Management District, "Executive Summary, Water Quality Management Strategy for Lake Okeechobee," 11 December 1981, 3, 11, Folder 16, Box 3, Marshall Papers.

[25] "Summary of Florida Sugar Industry's Presentation to Governor Bob Martinez and FDER Secretary Dale Twachtmann, Belle Glade, Florida," 11 April 1987, 19, File LOTAC I/LOTAC II, Box 18060, SFWMDAR.

[26] *The Palm Beach Post*, 29 July 1985.

[27] South Florida Water Management District, "Executive Summary, Water Quality Management Strategy for Lake Okeechobee," 11 December 1981, 12, Folder 16, Box 3, Marshall Papers.

[28] Terry Cole, Assistant Secretary, FDER, to John R. Maloy, Executive Director, SFWMD, 15 June 1982, Folder 16, Box 3, Marshall Papers.

[29] Colonel Robert M. Brantly, Executive Director, FGFWFC, to Victoria Tschinkel, Secretary, FDER, 18 May 1982, Folder 32, Box 2, Marshall Papers.

[30] John C. Jones, Executive Director, Florida Wildlife Federation, to Victoria Tschinkel, Secretary, FDER, 16 April 1982, Folder 16, Box 3, Marshall Papers.

[31] Paul C. Parks, Ph.D., Florida Chapter of the Sierra Club, to William Buzick, Deputy Director, Division of Permitting, FDER, 2 May 1982, Folder 16, Box 3, Marshall Papers.

[32] Parks to Buzick, 2 May 1982.

[33] Parks to Buzick, 1 June 1982, Folder 15, Box 3, Marshall Papers.

[34] "Summary of Florida Sugar Industry's Presentation to Governor Bob Martinez and FDER Secretary Dale Twachtmann, Belle Glade, Florida," 11 April 1987, 19.

[35] Paul C. Parks, Ph.D., Florida Defenders of the Environment, to Nancy Roen, General Development Corp., 5 August 1987, File LO Major Programs—Correspondence, Background, LOSAC, Box 18060, SFWMDAR; William F. Ritter and Adel Schirmohammadi, eds., *Agricultural Nonpoint Source Pollution: Watershed Management and Hydrology* (Boca Raton, Fla.: Lewis Publishers, 2001), 136.

[36] Appendix I in South Florida Water Management District, "Executive Summary, Water Quality Management Strategy for Lake Okeechobee," 11 December 1981, Folder 16, Box 3, Marshall Papers.

[37] John R. Wodraska, Executive Director, SFWMD, to Jim Smith, Chief of Staff, Office of the Governor, 4 May 1987, File LO Programs—Correspondence, Background, LOSAC, Box 18060, SFWMDAR.

[38] The success of the BMP program was not unusual. A recent study on agricultural pollution and watershed management states: "The main approach used to minimize pollution resulting from agricultural activities is implementation of Best Management Practices (BMPs). . . . Although cost-share incentives and some regulations are used, current nonpoint pollution abatement programs rely mostly on voluntary implementation of management practices." Ritter and Shirmohammadi, eds., *Agricultural Nonpoint Source Pollution*, 259.

[39] Bob Graham to Victoria J. Tschinkel, Secretary, 26 August 1985, File Lake Okeechobee File #1, Box 06591, SFWMDAR; Paul C. Parks, Ph.D., Florida Defenders of the Environment, Inc., to Nathaniel P. Reed, 3 February 1986, ibid.

[40] See Table 1, Major Algal Bloom Events on Lake Okeechobee, 1970-1988, in South Florida Water Management District, "Draft Interim Surface Water Improvement and Management (SWIM) Plan for Lake Okeechobee, 10 October 1988, File Okeechobee S.W.I.M. Plan (SFWMD)—1988, Box 1, S1497, Department of Agriculture and Consumer Services, Surface Water Improvement and Management Plan Files, FSA.

[41] Joseph D. Farish, Jr., 29 July 1985, File Lake Okeechobee File #1, Box 06591, SFWMDAR.

[42] Graham to Tschinkel, 26 August 1985.

[43] Patricia A. Bidol, Ph.D., Executive Program Director, SFWMD, "Lake Okeechobee: Proposed Solutions and Associated Costs," Presentation to the Florida House Committee on Natural Resources, 21 January 1987, File Lake Okeechobee, Box 06591, SFMWDAR; South Florida Water Management District, "Summary Sheets: Lake Okeechobee Protection Options," February 1987, ibid.

[44] Paul C. Parks to Nancy Roen, 5 August 1987, File LO Major Problems—Correspondence, Background, LOSAC, Box 18060, SFWMDAR.

[45] South Florida Ecosystem Restoration Task Force, Science Subgroup, "South Florida Ecosystem Restoration: Scientific Information Needs," 1996, 147, available at <http://everglades.fiu.edu/taskforce/scineeds/sub2.pdf> (30 July 2004.)

[46] Executive Order 86-150, 23 August 1986, File Lake Okeechobee #2, Box 06591, SFWMDAR. Estus Whitfield, the governor's lead advisor on the environment, crafted the executive order with input from environmental groups and state agencies. See letters in this file.

[47] Arthur G. Linton, Federal Facilities Coordinator, Enforcement Division, EPA to James L. Garland, Chief, Engineering Division, Jacksonville District, COE, 29 August 1980, File 1110-2-1150a (C&SF Martin County) Project General 1980, Box 1, Accession No. 077-01-0023, RG 77, FRC; T. J. Griffiths, Florida Citrus Mutual to Corps of Engineers, 12 September 1980, ibid.; Bradley J. Hartman, FGFWFC, to Garland, 16 October 1980, and Garland to Nathaniel P. Reed, 23 October 1980, ibid.

[48] Patricia A. Bidol, Ph.D., Executive Program Director, SFWMD, "Lake Okeechobee: Proposed Solutions and Associated Costs," Presentation to the Florida House Committee on Natural Resources, 21 January 1987, File Lake Okeechobee, Box 06591, SFWMDAR; South Florida Water Management District, "Summary Sheets: Lake Okeechobee Protection Options," February 1987, ibid.

[49] "Summary of Florida Sugar Industry's Presentation to Governor Bob Martinez and FDER Secretary Dale Twachtmann, Belle Glade, Florida," 11

April 1987; Patricia A. Bidol to George H. Wedgeworth, Sugar Cane Growers Coop, 6 March 1987, File LOTAC I/LOTAC II, Box 18060, SFWMDAR.

[50] "Summary of Florida Sugar Industry's Presentation to Governor Bob Martinez and FDER Secretary Dale Twachtmann, Belle Glade, Florida," 11 April 1987, 1-5, 51.

[51] "Summary of Florida Sugar Industry's Presentation to Governor Bob Martinez and FDER Secretary Dale Twachtmann, Belle Glade, Florida," 11 April 1987, 17-23, 38.

[52] "Summary of Florida Sugar Industry's Presentation to Governor Bob Martinez and FDER Secretary Dale Twachtmann," 43.

[53] Quotations in "Statement by Governor Bob Martinez Concerning SWIM Bill," 4 May 1987, File SWIM Bill 5/4/87, Box 1, S1401, Executive Office of the Governor Subject Files, 1987-1988, FSA; see also Alec Wilkinson, *Big Sugar: Seasons in the Cane Fields of Florida* (New York: Alfred A. Knopf, 1989), 239.

[54] Quoted in South Florida Water Management District, "Draft Interim Surface Water Improvement and Management (SWIM) Plan for Lake Okeechobee, 10 October 1988, File Okeechobee S.W.I.M. Plan (SFWMD)—1988, Box 1, S1497, Department of Agriculture and Consumer Services, Surface Water Improvement and Management Plan Files, FSA.

[55] Edward A. Fernald and Elizabeth D. Purdum, eds., *Water Resources Atlas of Florida* (Tallahassee: Institute of Science and Public Affairs, Florida State University, 1998), 164.

[56] South Florida Water Management District, "Draft Interim Surface Water Improvement and Management (SWIM) Plan for Lake Okeechobee, 10 October 1988, 18-19, File Okeechobee SWIM. Plan (SFWMD)—1988, Box 1, S1497, Department of Agriculture and Consumer Services, Surface Water Improvement and Management Plan Files, FSA. See also Daniel E. Canfield, Jr., and Mark V. Hoyer, "The Eutrophication of Lake Okeechobee," *Lake and Reservoir Management* 4 (No. 2 1988): 91-99.

[57] South Florida Water Management District, "Draft Interim Surface Water Improvement and Management (SWIM) Plan for Lake Okeechobee, 10 October 1988, Executive Summary 18-19, File Okeechobee SWIM Plan (SFWMD)—1988, Box 1, S1497, Department of Agriculture and Consumer Services, Surface Water Improvement and Management Plan Files, FSA.

[58] South Florida Water Management District, "Draft Interim Surface Water Improvement and Management (SWIM) Plan for Lake Okeechobee," Executive Summary 1-3.

CHAPTER 10

Envelopes of Protection: Land Acquisition Programs in Florida, 1980–1990

AT THE BEGINNING OF THE 1980s, many Floridians were concerned with the continuing growth of the state and the ever-increasing encroachment that this produced on South Florida's fragile ecosystem. To solve these problems, Governor Bob Graham, in his Save Our Everglades program, called for increased state and federal involvement in land acquisition efforts. Florida already had an ambitious purchasing program, but Graham and others believed that it needed to be expanded and strengthened, and they instituted new measures accordingly. Especially concerned with the Florida panther and its habitat in the Big Cypress and Fakahatchee Strand, Graham and Florida's congressional delegation proposed that the state—with federal help—acquire lands to add to the Big Cypress National Preserve, and Congress passed this measure in 1988. This was a significant achievement in the 1980s, especially given the negative attitude that the Reagan administration and its Interior Department had about land purchases. However, Governor Robert "Bob" Martinez, who succeeded Graham in 1987, believed that Florida needed to go even further, and he proposed Preservation 2000 in 1990, a massive funding effort for environmental land acquisition. By the early 1990s, Florida had the most impressive land acquisition program of any state, and it had secured vital territory in the Big Cypress area.

In the early 1980s, Florida already had a few ways to purchase environmentally endangered land. In 1963, state legislators amended the state constitution, authorizing officials to issue revenue bonds in order to acquire lands, water areas, or other resources in the interest of recreation or conservation. Income from these bonds was placed in a Land Acquisition Trust Fund.[1] The state empowered itself further in 1972 with the passage of two significant pieces of legislation. As we have already noted, the Florida Environmental Land and Wa-

ter Management Act allowed the state to designate regions as areas of critical state concern. The state also passed the Land Conservation Act, which established the Environmentally Endangered Lands Bond Issue, whereby the state could issue $200 million in bonds so that it could purchase, in the words of Graham, "environmentally significant and threatened lands."[2] Even though the Land Acquisition Trust Fund enabled the state to buy land for conservation purposes, some Florida officials, including Estus Whitfield, who served as an environmental adviser to Graham, saw Environmentally Endangered Lands as "the first major state land-acquisition program in Florida."[3]

The next important piece of legislation came in 1979 when the state legislature created the Conservation and Recreation Lands Program (also known as CARL). The Florida Department of Natural Resources called this plan a "direct successor to the Land Conservation Act."[4] It created a priority list for land purchases, administered by a Land Selection Committee. This committee had to follow certain procedures before placing lands on the final priority list, including holding public meetings and comparing and analyzing the selections. Each July, the committee presented its final priority list to the state cabinet for approval, and the state then worked to purchase the lands on that list.

Despite these measures, enormous growth in South Florida continued to threaten the environmental health of the region, especially the Everglades. Between 1970 and 1980, the population of Florida jumped from 6.8 million to 9.7 million, and it would escalate to 12.9 million by 1990. Demographers estimated that as many as 1,000 new residents came to Florida every day. Much of this growth occurred in South Florida, where areas such as Fort Lauderdale-Hollywood grew by 64.2 percent between 1970 and 1980. In addition, tourism was increasing, as 36 million people visited the state in 1980 alone. The tremendous growth had dire implications for South Florida's environment. "We're going to see enough water for the people," Patricia Dooris of the Southwest Florida Water Management District related, "but we're not going to have enough water to maintain the ecosystem that people expect in Florida."[5] Meanwhile, the sugar industry continued to expand in the EAA, creating pressure for more agricultural land and increasing the political influence of sugar growers. One 1984 publication noted that growers planted 349,000 acres to cane sugar in the EAA, a value of $600 million. This made Florida the largest sugar producing state in the United States. Likewise, farmers in Dade County were producing 75 percent of the country's winter vegetables and 95 percent of its limes.[6] The expanding agricultural industry and the increasing population meant the development of more and more land, leading to growing encroachment on Everglades National Park and Big Cypress National Preserve.

Workers harvesting vegetables in the EAA. (South Florida Water Management District)

The development especially affected wildlife. By the mid-1980s, some estimated that approximately 90 percent of wading birds in the Everglades had disappeared, dropping the population from more than 2.5 million in the 1930s to 250,000. The alligator was similarly imperiled, as were 13 other endangered species, including the Florida panther (placed on the Interior Department's first endangered species list in 1967), whose traditional habitat north of Big Cypress was jeopardized by citrus growers building orchards in the area. The problem for the panther (*Puma concolor coryi*) was that it needed around 300 square miles of land to hunt, and development intruded on that territory. State Highway 84, also known as Alligator Alley, a 76-mile-long roadway that crossed the Big Cypress National Preserve, the Fakahatchee Strand, and Water Conservation Area No. 3, bisected the panther's habitat as well, and the highway death rate for the animal surpassed its reproduction rate.[7]

In order to protect wildlife, as well as to preserve the quality of water flowing into Everglades National Park, restrictions on development or outright land purchases were necessary. But, as discussed previously, the Reagan administration did not place a high priority on environmental protection in the 1980s, and it weakened existing regulations and programs through budget cuts. The reduction in funding meant that little federal support was forthcoming for land purchases. As one periodical noted in 1985, the NPS "has not been buying land of late."[8]

Because of the federal government's attitude, Governor Graham and the state took even more responsibility to ease Florida's environmental stress. In 1981, the Florida legislature passed a bill to implement the Save Our Rivers program, which used revenue from a documentary stamp tax to create the Water Management Lands Trust Fund, administered by the Department of Environmental Protection. With this money, the state's five water management districts could purchase lands necessary for water management, water supply, and water conservation, following five-year plans that each district would develop. Also in 1981, the legislature established the Save Our Coasts program, which expanded the Land Management Trust Fund so that coastal lands could be acquired and preserved.[9] To coordinate these different programs, the state developed the Florida Statewide Land Acquisition Plan, thereby providing "a long-range strategy for the primary state-level acquisition programs."[10]

When Graham issued his Save Our Everglades program in 1983, land acquisition for environmental preservation was a big part of the program. Two areas especially were highlighted: Big Cypress, and the Holey Land and Rotenberger tracts (which bordered Conservation Area No. 3). Graham proposed that the state use Holey Land and Rotenberger as a wildlife buffer against agriculture and development, but one of the problems was that the Rotenberger Tract was a part of the Seminole Indian's state reservation and the Seminole were unwilling to agree to the flooding of this land. Negotiations over this area spilled into settlement talks over a lawsuit that the Seminole had introduced against the state in 1974, charging that Florida had never adequately compensated them for the flooding of their land in Conservation Area No. 3. After several years of negotiation, the two sides finally reached an agreement in September 1986. According to this settlement, the state would pay over $11 million to the Seminole for the Rotenberger Tract, the title and easement to other land flooded by Conservation Area No. 3, and for compensation for past projects conducted in Conservation Area No. 3A. Congress ratified this agreement in December 1987 under the Seminole Indian Land Claims Settlement Act, allowing the state of Florida to use the Rotenberger Tract for its buffer zone.[11]

Graham and state officials also focused on Big Cypress land acquisition, fearing that growth was adversely affecting both the Florida panther and the Big Cypress ecosystem. In order to obtain more information on these issues, he created the Big Cypress Area

Management Task Force, composed of representatives from Collier County, the SFWMD, Florida Department of Environmental Regulation, Florida Game and Fresh Water Fish Commission, Florida Department of Natural Resources, Florida Department of Transportation, the NPS, and the FWS. The governor instructed this committee "to review the known information concerning present and future access and uses in the Big Cypress Area" and to complete a report outlining the "environmentally sensitive areas" and what specific management actions were necessary.[12] In February 1983, the task force issued its conclusions. A "basic conflict" existed between "protection of endangered species and other uses of the area," it declared, such as hunting and the utilization of off-road vehicles. The task force also noted that plans were in the works to make Alligator Alley a part of Interstate 75, and it feared that this upgrade would "heavily impact the Preserve."[13]

The same effects could be seen in the Fakahatchee Strand, a tract of land located at the western end of Big Cypress Swamp and containing several watercourses and ponds, as well as hammock forests and over 45 species of orchids.[14] The state had designated the strand as a state preserve in 1974, and the Department of Natural Resources had attempted to acquire all of the acreage within the preserve. By 1983, it had purchased approximately 2/3 of the land, but it estimated that over 5,000 landowners still held tracts. According to an official

with the Florida Natural Areas Inventory (a project jointly operated by The Nature Conservancy and the Florida Department of Natural Resources to identify land and water areas in need of protection), it was essential to acquire these remaining lands, as well as three other parcels outside of the preserve's boundaries, in order to manage and protect the area effectively.[15]

Fakahatchee was especially important because it was one of the primary habitats of the Florida panther. Therefore, both the FWS and the Florida Game and Fresh Water Fish Commission examined how the panther could best be preserved in the area. Whatever the two agencies decided, it was clear that some action had to be taken, as the Fakahatchee was threatened with agricultural development and drainage projects. "It would be counterproductive to try and halt all development in the area," the Florida Natural Areas Inventory reported, but "development must be directed away from the most sensitive areas."[16]

Meanwhile, the NPS studied the question of how to protect more land in Big Cypress. In 1979, the Service had developed a land acquisition plan for the area, but in May 1982, the Interior Department had issued a policy statement that required any purchase programs to be either revised or replaced. Accordingly, in April 1983, the NPS announced that it was beginning a "land protection planning process" for the region. This included deciding what lands needed

Fakahatchee Strand Estuary. (South Florida Water Management District)

to be held in public ownership, as well as examining "the means of protection available to achieve the purpose of the Big Cypress as established by Congress."[17] The NPS noted that although 95 percent of the Big Cypress National Preserve had been purchased, over 550 tracts remained either in private or non-federal ownership. In order to maintain the preserve, the NPS would either have to acquire such lands or develop ways to manage them in accordance with the preserve's purpose. Robert L. Kelly, president of the Tropical Audubon Society, emphasized how important it was to complete the purchase of Big Cypress Preserve, stating that it was not only "an important area for several endangered species," but it also "protect[ed] the water supply of the western portion" of Everglades National Park.[18]

The importance of Big Cypress to endangered species was emphasized as state and federal officials continued to study the Florida panther problem. In October 1982, Graham and the state cabinet issued a one-year moratorium on oil and gas leasing within the Big Cypress National Preserve so that such practices would not "further compromise the panther's already tenuous survival" (only between 20 and 30 panthers still existed).[19] In addition, the Florida Game and Fresh Water Fish Commission tracked panther movement and habitat, and it reported in July 1983 that three "elements of public use" in Big Cypress especially threatened the panther: the utilization of access points and mineral roads, new kinds of off-road and all-terrain vehicles, and "a rapid increase" in recreation.[20] To combat these problems, the commission, together with the NPS, proposed certain management actions, such as requiring recreational use permits and annual vehicle registration, and supervising deer and hog hunting more closely (since those animals constituted the panther's main prey). The commission would also continue to study the panther, since any decision or action affecting the management of Big Cypress needed to be "based on sound logic and scientific knowledge."[21]

But Governor Graham decided that purchasing more land, rather than better management of the preserve, was needed, and he included acquisition of Big Cypress land and related regions as priorities in his Save Our Everglades program, announced in August 1983. Graham explained that development threatened the Florida panther, and he called on the federal government to purchase 70,000 acres of Big Cypress Swamp to provide protection. He also recommended that the state acquire the Fakahatchee Strand in order to forestall development in the panther's main habitat. The governor related that both Big Cypress and Fakahatchee held "immense ecological value," as they contained "some of the most diverse plant and animal communities in North America." At least four endangered animal species lived in the regions—the panther, wood stork, peregrine falcon, and red-cockaded woodpecker—as did 15 threatened plant species. Yet extensive development jeopardized this rich and fragile ecosystem. Therefore, Graham designated the Fakahatchee Strand as a "high priority for acquisition under the CARL program,"

and he lobbied the NPS "to increase its efforts to complete the acquisition of the Big Cypress National Preserve."[22]

Graham also asked President Reagan for federal help, informing the president that one of "the issues which must be resolved" was "completion of the acquisition of the Big Cypress National Preserve." This could not be done, Graham continued, without the cooperation of the Interior Department. "We pledge to work with you in such efforts," Graham stated, "with a special emphasis on the protection of Everglades National Park and the Big Cypress National Preserve."[23]

Before the Reagan administration could respond, Graham began formulating a land acquisition plan centered on the transformation of Alligator Alley into part of Interstate 75. Through consultations with federal officials from the Interior and Transportation departments, he reached an agreement that the restructuring of the highway be formulated so that the state could create a buffer zone around Everglades National Park, thereby protecting its resources and creating "permanent habitat for the Florida panther and other rare and endangered species."[24]

Under this proposal, the state and the federal government would acquire 165,000 acres in the Big Cypress Swamp and Fakahatchee Strand. The majority of this acreage, approximately 127,738 acres, was located northeast of the existing preserve, adjacent to the Miccosukee Indian Reservation, while the other 37,010 acres consisted of the northern part of Fakahatchee Strand. As a state news release reported, the land contained "wetlands, cypress swamp and hardwood hammock," as well as "a diversity of rare and endangered plants and animals including the panther, the bald eagle, and native orchids."[25] A large chunk of this acreage—mostly owned by Collier Enterprises and the Barron Collier Company—would be damaged by the highway expansion, necessitating damage payments by the Department of Transportation (90 percent) and the state (10 percent). State officials proposed that this compensation be used to reduce the total cost of acquisition, and that the state (20 percent) and the Interior Department (80 percent) assume the rest of the charges, with the state's contribution coming from Conservation and Recreation Lands Program funds. In addition, Graham proposed that the Department of Transportation build panther crossings into Alligator Alley, and that it design the reconstruction "to correct hydrologic problems in the Everglades."[26] On 18 April 1984, Graham announced this plan and asked Congress to approve it.

Over the next year, Graham met with Florida's congressional delegation to develop the necessary legislation, and in January 1986, U.S. Representative Thomas F. Lewis, a Republican from Palm Beach, introduced into Congress H.R. 4090, a bill to authorize additions to the Big Cypress National Preserve. U.S. Senator Lawton M. Chiles, Jr., a lifelong Florida Democrat, submitted a companion measure to the Senate (S. 2029), showing that, once again, environmental concerns in Florida were largely bipartisan. The bills proposed that the federal government add approximately 128,000 acres to the Big Cypress

National Preserve, to be known as the Big Cypress National Preserve Addition. These lands were necessary, the bills continued, in order to "limit development pressure on lands which are important both in terms of fish and wildlife habitat . . . and of wetlands which are the headwaters of the Big Cypress National Preserve."[27] The bills did not go into great detail about how the lands would be purchased, delineating only that the federal government would not pay more than 80 percent of the total cost (meaning the total acquisition costs minus any charges incurred by the Federal Highway Administration or the Florida Department of Transportation in damage payments).

In May 1986, the House Subcommittee on National Parks and Recreation of the Committee on Interior and Insular Affairs held a hearing on H.R. 4090. Several individuals testified in favor of the acquisition, including Florida's congressional delegation, James E. Billie, chairman of the Seminole Tribe of Florida, and Graham. As Graham stated, the acquisition would establish "an envelope of protection around Everglades National Park," preventing more Big Cypress acreage from becoming "citrus groves and subdivisions."[28] The purchase would also help preserve the Florida panther and other endangered animal and plant species.

Representatives from environmental organizations provided their support at the hearing. Paul C. Pritchard of the National Parks and Conservation Association told the committee that the Everglades Coalition had been reconstituted, in part to fight for "the addition of these critical environmental lands to the Big Cypress National Preserve," and he reported that the Sierra Club, Florida Audubon Society, Friends of the Everglades, and Florida Defenders of the Environment, among others, all backed the acquisition. Almost all of those who testified implored Congress to act quickly while the reconstruction of Alligator Alley was occurring, in order to minimize costs to both the federal government and the state of Florida.[29]

But not all were in favor of the acquisition. Benjamin G. Parks of the National Inholders Association protested the measure, stating that it left too many questions for property owners in the area. Claiming that approximately 1,000 litigation cases from the original establishment of Big Cypress National Preserve were still pending in court, Parks wondered whether an additional "4,000 small landowners" would be "left to battle in court for years . . . in order to receive a fair price on the property." Parks admitted that the legislation was

Florida panther. (South Florida Water Management District)

Big Cypress National Preserve. (South Florida Water Management District)

"well intended," but he insisted that this solution to "an alleged wild-life problem" would create "a very real problem to the people—the access to their property and recreational use of the preserve."[30] Many landowners agreed; James Humble, an avocado grower, had earlier related his displeasure with state and federal land acquisition efforts to the *U.S. News & World Report*, saying that "in the fervor for environmentalism, a basic property right is being run over."[31]

More startling, however, was the opposition expressed by P. Daniel Smith, deputy assistant secretary of the interior for fish, wildlife, and parks, at the hearing, especially since Graham had testified that Florida had worked closely with the Interior Department in the development of the acquisition strategy. According to Smith, the department could not "support the legislation based on current program and budgetary priorities" because it could not spend $40 million (Smith's estimate of the costs) for lands that "do not appear to be essential for purposes of the existing Big Cypress National Preserve." Smith claimed that no one had ever explored "the majority of the tract" to determine panther occupancy—Game and Fresh Water Fish Commission efforts notwithstanding—and he claimed that no panthers actually lived in the proposed acreage, since their primary habitat was located on the Fakahatchee Strand. The Interior Department was already involved in a process to acquire Fakahatchee Strand, Smith continued, and it was examining less expensive ways, such as exchanges, to acquire more land in South Florida.[32]

In some ways, Smith's testimony was accurate—the Fakahatchee Strand did constitute the main panther habitat, and most of the land to be purchased was not in the strand. However, Smith ignored the fact that panthers used a wide expanse of territory for hunting, and that the acquisition would protect such larger areas.[33] Another problem with Smith's testimony was that it implied that panther protection was the only reason why the legislation was necessary. While that was certainly an important reason for the measure, and perhaps even the driving force behind it, the preservation of water supply and water quality for South Florida was also a large reason why state officials wanted the land. Big Cypress National Preserve was created in 1974 to secure high quality water for Everglades National Park, and the addition would further that goal. Likewise, as Representative Lewis explained to his supporters, "this legislation is needed to ensure that South Florida's water supply will keep pace with its population increases," mainly because "the growing urban population of South Florida" was "dependent" on wetlands such as those in Big Cypress Swamp "as recharge sources for drinking water."[34]

Indeed, Representative John F. Seiberling (D-Ohio) believed that Smith was just throwing up a smokescreen to hide the Reagan administration's distaste for land purchases. "I am disappointed in the way the administration tempers its evaluation of things like this by the policy and ideology that is currently in vogue," Seiberling declared, charging that Reagan had his priorities "screwed up." The administration continually displayed an unfavorable reaction to

"anything but the military," Seiberling continued, even though the Florida situation constituted "a very serious problem." "We ought to have a more cooperative approach from the Administration," he concluded.[35]

When the Senate held a hearing on Chiles' bill in September 1986, William P. Horn, who also served as assistant secretary for fish, wildlife, and parks, made the same declarations as Smith. "We are aware that the lands covered by this legislation would be a desirable addition to the Big Cypress National Preserve," he stated, but "a distinction must be drawn between desirable lands and critical lands." The cost of the acquisition was prohibitive as well, Horn said, making it impossible for the Interior Department to support the measure.[36]

But Horn explained that another possibility existed to acquire at least some of the land: an exchange. According to Horn, the Interior Department proposed to give Collier Enterprises and Barron Collier Corporation federal lands in Phoenix, Arizona, currently housing a Navajo Indian school and worth $100 million, in exchange for 115,000 acres owned by the Colliers in South Florida and a payment of $50 million. The lands provided by the Colliers would consist of 70,000 acres to be added to the Big Cypress National Preserve; 16,000 acres to be used in the development of a Florida Panther National Wildlife Refuge; and 20,000 acres to establish the Ten Thousand Islands National Wildlife Refuge. If the exchange worked, Horn continued, the state of Florida would have to acquire only 57,000 acres to equal what it wanted the Interior Department to purchase, and the state already had the money necessary to do this under its Conservation and Recreation Lands Program program. Horn claimed that the land gained through the exchange would "constitute a significant addition to both the Park and Refuge systems . . . at no direct cost to the federal taxpayer," something that the Reagan administration fully supported.[37]

Nathaniel Reed, who was representing the environmental community, could see no real objections to the Interior Department's plan, other than the difficulty of its execution. The exchange "has got some real politics and . . . some very, very difficult things to overcome in Arizona," he related, but "it is doable, perhaps." Yet Reed also advised Congress to pass the proposed legislation anyway as a sort of "safety" measure in case an exchange could not be effected.[38] Steven Whitney, director of the national parks program for The Wilderness Society, agreed, explaining that "the exchange . . . would not be foreclosed by the passage of this legislation."[39]

Accordingly, the House of Representatives approved Lewis's bill in July 1986, but the Senate took no action before the adjournment of Congress, in part, according to one source, because of "Administration opposition to all discretionary federal land purchases."[40] Lewis and Chiles planned on reintroducing the measures in the subsequent Congress, believing that "the level of support" for the original bills indicated "a strong possibility" that they could pass.[41] Accordingly, in January 1987, the two submitted S. 90 and

Pine prairie in the Big Cypress National Preserve. (South Florida Water Management District)

H.R. 184, which, for the most part, were no different than the previous measures, except that they now proposed that 136,000 acres be acquired instead of 128,000. The new bills also specified that the Seminole Indians would maintain their traditional hunting, fishing, and trapping rights in the addition, just as they did in the original preserve.[42]

In February 1987, the Senate Subcommittee on Public Lands, National Parks and Forests held a hearing on the reintroduced bills. As in the 1986 hearings, many of the same individuals and organizations testified in favor of the legislation. Graham, who by now had become a U.S. Senator for Florida, claimed that not only was the addition necessary to preserve endangered wildlife such as the panther, it was also "crucial to the success of the Save our Everglades initiative, and to the maintenance" of Everglades National Park.[43] Florida Governor Bob Martinez declared his support for the addition as well, stating that "timely authorization and funding of federal acquisition" would serve several purposes, such as protection of Everglades National Park and southwest Florida's water supply, preservation of the Florida panther and other endangered species, and more recreational opportunities.[44] In addition, environmental groups continued to support the legislation, although some now asked the Senate to amend the bill so that the entire Fakahatchee Strand could be added to the preserve.[45]

The Interior Department, however, maintained its opposition to the measure; Horn testified that unless the bill was "amended to provide for acquisition by exchange," the department would not support it. "Our fundamental problem with S. 90 has nothing to do with the resources," Horn insisted, but everything to do with fiscal conservatism. He reiterated that the Arizona/Florida land exchange proposal was still the best method to pursue, and he claimed that the department was "in the process of completing negotiation" of such an exchange. Upon further questioning, Horn admitted that the interested parties in Arizona had not agreed to the exchange, but he believed that "the outlook [was] entirely positive."[46]

Whether because of the Interior Department's intransigence or not, no federal legislation was forthcoming in the summer of 1987. The major holdup came from the Senate, as the House had passed the legislation in March. In December, the Senate finally considered S. 90, with Chiles and Graham offering several amendments. First, the senators increased the amount of acreage to be acquired to 146,000. Second, they included a new section in the legislation, specifically allowing oil and gas exploration, development, and production in the area under certain terms. Production companies had to obtain a permit from the NPS before conducting any activities in the addition, for example, and the secretary of the interior would have to establish rules for such activity largely in conformance with regulations in "similar habitats or ecosystems within the Big Cypress National Preserve."[47] No debate ensued over these amendments, and they readily passed, as did the entire measure.

When the bill went back to the House, Representative Bruce Vento of Minnesota noted that the legislation would not affect the land exchange discussions. However, Vento also explained that the "Arizona-Florida land exchange has proven to be quite complex." He was not sure whether it would come to fruition, but he encouraged the House to pass the amended bill anyway, stating that the government should not "delay the addition of critical lands to the Big Cypress National Preserve on the basis of a land exchange that may or may not come about."[48] The Everglades Coalition agreed, petitioning Congress at the coalition's third annual conference in January to approve the measure.[49] Accordingly, the House passed the bill, and President Ronald Reagan signed it into law in April 1988, although he insisted that the land exchange proposal be pursued and executed as a condition of his approval.[50]

Only a few days after Reagan signed the act, Representative Morris Udall of Arizona introduced the Arizona-Florida Land Exchange bill into Congress. This measure, Udall explained, would allow the transfer of 118,000 acres owned by the Collier family in South Florida (valued at $45.1 million) to the federal government in exchange for 68 acres in downtown Phoenix and a payment of $34.9 million. That money would be used to establish an Indian education trust fund to compensate the Navajo for the loss of their boarding school, while the Collier's land would be added to the Big Cypress National Preserve and used to create the Florida Panther National Wildlife Refuge.[51] Governor Martinez of Florida strongly supported this bill, calling it "a unique opportunity to acquire over one-half of the authorized Big Cypress Addition relatively quickly," and he urged Florida's congressional delegation to work hard for its passage.[52]

For the next several months, this measure was debated in both chambers of Congress. Various senators and representatives raised objections over the valuation of the land and the effect that the closure of the Navajo school would have on the Indians—some, such as Representative Sidney R. Yates of Illinois, even characterized the proposal as "a cozy, private, preferential deal" between the secretary of the interior and the Collier interests.[53] Many, however, including Lewis, Graham, and Chiles, advocated the measure as a win-win situation for both Florida and Arizona. Likewise, in a hearing held on the bill in July 1988, numerous environmentalists and state officials favored the acquisition. Finally, on 18 November 1988, the measure, known as the Arizona-Florida Land Exchange Act, became law.[54]

Yet the authorization did not prevent problems that developed in the execution of the exchange. For one thing, Collier interests and the city of Phoenix could not reach agreement on the use of the exchanged land. Barron Collier wanted to create 7.7 million square feet of commercial and residential space on the acreage, while the city wanted a 90-acre park. For another, a recession in the first part of the 1990s devalued the land from $80 million to between $25 and $35 million, causing the Collier interests to declare that they were no longer interested in the exchange. By 1991, the agreement was still undecided, and the federal government had made no appropriations for the purchase of the Big Cypress land.[55] However, in December 1996, the land exchange finally occurred, meaning that by the end of the 1990s, 146,000 acres had been added to the Big Cypress National Preserve, making a total of 700,000 acres under protection.[56] In addition, 26,400 acres in the northern portion of the Fakahatchee Strand was preserved as the Florida Panther National Wildlife Refuge, administered by the FWS.

Some environmentalists, however, believed that the protection was not adequate. For one thing, hunting and off-road vehicles were still permitted in the Big Cypress National Preserve, even though earlier state studies had noted the threat that they posed to wildlife. For another, original landowners were permitted to maintain their mineral rights in the land, and oil and gas drilling could still occur, albeit under NPS supervision. To many, these concessions, granted at the bequest of Congress due to the pleadings of groups such as the National Rifle Association, meant that Big Cypress National Preserve was a "park service stepchild" and that the NPS could not adequately protect the region's ecology.[57]

Regardless, the state of Florida had successfully obtained the means to acquire Big Cypress land. Yet state officials were not done. Governor Martinez, who many environmentalists had believed would work against ecological concerns and the Save Our Everglades program, actually accelerated environmental land acquisition efforts during his one term as governor. In the late 1980s, he ensured that Florida's Conservation and Recreation Lands Program fund had $43 million for land acquisition, and he proposed that over the next nine years, the state enhance the fund by $200 million.[58]

As the 1990s approached, Martinez developed a plan to generate even more money for land acquisition, in part because, even with the efforts that the state had already made, much of Florida's rich ecology still remained in danger. In 1990, a state commission investigating environmental concerns reported that by the year 2020, another three million acres of wetlands and forests would be lost to development. It was also estimated that, according to the Department of Environmental Protection, "about 19 acres per hour of forest wetland and agricultural land was being converted for urban uses." Many Floridians were concerned with these facts; a November 1989 poll indicated that 88 percent of Florida residents wanted the state to devote more attention to the environment. Aware of these trends, Martinez proposed—and the Florida legislature passed—a huge land acquisition program in 1990 known as Preservation 2000.[59]

Under Preservation 2000, the state increased the tax on real estate documents to fund an additional $300 million in bonds every year. With such an arrangement, Florida would have a $3 billion land preservation fund by the year 2000—more than the federal government spent on environmental land acquisition efforts. Martinez justified such a huge amount by saying that Floridians had "an important choice to make: We can buy up environmentally sensitive lands

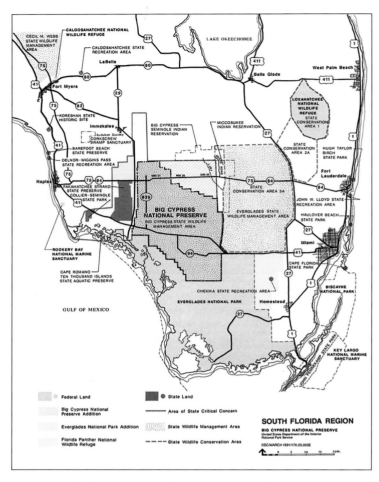

Map of the Big Cypress National Preserve and Addition. (National Park Service, *Big Cypress National Preserve, Florida: General Management Plan, Final Environmental Impact Statement, Volume 1,* 5)

that would otherwise be lost to future generations, or we can let a golden opportunity slip by."[60] Newspaper editorials called the program "staggering," "unprecedented," and "one of the most significant environmental initiatives in the past two decades," and environmentalists were pleased as well. "This, I think, is going to change the face of Florida more than any single thing I can think of," Nathaniel Reed noted.[61] Others agreed. Ernest "Ernie" Barnett of the Department of Environmental Protection claimed that Preservation 2000 was the shining environmental jewel in Martinez's administration, and that it established a program that spent more per year than the federal government and many small countries on land acquisition.[62]

Yet Preservation 2000 was merely the culmination of land acquisition programs that the state developed in the 1960s and 1970s. Always one of the most ambitious states in terms of environmental land purchases, the state increased its efforts in the 1980s with the establishment of the Conservation and Recreation Lands Program, the Save Our Rivers and Save Our Coasts legislation, and the Save Our Everglades plan. This focus on land acquisition was necessary in the 1980s because of the Reagan administration's discouragement of federal land purchases. In the words of one NPS officer, "it was administration policy that they didn't want to be expanding parks that they'd have to pay for."[63] Even though all sides agreed that additional lands in Big Cypress Swamp and in the Fakahatchee Strand were necessary for preservation, their acquisition occurred only after the Interior Department negotiated an exchange—not a purchase—between the Collier interests and the federal government. At the same time that the state fought for Big Cypress acquisition, however, an even bigger battle was occurring over another area in need of protection: the East Everglades. The fight over that land—largely between the NPS, the Corps, and agriculturists, hunters, and others interested in the region—would dwarf the Big Cypress difficulties.

Endnotes

[1] See Supreme Court of Florida, Opinion No. 76,984, 20 December 1990, Florida Supreme Court Briefs & Opinions, College of Law Library, Florida State University, Tallahassee, Florida <http://www.law.fsu.edu/library/flsupct> (15 February 2006). In 1968, a new constitution was adopted in Florida that prohibited the issuance of any of these bonds, but Senate Joint Resolution No. 292 rescinded that prohibition and allowed the Land Acquisition Trust Fund to continue.

[2] Senator D. Robert Graham, "A Quiet Revolution: Florida's Future on Trial," *The Florida Naturalist* 45 (October 1972): 149.

[3] Whitfield interview, 11.

[4] Florida Department of Natural Resources, "Statewide Land Acquisition Plan and Procedure for Individual Project Design and Acquisition Phasing," 20 November 1984, 2-3, Appendix 4, copy in File Department of Natural Resources Statewide Land Acquisition Plan, 1984-85, Box 02161, SFWMDAR.

[5] As quoted in "Florida's Growth Straining Fragile Groundwater," *Engineering News Record* 216 (3 January 1985): 26.

[6] Hansen, "South Florida's Water Dilemma," 17, 20; Land Acquisition Selection Committee for the Board of Trustees of the Internal Improvement Trust Fund, *Florida Statewide Land Acquisition Plan* (Tallahassee, Fla.: Board of Trustees of the Internal Improvement Trust Fund, 1986), 1 [hereafter referred to as *Florida Statewide Land Acquisition Plan*]; Southwest Florida Regional Planning Council, "Demographics" <http://www.swfrpc.org> (16 December 2005).

[7] Moreau, "Everglades Forever?" 72-74; James Carney, "Last Gasp for the Everglades," *Time* 134 (25 September 1989): 26.

[8] Yates, "Saga of the Glades Continues," 38.

[9] Florida Department of Natural Resources, "Statewide Land Acquisition Plan and Procedure for Individual Project Design and Acquisition Phasing," 3.

[10] *Florida Statewide Land Acquisition Plan,* 2. It appears that the need for a comprehensive plan was first addressed in 1984. See Florida Department of Natural Resources, "Statewide Land Acquisition Plan and Procedure for Individual Project Design and Acquisition Phasing."

[11] Seminole Indian Land Claims Settlement Act of 1987 (101 Stat. 1556); Kersey, "The East Big Cypress Case, 1948-1987," 466-474.

[12] Quotations in Big Cypress Area Management Task Force, "Report to Governor and Members of the Cabinet," 17 February 1983, 3-4, File Big Cypress Area Mgmt Task Force Report, 1983-85, Box 02186, SFWMDAR;

see also Rob Magee, State of Florida Department of Community Affairs, Division of Local Resource Management, to Nancy Linnan, 5 July 1983, Folder 30, Box 25, Series II, Marjory Stoneman Douglas Papers, Manuscript Collection 60, Archives and Special Collections, Otto G. Richter Library, University of Miami, Miami, Florida [hereafter cited as Douglas Papers].

[13] Big Cypress Area Management Task Force, "Report to Governor and Members of the Cabinet," 29-30.

[14] Senate Committee on Energy and Natural Resources Subcommittee on Public Lands, National Parks and Forests, *El Malpais National Monument and Big Cypress National Preserve: Hearing Before the Subcommittee on Public Lands, National Parks and Forests of the Committee on Energy and Natural Resources, United States Senate,* 100th Cong., 1st sess., 1987, 74 [hereafter referred to as *El Malpais National Monument and Big Cypress National Preserve*].

[15] Steve Gatewood, Florida Natural Areas Inventory, to Jim McKinley, FLTNC, 6 June 1983, Folder 36, Box 25, Series II, Douglas Papers.

[16] Gatewood to McKinley, 6 June 1983.

[17] "National Park Service Initiates Land Protection Planning for Big Cypress National Preserve," National Park Service News Release, 25 April 1983, Folder 2, Box 23, Series II, Douglas Papers.

[18] Robert L. Kelley, "The Future of Everglades National Park," Folder 27, Box 25, Series II, Douglas Papers.

[19] "Panther Preservation in the Big Cypress National Preserve: A Discussion of the Issues," March 1985, File PRO Everglades Panther, Box 21213, SFWMDAR.

[20] "Florida Panther Recovery: A Status Report to The Governor and Cabinet, July 7, 1983," File PRO Everglades Panther, Box 21213, SFWMDAR.

[21] "Florida Panther Recovery: A Status Report to The Governor and Cabinet, July 7, 1983."

[22] "Issue Paper: Save Our Everglades," 9 August 1983, *9-10, copy in Folder 16, Box 2, Marshall Papers. The Save Our Everglades program also called for land acquisition efforts in East Everglades; this will be dealt with in another chapter.*

[23] Bob Graham, Governor, to Honorable Ronald Reagan, 8 August 1983, Folder 16, Box 2, Marshall Papers.

[24] "Save Our Everglades: Restoring Everglades National Park," 18 April 1984, 2, Folder 16, Box 2, Marshall Papers.

[25] "Save Our Everglades: Restoring Everglades National Park," 2-5.

[26] "Save Our Everglades: Restoring Everglades National Park," 2-5.

[27] Quotations in House Committee on Interior and Insular Affairs Subcommittee on National Parks and Recreation, *Additions to the National Park System in the State of Florida: Hearings Before the Subcommittee on National Parks and Recreation of the Committee on Interior and Insular Affairs, House of Representatives,* 99th Cong., 1st and 2d sessions, 1985 and 1986, 105-107 [hereafter cited as *Additions to the National Park System*]; see also Bob [Graham], Governor, to Mrs. Marjory Stoneman Douglas, Friends of the Everglades, 25 February 1985, Folder 85, Box 30, Series II, Douglas Papers.

[28] *Additions to the National Park System, 164.*

[29] *Additions to the National Park System, 187-188.*

[30] *Additions to the National Park System, 173-174.*

[31] As quoted in Ronald A. Taylor, "Saving a Fountain of Life," *U.S. News & World Report* 100 (24 February 1986): 64.

[32] *Additions to the National Park System,* 151-152.

[33] For a good discussion of panther movements in the Big Cypress area, see Levin, *Liquid Land,* 99-117.

[34] "Big Cypress Addition: An Attempt to Preserve Water, Life in 'Glades," *Congressman Tom Lewis' Florida Environment Report,* n.d.

[35] *Additions to the National Park System,* 158-159.

[36] Senate Committee on Energy and Natural Resources Subcommittee on Public Lands, Reserved Water and Resource Conservation, *Additions to the Big Cypress National Preserve; Establishing the San Pedro Riparian National Conservation Area; Designating the Horsepasture River as a Component of the National Wild and Scenic Rivers System; and Amending FLPMA: Hearing Before the Subcommittee on Public Lands, Reserved Water and Resource Conservation of the Committee on Energy and Natural Resources, United States Senate,* 99th Cong., 2d sess., 1986, 99-103 [hereafter referred to as *Additions to the Big Cypress National Preserve*].

[37] *Additions to the Big Cypress National Preserve,* 99-103.

[38] *Additions to the Big Cypress National Preserve,* 162.

[39] *Additions to the Big Cypress National Preserve,* 162.

[40] "Big Cypress Expansion Legislation Fact Sheet, January 1987," File Washington Office Big Cypress, Box 2, S1401, Executive Office of the Governor Subject Files, 1987-1998, FSA.

[41] "Big Cypress Addition: An Attempt to Preserve Water, Life in 'Glades."

[42] *El Malpais National Monument and Big Cypress National Preserve,* 20-30, 41.

[43] *El Malpais National Monument and Big Cypress National Preserve,* 39-42.

[44] Bob Martinez, Governor, to Senator Dale Bumpers, Chairman, Senate Subcommittee on Public Land, National Parks and Forests, 12 February 1987, in *El Malpais National Monument and Big Cypress National Preserve,* 43-44.

[45] *El Malpais National Monument and Big Cypress National Preserve,* 356. At a January 1987 meeting of the Everglades Coalition, the organization encouraged the Reagan administration to support the new measures, stating that "all available methods including possible land exchanges, should be considered to bring these critical lands under public ownership." "Statement of Everglades Coalition Members, Lake Wales, Florida, January 17, 1987," File 10-1-7a (Kissimmee River-Lake Okeechobee, FL) 12222, Box 25, Accession No. 077-96-0033, RG 77, FRC.

[46] *El Malpais National Monument and Big Cypress National Preserve,* 67-81.

[47] *Congressional Record,* 100th Cong., 1st sess., 11 December 1987, 133:17971.

[48] *Congressional Record,* 100th Cong, 2d sess., 1 March 1988, 134:528.

[49] "Coalition Outlines Plan to Halt Everglades Decline," *National Parks Magazine* 62 (March/April 1988): 9.

[50] Act of 29 April 1988 (102 Stat. 443); Ronald Reagan, "Statement on Signing the Big Cypress National Preserve Addition Act," 29 April 1988, Ronald Reagan Presidential Library Archives <http://www.reagan.utexas.edu/archives> (20 December 2005).

[51] *Congressional Record,* 100th Cong, 2d sess., 3 May 1988, 134:2897.

[52] Bob Martinez, Governor, to The Honorable Lawton Chiles, United States Senate, 6 July 1988, File Big Cypress Land Swap, Box 88-02, S1331, Executive Office of the Governor, Brian Ballard, Director of Operations, Subject Files, 1988, FSA.

[53] *Congressional Record, 100th Cong, 2d sess., 14 July 1988, 134:5764.*

[54] Act of 18 November 1988 (102 Stat. 4571); see also *Congressional Record,* 100th Cong, 2d sess., 27 July 1988, 134:5895.

[55] "Everglades Land Swap Imperiled," *National Parks Magazine* 65 (September/October 1991): 10-11.

[56] See "Save Our Everglades: A Status Report by the Office of Governor Lawton Chiles," 30 June 1996, 9, copy provided by SFWMD.

[57] See, for example, Levin, *Liquid Land,* 101-102.

[58] "Past Accomplishments," File Governor's Environmental Initiatives, Box 1, S1401, Executive Office of the Governor Subject Files, 1987-1988, FSA.

[59] Florida Department of Environmental Protection, "History of Florida's Conservation Efforts" <http://www.dep.state.fl.us/lands/acquisition/P2000/BACKGRND.htm> (20 December 2005).

[60] As cited in "Florida's Preservation 2000," *St. Petersburg Times,* 25 January 1990.

[61] Reed and editorial quotations in "Florida's Preservation 2000," *St. Petersburg Times,* 25 January 1990; see also Florida Department of Environmental Protection, "History of Florida's Conservation Efforts."

[62] Ernest "Ernie" Barnett interview by Matthew Godfrey, 23 September 2004, Tallahassee, Florida, 3-4.

[63] As cited in "Everglades Land Swap Imperiled," 10.

CHAPTER 11

Brewing Storm: Development, Water Supply, and the East Everglades

ECOLOGICAL PROBLEMS IN EVERGLADES NATIONAL PARK were getting worse fast. That was the consensus among science experts at the beginning of the 1980s. The good news was that scientific evidence increasingly pointed to poor water management as the underlying cause of most of the Everglades' biological decline. If water deliveries to the park could be rectified, it followed, the Everglades might be saved. But scientific studies indicated that it was not enough simply to guarantee minimum quantities of water to the park. Rather, water management had to be modified so that water entering the park was distributed more nearly in the historic pattern of sheet flow. Moreover, the timing of water deliveries and duration of inundation—what was called "hydroperiod"—had to parallel the natural rainfall pattern.

Even as scientific understanding of the Everglades ecosystem improved, engineering solutions for modified water management became more difficult. Most of the sheet flow into Everglades National Park came through two broad sloughs: Shark River Slough and Taylor Slough. The entrance to both sloughs was an area bordering the east edge of the park that had remained practically uninhabited until recent years. By the early 1980s, it was lightly populated and portions were under cultivation. Initially, efforts to modify water deliveries to the park through this area focused on re-engineering options that would balance the park's water supply requirements with the flood control needs of these area residents. By the end of the decade, those options no longer appeared realistic. In certain portions of this hotly contested area, the protection of park values was incompatible with flood control. Increasingly, water managers believed it was necessary to buy out the landowners and change the use of the land. This thinking culminated in the Everglades National Park Protection and Expansion Act of 1989.

The East Everglades is the name given to an area bordering the east side of Everglades National Park. The East Everglades area in the 1980s (until a portion was added to the national park in 1989) encompassed some 153,600 acres or approximately 242 square miles.[1] It included the headwaters of Shark River Slough (until 1989) and Taylor Slough. Shark River Slough, the larger slough at approximately 25 miles wide, gathers some of its waters from north of the park in Conservation Area No. 3B; that portion of the slough that runs through the East Everglades area is called Northeast Shark River Slough. Taylor Slough drains approximately 40 square miles southeast of Shark River Slough. Sawgrass marshlands predominate throughout this area, while hardwood hammocks, or tree islands, occur on higher elevations.[2]

In terms of hydrology and biology the area is part of the Everglades ecosystem; in terms of land use and ownership it constitutes the farthest limits of Dade County's urban/rural interface abutting the park. Its boundaries in the 1980s were the Tamiami Canal on the north, the national park on the west and south, and Levee 31 and Canal 111 on the east. The Tamiami Canal and Levee, or L-29, it will be recalled, formed the southern edge of Conservation Area No. 3 and was completed by the Corps in 1963. L-31 was a southern extension of the eastern perimeter levee, while C-111 was at the southern end of this system and dated from the mid-1960s.[3]

Although the land was mostly in private ownership, it remained largely uninhabited until the 1970s. Farther south, in the C-111 basin, it was still uninhabited in the 1980s, inundated by water during much of the rainy season. Early in the twentieth century the state had offered these lands for 25 cents an acre; speculators had purchased them in the 1920s and sold them as bonus property for

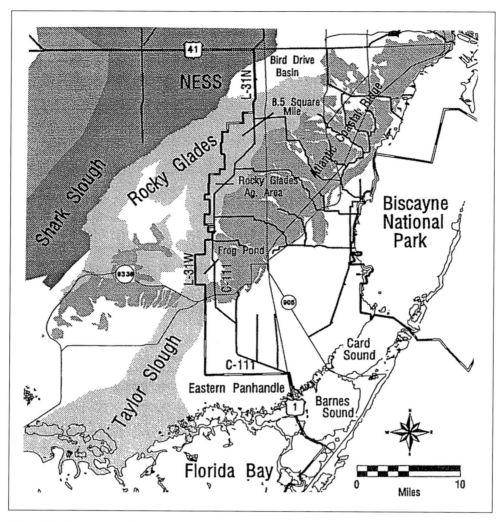

Map of the East Everglades area. (Everglades National Park, "Seepage Control in Western Dade County," 1994)

the buying of land elsewhere. A generation passed and the worthless parcels of waterlogged property were sold and resold. The water conservation areas were created and portions of the East Everglades began to dry out. Even then, most of the area remained too wet to inhabit. Under the 1948 and 1954 congressional authorizations of C&SF Project works, the Corps had received approval to construct L-31, running southwest from L-29, in order to provide flood and salinity protection to the area. As proposed, L-31 would consist of a northern portion (L-31N) and a western portion (L-31W); the western component would encompass a 5,000-acre area known as the Frog Pond, located at the head of Taylor Slough. Farmed as early as the 1940s, the Frog Pond had agricultural potential, but had attracted only a few vegetable growers to this point. In the 1960s, the Corps aligned the different L-31 parts, completing construction of L-31W in the early 1970s. These engineering works, combined with severe drought in 1971, exposed more ground, and by the mid-1970s the situation changed rapidly as people moved in to build residences and raise crops, attracted by the comparatively low price of land in this area of southwest Dade County.[4]

The government of Dade County was ambivalent about development of this area. In the early 1960s, the Metro-Dade County Commission supported a water control plan that would permit agricultural use during the dry season. The Southwest Dade Project received congressional authorization in 1965, but after considerable planning by the Corps it was never built. County support for the project waned as the NPS began to voice opposition, fearing that the project would complicate efforts to convey water to Taylor Slough and the southeast corner of the park. County officials were even less enthusiastic about agricultural development in Southwest Dade County following the drought of 1971, which heightened concern about saltwater intrusion into the Biscayne aquifer. Nevertheless, Dade County classified nearly the entire area as agricultural in its land-use plan, and it supported construction of canals and levees north and west of the town of Homestead, which served to drain that area for crop production. It also passed zoning ordinances with stringent performance criteria that, in the minds of some residents, conveyed a public commitment to flood protection.[5]

The Corps, too, saw problems with development of the East Everglades. A farming practice known as "rock plowing" was particularly harmful to the environment. Farmers, when clearing the land for planting, would plow up the limestone substratum to augment the thin layer of topsoil. This limestone was very porous and acted as a filter as water percolated into the Biscayne aquifer. The Corps maintained that rock plowing was harmful to the aquifer. It discouraged the practice through its Section 404 permitting program, although by this time the Frog Pond area already bore the scars of rock plowing and the practice continued elsewhere to a limited extent.[6]

The Corps also required homeowners in the East Everglades to apply for Section 404 permits as they were filling in wetlands to improve their home sites. The agency notified some 50 residents that they must obtain permits. After residents ignored repeated orders by the Corps to desist, the U.S. Attorney's Office brought suit against five offenders. U.S. District Judge James W. Kehoe ruled in favor of the government, persuaded by the testimony of eight government experts that dumping in the Everglades lowlands threatened the environment. One of the defendants, Russell Carter, formed a local group in defense of property rights and posted an anti-government sign in his yard. Another landowner threw "a pesky government bureaucrat" into a pond housing his pet alligator, while still others told the media that "if we don't get justice through the courts we'll get it with our guns." Fairly or not, the homesteaders in the East Everglades were gaining a prickly reputation.[7]

In August 1981, Tropical Storm Dennis soaked the East Everglades with three days of torrential rain. Row crops of tomatoes and malanga disappeared under water. Avocado, lime, and mango groves were ruined. Roads around Florida City (between Homestead and the national park) remained impassable for three weeks following the storm. Farmers and homeowners angrily confronted SFWMD employees, accusing the district of deliberately ignoring their plight. Someone threw a pipe bomb into one of the canal structures on the edge of the East Everglades, though it failed to explode.[8]

If the county government had been dubious of this community before, it now took definite action to curtail further growth in the area, which soon became known as the 8.5 Square Mile Area. This moniker arose from the passage of a zoning ordinance on 27 October 1981 by the Board of County Commissioners stating that within the region, located in the East Everglades area west of the levees separating the park and the Miami suburbs, the county would permit only a maximum of one dwelling unit per 40 acres for residential use or one dwelling unit per 20 acres in conjunction with agriculture, replacing the existing one house per 5 acres regulation. The ordinance also required notification of property purchasers and individuals seeking building permits that the county and the SFWMD had no drainage plan for the area. The Governing Board of the SFWMD endorsed the ordinance at a special meeting eight days prior to its enactment.[9]

The county aimed to slow growth in the 8.5 Square Mile Area in order to protect Miami's water supply, but it had difficulty enforcing the ordinance. Some property owners openly defied the ordinance, building homes without permits. Others worked the system, obtaining a permit to build seasonal housing for migrant farm workers, for example, and subsequently expanding it into a second permanent dwelling, thereby getting two residential dwellings onto 40 acres. Many of the landowners were Cuban refugees who were not familiar with the permitting process. Indeed, many had already paid too much for land that had been cynically advertised as "waterfront property" and they perceived the zoning restrictions and absence of flood security as added injustices.[10]

Settlement of the East Everglades, and specifically of the 8.5 Square Mile Area, encroached on the national park, resulting in loss of wetlands and wildlife habitat and further interfering with the natural sheet flow of water into the park through Northeast Shark River Slough and Taylor Slough. As homesteaders in the East Everglades drained their land, it set in motion the familiar train of disturbances: marsh fires, soil subsidence, and invasion of exotic species such as Australian pine and melaleuca. To have this occurring on the park's doorstep was especially harmful to fauna and flora. Many wildlife species used the upper sloughs within the East Everglades for nesting, feeding, foraging, and cover. It was estimated that the East Everglades had at one time provided 35 percent of South Florida's wood stork feeding grounds, but water level manipulation rendered the area unsuitable for this species during its crucial nesting period. Development pressure also caused a reduction in incidence of 12 rare, endemic plant species, which appeared to have a detrimental effect on the biological diversity and productivity of the flora in the adjoining portion of the park.[11]

Park officials viewed these developments with growing concern. John M. Morehead, superintendent of Everglades National Park, welcomed the Dade County ordinance as a sign of "exceptional foresight" on the part of local officials, but it was the Corps, a sister federal agency, that held the key to improved water management in the area. In a long letter to Colonel Alfred Devereaux, Jacksonville District Engineer, in September 1982, the superintendent expressed gratitude that the Corps was studying water deliveries through the Shark River Slough and he urged the Corps to "examine ways to rejoin the historical hydrological equilibrium between the east Everglades, the Water Conservation areas, and Everglades National Park."[12]

Morehead stressed the importance of Shark River Slough for wildlife habitat, and explained that loss of wetlands within the East Everglades as a result of development or hydrological change would substantially reduce wildlife populations in the park. "We believe that this condition must be reversed," he noted. "Everglades National Park and the east Everglades are hydrologically connected and should be treated as one hydrological unit." In other words,

Everglades National Park provided habitat for rare birds such as the Snowy Egret. (Everglades National Park)

Morehead urged a management approach that did not stop at the park boundary. In the superintendent's opinion, the federal interest in preserving Everglades National Park justified federal action outside the park. "There can be little doubt," Morehead stated, "that the future of Everglades National Park is intimately tied to the future of the Shark Slough within the east Everglades."[13] This was a bold position that reflected recent changes in NPS thinking about how to approach "external threats" to national parks.

Morehead explained that the paramount need of the park was to obtain a more natural flow of water based upon rainfall in the drainage north of the park. Currently, water deliveries were based on a minimum monthly delivery schedule established under P.L. 91-282 (1970). A new approach was required. For one thing, the 260,000 acre-feet minimum allocation for Shark River Slough was based on median flows through Tamiami Trail culverts during the period 1940-1962. It had to be recognized, Morehead wrote, that these flow data were below pre-drainage era levels since six major Everglades drainage canals were already operational by that time.[14]

A revised water delivery schedule must also take into account natural fluctuations in rainfall within each season, Morehead ar-

gued. The current schedule was based on average monthly flows. The resulting monthly breakdown of the schedule provided for peak flows in October and minimum flows in April and May. While the current schedule did provide for some deviations in time of drought or high water, the magnitude and timing of the deviations were determined by urban and agricultural water supply needs rather than the park's ecological needs. Morehead wanted a schedule that would allow fluctuations "in synchrony with the natural system; a system to which the slough's animal and plant populations have become adapted over millennia."[15]

Morehead had been arguing these points for two years. He had written a similar letter to Colonel Devereaux's predecessor, Colonel James W. R. Adams, in July 1980, in response to proposed changes in regulation water levels for Lake Okeechobee and the water conservation areas. The concern in the summer of 1980 was that Everglades National Park could face a water shortage if there was a "drawdown" of water levels in these other areas. The SFWMD and the Corps were making plans to lower the water level in the conservation areas for the purpose of giving vegetation (and the deer population) a chance to rebound under drier conditions. But the park's concerns went beyond questions of water shortage. The regulation levels tended to flatten out extreme high and low water events. While this certainly benefited agriculture and the urban populations, it harmed the park. "The trouble biologically with all such modulations is that in time the modulation reduces animal and plant diversity and favors only those few species that happen to be adapted to the modulation," Morehead cautioned. "In an ecosystem like the Everglades marsh that isn't particularly diverse in species to begin with, reduction in population and loss of species can happen dramatically and rapidly." Morehead posed a series of technical questions to Colonel Adams concerning the regulation schedules for the conservation areas. His last question was: "What modifications could be done to make the system react more fairly to systemwide rainfall?"[16]

The problem was deceptively complex. The C&SF Project was unique among the Corps' flood control projects in that it covered vast areas of impounded waters moving very slowly over a nearly imperceptible gradient. From an operational standpoint it took weeks, even months, to move water from one end of the system to the other. The Corps followed a schedule of water releases for each area that was more weighted toward water supply and flood control, although

other purposes such as fish and wildlife benefits were also taken into account. Maximum water levels were set according to normal rainfall patterns, meaning that when water levels rose above the maximum allowed in the regulation schedule the Corps had a responsibility to open gates and move water out of the area, if only as a hedge against flooding in case of abnormally high rainfall perhaps two or three months in the future. Colonel Adams readily admitted that the system was imprecise and subject to the caprice of nature. "We're in a situation where science has gone about as far as it can in predicting Mother Nature but she still has the last card to play," he told an interviewer in 1981.[17] Not surprisingly, while the park superintendent focused on fluctuations in nature, the District Engineer concerned himself with weather extremes.

The winter of 1982-1983—an El Niño event—produced the kind of freakish weather that Colonel Adams warned about. Heavy rains began in October and continued through the winter—normally Florida's dry season—culminating in a 60-day, 20-inch deluge in January and February. The SFWMD made emergency releases from Lake Okeechobee to the St. Lucie and Caloosahatchee estuaries but that was not enough. The Corps opened the floodgates in

Gate structure at S-12C. (South Florida Water Management District)

the Tamiami Canal and levee along the northeast boundary of Everglades National Park, and from October 1982 through February 1983 the park received three-and-a-half times its minimum quota.[18] In the month of February, when the minimum quota for releases through these structures was 9,000 acre-feet, the park received a whopping 88,000 acre-feet.[19] While the C&SF Project afforded admirable flood protection to sugar cane fields in the EAA, winter crops in Dade County, and all the coastal cities in South Florida, the Corps succeeded only by dumping much of the excess water on the park. Ironically, at the same time that the four open gates in L-29 were disgorging water into the park at the rate of approximately five billion gallons per day, engineers and hydrologists in the Corps and the SFWMD were beginning to consult with the park's chief scientist, Gary Hendrix, on how to develop a rainfall-driven water delivery schedule. This was hardly the type of deviation from average monthly flows that Superintendent Morehead had in mind.

Indeed, the dumping of excess water on the park caused severe ecological damage. According to Jim Kushlan, a park wildlife biologist, the practice destroyed alligator nests and disrupted feeding patterns of the woodstork.[20] Morehead agreed. "Just as soon as the birds and gators would get their nests settled, they'd get blown away by waves of water," he related. These releases destroyed the "natural wet-and-dry rhythm" that had characterized water flow in the Everglades before drainage began, harming the lifeways of both flora and fauna.[21]

Insisting that floodwaters were causing grave harm to the park, Morehead and Hendrix requested an emergency meeting with the SFWMD. On 10 March 1983, Hendrix arrived alone in West Palm Beach and presented a seven-point plan to the SFWMD's Governing Board. He began by saying that in the past few months the park had been assessing the effects of the water delivery schedule that had been in place since 1970 and the park staff had concluded that most of the degradation to ecological values in the park had occurred from excess water in the dry season. (This was the opposite, of course, from the longstanding perception that the Everglades was dying of thirst.) Both he and the superintendent believed that without some "urgent measures" the park could not "sustain much of its resources for very long."[22]

The first four points in the seven-point plan aimed at undoing the fragmentation of Conservation Area No. 3 and restoring sheet flow to the park. The plan called for filling in the L-28 canal, which ran north and south down the boundary of Conservation Area No. 3A and Big Cypress National Preserve; filling in the L-67 canal extension and removing the levee; rededicating Conservation Area No. 3B for water storage and sheet flow; and redistributing water deliveries from Conservation Area No. 3A along the whole length of the Tamiami Canal from L-28 to L-30. Collectively, these four actions would redistribute water flowing into the park from the confined area around the four floodgates in the Tamiami

Canal to several historic drainages south of Big Cypress as well as the Northeast Shark River Slough. The fifth item called for a water quality monitoring program. The sixth was a request that the Corps and SFWMD defer any implementation of new drainage districts until impacts to the park were fully considered. The seventh and final point was to field test a new water delivery schedule starting as soon as possible.[23]

At the end of Hendrix's presentation the Governing Board recommended that the SFWMD's executive director, John "Jack" Maloy, report to the board in a month with a studied response. When board member Jeanne Bellamy asked Maloy for his off-the-cuff reaction, he said that he was "overwhelmed." "Oh, you're never overwhelmed," Bellamy prompted. "I'll tell you one thing," Maloy replied. "This is a real test of whether the organization is . . . a regular bureaucratic organization or something different." A request of this scope, Maloy explained, would normally take the organization three years just to study it. "By then the Park will be a desert," Bellamy cut in. Maloy noted that most of the points Hendrix brought with him had been discussed already with district engineers, but this was a lot to consider all at once.[24]

In a letter to Marjory Stoneman Douglas and 16 other "Everglades watchers," Nathaniel P. Reed, a member of the governing board, offered further commentary on the meeting, characterizing the Seven Point Plan as a "bombshell." Reed seemed most surprised that Morehead, who enjoyed an exceptionally close working relationship with the board, had chosen to send a messenger instead of appearing himself. "When the District and the Park have had a problem during Morehead's tenure, his presence, his explanation, his superb ability at negotiation have made the Board willing to find some area of cooperation," Reed explained. After Hendrix made his proposal, Reed continued, the Corps representative, Carol White, "appeared to have apoplexy." The emergency ploy was even a bit worrisome to Reed, who wondered if "emergency actions" were appropriate or wise. He noted that such Everglades experts as Art Marshall and Johnny Jones had "expressed sincere reservations" about implementing some of the park's proposals without further study.[25]

As a former assistant secretary of the interior, Reed offered his own analysis of the park's Seven Point Plan. "*The Park's request represents a major change in attitude*," he headlined. "The new approach may be the result of the flood conditions inundating the Park or may reflect the Superintendent's view that as the Department of the Interior and the National Park Service are not actively defending the Park's integrity [then] the local representatives must declare the present state of affairs an 'emergency' requiring 'emergency' measures."[26]

Following the meeting, Maloy acted decisively. He redeployed staff to evaluate the Seven Point Plan at lightning speed, and then he called an "emergency meeting" of the Governing Board for 5 April. Defining the situation as an "emergency" gave him authority to issue an order without prior notice. He had his legal staff prepare a draft

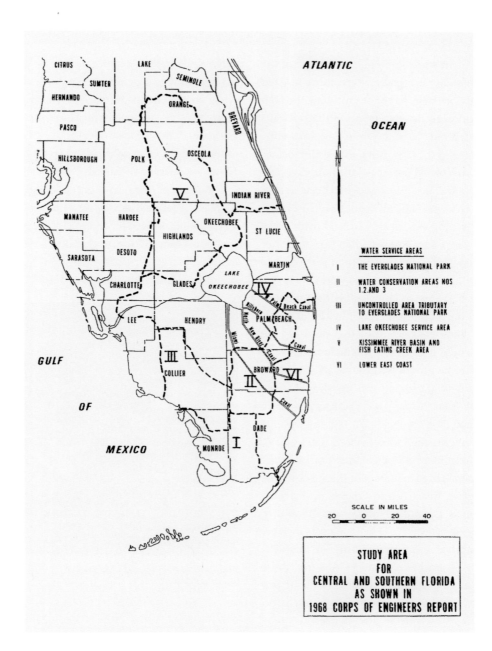

CITRUS — SUMTER — HERNANDO — PASCO — HILLSBOROUGH — POLK — LAKE — SEMINOLE — ORANGE — OSCEOLA — BREVARD — ATLANTIC OCEAN — MANATEE — HARDEE — INDIAN RIVER — OKEECHOBEE — ST LUCIE — HIGHLANDS — SARASOTA — DESOTO — MARTIN — CHARLOTTE — GLADES — LAKE OKEECHOBEE — PALM BEACH — LEE — HENDRY — BROWARD — COLLIER — DADE — GULF OF MEXICO — MONROE

V — IV — III — II — VI — I

Hillsboro Canal — West Palm Beach Canal — Miami Canal — New River Canal — North New River Canal — Canal

WATER SERVICE AREAS

I   THE EVERGLADES NATIONAL PARK

II   WATER CONSERVATION AREAS NOS 1,2,AND 3

III   UNCONTROLLED AREA TRIBUTARY TO EVERGLADES NATIONAL PARK

IV   LAKE OKEECHOBEE SERVICE AREA

V   KISSIMMEE RIVER BASIN AND FISH EATING CREEK AREA

VI   LOWER EAST COAST

SCALE IN MILES
20   0   20   40

STUDY AREA
FOR
CENTRAL AND SOUTHERN FLORIDA
AS SHOWN IN
1968 CORPS OF ENGINEERS REPORT

Study area of the water supply restudy. (U.S. Army Corps of Engineers, Jacksonville District, *Central and Southern Florida Water Supply: Reconnaissance Report*, 1979)

emergency order with findings of fact and conclusions of law for the Governing Board's approval.[27] At the Governing Board meeting on 5 April, Maloy duly reported on the SFWMD's technical response to the Seven Point Plan. He recommended support for all seven measures. If the board approved, the district would approach the Corps and the NPS and ask for financial help, recognizing the federal interest in protecting the national park. Maloy presented the board with the draft emergency order, which the board adopted.[28]

Representatives of several environmental groups also attended the emergency meeting and expressed support for the Seven Point Plan. Peter Mott, president of the Florida Audubon Society, correctly noted that the seventh point, calling for a field test of a new water delivery schedule (and the abandonment of the present congressionally mandated one) would require an act of Congress. He predicted a "united front" on this matter in the coming year.[29] Two other attendees, Michael Hevener, executive director of the Dade County Farm Bureau, and William Earl, counsel for that organization, spoke on behalf of 5,800 farmers and 7,000 farm workers of Dade County. They worried that the Seven Point Plan would cause flooding, wreck crops, and damage private property. They wanted an EIS "for any structural changes that would affect the farming interests in south Dade County."[30] Immediately following the meeting, Earl sent a letter by courier to Colonel Devereaux, commander of the Jacksonville District, requesting an EIS.[31]

As the Corps and the SFWMD entered discussions about implementing the Seven Point Plan, it became clear that the Corps had serious misgivings. The main sticking point was the fate of the residential development in the East Everglades. Devereaux insisted that the Corps would not support steps to restore full flow through Northeast Shark River Slough until its study showed that area inhabitants would not be flooded out. Others were less sympathetic. Charles L. Crumpton, a member of the Governing Board and former Dade County planner, observed that 50 percent of the houses and 90 percent of the mobile homes in the East Everglades had been erected without building permits.[32] No state officials would say so explicitly, but to Devereaux the meaning was clear: "To heck with these people. Just flood them out. Then they'll move. Then they'll get out of there." Devereaux disagreed with such a position, in part because he believed that, even though the property owners were "operating at their own risk in that area," the federal government should and could not "deliberately flood somebody, or increase the risk of flooding, without compensation." Devereaux admitted that restoring the flow to Shark River Slough would not cause immediate flooding of landowners, but it would raise the groundwater table, elevating the possibility of a flood. "I just did not personally feel, nor did my superiors feel," he later recollected, "that the Corps of Engineers could be party to anything that would do that."[33]

To break this impasse, Congressman John Seiberling (D-Ohio), chairman of the House Subcommittee on Public Lands and National Parks, visited South Florida with three members of the subcommittee—James Weaver (D-Oregon), Bruce Vento (Democrat-Farmer-Labor-Minnesota), and Thomas Lewis (R-Florida)—at the end of April. Their three-day tour ended at SFWMD headquarters in West Palm Beach, where Seiberling had some stern words for the Jacksonville District commander. "I am here because I was told there is an emergency by a lot of experts," Seiberling said. "It's your position there is no emergency?" Devereaux responded, "There is no emergency right now. No sir." Everyone in attendance—the four congressmen, the Governing Board members, SFWMD staff, and park staff—could not believe the colonel's words. Seiberling asked incredulously, "There is a congressional subcommittee here because there's an emergency . . . What does it take to prove it?" Congressman Lewis demanded, "What makes you think we don't have an emergency? What does it take—the East Coast sliding off?"[34] The colonel remained impassive.

Devereaux probably held his ground on this point because he was operating under a different code of authorities than the SFWMD and the NPS. He later explained to an interviewer, "I couldn't use emergency measures, because emergency measures can't be used for environmental purposes."[35] The Corps' authorities to deal with an emergency came from Public Law 84-99, first passed in June 1955 and amended several times since. According to this law, when flooding, hurricanes, or drought constituted an emergency, the Corps could engage in any action "which is essential for the preservation of life and property," such as strengthening existing flood control struc-

tures, constructing temporary levees, clearing channels, removing debris and wreckage once a flood had receded, and providing clean water to regions in need. Nowhere in the act did it authorize the Corps to take emergency measures for environmental preservation purposes.[36] By contrast, Florida state law explicitly allowed Maloy authority to protect wildlife and fish if he and the Governing Board found an emergency existed.

Nathaniel Reed, who sat at Seiberling's side in the meeting at SFWMD headquarters, by now had resolved any doubts in his own mind about the need for emergency measures. Reporting on the congressional tour to Richard Davidge, a Watt loyalist who occupied the assistant secretary position that Reed himself had held in the Nixon and Ford administrations, he wrote: "To everyone's astonishment,

Flooding in East Everglades area. (U.S. Army Corps of Engineers, Jacksonville District)

Col. Devereaux, the District Engineer, was totally uncooperative." It was not clear to Reed if the colonel was acting under orders or on his own initiative, but he had never seen such a hard line in his 30 years of involvement with the Everglades. "Apparently, the District staff is in an upheaval. The older in service staff members are rear guarding and resist any changes. The younger staff [members] want to solve the ongoing Everglades crisis and agree to be innovative." Reed ended his letter to Davidge with a warning and a plea. "The general perception is that the Administration has written off Everglades Park. I urge you to give this issue priority."[37]

Throughout 1983, the SFWMD strove for rapid implementation of the Seven Point Plan, including opening S-333 in April to allow water to flow to the eastern third of Everglades National Park. The Corps, meanwhile, took a more deliberative approach. Those different approaches were evident in how each agency dealt with challenges to the plan from farmers and property owners. Perhaps the most serious challenge involved efforts by residents of the 8.5 Square Mile Area to have the restrictive county zoning ordinance lifted. Spurred on by droughts in the 1970s that convinced many that flooding would not be a serious problem, the area had grown into a community of approximately 800 persons, who had constructed several hundred residences and agricultural structures to serve the region's numerous plant nurseries and farms.[38] As we have seen above, Tropical Storm Dennis debunked the flooding myth, making some residents clamor for a government-sponsored drainage plan and better flood protection, while others merely wanted to subdivide their land and cash out.

Underpinning the zoning ordinance was a county ruling that the East Everglades was an "Area of Critical Environmental Concern." When Maloy learned that the Dade County Board of Commissioners was considering a repeal of that ruling, he acted swiftly and decisively to move the issue up to the state level. On 22 June, he appealed to Dr. John M. DeGrove, secretary of the Department of Community Affairs (the state land planning agency), to initiate the process of designating the East Everglades an "Area of Critical State Concern."[39] Governor Bob Graham established the Everglades National Park/East Everglades Committee on 7 February 1984—a major step in the designation process and a strong indication that the state would likely assert control if Dade County backed off its own environmental protection plan.[40]

The Corps, meanwhile, contended with a legal challenge, which would eventually become known as *Kendall v. Marsh*, after the Dade County Farm Bureau filed suit in U.S. District Court on behalf of East Everglades farmers and property owners. Concerned that knocking gaps into L-67 would flood 80,000 acres of vegetable and fruit farms, the farmers sought an injunction that would prevent the Corps from modifying any structures in the C&SF Project until it completed an EIS. The farm bureau contended that the removal of levees was not an emergency procedure and that agriculturists would "suffer substantial and irreparable harm" from "higher ground water and increased flooding danger" if the Corps was allowed to proceed.[41] Wanting to forestall litigation, and without any authority to implement the Seven Point Plan, the Corps moved cautiously on any elements that might result in flooding of crops and homes in the East Everglades. The Corps also arranged for meetings with the farmers without notifying or consulting the SFWMD, much to the dismay of Maloy. Despite these efforts, the suit continued into 1985.[42]

The Corps' deliberative approach frustrated the park superintendent, state water managers, and environmentalists, all of whom wanted prompt action and believed that the Corps should move ahead undeterred by the threat of lawsuits. The Corps was already on record concerning the first four points in the Seven Point Plan—the modifications to L-28 and L-67 and the redistribution of waters in Conservation Area No. 3A and 3B—but the report was still in draft. To implement those measures immediately would be to circumvent the standard process of sending project proposals up through the Board of Engineers and Congress. As Colonel Devereaux later explained to an interviewer, the park, by declaring an emergency

and getting members of Congress involved, "put an extraordinary amount of political heat on the Corps to implement these things as rapidly as we could." It placed Devereaux in a tenuous position because he did not have legal authority to expend funds for the actions that the park and SFWMD wanted done.[43]

The Corps was relatively receptive to the first action: modifications to the western levee, or L-28, in Conservation Area No. 3A. It was the least controversial action because it did not affect agricultural interests in Dade County. The Corps modified this levee so as to divert waters entering 3A back into Big Cypress National Preserve, from which they flowed to the western side of Everglades National Park. To accomplish this it breached the L-28 tie-back levee, installed culverts connecting the inside and outside canals on either side of the L-28 levee, and put plugs in the lower collector canal.[44] This work was completed in March 1984.

The second point in the plan, removal of the L-67 extension, was more problematic, as it required the removal of structures already built. The Corps finally agreed to take more modest measures. It would install two control culverts or "plugs" in the canal in order to add resistance to its flow, forcing some of the water to move to the west. It also discussed putting gaps in the last four miles of the levee. When Nathaniel Reed heard of this he wondered if the park was backing off its request to have the entire canal and levee removed. Morehead informed Maloy, "We go along with these gaps only because it is action of some sort."[45] However, the Corps did not actually place gaps into L-67 until another crisis arose a few years later over the status of the Cape Sable seaside sparrow.

When it came to redistributing the waters in Conservation Area No. 3A and 3B in order to restore more sheet flow into the park, the Corps and the SFWMD disagreed about what to do. All the water entering the park from Conservation Area No. 3A came through a set of four gates spaced along the Tamiami Canal called S-12A, S-12B, S-12C, and S-12D. Most of the water came through S-12D. In order to spread this inflow into the park the Corps closed S-12D, forcing more water through the other three gates. The SFWMD argued that this was a half-measure. It proposed to use S-333 and divert water from Conservation Area No. 3A into the Tamiami Canal, where it would flow east and then south through a series of 53 culverts under U.S. Highway 41, thereby feeding into the Northeast Shark River Slough. The Corps maintained that this would be a misuse of S-333 as it would likely flood out residents in the East Everglades area. The disagreement became bitter as state water managers tried to assert their prerogative to operate the C&SF Project as they saw fit, while the Corps insisted that the interests of property owners in the East Everglades must come first.[46]

Maloy raised the dispute over S-333 with the Department of Environmental Regulation, threatening to make control of the C&SF Project into a states rights issue. Colonel Devereaux offered to meet with the governor.[47] Finally, in January 1984, the Corps and the SFWMD reached a compromise; the Corps consented to new operating criteria for S-333 and water began to flow through this gate into Northeast Shark River Slough. The operating criteria were to be incorporated into the field test of a new water delivery schedule for the park (the seventh item in the Seven Point Plan).

The last point in the Seven Point Plan had to be addressed by Congress. Congressman Dante Fascell (D-Florida) introduced a bill in the House authorizing the Secretary of the Army to modify the water delivery schedule for the park. The measure was incorporated into a supplemental appropriations act for 1984, enacted in November 1983. The law provided for a two-year field test to begin immediately and authorized the Secretary of the Army to acquire farmlands that would be subject to flooding and to construct flood protection works for homes in the area. It provided $10 million for land acquisition.[48]

The Corps regarded this law with a great deal of skepticism—it could not possibly do all these things in the two-year timeframe that the law required—but by January 1985 it had prepared a "General Plan for Implementation of an Improved Water Delivery System to the Everglades National Park." The plan set out a strategy for how the Corps would comply with P.L. 98-181. It required an innovative,

expedited process, for the law had already circumvented the usual steps in which the Corps reported to Congress with a reconnaissance study, followed in a few years by a feasibility study. Instead, the Corps would proceed straight to the preparation of a General Design Memorandum, and, concurrently with that effort, it would prepare an EIS and conduct a "limited field test." The field test would "not significantly impact residential or agricultural interests."[49]

The field test rested on a compromise agreement that the Corps had worked out with the park, the SFWMD, and area farmers during the preceding year. The farmers, in their lawsuit against the Corps, raised two demands. The first, as already noted, was to delay additional water releases into the Northeast Shark River Slough until the Corps had prepared an EIS. The second demand was that the Corps should continue its annual fall drawdowns of water levels in the Frog Pond to assist fall planting. The park believed that the fall drawdowns, which flushed water through the L-31W and C-111 canals into Barnes Sound, sucked water out of the park as well. On the recommendation of the SFWMD, and to head off litigation, the park agreed not to object to the fall drawdowns for one year if, in turn, the farmers agreed not to oppose water releases into Northeast Shark River Slough.[50]

Gate structure at S-333. (South Florida Water Management District)

By the time the field test was set to begin, it came under the purview of the Everglades National Park/East Everglades Committee, established by Governor Graham in February 1984. This committee was also called the 380 Committee because it was formed according to Chapter 380 of the Florida statutes for the purpose of recommending whether the East Everglades should be designated an Area of Critical State Concern. Governor Graham charged this committee with finding consensus among the many disparate agencies and competing interests that had locked horns over water management in the East Everglades. The committee included federal, state, regional, local, tribal, and non-government representatives. The Miccosukee Tribe was represented on the committee, as were environmentalists, Dade County businessmen, East Everglades residents, and farmers. Colonel Devereaux sat on the committee for the Corps, and Superintendent Morehead represented the NPS, while the Florida Game and Fresh Water Fish Commission participated on behalf of fish and wildlife interests.[51]

After more than a year of study, the committee submitted an "implementation plan" to the governor. The plan proposed a three-part strategy for improving water management in the area. The first part of the strategy was to establish an "iterative testing process." (This corresponded with the "field test" authorized by Congress.) Incremental changes to existing structures and operating procedures would be introduced and analyzed to determine best water management practices. The process, of course, would involve collaboration by the Corps, the SFWMD, and the NPS. In the second part of the strategy, the committee formed the Southern Everglades Technical Committee, a subgroup of hydrologists and ecologists who would review the analysis on the iterative testing process and recommend changes. Recognizing that this group's recommendations could be controversial, the third part of the strategy was to impose a conflict resolution process for solving, or even mediating, disputes as they arose.[52]

The committee's implementation plan focused, appropriately enough, on fashioning a workable administrative process. But it also provided a consensus-based view of the myriad land and water management problems that beset the park and the East Everglades. Significantly, the 90-page report recognized the ecological importance of restoring sheet flow through Northeast Shark River Slough, as well as the adverse impacts of certain farming practices, including rock plowing, on water quality. The committee also recommended that the 8.5 Square Mile Area be provided with flood protection adequate to protect the community from a one-year-in-ten flood.[53]

The considerable time and commitment that went into the Everglades National Park/East Everglades Committee sowed good will among the many parties, and it produced about three years of concerted effort at building consensus. The Corps and the SFWMD began making field tests of water flows into the East Everglades in early 1984 and continued making them through the following year and into the next under the committee's watchful eye. Near the end of 1985 the Corps began making controlled releases of water south of the Tamiami Canal to simulate natural sheet flow in response to rainfall, while the SFWMD used the field tests to refine its hydrological computer model.[54]

Meanwhile, the NPS initiated studies of aquatic vegetation where the sheet flow was tentatively being restored for the purpose of measuring water quality. These studies showed alarming results. Water flowing into the park from Conservation Area No. 3 was so laden with nutrients from the agricultural areas that it was altering plant life in the park. Both the Corps and the SFWMD had a growing body of data on water quality based on water sample analysis. According to an interagency memorandum of agreement on water quality executed in February 1984 (pursuant to the fifth point in the Seven Point Plan) the Corps collected samples of surface water at specific locations and tested them for pesticide residues and trace metals, providing data to the SFWMD and the NPS on a monthly basis. The SFWMD had a similar responsibility.[55] Experts all over South Florida recognized that agriculture was loading nutrients into an ecosystem that was naturally nutrient-deficient; the spread of cattails through the water conservation areas provided proof. What they did not yet know was the extent to which nutrient "dosing" (or the addition of nutrients to the area) was affecting the ecology of Everglades National Park.[56]

Amid this synchronous hum of activity by the three agencies, a turnover of leadership occurred: Colonel Charles Myers III relieved Devereaux of command over the Jacksonville District, John R. "Woody" Wodraska replaced Maloy as executive director of the SFWMD, and Michael Finley took the place of Morehead as superintendent of Everglades National Park. The new leadership, coupled with a perception on the part of environmentalists that changes in water management were occurring too incrementally, led to renewed disagreement over how to implement a new water management regime in the East Everglades.[57]

Superintendent Finley brought a new edge to the park's demands. Finley was a rising young star in the NPS, and Everglades National Park was a difficult post. The director of the NPS, William Mott, met Finley at National Airport in Washington, D.C., for what Finley thought was an interview. Instead, Mott simply told Finley he wanted him to go down to Florida and do what he could for the Everglades. Finley arrived in June 1986, and it did not take him long to decide that the Everglades were in "great jeopardy," that this was a "system approaching collapse." He quickly came to appreciate that the causes and the politics were complex. The Seven Point Plan and the Supplemental Appropriations Act of 1984 notwithstanding, Finley took the view that "the park was not at the table nor taken seriously by any of the water management agencies." His job, he believed, "was to get the park taken seriously."[58]

Sawgrass, once abundant, became displaced in areas with high nutrient levels. (South Florida Water Management District)

Despite Finley's efforts, the SFWMD altered its position on the East Everglades following a 1987 change in state governors from the Democrat Graham to the Republican Bob Martinez. Whereas Graham had appointed a number of champions of Everglades National Park to the Governing Board of the SFWMD, the new governor returned the membership of the board to a more pro-business, pro-agriculture orientation. Woody Wodraska, executive director of the SFWMD since 1985, responded to the new dynamic—as did farmers. Although the SFWMD continued to support new approaches to water management in the East Everglades, after 1986 it leaned more toward agricultural interests.

The breach between the SFWMD and the park occurred over the Frog Pond, which had begun to attract interest in the mid- to late 1970s when drier conditions encouraged much more intensive use. Farmers began planting tomato crops in the area as soon as standing water receded in the fall. After a winter harvest and the coming of summer rains, the Frog Pond once more filled with water. With the

return of wetter weather in 1982, the Corps and the SFWMD began operating the L-31W canal as a means to prevent these tomato crops from being flooded in the fall and winter. The park considered this use of the canal inappropriate, since it had been built as part of the South Dade Conveyance System for the purpose of getting water to Taylor Slough and the southeast corner of the park. In 1984, the park consented to this use of the L-31W canal for one year in return for the farmers' permission to allow field tests in Shark River Slough. The SFWMD renewed this arrangement with the farmers for two years following. In May 1987, the Governing Board passed a resolution calling for a phase-out of the use of L-31W by 1990, and construction of an internal drainage system so that excess water in the Frog Pond during the winter growing season would drain east. The purpose was to ensure there would be "no net reduction in farmable acreage in the Frog Pond."[59]

It seemed to Superintendent Finley that neither the SFWMD nor the Corps was following through on earlier commitments to restore natural flow to the Taylor Slough. Draining the water to the east would still leave Taylor Slough in short supply. Moreover, draining the Frog Pond to the east depended on lowering the water level in the C-111 canal, with consequences for the southeast corner of the park and Florida Bay. The water level in the C-111 basin was normally maintained by a gated culvert structure or "plug" (S-197) in the lower end of the C-111 canal, which the Corps had added to the project in the 1960s as a result of a lawsuit by the National Audubon Society. Occasionally the Corps removed this plug to provide flood relief for the C-111 basin. It had done so in 1981, 1982, and 1985. Overruling the park's objections, the Corps removed the plug again in 1988. For eight days, freshwater discharged in massive quantities through the C-111 canal into Barnes Sound and Florida Bay, with deadly consequences for the saltwater marine life.[60]

For all of these reasons, Finley believed the NPS must take a separate road in order to get acceptable water management. "My view," Finley recalled in an interview, "was that this was going to have to be forced either by public opinion and politics or by the courts. Individual agency action wasn't going to do it—they either didn't have the guts or the ability to do it."[61] One surprise, however, was Governor Martinez's strong support for his predecessor's "Save Our Everglades" program. When Martinez was elected governor in November 1986, the environmental community was dubious. The Everglades Coalition immediately invited the governor-elect to address the coalition's second annual conference in January. Meanwhile, Governor and Senator-elect Bob Graham communicated with Martinez about the importance of sustained gubernatorial focus on the federal-state agenda for Everglades restoration. According to one administrative official, it was Graham's intention to present the new governor "with early opportunities to work visibly and productively with the Congressional Delegation on Save Our Everglades issues."[62] Graham had built strong public support in Florida for Everglades restoration, and the outgoing governor suggested that Martinez would be wise to embrace this popular agenda. Should he do so, it would "help establish a positive climate for dealing with the Congressional Delegation on other issues of interest to the Governor."[63] Martinez took this bait. After taking office, he quickly positioned himself to lead a number of Save Our Everglades program initiatives, and he retained Graham's Save Our Everglades program coordinator, Estus Whitfield, on his staff.

In 1987, Martinez took important steps for expanding Big Cypress National Preserve, improving the water quality of Lake Okeechobee, and accomplishing restoration of the Kissimmee River. If there was one thing that distinguished Martinez's overall approach to saving the Everglades from that of Graham's, it was the Republican governor's emphasis on just compensation for private property takings. "The key to protecting and restoring the Everglades is land acquisition," Martinez announced toward the end of 1987.[64] At Martinez's urging, the state legislature increased the Conservation and Recreational Lands fund by $200 million over the next nine years.

With regard to the East Everglades, Martinez announced on 22 January 1988 a federal-state initiative to acquire approximately 70,000 acres in public ownership. Two months later, the governor established the East Everglades Land Acquisition Task Force, with members drawn from federal, state, and local government, as well as the environmental community and private landowners. The task force's job was to evaluate the feasibility of joint federal/state acquisition of the land, and to develop a plan for acquiring, managing, and protecting it. In particular, the task force was to report to the governor in six months as to whether the state of Florida ought to support federal legislation to expand Everglades National Park in this controversial area.[65]

The task force made its report to the governor on 1 October 1988. It recommended three areas for inclusion in Everglades National Park: first, the Northeast Shark River Slough, containing 70,740 acres; second, the state-owned East Everglades Wildlife and Environmental Conservation Area, containing 34,560 acres; and third, an area between the wildlife sanctuary and the L-31 canal, containing about 2,300 acres. Five other tracts, it stated, should not be included: the area between the L-31 canal and Krome Avenue (the outskirts of Homestead), the 8.5 Square Mile Area, the developed agricultural area south of it, the Frog Pond, and an area south of

The Frog Pond agricultural area. (U.S. Army Corps of Engineers, Jacksonville District)

the Frog Pond known as the Aerojet lands. However, most of the Aerojet lands were newly acquired by the SFWMD using Save Our Rivers program moneys and the task force suggested this area might be added to the park at a later time. It proposed that the lands be acquired using the federal land acquisition process. It suggested that hunting should be prohibited and airboat use should be phased out in the additions to the park. It also recommended that field tests of modified water delivery to the park, currently set to expire in January 1989, should be continued "until the land acquisition is accomplished and the permanent water delivery program proposed in the Corps of Engineers General Design Memorandum begins."[66]

Soon after the committee made its recommendations, Congressman Dante Fascell introduced legislation expanding Everglades National Park in the House, while Senator Bob Graham and Senator Connie Mack III (R-Florida) co-sponsored similar bills in the Senate. Mack, a former member of the House, had been elected as Florida's junior senator in November. The bipartisan showing by Florida's two senators helped the bill's prospects. Also important was the election of George H. Bush as president. In his political campaign, Bush had promised to be "the environmental president," a pledge environmentalists regarded with skepticism. Yet it did seem that Bush was genuinely more interested in protecting ecological values than President Ronald Reagan. A few weeks prior to his inauguration, President-elect Bush went sport fishing in the Florida Keys, and, through a prior arrangement, Superintendent Finley boarded Bush's boat for a 20-minute chat. At the end of the conversation, Bush indicated that he would support the park expansion bill provided that it was bipartisan.[67]

There was little outright opposition to the legislation. Fiscal conservatives were concerned about the cost of land acquisition—an estimated $32 million according to the NPS or $70 million according to the Corps. Sportsmen's groups wanted the area added to the national wildlife refuge system rather than the park. Dade County farmers had reservations about the modified water delivery plan, but they generally wanted a horse trade: restoration of sheet flow to the park for greater flood protection in nearby agricultural areas. The SFWMD backed the legislation with the proviso that the bill should be amended to recognize the multi-purpose nature of the C&SF Project.[68] These were the main outlines of the demands for making the legislation bipartisan and acceptable to all interests.

After extensive amendment of the bill in committee, Congress enacted it in November 1989. The purpose of the act was first, to increase protection and "to enhance and restore the ecological values, natural hydrologic conditions, and public enjoyment" by adding certain lands to the park; and second, to assure that the park was "managed in order to maintain the natural abundance, diversity, and ecological integrity of native plants and animals . . . as a part of their ecosystem."[69] The act provided specific steps for modifications to the C&SF Project, and directed the Corps to complete a General Design

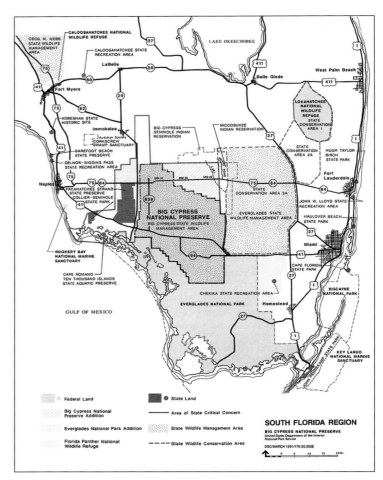

1989 Additions to Everglades National Park. (National Park Service, *Big Cypress National Preserve, Florida: General Management Plan, Final Environmental Impact Statement, Volume 1*, 5)

Memorandum entitled "Modified Water Deliveries to Everglades National Park." The study was to include flood protection, if warranted, for the 8.5 Square Mile Area and the adjacent agricultural region. With regard to the C-111 basin, the General Design Memorandum was to "take all measures which are feasible and consistent with the purposes of the project to protect natural values associated with Everglades National Park."[70]

The law stated that construction of modifications to the C&SF Project were justified by environmental benefits and did not require further economic justification. Thus, the General Design Memorandum would not be subject to the Corps' usual cost-benefits analysis. Funds for the so-called Modified Water Deliveries project would subsequently come out of Interior Department appropriations acts, since this was national park legislation.

The law defined project purposes generally, but it stopped short of declaring that the project was multi-purpose, as Wodraska had requested in his testimony. Nothing relating to the Modified Water Deliveries was to be "construed to limit the operation of project facilities to achieve their design objectives, as set forth in the Congressional authorization and any modifications thereof." Significantly, the language in the House version of the bill asserted the interests of the park. In the bill passed by the House on 7 November, this subsection read as follows: "Nothing in this Act shall be construed to limit operation of project facilities to achieve their original design objectives . . . provided, however, that the project shall be operated to maximize the restoration of natural hydrologic conditions within Everglades National Park . . . and any modifications thereto, must receive the written concurrence of the National Park Service."[71] The Senate amended the House bill, eliminating this proviso, and the House concurred in the Senate amendment. The result of the Senate amendment was to maintain the possibility that structures such as the C-111 canal could be built for environmental purposes and then operated for other uses.[72]

Enactment of the Everglades National Park Protection and Expansion Act of 1989 was a victory for the park and environmentalists. It provided a roadmap for the SFWMD, the Corps, and the NPS to work together in resolving land use and water management issues in the East Everglades where conflicts were longstanding. However, park officials and environmentalists worried that the legislation was too little too late. The law addressed the problems of quantity, timing, and distribution of water deliveries to the park, but by 1989 the focus of environmental concern was already shifting elsewhere: to the protection of water quality. The problem of excessive phosphorus entering the Everglades and altering the aquatic life was rooted not in the East Everglades but in the sugarcane fields farther north and the heavily urbanized coastal area to the northeast.

Yet the 1989 act was also a triumph for Florida politicians who believed that bipartisanship and increased federal support were the key ingredients to shaping a brighter future for South Florida. Embedded in the notion of increased federal support was the expectation of greater federal-state cooperation. But by the time the act was passed, the issue of water quality had reached the point of litigation, and the lawsuit that followed would become one of the most divisive events in the history of South Florida water management.

Endnotes

[1] Kathleen Shea Abrams, et al., "The East Everglades Planning Study," in *Collaborative Planning for Wetlands and Wildlife: Issues and Examples*, Douglas R. Porter and David A. Salvesan, eds. (Washington, D.C.: Island Press, 1995), 226.

[2] Abrams et al., "The East Everglades Planning Study," 226.

[3] Hansen, "South Florida's Water Dilemma," 18; Light and Dineen, "Water Control in the Everglades," 68.

[4] Statement of Nathaniel Pryor Reed in House Committee on Interior and Insular Affairs Subcommittee on National Parks and Public Lands, *Everglades National Park Protection and Expansion Act of 1989: Hearing before the House Subcommittee on National Parks and Public Lands of the Committee on Interior and Insular Affairs*, 101st Cong., 1st sess., 1989, 63;

Robert Johnson interview by Theodore Catton, 16 July 2004, 5-6 [hereafter referred to as Johnson interview]; Abrams et al., "The East Everglades Planning Study," 231-232; Dade County Planning Department, "Water Control Facilities in and Around the East Everglades Area," 18 September 1981, File Dade County East Everglades 1958-83 Resolutions/Ordinances/General, Box 02172, SFWMDAR; Walter A. Gresh, Regional Director, to District Engineer, 9 May 1963, File CE-SE Central and Southern Florida FCP, South Dade County, FWSVBA; University of Florida School of Natural Resources and Environment, "Water Management Issues Affecting the C-111 Basin, Dade County, Florida: Hydrologic Sciences Task Force Initial Assessment Report," 6 June 1997, available at <http://snre.ufl.edu/publications/c111. htm> (27 April 2006).

[5] Dade County Planning Department, "Water Control Facilities in and Around the East Everglades Area"; A. J. Salem, Acting Chief, Planning Division to Henry Iler, Senior Planner, Metro Dade Planning Department, 1 September 1981, File 1110-2-1150a (C&SF Southwest Dade County) Project General 1965 Authority, Box 10, Accession No. 077-96-0037, RG 77, FRC; Abrams et al., "The East Everglades Planning Study," 235.

[6] Lieutenant Colonel Robert J. Waterston III, interview by George E. Buker, 12 April 1984, Jacksonville, Florida, 24, transcript in Library, Jacksonville District, U.S. Army Corps of Engineers, Jacksonville, Florida; Abrams et al., "The East Everglades Planning Study," 236-237.

[7] Quotations in "Florida's Battle of the Swamp," *Time* 118 (24 August 1981): 41; see also *The Miami Herald*, 25 December 1981.

[8] East Everglades Task Force, untitled 27-page memorandum, no date, File East Everglades, Box 15746, SFWMDAR.

[9] M. R. Stierheim, County Manager to Joan Hagan, Secretary, Florida Department of Veteran and Community Affairs, 9 November 1981, File East Everglades, Box 15746, SFWMDAR.

[10] Johnson interview, 6; Abrams et al., "The East Everglades Planning Study," 236.

[11] Abrams et al., "The East Everglades Planning Study," 229-230.

[12] John M. Morehead, Superintendent, to Colonel Alfred B. Devereaux, District Engineer, 17 September 1982, File Everglades National Park 1958-88 General/Resolutions/Agreements, Box 02161, SFWMDAR.

[13] Morehead to Devereaux, 17 September 1982.

[14] Morehead to Devereaux, 17 September 1982.

[15] Morehead to Devereaux, 17 September 1982.

[16] John M. Morehead, Superintendent, to Colonel James W. R. Adams, District Engineer, 7 July 1980, File 1110-2-1150a (C&SF) Conservation Areas June 1980—December 1982, Box 12, Accession 077-96-0038, RG 77, FRC.

[17] Colonel James W. R. Adams, interview by George E. Buker, 18 November 1981, Jacksonville, Florida, 33, transcript in Library, Jacksonville District, U.S. Army Corps of Engineers, Jacksonville, Florida.

[18] Hansen, "South Florida's Water Dilemma," 41.

[19] South Florida Water Management District, "Order No. 83-10," File Everglades National Park Relief Plan—General/Agreement 1983-85, Box 02161, SFWMDAR.

[20] Steve Yates, "Marjory Stoneman Douglas and the Glades Crusade," *Audubon* 85 (March 1983): 122.

[21] Moreau, "Everglades Forever?" 73.

[22] Verbatim Excerpt from Workshop Meeting—Everglades National Park 7-Point Plan, 10 March 1983, no file name, Box 15747, SFWMDAR.

[23] Verbatim Excerpt from Workshop Meeting—Everglades National Park 7-Point Plan, 10 March 1983.

[24] Verbatim Excerpt from Workshop Meeting—Everglades National Park 7-Point Plan, 10 March 1983.

[25] Nathaniel P. Reed to Marjory Stoneman Douglas et al., 14 March 1983, Folder 29, Box 2, Marshall Papers.

[26] Reed to Douglas et al., 14 March 1983 (emphasis in original).

[27] South Florida Water Management District News Release, 29 March 1983, File Everglades National Park Relief Plan—General/Agreement 1983-1985, Box 02161, SFWMDAR.

[28] "Minutes of an Emergency Meeting of the Governing Board of the South Florida Water Management District," 5 April 1983, File Everglades National Park Relief Plan—General/Agreement 1983-1985, Box 02161, SFWMDAR.

[29] "Minutes of an Emergency Meeting of the Governing Board of the South Florida Water Management District," 5 April 1983.

[30] "Minutes of an Emergency Meeting of the Governing Board of the South Florida Water Management District," 5 April 1983.

[31] William L. Earl to Colonel Alfred B. Devereaux, District Engineer, 5 April 1983, File Everglades National Park Relief Plan—General/Agreement 1983-1985, Box 02161, SFWMDAR.

[32] "Lawmaker: Everglades Faces Crisis," *The Palm Beach Post*, 1 May 1983.

[33] Devereaux interview, 44.

[34] As reported in "Lawmaker: Everglades Faces Crisis," *The Palm Beach Post*, 1 May 1983; see also Nathaniel P. Reed to Rick Davidge, Assistant Secretary of the Interior for Fish, Wildlife & Parks, 2 May 1983, Folder 25, Box 2, Marshall Papers.

[35] Devereaux interview, 45.

[36] Public Law 84-99, Emergency Flood Control Work.

[37] Reed to Davidge, 2 May 1983.

[38] Abrams et al., "The East Everglades Planning Study," 234-235; "WMD Director: Corps Blocking Everglades Plan," *The Palm Beach Post*, 14 October 1983.

[39] John R. Maloy, Executive Director to Mayor and Board of County Commissioners, 16 June 1983, and Maloy to Dr. John M. DeGrove, Secretary, Department of Community Affairs, 22 June 1983, File Dade County East Everglades 1958-83 Resolutions/Ordinances/General, Box 02172, SFWMDAR.

[40] Abrams et al., "The East Everglades Planning Study," 238-239.

[41] Quotation in *Dade County Farm Bureau v. John O. Marsh, Jr., et al.*, Case No. 83-1210, Emergency Motion for Temporary Restraining Order and Other Emergency Relief, 3-4, copy provided by James Vearil, Senior Project Manager, RECOVER Branch, Programs and Project Management Division, U.S. Army Corps of Engineers, Jacksonville District, Jacksonville, Florida; see also "Cite East Everglades as Area of Concern, Agency Asks Graham," *Fort Lauderdale Sun-Sentinel*, 11 June 1983.

[42] "WMD Wants State to Run East Everglades," *The Palm Beach Post*, 11 June 1983; "Cite East Everglades as Area of Concern, Agency Asks Graham," *Fort Lauderdale Sun-Sentinel*, 11 June 1983; "Farmers' Suits May Delay Plan," *The Evening Times*, 14 October 1983.

[43] Devereaux interview, 40.

[44] Devereaux interview, 40-41; "Plan of Action Approved by SFWMD Governing Board on April 5, 1983 Designed to Improve Water Deliveries to Everglades National Park," File Dade County East Everglades 1958-83 Resolutions/Ordinances/General, Box 02172, SFWMDAR.

[45] John M. Morehead, Superintendent to Jack R. Maloy, Director, July 12, 1983, File East Everglades, Box 15746, SFWMDAR. See also Devereaux interview, 42-43.

[46] Devereaux interview, 43-45; "Plan of Action Approved by SFWMD Governing Board on April 5, 1983 Designed to Improve Water Deliveries to Everglades National Park," File Dade County East Everglades 1958-83 Resolutions/Ordinances/General, Box 02172, SFWMDAR; "WMD Director: Corps Blocking Everglades Plan," *The Palm Beach Post*, 14 October 1983. Ironically, in later years, the Corps operated S-333 instead of the SFWMD because of district fears about flooding and possible legal action.

[47] Victoria J. Tschinkel to Stanley W. Hole, 29 November 1983, File Everglades National Park Relief Plan—General/Agreement 1983-85, Box 02161, SFWMDAR.

[48] Act of 30 November 1983 (97 Stat. 1153).

[49] U.S. Army Corps of Engineers, "General Plan for Implementation of an Improved Water Delivery System to Everglades National Park," January 1985, copy in South Florida Water Management District Reference Center, West Palm Beach, Florida.

[50] Abrams et al., "The East Everglades Planning Study," 238. The arrangement was continued in 1985 and 1986. See "Frog Pond Summary," 26 May 1987, File Everglades National Park NESRS Frog Pond, Box 15746, SFWMDAR.

[51] Abrams et al., "The East Everglades Planning Study," 240-241; Devereaux interview, 34.

[52] Quotations in Everglades National Park/East Everglades Resource Planning and Management Committee, "Everglades National Park/East Everglades Resource Planning and Management Implementation Plan," 18 April 1985, File East Everglades, Box 15746, SFWMDAR, 7-14; see also Abrams et al., "The East Everglades Planning Study," 245-247.

[53] Everglades National Park/East Everglades Resource Planning and Management Committee, "Everglades National Park/East Everglades Resource Planning and Management Implementation Plan," 25-27, 53-63.

[54] SFWMD, "District Proposal will Help Restore Natural Water Flow to Park," 11 December 1985, File Pro Everglades SOE, Box 21213; "Agreement," 1 October 1987, File Everglades National Park NESRS Frog Pond, Box 15746, SFWMDAR. The field tests were reauthorized by Congress in December 1985 to continue through 1 January 1989.

[55] Memorandum of Agreement among the Army Corps of Engineers, the South Florida Water Management District, and the National Park Service for the Purpose of Protecting the Quality of Water Entering Everglades National Park," 8 February 1984, File Everglades National Park, Box 15754, SFWMDAR.

[56] Finley interview, 1.

[57] T. MacVicar, Deputy Director to J. R. Wodraska, Executive Director, 25 November 1985, and Alice Wainwright, Coordinator, Southeast Florida Chapters, National Audubon Society to Colonel Charles Myers, District Engineer, 21 March 1986, File Everglades National Park Relief Plan—General Agreement 1983-85, Box 02161, SFWMDAR.

[58] Michael Finley interview by Matthew Godfrey, 20 October 2004, 2 [hereafter referred to as Finley interview].

[59] "Frog Pond Summary," 26 May 1987, and D.W. Edwards, GenCorp Realty Company to John Wodraska, South Florida Water Management District, 21 May 1987, File Everglades National Park NESRS Frog Pond, Box 15746, SFWMDAR.

[60] Finley interview, 2; Herndon interview, 30-31; Light and Dineen, "Water Control in the Everglades: A Historical Perspective," 70; Abrams et al., "The East Everglades Planning Study," 236.

[61] Finley interview, 2-3.

[62] Dave Johnson, Office of the Governor, to Mac Stipanovich, 25 November 1986, File Pro Everglades SOE, Box 21213, SFWMDAR.

[63] Johnson to Stipanovich, 25 November 1986.

[64] "Governor Issues Everglades Statement," 17 November 1987, File Pro Everglades SOE, Box 21213, SFWMDAR.

[65] East Everglades Land Acquisition Task Force, "A Report to Governor Bob Martinez," 1 October 1988, ii, File Everglades, Box 88-02, S1331, Executive Office of the Governor, Brian Ballard, Director of Operations, Subject Files 1988, FSA.

[66] East Everglades Land Acquisition Task Force, "A Report to Governor Bob Martinez," 1 October 1988, iii-vii.

[67] Finley interview, 6.

[68] Senate Committee on Energy and Natural Resources Subcommittee on Public Lands, National Parks and Forests, *Everglades National Park Protection and Expansion Act of 1989: Hearing before the Subcommittee on Public Lands, National Parks and Forests of the Committee on Energy and Natural Resources*, 101st Cong., 1st sess., 1989, 3-7, 79; House Subcommittee on National Parks and Public Lands, *Everglades National Park Protection and Expansion Act of 1989*, 59, 117.

[69] Everglades National Park Protection and Expansion Act of 1989 (103 Stat. 1946).

[70] Everglades National Park Protection and Expansion Act of 1989 (103 Stat. 1946, 1949).

[71] Everglades National Park Protection and Expansion Act of 1989 (103 Stat. 1946, 1949).

[72] "Everglades National Park Protection and Expansion Act," no date, File Everglades National Park, Box 15754, SFWMDAR.

12   The "Ultimate Hammer": Dexter Lehtinen's Lawsuit

LITIGATION WAS PART OF THE POLITICAL MIX in South Florida water management as early as the nineteenth century. But when the United States brought suit against the SFWMD in 1988 it raised litigation to a new level, initiating one of the largest environmental lawsuits in American history. The suit pitted federal and state agencies against each other, pushed agricultural organizations to harden their position against environmental remediation, incited environmental organizations to vilify Big Sugar, and alienated the people who were nearest to the geographic center of it all, the Miccosukee Tribe. For all of the turmoil that it caused, however, the suit raised awareness and compelled action. It laid the foundation for the broad consensus approach that would triumph at the end of the century in Congress's billion-dollar blessing of the Comprehensive Everglades Restoration Plan. To people who worked on Everglades issues and were inured to litigation, the suit that began in 1988 would long be known as "the Big One," or simply as "Dexter Lehtinen's lawsuit."[1]

Dexter Lehtinen, raised in Homestead, Florida, in the 1950s, knew the Everglades as a place of tranquility and boyhood innocence. During the Vietnam War, Lehtinen volunteered to serve in the U.S. Special Forces as a paratrooper and ranger. Gravely wounded while leading his platoon on reconnaissance during the invasion of Laos in 1971, he bore a deep scar on his left cheek afterwards—a "trademark," journalists would later write, of his fiery, combative public persona. Returned from the war, he went to Stanford Law School and graduated at the top of his class. In the 1980s, he entered Florida politics, serving one term in the House and one in the Senate. As a state senator, Lehtinen switched from the Democratic to the Republican Party after marrying a Republican colleague—Ileana Ros—thereby attracting the attention of Republicans at the national level. In 1988, he was appointed the U.S. Department of Justice's top attorney in South Florida. The Reagan administration picked Lehtinen for the prominent position of U.S. attorney in Miami because they saw a man who would increase efforts in the drug war. Said former Associate Attorney General Frank Keating, he was "the brightest, toughest, meanest scrapper we could find." Lehtinen immediately grabbed attention by trying to assume the lead role in prosecuting former Panamanian dictator and drug lord General Manuel Noriega. Lehtinen further made news by carrying a plastic AK-47 as a symbol of his aggressive attack on drugs and by publicizing his office's new motto, "No Guts. No Glory."[2] He received the nickname "Machine Gun."[3]

Lehtinen was also passionate, if less demonstrative, about protecting the environment. Soon after taking office he arranged a meeting with Michael Finley, the superintendent of Everglades National Park, who, since his arrival in 1986, had become very concerned about the quality of water entering the park. The problem, as Finley discovered, was that EAA farmers—primarily sugar growers—used nitrate and phosphate fertilizers to stimulate their crops, and these nutrients became absorbed in the runoff that ultimately flowed into the water conservation areas and then into the park. Because of the influx of nutrients, the water conservation areas (especially Loxahatchee National Wildlife Preserve, which adjoined the EAA) and the canals transmitting the water were choked with cattails and algae that prevented sunshine from reaching underwater plants, creating stagnant, oxygen-depleted waterbodies. Although Everglades National Park had so far experienced few of these problems, Finley realized that it was only a matter of time. "It's like a cancer," he told *Time* magazine, "and the cancer is moving south."[4]

After meeting together in 1988, both Lehtinen and Finley saw an opening to combat this agricultural pollution of South Florida waters. The state, under its five water management districts (including the SFWMD), was chiefly responsible for regulating water quality. Since the water entering the conservation areas and the park was, in the opinion of Lehtinen, Finley, and other park officials, of poor quality, the state had obviously failed to fulfill its mission, opening itself to litigation for damages done to Loxahatchee National Wildlife Refuge and Everglades National Park.[5]

Lehtinen and Finley relied on the work of Ron Jones, a microbiologist at Florida International University, for their evidence. Jones, described by one journalist as "a nerdy young [man] who was a devout adherent of an Amish-style sect called Apostolic Christianity, and believed God had sent him to Florida to save the Everglades," conducted studies that convinced him that any phosphorus amounts over 10 parts per billion would destroy the Everglades ecosystem by, among other things, transforming sawgrass swaths into areas choked with cattails—"the markers on the grave of the Everglades," according to Jones.[6] Phosphorus also killed periphyton, a food source for fish and snails that are then consumed by birds, disrupting the food chain. Yet phosphorus-rich runoff continued to pour into the Everglades, making it oligotrophic and poisoning it to death. Only by reducing phosphorus amounts to 10 parts per billion, Jones argued, could any healing begin.[7]

Periphyton. (South Florida Water Management District)

When the SFWMD released a first draft of its SWIM plan for protecting the water quality of Lake Okeechobee, Lehtinen and Finley had a clear target for their lawsuit. Although there was no direct federal interest in Lake Okeechobee, the SWIM plan clearly had ramifications for waters draining into Loxahatchee National Wildlife Refuge and Everglades National Park, two federal areas. In Lehtinen's and Finley's view, the draft SWIM plan would not reduce phosphorus levels quickly or drastically enough to protect the fed-

eral areas from the contaminated sheet flow emanating from Lake Okeechobee. Therefore, the lawsuit would ask the U.S. district court in Miami to maintain its jurisdiction until the state agencies developed an adequate plan. In other words, the suit would force the state to take a tougher stand against polluters, particularly the sugar industry.[8]

Finley had been searching for solid ground for a lawsuit against the state for the previous two years, consulting with legal counsel in the Natural Resources Defense Council and the Sierra Club, and later assigning members of his staff to develop causes for action. But it was Lehtinen who finally crafted the complaint. Legal scholar William H. Rodgers, Jr., has written that the lawsuit, entitled *United States v. South Florida Water Management District and Florida Department of Environmental Regulation, et al.*, was "brilliantly conceived" and "one of the most creative contributions in the history of modern environmental law."[9] The complaint contained five counts. The first and second counts held that the damage to natural vegetation in the Loxahatchee National Wildlife Refuge and Everglades National Park—which the state was allowing to happen by not enforcing water quality regulations—violated state law and the public trust doctrine because it was destroying federal property. The third count alleged a breach of contract: the National Park Service had contracted with the SFWMD to have water of a certain quality delivered to the park and the SFWMD had not complied. The fourth count maintained that the excessive water-born nutrients entering the park constituted a nuisance under common law and riparian water rights, while the fifth held that the state's actions violated the National Park Service Organic Act, which provided that parks would be preserved in an unimpaired condition for future generations.[10]

The strength of the lawsuit was that it claimed that the state failed to enforce its own water quality standards, in particular the narrative standard for high quality waters as defined in the Florida Administrative Code. For so-called Class III waters, the code stated that "in no case shall nutrient concentrations of a body of water be altered so as to cause an imbalance in natural populations of aquatic flora or fauna." Although experts disagreed on the precise causes for the changes in natural vegetation, water quality was clearly involved. Thus, in the eyes of many environmentalists, Lehtinen's nuisance theory was practically irrefutable, and the litigation came to focus on nutrient loading as the keystone pollutant that altered natural conditions in both the refuge and the park.[11]

But to state lawyers and administrators, a bitter irony existed in the lawsuit: the C&SF Project—the pollution delivery system—was largely a federal project. As Keith Rizzardi, an attorney for the SFWMD, later wrote, "The federal government sued the State of Florida and the Water Management District for the consequences of operating the flood control project that the United States had helped to design and build."[12] The lawsuit simply sidestepped the federal interest in the C&SF Project, focusing instead on the federal interest in conservation lands. Lehtinen's client in this case was the Department of the Interior, not the Corps of Engineers.

In a similar vein, the agricultural interests declared that the state had developed its water quality standards under the aegis of a federal statute, the Clean Water Act, in cooperation with the federal enforcing agency, the Environmental Protection Agency. There was no legal precedent, they observed, for using the Clean Water Act to control non-point-source pollution. Since the Biscayne aquifer lay just beneath the ground surface in South Florida, non-point source pollution was ubiquitous in that region. Agricultural interests contended that the Clean Water Act did not create a federal right to sue the state over how it was managing non-point-source pollution, but Lehtinen's litigation took the opposite view, one of the first lawsuits to do so.[13]

Lehtinen filed the lawsuit on 11 October 1988, one day after the SFWMD released its draft SWIM plan for Lake Okeechobee. The SFWMD acknowledged in the plan that phosphorus levels in the lake had increased by more than two and a half times since the early

An employee of the SFWMD conducting sampling for water quality studies. (South Florida Water Management District)

1970s, and it recommended that the phosphorus concentration be reduced by at least half. According to the lawsuit, this was not good enough. Phosphorus levels in Lake Okeechobee had reached approximately 120 parts per billion (ppb), and ran as high as 200 ppb in the runoff from the EAA. By contrast, ambient levels of phosphorus in park waters were about 10 ppb. The lawsuit therefore highlighted the need for an Everglades SWIM plan in order to reduce nutrient levels to an amount that would not harm park resources.[14]

Lehtinen had other reasons for filing the lawsuit when he did. According to Finley, he and the U.S. attorney waited for Governor Marti-

nez to endorse the proposed Everglades National Park Protection and Expansion Act, anxious that the litigation should not derail that effort. Perhaps, too, Lehtinen waited because he doubted whether the Reagan administration would support such a headlong legal battle with the sugar industry in Florida. By October, Vice President George H. W. Bush was in the final heat of his presidential campaign, castigating the Democratic Party nominee, Massachusetts Governor Michael S. Dukakis, for his failure to clean up Boston Harbor. The U.S. Justice Department would hardly be able to back away from a lawsuit aimed at protecting the Everglades. Regardless, Lehtinen filed the lawsuit without consulting his superiors at "Main Justice" in Washington.[15]

Because Finley had been working closely with Governor Martinez on the matter of expanding the boundaries of Everglades National Park, the superintendent wanted to maintain a good relationship. Therefore, immediately after Lehtinen filed the suit, Finley telephoned Martinez so the governor would not have to discover the action in the newspapers. Finley tried to inform Martinez gently, using the bad-news, good-news formula. "What could possibly be the good news?" the governor responded when he was told that his state and the water management district were being sued by the United States. The good news, Finley replied, was that the suit did not name the governor personally.[16]

Martinez issued a statement on the lawsuit the following day. He listed various initiatives he had taken as governor for the protection of Florida's environment. He was proud of what his administration had accomplished, he said, and it would do more in the future. "While I have not seen the federal lawsuit and cannot comment on it at this time," he said, "I welcome the efforts of anyone who chooses to join in our efforts to protect one of the world's unique environmental resources."[17]

Despite Martinez's spirit of turning the other cheek, the litigation was politically charged from the outset, and it grew more politicized as various interest groups lined up on either side. The governing board of the SFWMD immediately hired outside counsel to assess the implications of the lawsuit. Vice Chairman James Garner persuaded Governor Martinez that he should request the Department of Justice to drop the suit. They flew to Washington and met with Attorney General Richard Thornburgh. According to another board member, Nathaniel Reed, who strongly opposed this move, Thornburgh told the governor, "I do not force my U.S. attorneys to drop lawsuits." If Martinez felt that the state was being unjustly sued, Thornburgh continued, he should prepare a good defense. Reed recollected that the lawsuit divided the SFWMD's governing board, as members like Reed contended that the district needed to listen more assiduously to its own scientists and agree to more stringent pollution controls, while others urged the state to spend enormous sums on legal defense so as to defeat the lawsuit without taking any action.[18] "There has to be a change," Reed insisted, while board member Doran Jason retorted, "If [Lehtinen] wants to fight, let's go ahead."[19]

Meanwhile, the U.S. Army Corps of Engineers' reaction to the lawsuit was mixed. Colonel Terrence "Rock" Salt, who became District Engineer of the Jacksonville District in 1991, claimed that the lawsuit was useful for bringing about stronger environmental protections, but he also recognized the unprecedented strain it placed on the Corps' historic partnership with the SFWMD. The action put the Corps between the SFWMD and the National Park Service, two agencies with which it had long enjoyed close, if sometimes contentious, relationships. The Corps staff was conflicted about the litigation, with some division managers approving it and others opposing it. Legal counsel in the Jacksonville District were cautiously supportive, supplying documents upon request by the Justice Department, preparing its experts for deposition, but never offering advice on litigation strategy.[20]

In the Justice Department, the lawsuit was not given high priority, and many attorneys were doubtful that Lehtinen could win the case. His legal arguments involving the Clean Water Act were unprecedented. Moreover, without strong backing from Washington, Lehtinen and his staff attorneys in Miami were soon outgunned. While the federal government assigned relatively few lawyers to the case, the state began to spend millions of dollars on legal fees. In the words of one publication, it "responded to the suit by hiring the most expensive lawyers it could find," eventually expending approximately $6 million.[21] In addition, the court granted the Florida Sugar Cane League and other agricultural interests intervention in the case in January 1991, allowing the sugar industry to supplement state efforts with its financial resources. The industry hired high-priced law firms in Miami, and these attorneys began to accumulate deposition after deposition of interminable testimony taken from experts on both sides. By the early 1990s, the lawsuit rivaled the litigation surrounding the *Exxon Valdez* oil spill as the most expensive environmental litigation ever seen.[22]

U.S. Senator Lawton Chiles (D-Florida) made the litigation expense a campaign issue when he ran for Florida's governorship in 1990. Chiles argued that the millions of dollars Governor Martinez was spending on legal fees would be better spent on working with the federal agencies to solve the problem. Chiles promised not only to settle the lawsuit, but he also declared that cleanup of the water flowing into the Everglades would be his top environmental priority. In the November election, Chiles defeated Martinez, but it is unclear how much of a deciding factor the Everglades lawsuit played in the outcome. Nevertheless, in fulfillment of his campaign promise, Chiles made settlement of the Everglades lawsuit his "Number 1 Environmental Priority," assigning Carol Browner, secretary of the Department of Environmental Regulation, to oversee the negotiations.[23]

Encouraged by the change in administration, a number of environmental organizations began to urge a negotiated settlement, and commenced work in that direction.[24] Also influential was Richard Stewart, assistant attorney general for the Bush administration, who had formerly worked as a lawyer specializing in environmental lawsuits against copper smelters. Stewart, described by one observer as "pompous, well organized, and conniving," in contrast to Lehtinen, who was "down-to-earth, frantic, and candid," organized federal agencies responsible for the South Florida ecosystem and got them to submit unified comments on the Everglades SWIM Plan developed by the SFWMD, decrying the destruction that had taken place to the environment.[25] This united front helped convince Governor Chiles that continuing a defense in the lawsuit was fruitless. Accordingly, on 20 May 1991, in a bit of political theater that Everglades hands would recount for years afterwards, Governor Chiles walked into the federal courthouse in Miami and appealed directly to Judge William Hoeveler to end the litigation. "I am ready to stipulate today that water is dirty," Governor Chiles declared. "I am here and I brought my sword. I want to find out who I can give that sword to and I want to be able to give that sword up and have our troops start the reparation, the clean up . . . . We want to surrender. We want to plead that the water is dirty. We want the water to be clean, and the

Governor Lawton Chiles, who "surrendered his sword." (The Florida Memory Project, State Library and Archives of Florida)

question is how can we get it the quickest."[26] A few weeks later, the Florida Department of Environmental Regulation filed papers with the court agreeing that water going into the conservation areas and into Everglades National Park contained excessive amounts of nutrients. Department Secretary Carol Browner explained why both Chiles and the state took these actions. "The real challenge for everyone concerned," she noted, "is to stop pointing fingers to prove who is at fault and get on with the cleanup."[27]

Although environmentalists lauded Chiles and the state, some in the sugar industry were not pleased, especially since they believed that the state had a sound defense against Lehtinen's allegations. Chiles did not "want to have an albatross of a lawsuit, so he waltzed into federal court [and] surrendered his sword," Barbara Miedema, vice-president of communications for the Sugar Cane Growers Cooperative of Florida, stated in her characterization of the situation. This action, according to George Wedgworth, founder and president of the Sugar Cane Growers Cooperative, "forfeited our interests."[28]

The sugar industry's preferences notwithstanding, Chiles' action set in motion a more intense period of negotiations, and in July 1991, the Florida Department of Environmental Regulation, the SFWMD, and the U.S. Department of Justice reached a settlement. In the resulting 30-page "Settlement Agreement," a landmark document, the parties defined the problem, articulated a set of remedial solutions, and specified dates in the future by which certain goals had to be met. It began with a set of definitions, including item "F," which defined "imbalance in natural populations of flora and fauna" as "situations when nutrient additions result in nuisance species." Such circumstances included

> replacement of native periphyton algal species by more pollution-tolerant algal species, loss of the native periphyton community or, in advanced stages of nutrient pollution, native sawgrass and wet prairie communities giving way to dense cattail stands or other nutrient-altered ecosystems, which impair or destroy the ability of the ecosystem to serve as habitat and forage for higher trophic levels characteristic of the Everglades.[29]

With "imbalance" of natural systems defined, the document proceeded to describe the problem, drawing a link between the phosphorus-loaded water flowing out of the EAA and the nutrient-lean (oligotrophic) natural condition of the Everglades ecosystem. The following statement carried unusual weight because it was prefaced by "the Parties agree" and it concluded with the freighted term "imbalances":

> Excess phosphorus accumulates in the peat underlying the water, alters the activity of microorganisms in the water, and disturbs the natural species composition of the al-

gal mat (periphyton) and other plant communities in the marsh. These disturbed communities deplete the marsh of oxygen, and, ultimately, result in native sawgrass and wet prairie communities being replaced by dense cattail stands or other nutrient-tolerant ecosystems. The ability of the ecosystem to serve as habitat and forage for the native wildlife is thereby greatly diminished or destroyed. These changes constitute imbalances in the natural populations of aquatic flora and fauna or indicators of such imbalances.[30]

Following the sections on definitions and background, the document contained 20 more numbered paragraphs, of which three were especially important. In Paragraph 7, the parties agreed that phosphorus concentrations in waters entering Everglades National Park would be reduced to amounts that would prevent an imbalance of flora and fauna. In general, the objective was to obtain prescribed concentration limits for Shark River Slough and Taylor Slough in two stages, with "interim concentration limits" met by 1 July 1997 and "long-term concentration limits" by 1 July 2002. Target levels were tied to "baseline" amounts measured in 1978 and 1979. These levels, expressed in parts per billion (ppb), were set forth in Appendix A of the Settlement Agreement. The amounts varied to take into account wet and dry cycles, but reflected an overall target of about 10 ppb. Paragraph 8 of the Settlement Agreement established similar goals for water discharged from the EAA into the Loxahatchee National Wildlife Refuge. Target levels for this area were set forth in Appendix B.[31]

Paragraph 10 committed the SFWMD to develop stormwater treatment areas (STAs). The agreement identified STAs as "the primary strategy to remove nutrients from agricultural runoff." Construction and operation of these giant water filtration plants would constitute the primary remedial action, and, as such, they would become the focus of much further debate over the next decade. The district was to purchase land for the STAs, design the structures, and build them (the agreement was later amended to commit the Corps to this task as well). Initially, the SFWMD was to construct four STAs, and if these did not sufficiently reduce phosphorus concentrations coming from the EAA, the district would acquire more acreage and build additional facilities. The location and size of the four STAs and the basins that each STA would serve were stipulated in a table, with further specifications detailed in Appendix C. In addition to the STAs, the Florida Department of Environmental Regulation agreed to regulate agricultural discharges by a regulatory permit system. The STAs and the permits together were expected to reduce phosphorus loading by 80 percent.[32]

But the Settlement Agreement was not the only result of Lehtinen's lawsuit. In May 1991, the Florida legislature had also passed unanimously the Marjory Stoneman Douglas Everglades Protection Act, which specifically dealt with water quality in the conservation

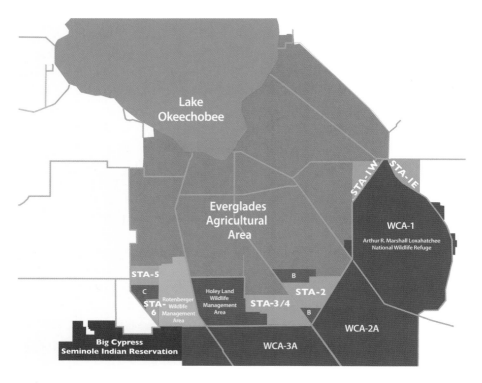

Stormwater Treatment Areas and Water Conservation Areas. (South Florida Water Management District)

areas and Everglades National Park. The law declared that it was the state's imperative to preserve and restore the Everglades Protection Area, which it defined as the Loxahatchee National Wildlife Refuge, the other water conservation areas, and the park, and it required the SFWMD to develop specific programs to protect and restore the Everglades. In addition, the act mandated tougher objectives for incorporation into the draft Everglades SWIM plan, including the development of STAs and the implementation of a permit system for discharges into waters managed by the district.[33]

In February 1992, Judge Hoeveler approved the Settlement Agreement, entering it as a consent decree. The judge noted that its "ambitious plan" essentially implemented what the state had set forth in the Marjory Stoneman Douglas Everglades Protection Act. Indeed, the only real differences were that the agreement delineated additional specificity for schedules and it imposed an administrative process rather than a result.[34] This administrative process was based on interagency cooperation and consensus, achieved through a Technical Oversight Committee. This committee consisted of five members representing Everglades National Park, the Loxahatchee National Wildlife Refuge, the Florida Department of Environmental Regulation, the SFWMD, and the Corps. It had the responsibilities of planning, reviewing, and recommending all research pursuant to the Settlement Agreement, and it was supposed to operate under a consensus approach, defined as a four out of five majority. In the absence of a consensus decision, parties could seek arbitration.[35]

Although the entering of the Settlement Agreement as a consent decree supposedly ended the litigation, it continued, in large part be-

cause some entities were not happy with the arrangement. The Florida Cane Sugar League and other agricultural interests, for example, appealed the court order approving the settlement. Likewise, in the spring of 1992, following the SFWMD's Governing Board's approval of the final Everglades SWIM Plan (which, to no one's surprise, mirrored the requirements in the Everglades Protection Act and the Settlement Agreement), more than 30 agricultural cooperatives and corporations brought suit against the SFWMD. Several of these entities, mostly representing the sugar industry, petitioned for administrative proceedings to determine the legality of the Everglades SWIM Plan. The petitioners argued that the SFWMD, in refusing to disclose technical information that had been used in the settlement process and in developing the Everglades SWIM Plan, had violated the Florida Administrative Procedures Act. The petitions went to the Division of Administrative Hearings, which consolidated them into three cases. The Florida Department of Environmental Regulation, the United States, the Miccosukee Tribe, and certain environmental organizations moved to intervene in the litigation, and the Florida Division of Administrative Hearings granted all these motions for intervention.[36] As Carol Browner, secretary of the Department of Environmental Regulation derisively explained, "We get sued every day by sugar. I call it 'suit du jour.'"[37]

Referring to these challenges, Deputy Assistant Attorney General Myles Flint later explained to Congress that "relaxed rules of evidence and procedure and a plenary grant of jurisdiction governed these proceedings," so that despite efforts by the state and federal agencies to stand by the Settlement Agreement and Consent Decree, "the administrative challenges became protracted and complicated, with voluminous discovery." Not only did this renewed litigation cause further delays and expense, it threatened to undermine the consensus approach fashioned in the Settlement Agreement and Consent Decree as agricultural interests took one side while the Miccosukee Tribe and environmental organizations closed ranks on the other.[38]

Indeed, even though the Miccosukee had not participated in the water quality suit, the tribe, whose reservation lands were affected by quality issues, still had an interest in the proceedings. Lehtinen and his staff had carefully framed the lawsuit so that it neither embraced nor prejudiced tribal interests, but once a settlement was reached the tribe did not want to be left out of the remediation effort. It therefore filed a motion to intervene in the case and attain status as a party to the Settlement Agreement. U.S. attorneys, however, were concerned that the tribe's move might jeopardize the agreement. Following negotiations, the tribe withdrew its motion to intervene in return for a Memorandum of Agreement with three Interior Department agencies: the NPS, the FWS, and the Bureau of Indian Affairs. This memorandum, dated 1 November 1991, pledged that the Interior Department would provide the tribe with results and data of all studies relating to water quality in the Everglades, allow the tribe to attend Technical Oversight Committee meetings as an interested non-member, and consult with the tribe on the Department's position prior to such meetings. For its part, the tribe agreed to give the Department notice before taking any further actions in court with regard to the Settlement Agreement.[39]

Meanwhile, even though agricultural interests continued fighting the Settlement Agreement and Consent Decree in court, Lehtinen's role in the lawsuit had just about run its course. His superiors in Washington had lost patience with his renegade spirit, while many of his staff attorneys in Miami had had enough of his autocratic management style. More importantly, the Justice Department now wanted to preserve the fragile consensus that the Settlement Agreement produced, even though it was seemingly teetering on a precipice. Toward the end of 1992, Lehtinen quit his office as U.S. attorney in Miami, leaving behind a staff that was deeply divided and isolated from the rest of the Justice Department.[40]

Lehtinen was far from through with Everglades litigation, however. Less than a year after resigning from the Justice Department, he went to work for the Miccosukee Tribe. As the tribe's attorney, Lehtinen would file suit against the United States in 1995, initiating another phase in the Everglades litigation. For Lehtinen's detractors, the volatile attorney's new championship of the Miccosukee Tribe appeared self-serving, perhaps even vengeful. "You have to be careful, because Dexter is like gasoline," complained one federal official.[41] But by then, Lehtinen was no longer calling the shots. The Miccosukee Tribe was making its own decisions and Lehtinen was merely its agent. He would continue to make himself heard on Ever-

Cattails, "the markers on the grave of the Everglades." (South Florida Water Management District)

glades issues, but henceforth he would be at the edge of the process rather than at the center of it, accusing the federal government of selling out the Everglades and his client, the Miccosukee Tribe, to the wealthy corporations that had an economic stake in polluting the waters of South Florida.[42]

In a similar way, Lehtinen's lawsuit continued. Although the Everglades Forever Act of 1994, discussed in Chapter 14, brought some resolve to the litigation—in that it appeased the sugar industry, which called it a "far better, more comprehensive solution than the settlement agreement"—later amendments to that act would be the subject of additional appeals and contentions. In the initial years of the twenty-first century, *U.S. v. South Florida Water Management District* remained active, although under the jurisdiction of Judge Federico A. Moreno.[43] To Michael Finley, this was a good thing. "The court still has jurisdiction," he stated in a 2004 interview, "which is the ultimate hammer over the state and the South Florida Water Management District."[44]

The lawsuit that Dexter Lehtinen instigated in 1988, then, was not a happy affair. It sharpened differences among all stakeholders in South Florida's water resources and drove wedges between federal and state agencies that had long labored to work cooperatively and share information with one another. As Estus Whitfield, environmental adviser to both Governor Martinez and Governor Chiles maintained, "the lawsuit set back the restoration efforts substantially" by "pitt[ing] everybody against everybody else." "That is not the formula for getting something done," Whitfield contended. "That is the formula for fussing and fighting and going nowhere."[45]

Yet in other ways, the lawsuit was a necessary instrument of change. The cost of litigation—both in monetary terms and in the toll it took on people's lives—drove many diverse interests to seek consensus as an alternative to fighting and gridlock. At the same time, it jarred Florida into taking action to restore water quality to the Everglades. "Without litigation," Nathaniel Reed contended, the SFWMD "never would have been able to persuade the taxpayers and the sugar industry that steps had to be taken to control the pollution of the Everglades marsh."[46] Indeed, the litigation brought about four specific actions that established a foundation for environmental mitigation efforts in the 1990s: the Marjory Stoneman Douglas Everglades Protection Act, the Everglades SWIM Plan, the Settlement Agreement, and the Consent Decree. Viewed in retrospect, the lawsuit was a major turning point in the long, complicated, and arduous transformation of the C&SF Project from a system designed primarily for flood control and irrigation to one bent toward ecosystem restoration and the preservation of a sustainable environment.

Endnotes

[1] William H. Rodgers, Jr., "The Miccosukee Indians and Environmental Law: A Confederacy of Hope," *Environmental Law Reporter* 31 (August 2001): 10918; John D. Brady interview by Theodore Catton, 12 May 2005, 1 [hereafter referred to as Brady interview].

[2] All quotations in Rebecca Wakefield, "Lehtinen for Mayor," *Miami New Times*, 22 May 2003; see also James Carney, "Last Gasp for the Everglades," *Time* 134 (25 September 1989): 27.

[3] Rodgers, "The Miccosukee Indians and the Environmental Law," 10918.

[4] As quoted in Carney, "Last Gasp for the Everglades," 27.

[5] Finley interview, 2.

[6] Both quotations in Grunwald, *The Swamp*, 284-285.

[7] Michael Grunwald, "Water Quality is Long-Standing Issue for Tribe," *The Washington Post*, 24 June 2002.

[8] Keith Rizzardi, "Translating Science into Law: Phosphorus Standards in the Everglades," *Journal of Land Use and Environmental Law* 17 (Fall 2001): 151; "Everglades Water Threatened," *Engineering News-Record* 221 (27 October 1988): 16.

[9] Rodgers, "The Miccosukee Indians and the Environmental Law," 10918.

[10] Finley interview, 4; U.S. Civil Complaint, "Everglades Litigation and Restoration" <http://exchange.law.miami.edu/everglades> (29 August 2005).

[11] Rodgers, "The Miccosukee Indians and the Environmental Law," 10918. Florida Administrative Code quoted in "Settlement Agreement," *United States v. South Florida Water Management District*, Case No. 88-1886-CIV-Hoeveler, 4 January 1995, 3.

[12] Rizzardi, "Translating Science into Law," 150-151.

[13] Brady interview, 2.

[14] South Florida Water Management District, "Draft Interim Surface Water Improvement and Management (SWIM) Plan for Lake Okeechobee," 10 October 1988, 19, File Okeechobee S.W.I.M. Plan (SFWMD)—1988, Box 1, S1497, Department of Agriculture and Consumer Services, Surface Water Improvement and Management Plan Files, FSA, 19; Rizzardi, "Translating Science into Law," 151; "Everglades Water Threatened," 16.

[15] Grunwald, "Water Quality is Long-Standing Issue for Tribe."

[16] Finley interview, 4.

[17] "For Your Information," 12 October 1988, File Everglades, Box 88-02, S1331, Executive Office of the Governor, Brian Ballard, Director of Operations, Subject Files 1988, FSA.

[18] Nathaniel Reed interview by Julian Pleasants, 18 December 2000, 22, Everglades Interview No. 2, Samuel Proctor Oral History Program, University of Florida, Gainesville, Florida [hereafter referred to as Reed interview].

[19] As quoted in Carney, "Last Gasp for the Everglades," 27.

[20] Brady interview, 2.

[21] Norman Boucher, "Smart as Gods: Can We Put the Everglades Back Together Again?" *Wilderness* 55 (Winter 1991): 18.

[22] Brady interview, 2; Michael Satchell, "Can the Everglades Still Be Saved?" *U.S. News & World Report* 108 (2 April 1990): 24; Wedgworth and Miedema interview, 5.

[23] Jim Lewis, "Key Environmental Accomplishments of the Chiles-McKay Administration," 12 December 1991, File Environmental Issues, Box 5, S1824, Executive Office of the Governor Subject Files, 1991-1996, FSA.

[24] "Everglades Lawsuit and Cleanup," undated memorandum, File Environmental Issues, Box 5, S1824, Executive Office of the Governor, Subject Files, 1991-1996, FSA. The environmental organizations were the Wilderness Society, Florida Audubon Society, and Environmental Defense Fund.

[25] Rodgers, "The Miccosukee Indians and Environmental Law," 10923.

[26] As quoted in John J. Fumero and Keith W. Rizzardi, "The Everglades Ecosystem: From Engineering to Litigation to Consensus-Based Restoration," *St. Thomas Law Review* 13 (Spring 2001): 674; see also Grunwald, *The Swamp*, 290-291.

[27] "DER Stipulates to Majority of Disputed Facts in Everglades Lawsuit," Florida Department of Environmental Regulation Press Release, 10 June 1991, File Environmental Regulation, Box 5, S1824, Executive Office of the Governor Subject Files, 1991-1996, FSA.

[28] Wedgworth and Miedema interview, 5.

[29] *United States of America, et al., v. South Florida Water Management District*, Settlement Agreement, 26 July 1991, "Everglades Litigation and Restoration" <http://exchange.law.miami.edu/everglades> (25 August 2005) [hereafter referred to as Settlement Agreement].

[30] Settlement Agreement.

[31] Settlement Agreement.

[32] Settlement Agreement.

[33] South Florida Water Management District, "Draft Surface Water Improvement and Management Plan for the Everglades: Planning Document," 24 September 1991, File Everglades SWIM Plan 1990-1991, Box 1, S1497,

Department of Agriculture & Consumer Services, Surface Water Improvement and Management Plan Files, FSA.

34 *United States of America, et al., v. South Florida Water Management District*, Consent Decree, 24 February 1992, File 1110-2-1150a Settlement Agreement, Case #88-1886-CIV-HOEVELER, Box 3977, JDAR.

35 Settlement Agreement.

36 SWIM Challenges, "Everglades Litigation and Restoration" <http://exchange.law.miami.edu/everglades> (29 August 2005). The following organizations petitioned: Sugar Cane Growers Cooperative of Florida; Roth Farms, Inc.; and Wedgworth Farms, Inc. (DOAH Case No. 92-3038); Florida Sugar Cane League, Inc,; United States Sugar Cane Corporation; and New Hope South, Inc. (DOAH Case No. 92-3039); Florida Fruit and Vegetable Association; Lewis Pope Farms; W. E. Schlecter & Sons, Inc.; and Hundley Farms, Inc. (DOAH Case No. 92-3040).

37 As quoted in Boucher, "Smart as Gods," 18.

38 House Committee on Natural Resources, Subcommittee on National Parks, Forests and Public Lands, *Oversight Hearing on the Land Use Policies of South Florida with a Focus on Public Lands and what Impact these Policies are Having*, 103rd Cong., 2d sess., 1994, 66-67.

39 "Memorandum of Agreement among the National Park Service, the Fish and Wildlife Service, and the Bureau of Indian Affairs, of the United States Department of the Interior, and the Miccosukee Tribe of Indians of Florida," 1 November 1991, File Everglades Mediation Miccosukee, Box 19706, SFWMDAR.

40 Rebecca Wakefield, "Lehtinen for Mayor," *Miami New Times*, 22 May 2003.

41 Stan Yarbro, "Dexter Lehtinen Makes His Stand with the Miccosukees," *The Daily Business Review*, 3 March 1995.

42 Dexter Lehtinen, "The Everglades Need Clean Water, Not Broken Promises," *The Miami Herald*, 12 November 1993.

43 Wedgworth and Miedema interview, 5-6.

44 Finley interview, 5.

45 Whitfield interview, 25.

46 Reed interview, 22.

CHAPTER 13 A Broader Perspective: Ecosystem Restoration Becomes National Policy

THROUGHOUT THE 1980S AND EARLY 1990S, Florida politicians lobbied Congress and the president of the United States for federal help in Everglades restoration. Governor Bob Graham appealed to President Ronald Reagan for federal assistance in support of the state's Save Our Everglades program. Congressman Dante Fascell pushed enactment of the law initiating experimental water deliveries to Everglades National Park—a federal incursion into state water rights that he and other Florida lawmakers regarded as a practical necessity. Graham, both as governor and as a U.S. senator, fought for a congressional directive for the Corps to restore the Kissimmee River. All of these initiatives required federal appropriations. The threshold question for these politicians was always whether or not there was a national interest. But the problems of ecological decline stemmed fundamentally from Florida's burgeoning population growth, critics contended, and it was the responsibility of the state to manage growth. Therefore, why should the federal government invest in Everglades restoration if the state ultimately controlled the outcome?

With that counterargument in view, Florida's entire congressional delegation vigorously pursued more federal protections for South Florida wild lands: establishment of Biscayne National Park in 1980, additions to Big Cypress National Preserve in 1988 and Everglades National Park in 1989, creation of Florida Keys National Marine Sanctuary in 1990, and designation of Dry Tortugas National Park in 1992. By the early 1990s, the federal interest in South Florida was manifestly huge, and Florida politicians pointed to the federal lands whenever they angled for more federal involvement in South Florida's water management. "We are right now on the edge of a severe water crisis," Congressman Clay Shaw, Jr., a Republican from Miami, declared to his fellow members of the House. "The Federal

Government, as the largest landowner . . . has the responsibility . . . of seeing to it that its investment is preserved and the water flow is preserved."[1]

As the complexity, scale, and cost of ecosystem restoration in South Florida grew, the threshold question for federal involvement subtly changed. The national politics of saving the Everglades turned a corner. Instead of "is this a federal interest?" the question became "is this a national priority?" The problem was not *if* the government should develop and implement a comprehensive plan for saving the Everglades from ecological collapse, but *how*. And the politicians speaking out for Everglades restoration were no longer just Florida politicians. Increasingly, political leaders from across the nation saw Everglades restoration as a test case for efforts to restore and protect other ecosystems at risk throughout the United States. They adopted the dire rhetoric that Graham, other Florida politicians, and environmentalists had used for more than a decade: Everglades National Park, one of the crown jewels in the national park system, was dying. As Representative George Miller, a Democrat from California, ominously observed at a field hearing in the Florida Keys in July 1993, "We are not prepared to de-designate, if you will, the Everglades, Yellowstone, or Yosemite" as areas needing federal protection.[2]

This new political framework began to take shape following the election of William J. Clinton to the United States presidency in November 1992. Despite Clinton's mixed record on the environment as governor of Arkansas, many environmentalists saw him as the "great green hope."[3] During the presidential campaign, Clinton made numerous pledges of increased federal support for environmental programs, such as enactment of a new Clean Water Act that would regulate nonpoint sources of pollution and real commitment to "no net loss" of wetlands (two matters of importance to South Florida). Clinton also rejected the Bush administration's position that environmental protection was adverse to economic growth. Rather, Clinton maintained, environmental cleanup efforts would create jobs and lead to a stronger economy based on sustainable development.[4] Florida lawmakers who wanted the federal government to get more involved in cleaning up the Everglades were encouraged by this rhetoric.

Clinton boosted his environmental credentials by selecting Albert Gore, Jr., as his vice-presidential running mate. Gore, a senator from Tennessee, was recognized as one of the leading thinkers on environmental policy in Congress; his book *Earth in the Balance* came out during the election year. In that work, Gore argued that environmental problems were the most urgent global challenge of the post-Cold War era, that the United States had a responsibility to

President Bill Clinton at a joint session of Florida's legislature. (The Florida Memory Project, State Library and Archives of Florida)

Vice President Albert Gore, Jr. (The Florida Memory Project, State Library and Archives of Florida)

lead the world community on environmental issues, and that President George H. W. Bush had failed to provide that leadership.[5] When Clinton was elected president, members of Congress who supported environmental issues expected presidential leadership in areas where it had been lacking over the past 12 years.

Floridians who desired a larger federal role in saving the Everglades had reason to be pleased, too, as President Clinton formed his administration. His nominee for attorney general was Janet Reno, a Florida native, who soon began overseeing the job of preparing a new settlement in Dexter Lehtinen's lawsuit. His choice for EPA administrator was Carol Browner, another Floridian, who had served under Governor Bob Martinez and Governor Lawton Chiles as chief of the state's Department of Environmental Protection.[6] Clinton's selection for secretary of the interior, Bruce Babbitt, a former governor of Arizona, was not as familiar to Floridians. Babbitt, however, was eager to dispel any concerns that he would focus inordinate attention on the West, and he quickly dove into the Everglades issues, making Everglades restoration his leading cause in the eastern United States.[7]

As the Jacksonville District of the Corps, the SFWMD, and other agencies in South Florida took measure of the new administration, they noted events occurring in the opposite corner of the country. Clinton and Gore, delivering on a campaign promise, convened a "forest summit" to break the deadlock over old-growth logging and protection of the northern spotted owl on national forests in Oregon and Washington state. The president and vice-president met with environmentalists and the timber industry in Portland, Oregon, in April 1993, and announced a forest plan the following July. Emblematic of Clinton's compromise approach to controversial issues, the plan allowed for a resumption of logging at set harvest levels for 10 years, designation of certain areas for habitat conservation, and federal assistance for retraining displaced timber industry workers in other jobs. While the forest plan was fundamentally a political compromise, it charted a course for the future by employing a rigorous and revolutionary new method called "ecosystem management."[8] The Clinton administration's early commitment to ecosystem management in such a highly charged atmosphere as that surrounding the northern spotted owl sent a powerful signal all the way from the Pacific Northwest to South Florida.

South Florida's resource managers had long practiced elements of ecosystem management before the term became fashionable in the late 1980s and early 1990s. Members of the Corps of Engineers applied principles of ecosystem management when they worked with Everglades National Park staff in restoring sheet flow to Shark River Slough for the purpose of protecting the park's flora and fauna. So, too, did NPS specialists who developed a fire management plan for Everglades National Park, scientists in the SFWMD who collected and analyzed water samples from Lake Okeechobee, and members of the Miccosukee Tribe who hunted, fished, and trapped in their usual and accustomed places within the Big Cypress National Preserve.

What was new in the 1990s was that resource management agencies began to adopt ecosystem management as an organizing principle for many of their disparate activities. With the advent of the Clinton administration in 1993, ecosystem management was elevated to national policy.

Amid a deluge of scientific papers examining ecosystem management as a concept, an essay by ecologist R. Edward Grumbine, published in the journal *Conservation Biology* in 1994, offered the most round and succinct appraisal of what it entailed.[9] Grumbine recognized ten dominant themes of ecosystem management, beginning with a "hierarchical context," or "systems perspective," for addressing environmental problems. A systems perspective meant that managers working on a problem at any one level or scale in the biosphere—whether they were focused on genes, species, populations, ecosystems, or landscapes—needed to seek connections between all levels in the system. A corollary or second theme of ecosystem management involved the need to define ecological boundaries at appropriate scales. In other words, managers had to recognize when it was necessary to seek environmental solutions across jurisdictional lines. In the case of South Florida, resource managers had long understood—but with growing clarity—that the ecological boundaries of concern to them encompassed the entire Kissimmee River-Lake Okeechobee-Everglades watershed, even extending to Florida Bay and the Florida Keys. Third, ecosystem management aimed to preserve "ecological integrity." Standards for maintaining this integrity varied, but generally they included conserving viable populations of native species and maintaining natural disturbance regimes. For example, in South Florida, a state game warden and a NPS scientist might have different objectives for maintaining ecological integrity, but they would agree that restoring the natural hydropattern and allowing for extremes of high water and drought were key elements in their work. Additional themes were associated with the scientific method—data collection, monitoring, adaptive management—and with institutional processes, such as interagency cooperation and organizational change. Finally, Grumbine emphasized that ecosystem management was a social construct: it recognized that "humans are fundamental influences on ecological patterns" and that "human values play a dominant role in ecosystem management goals."[10]

Many viewed ecosystem management as essentially a change of focus from the protection of single species to the conservation of whole systems, but Grumbine noted that this did not capture the full scope of the "seismic shift" in thinking that the new approach required. At base, ecosystem management was "an early stage in a fundamental reframing of how humans value nature." It was an alternative to "resourcism"—the premise long held by modern industrial societies that nature was a storehouse of raw materials awaiting exploitation by humankind. Ecosystem management recognized biodiversity as something with intrinsic value, or as one set of authors included in Grumbine's survey explained, it assumed that "living systems have importance beyond their traditional commodity and amenity uses."[11] Other authors whom Grumbine cited argued that ecosystem management required an ethical reorientation to nature, even a "rejection of humanism or anthropocentrism" in favor of "a biocentric embrace of all life," although not all proponents would accept this philosophy.[12] One of the central challenges of ecosystem management, Grumbine suggested, was to pursue the goal of ecological integrity within a sociopolitical framework still governed by values that supported resourcism. Distilling all of these factors into a working definition, Grumbine declared that "ecosystem management integrates scientific knowledge of ecological relationships within a complex sociopolitical and values framework toward the general goal of protecting native ecosystem integrity over the long term."[13]

When President Clinton came into office, the best example of ecosystem management in South Florida was what the Corps and the SFWMD were undertaking in the Kissimmee River restoration project. But the scale was limited; some resource managers had begun thinking more grandly. One reason for this was because major problems with Florida Bay had surfaced, and many believed that the water management regime in South Florida under the C&SF Project was to blame.

Florida Bay, a shallow triangular coastal lagoon located south of the southern Florida peninsula, extended south and east to the Florida Keys and west to the Gulf of Mexico. The unusual geography of the bay made it especially susceptible to changes in salinity. Exceedingly shallow (generally three to ten feet deep over most of its expanse), the bay's rate of evaporation relative to the volume of water was very high. In addition, mud banks covered considerable parts of the bay floor, moving like underwater sand dunes. Resting just below the surface of the water, the banks reduced the force of lunar tides and restricted the circulation of seawater into the bay. Fresh water flowed to the bay mainly through Taylor Slough (and, to a lesser degree, Shark River Slough), and this water mingled with gulf currents in the outer portion of the bay. The brackish waters supported rich communities of seagrasses, molluscs, crustaceans, and fish, and, in general, the seagrasses were more prolific where the waters of the bay mixed more freely with gulf waters.[14]

Scientists and environmentalists had been concerned about the bay for years. In the 1960s, a dearth of fresh water in Everglades National Park caused many to worry that Florida Bay's salinity would rise to dangerous levels, killing the shrimp and fish. This, in turn, harmed the shrimp and commercial fishing industries that depended on the bay for their livelihood. Additional concerns arose in the 1970s, and resident fishermen, such as Michael Collins, were the first to call attention to ecological changes in Florida Bay. A resident of the island community of Islamarada in the Florida Keys, Collins made a living taking wealthy clients out on his charter fishing boat around the Everglades, the Bahamas, and the bay. With other fishermen, Collins began observing changes in seagrass communities in Florida Bay, and in 1976 the Islamarada Fishing Guide Association sent him to Everglades National Park to consult with research scientists about possible causes. Not satisfied that the park was giving the problem adequate attention, Collins began to research the history of the C&SF Project on the theory that water diversions from the Everglades—particularly the construction of the C-111 canal—had reduced freshwater flows into Florida Bay, thereby altering the bay's estuarine characteristics.[15]

During the 1980s, Collins took his concerns to the SFWMD, and at the end of the decade Governor Bob Martinez appointed him chairman of the Resource Planning and Management Committee for the Keys Areas of Critical State Concern. According to Collins, that group's activity "was one of the first efforts I saw to get a number of government entities from different branches of government together to discuss resource management." The interagency cooperation was at the state and county level, rather than the federal level, and participants tried to define the ecological boundaries of the problem.[16]

Aerial view of Florida Bay. (Everglades National Park)

Turtle grass, or *Thalassia testinum*, the most abundant species of seagrass in Florida Bay, proved to be the canary in the coalmine. Fishing guides first observed that the turtle grass was spreading, colonizing the inner part of the bay, an indication that conditions were becoming more saline. In 1987, they began to see huge patches of turtle grass looking sick or dead. During the next four years, the seagrass die-offs spread over several hundred thousand acres. Floating mats of the decomposing matter blocked out sunlight, lowered the oxygen content in the water, and led to massive algal blooms. The normally crystal clear waters of Florida Bay became more turbid. As Collins told one journalist, "You should be able to read a newspaper

lying on the bottom in 10 feet of water."[17] In southwestern portions of the bay, increased turbidity and phytoplankton growth led to massive die-offs of sponges.[18]

By 1991, these conditions had reduced shrimp and fish harvests to record lows. Then, in November 1991, a huge algal bloom erupted in Florida Bay, spreading until, by the summer of 1992, it covered miles and miles of the bay, choking out sunlight, devastating sponge, shrimp, and fish populations, and creating a "dead zone" along the bay's western edge.[19] Observers, including commercial fishermen, Everglades National Park officials, and environmentalists, were horrified by the developments. "Florida Bay is falling apart like a rotting piece of cloth," Jay Zieman, a marine scientist with the University of Virginia, asserted. "This is a disaster on the same scale as the Yellowstone fires" (which, ironically, turned out *not* to be a disaster after all).[20] The bay was "becoming a huge dead zone," an editorial in *The Miami Herald*, declared. "Slime and algae cloud its once clear waters, where sea grass waved gently in the current."[21] The condition of the bay, Mike Robblee, chief of Everglades National Park's marine science section, related, showed that either the bay was "very sick" or it was "changing drastically." Whatever the situation, Robblee continued, "we need to sit up and take notice."[22]

Collins, who would later become a member of the SFWMD governing board, continued to assert that the cause of the devastation lay in the management of water in South Florida. "It was the drainage system that had been put in that was the problem," he averred.[23] Some environmentalists agreed. George Barley, an Orlando developer who was also an avid sports fisherman, part time summer resident of Islamorada in the Florida Keys, and chairman of the Florida Keys National Marine Sanctuary Advisory Council (created in 1990 by the Florida Keys National Marine Sanctuary and Protection Act, in part to deal with Florida Bay issues), became convinced that "the basic problem in Florida Bay is its fresh water has been taken away by a variety of means upstream."[24] Barley and others claimed that development in South Florida and the C&SF Project had drastically reduced how much fresh water flowed into the bay, creating an imbalance between the amounts of salt and fresh water that characterized a healthy estuary and making it more like the sea. Others, however, insisted that the problem came from an overabundance of nutrients resulting from runoff from the EAA and South Florida's urban areas.[25]

But the real dilemma was that no one could say with certainty what had caused the dramatic seagrass dieoff. Were the seagrass communities responding to nutrient loading similar to that occurring in Lake Okeechobee? Was Florida Bay receiving nitrogen and phosphorus coming all the way from the sugar cane fields? Or was it a problem of water supply and increased salinity? Was the sharp reduction of freshwater input from the C-111 basin causing more seawater to infiltrate and mix with the shallow waters in Florida Bay?[26] No one seemed to know. As Everglades National Park

Superintendent Richard Ring explained, park scientists had largely ignored Florida Bay since the 1960s in order to concentrate on mainland water issues. "Basic research that should have been done in the 1970s has not been done," Ring stated, noting that the park's research center did not have the funding to study the problem adequately.[27]

Realizing the severity of the situation, and hoping to prevent the bay's impending collapse, Barley used his position with the Florida Keys National Marine Sanctuary—and his friendship with President George H. Bush, an avid fisherman of Florida Bay waters—to warn public officials of the problems. He recruited a wealthy friend with a seaplane to give flight tours of the bay to any public official who was interested in having a look. At first county commissioners accepted the offer, then elected officials who came from outside the area. This sounded an alarm that was soon heard in Washington; in the words of Billy Causey, director of the Florida Keys National Marine Sanctuary, "the noise level started getting so loud that [Congress] couldn't help but . . . hear it."[28]

Even before Barley began publicizing the Florida Bay issue, the Corps of Engineers had recognized the need for increased coordination in South Florida between water management agencies in order to promote the overall environmental health of the region. Colonel Terrence "Rock" Salt, District Engineer of the Jacksonville District, for example, had proposed a review of the whole C&SF Project in 1991 with a view to developing a comprehensive framework for interagency coordination on water management issues in South Florida. He took Lieutenant General Arthur Williams, director of the Corps' civil works program, and Nancy Dorn, Assistant Secretary of the Army, on a helicopter tour of the Kissimmee River system, receiving their support to put the review study into the annual appropriation bill for the Corps' civil works program. Although the Water Resources Development Act of 1992 (WRDA-92) authorized the review study, it got lost in the frenzy of the Corps' emergency response to Hurricane Andrew, which struck South Florida in August 1992, and, as a result, the Bush administration did not allocate funds for a review. When Clinton came into office in January 1993, then, the idea of a comprehensive ecosystem restoration plan was embryonic and without a federal funding source.[29]

Florida environmentalists knew of the proposed study and wanted to see it funded through the Corps. After Colonel Salt became absorbed in the Hurricane Andrew disaster relief efforts, James "Jim" Webb of the Wilderness Society took the matter into his own hands and drafted the language for a congressional authorization.[30] Meanwhile, the Everglades Coalition produced its own restoration plan for the "Greater Everglades Ecosystem," influenced in part by the condition of Florida Bay. Not surprisingly, the coalition's plan called for restoration of "the essential features of the natural hydrology—the volume, depth, timing and distribution of water that once flowed through the system." It also sought a return of pristine water quality and enhancement of urban and agricultural water supplies.

Drawing upon ecosystem management concepts then in development for the "Greater Yellowstone Ecosystem," the plan further called for restored connectivity among wetland communities and use of biological indicator species to monitor the health of the ecosystem.[31]

After the 1992 presidential election, Florida environmentalists scrambled to reposition themselves and to establish links to the Clinton administration, even though some took a dim view of Clinton. According to Joseph Browder of the Audubon Society, those who considered themselves close to Clinton advised that the way to get his attention was to recast the Everglades restoration plan as a way to create jobs. "I had been getting reports by people who were supposedly in the know that we needed to turn this into a pork barrel program," Browder remembered in an interview. Browder himself heard the president-elect make an off-the-cuff remark at a gathering in Hilton Head, South Carolina, during the winter of 1992-1993, that the only people that mattered were those who invested money and created jobs. "It reinforced the feeling that this was going to be a tough slog," Browder recalled.[32]

Jim Webb of the Wilderness Society had other ideas. Webb knew Bruce Babbitt from his years in Arizona and he correctly recognized the new secretary of the interior as the key figure on Clinton's environmental team. Webb got Babbitt to come to Tallahassee, Florida, in January 1993 and give the keynote address to the annual conference of the Everglades Coalition. This was Babbitt's first public appearance after his confirmation. At the podium, Babbitt referred warmly to his two dinner companions, Colonel Salt and Richard Ring, superintendent of Everglades National Park, and promptly launched into a visionary speech about a Corps restudy of the whole ecosystem based on consultation with other federal agencies, input by a team of scientists, and political support from the highest levels. The audience cheered, applauding this bold new course.[33]

Babbit's resolve stiffened after paying a visit to Everglades National Park. His examination of the park left him "absolutely appalled," and Webb convinced him that drastic measures were needed, including the purchase of more private land to protect the park's boundaries. "We can't defend the Everglades—or Yellowstone—just at their boundaries," Webb noted. "We have to deal with the whole ecosystem."[34] Back in Washington, Babbitt put this plan into motion. Just as he had outlined in Tallahassee, the restoration effort would go forward simultaneously at three levels in the federal government: at the cabinet level in Washington, at the agency level with the coordination of key managers like Colonel Salt and Superintendent Ring, and at the field level with scientists in each agency participating on an interagency team. Cooperation would start at the cabinet level with a new interagency task force and flow down to the field level. Whatever emerged from this effort would be science-driven.

In attempting to implement this plan, Babbitt had other examples of interagency efforts providing advice on water resource

Secretary of the Interior Bruce Babbitt. (U.S. Department of the Interior Library)

management. In the 1960s, for example, the St. Paul District of the Corps initiated the Upper Mississippi River Comprehensive Basin Study, an interagency examination of the river that morphed into the Upper Mississippi River Basin Coordinating Committee in the 1970s. Consisting of representatives from the Corps and the Departments of Agriculture, Commerce, Health, Education and Welfare, Housing and Urban Development, Interior, and Transportation, as well as individuals from the EPA and the Federal Power Commission, this committee was specifically tasked with developing a plan to solve water and land resource problems on the Upper Mississippi River. For additional management of the Upper Mississippi, the Great River Environmental Action Team was formed in the late 1970s, made up of representatives from the Corps, the USGS, the EPA, the Soil Conservation Service, the Bureau of Outdoor Recreation, and the Department of Transportation. The team, also known as GREAT, had the responsibility of coordinating navigation and dredging on the Upper Mississippi River with other river uses, especially recreation and fish and wildlife management. Studies initiated by GREAT eventually led to congressional authorization of the Upper Mississippi River System Environmental Management Program in 1986, which, under the leadership of the Corps, specifically focused on enhancing and preserving environmental values on the Upper Mississippi River.[35]

In a similar way to these Upper Mississippi management committees, Babbitt established a cabinet-level task force for South Florida, composed of five assistant or under secretaries representing the Departments of Interior, Defense, Commerce, Agriculture, and Justice, and an assistant administrator representing the Environmental Protection Agency.[36] It would meet semi-annually. Although task force members would delegate most of the effort to the Interagency Working Group, such attention to an ecosystem by so many senior officers in the executive branch of government was unprecedented.

In the early 1990s, observers had begun making references to the "federal family" in South Florida, meaning the constellation of federal agencies involved in resource management. In welding this federal family into an interagency team, Babbitt's first task was to get together the several agencies in the Department of the Interior. These included the NPS, the FWS, the Bureau of Indian Affairs (BIA), and the USGS. Babbitt arranged a meeting of the Interior agencies in South Florida in April 1993 so that they could begin to develop a united vision for Everglades restoration. He sent his own science advisor, Thomas E. Lovejoy, as his representative. Lovejoy, a renowned conservation biologist, had recently gone to work for Babbitt to head up a new National Biological Survey, and on top of that effort Lovejoy plunged headlong into Everglades issues. At the April meeting Lovejoy encountered a general mood of optimism, although the representative from Everglades National Park sounded a discordant note when he insisted that the park did not want the USGS to conduct a hydrological survey in the park, preferring to have its own science staff do it.[37]

The focal point of this meeting was a composite satellite view of South Florida in which human development showed up in red and natural vegetation appeared in green. The satellite view was a remarkably clear expression of the extent of human manipulation of the natural environment and the hydrological pattern of flow from the headwaters of the Kissimmee River through Lake Okeechobee and the Everglades to Florida Bay. "You could see where the agricultural interests had encroached, and the way the water didn't flow unless somebody turned a valve somewhere," Lovejoy remembered in an interview. "You could see all the manmade structures, ditches, and dikes."[38] The satellite image was a fitting point of departure for the new interagency planning effort. Jurisdictional lines did not appear in the image, though the location of certain boundaries could be inferred from various hard edges separating red and green areas. More importantly, the image stimulated a holistic view or ecosystem perspective.

In June, Babbitt called the first meeting of the Interagency Working Group in Key Largo. Billy Causey, director of the Florida Keys National Marine Sanctuary, described this conference as "pivotal." The group's initial task was to define the extent of the ecosystem and agree upon some restoration objectives. "Never in my wildest imagination," Causey said, "did I expect all the people in that room to define the ecosystem as starting in the Kissimmee headwaters and coming all the way down to the Florida Keys."[39] However, since Florida Bay's condition was not improving, and, in many ways, was worsening, the group's definition was not surprising. As a panel of scientists later concluded, South Florida ecosystems had been "managed as if they were in isolation from one another," in many ways causing the freshwater problems that Florida Bay now faced. In their estimation, "it is clear that what is now needed is a broader perspective."[40] Accordingly, the group began coordinating several different Everglades project already underway, such as the C-111 Project (replumbing the East Everglades for better water flow to Everglades National Park) and the investigation of Florida Bay's problems, with the goal of improving the Everglades ecosystem as a whole.[41]

A satellite map of South Florida. (U.S. Army Corps of Engineers, Jacksonville District)

The Key Largo meeting also saw the emergence of some interesting group dynamics. Babbitt had insisted that each department send two—and only two—representatives to this initial meeting because he did not want an influx of Interior personnel. Moreover, he asked Assistant Secretary of the Interior George Frampton to co-chair the meeting with Deputy Under Secretary of Commerce Doug

Hall—a clear signal that the Commerce Department's National Oceanic and Atmospheric Administration (NOAA) had a role in Everglades restoration as the managing agency of South Florida's coastal waters. Frampton and Hall effectively led the group, displaying a new confidence that national park interests would get their due. Representatives of the Miccosukee and Seminole tribes and some state officials attended the meeting as well, but were not invited to sit at the table. Instead, they sat mutely against the back wall. This peculiar seating arrangement struck some participants as imperious on the part of the federal government. Colonel Salt showed up with Jimmy Bates, the senior civilian in the civil works directorate of the Corps' headquarters division, plus four others, all in Army uniform. This military escort was contrary to Babbitt's instruction that exactly two people attend for each department. "We all kind of bristled," Causey remembered. "We started counting heads." However, Salt, a large, square-shouldered man whom everyone knew as "Rock," quickly put everyone at ease with his disarming and enthusiastic manner, and he began to act as the group facilitator. "We could see it was a new era for the Corps," Causey recalled. "We had had some good colonels but Rock was here to get the work done."[42]

Colonel Salt was undoubtedly the right man in the right place at the right time, another one of the many fortunate circumstances that propelled Everglades restoration to a national priority status during the Clinton administration. Salt's consensus-based leadership style was atypical of a commanding officer.[43] He was deeply interested in ecosystem restoration. Earlier in his career he had been assigned to the Corps' Walla Walla District in the Pacific Northwest where he worked on mitigating the impacts of Columbia-Snake River dams on anadromous fish runs, and on other efforts to restore habitat for endangered salmon. He also had the backing of leaders in the Corps who wanted to move the organization in a "greener" direction, notably Lieutenant General Henry Hatch, Chief of Engineers from 1988 to 1992. When Salt was selected for the Jacksonville District command, he went to G. Edward "Ed" Dickey, the Acting Secretary of the Army for Civil Works who had contributed to the development of the "Principles and Guidelines" in 1983, by which the Corps evaluated the federal interest in proposed environmental projects. Salt asked Dickey bluntly if the Corps was serious about Kissimmee River restoration and Everglades modified waters projects. "Oh, yes," Dickey replied, but the colonel must do two things: demonstrate that the project was in the federal interest, and show that it was deserving of high priority in the nation. Salt focused on those problems when he represented the Corps in the Interagency Working Group and when he initiated the restudy of the C&SF Project. Ultimately, he had to prove to his superiors in Washington that the federal interest in ecosystem restoration in South Florida was more compelling than competing initiatives contemplated in regions such as California or the Mississippi Valley.[44]

Salt understood the need to follow and respect the internal process of the Corps even as that process began to get short-circuited by Washington politics. One significant consequence of Babbitt's initiative in creating a federal task force was that the Jacksonville commander communicated directly with the Army's task force representative, Acting Assistant Secretary Dickey. The normal chain of command in the Corps of Engineers ran from the Secretary of Defense, the Secretary of the Army, and the Assistant Secretary of the Army to the Chief of Engineers at Corps headquarters, then to the Division Engineer, and then to the District Engineer. Direct communications between Dickey and Salt, which grew increasingly frequent, bypassed headquarters and the division. During Salt's command the task force's impact on the Corps' organizational structure did not produce significant tensions or repercussions, but by the end of the Clinton administration it would.[45]

Colonel Terrence "Rock" Salt, District Engineer of the Jacksonville District. (U.S. Army Corps of Engineers, Jacksonville District)

At the same time that Babbitt initiated the creation of a federal task force on ecosystem restoration in South Florida, he pushed the Corps to commence an immediate comprehensive review of the C&SF Project. If the seeds of this restudy were already sown before Babbitt came into office, it was undoubtedly Babbitt's energy that caused the project to germinate. As Salt remembers, he received a "frantic call from Ed Dickey" in April 1993. Did Salt know anything about a restudy, Dickey inquired. The next day Dickey called him again, this time relating that the administration wanted the Corps to begin a restudy immediately using existing funds. Next, General Roger F. Yankoupe, Division Engineer of the South Atlantic Division, phoned Salt, telling him to bring his chief planner to Atlanta to get the restudy started. With the help of John Rushing, Chief of Planning in the South Atlanta Division, Salt moved the project expeditiously "through the stovepipes in the Corps." Initially, Salt and others thought the study would be funded out of the Corps' general investigations account, but

Rushing had another idea. "By calling it a review study [we] could use construction dollars, which were an order of magnitude greater than [general investigations] dollars," Salt later explained. "By putting it into that account we were able to initiate a $2 million reconnaissance study that was unprecedented in terms of size."[46]

By June 1993, the "Restudy" (as it was now officially called) had assumed national importance. Jimmy Bates, Deputy Director of Civil Works, instructed Salt to select his planning team carefully and assemble the best talent the Corps had. With such strong backing at the highest Corps level, it was no wonder that Salt exuded confidence at the initial gathering of the Interagency Working Group.[47]

Salt tapped Stuart Appelbaum, chief of the Jacksonville District's Flood Control and Floodplain Management Planning Group, to head the Restudy. Appelbaum, who had worked on the Kissimmee River restoration plan, had contemplated how he would run the C&SF Project review study since its first discussion in 1992, influenced by the mentoring of Mann Davis, who had headed the District's 1980 water supply study. Because that examination had been less than a stellar success, Davis had determined that the Corps needed to improve the way it conducted the study, and he transmitted some of these ideas to Appelbaum. Appelbaum therefore decided that the Restudy would have to involve the public and be interdisciplinary and interagency. Most importantly, people had to perceive it as something new and different. In order to accomplish these purposes, Appelbaum co-located all of the team—all disciplines, all agencies—in one room. His organizational model was the Skunkworks operation in the Lockheed Corporation. As Appelbaum explained, "You give them their own status off on the side; they are no longer working for the same organization, but they're kind of a unique, standalone organization; you let them go solve tough problems." By late summer Appelbaum had a team of 12 people and a room in the basement of the Jacksonville District affectionately known as "the cave." His oft-repeated instruction to his team members was that they leave their agency hats at the door. One wag brought in 12 hats with a generic "agency" logo printed on each one. The team began to form a group identity.[48]

By the end of summer it was clear that the Restudy would serve as the vehicle for developing a comprehensive Everglades restoration plan. The Task Force and the Interagency Working Group would provide oversight. In September 1993, the second meeting of the Interagency Working Group occurred in Orlando. Ed Dickey attended with Salt. Talking about the Restudy, Dickey told the group that the other agencies must decide what they wanted restored, and then the Corps would draw up the engineering plans. This was a familiar refrain, but never in history had such an invitation involved so many agencies and so much area. Indeed, it was now evident that the scope of the Restudy would exceed the geographic limits of the C&SF Project.[49]

The Orlando meeting produced an interagency agreement on South Florida ecosystem restoration, which formally established the

South Florida Ecosystem Restoration Task Force. The agreement declared that the South Florida ecosystem encompassed the Kissimmee watershed, Lake Okeechobee, the Big Cypress Basin, the Everglades, Florida Bay, and the Florida Keys. It listed the many federal interests in the area. These were not limited to federal lands, but also included the C&SF Project and the enforcement of environmental laws such as the Clean Water Act, the Clean Air Act, the Endangered Species Act, and others. The purpose of the Task Force was "to coordinate the development of consistent policies, strategies, plans, programs, and priorities for addressing the environmental concerns of the South Florida Ecosystem." The agreement acknowledged the need for coordination with state, local, and tribal governments, as well as with member agencies. Specific goals of the task force were to agree on federal objectives for ecosystem restoration; to promote an ecosystem-based science program; to support the development of "appropriate multi-species recovery plans for threatened and endangered species" (a careful effort to move from single-species management to the conservation of whole systems); and to help expedite projects aimed at ecosystem restoration.[50]

The interagency agreement also formally established the Interagency Working Group. It was to be composed of Florida-based representatives of the following federal agencies: NPS, FWS, USGS, BIA, National Biological Survey (Department of the Interior); NOAA (Department of Commerce); Soil Conservation Service (Department of Agriculture); U.S. Attorney for the Southern District of Florida (Department of Justice); EPA; and U.S. Army Corps of Engineers (Department of the Army). The Working Group was to prepare recommendations in the form of an integrated plan one year from the first meeting of the Task Force, and update this document annually thereafter. Other responsibilities included developing an integrated financial plan, an ecosystem-based science program, and public outreach efforts. The Working Group was also charged with identifying and resolving interagency differences concerning ecosystem restoration, and it was empowered to establish subgroups.[51]

Yet some groups—most notably the Miccosukee Indians—believed that they had been intentionally excluded from both the South Florida Ecosystem Restoration Task Force and the Interagency Working Group, despite their obvious interests in Everglades restoration. Indeed, both the Seminole and the Miccosukee were intensely interested in water quality and restoration issues, especially since the quality of water entering Conservation Area No. 3 directly affected their lands, and had developed water rights compacts in the 1980s and 1990s to protect their interests. Having expressed this concern in the past, both the Seminole and the Miccosukee expected at least some kind of a role in ecosystem restoration efforts. When no formal position was offered, the Miccosukee protested, spurred on by Dexter Lehtinen, who they had hired as their attorney. In 1994, for example, the tribe sued the federal government, charging that it had been unfairly excluded from a meeting where SFWMD and Florida

Department of Environmental Protection scientists had met with federal scientists. Although Truman E. "Gene" Duncan, Jr., head of the Miccosukee water management division, attended the meeting, he alleged officials ejected him from the gathering. Jay Ziegler, spokesman for the Interior Department, did not dispute the charge, but said that the reason for the action was so that federal authorities could discuss President Clinton's upcoming budget. The Miccosukee disagreed; Angel Cortinas, one of their attorneys, insisted that the Indians were "being excluded from the discussions that affect the tribe's interest."[52]

Yet the Task Force and the Working Group did not maliciously prevent the inclusion of the Miccosukee; instead, the Federal Advisory Committee Act, which authorized the creation of organizations such as the Task Force, precluded non-federal interests from actively sitting on federal committees. Non-federal groups could attend meetings, but could not participate in any decision-making. As explained in Chapter 18, not until 1995 would Congress remedy this situation by amending the Federal Advisory Committee Act. Until then, Task Force and Working Group officials believed there was nothing they could do.

A field of sawgrass, one of the dominant plants of the pre-drainage Everglades. (The Florida Memory Project, State Library and Archives of Florida)

The Miccosukee action indicated that the consensus approach that Secretary Babbitt was trying to produce with the working groups was not entirely successful, but the Task Force and Working Group continued their operation. In order to carry out Babbitt's vision of science lying at the heart of the restoration efforts, the Working Group established a Science Subgroup, and in November 1993, this subgroup completed its initial report, "Federal Objectives for South Florida Restoration." This document foresaw the outcome of ecosystem restoration as follows:

The idealized goal for the natural areas of South Florida is to restore to predrainage conditions the landscape-scale hydrologic and ecologic structure and function in order to reinstate ecological integrity and sustainable biodiversity. The goal is an ecosystem that is resilient to both chronic stresses and catastrophic events with as little human intervention as possible.[53]

The report also presented more specific restoration objectives and measurable success criteria for the entire region and nine sub-regions. In each case, it described three levels of protection based on the amount of developed area that would be restored to wetlands. The Science Subgroup termed the levels of protection at either end of this continuum as "constrained" and "unconstrained" options, while the level of protection in the middle was termed the "incremental" choice. The point of this presentation was to show that for each increment of developed area restored to wetlands, the social and economic costs rose while the environmental risk fell. Put another way, if ecosystem restoration did not go far enough, it would entail a high risk of failure.

When the Working Group released this report, controversy ensued. The "unconstrained" option of complete restoration of all wetlands, which the Science Subgroup described only for purposes of framing the "incremental" option, inflamed certain stakeholders—and with good reason. Under this option, the report graphically showed one swath of restored wetlands obliterating a small city north of Tampa Bay, while also displaying an immense area of restored wetlands completely engulfing the EAA. As if these visual images were not provoking enough, the Science Subgroup's choice of terminology seemed strangely aggressive: to say that the presence of communities and farms was "constraining" sent the wrong public message. The Task Force, the Working Group, and the Science Subgroup were all chastened by the public reaction, which served as a healthy reminder to them that ecosystem management was fundamentally a social endeavor.[54] As Grumbine would write less than a year later in his timely synopsis of ecosystem management, "human values play a dominant role in ecosystem management goals."[55]

Thus far, Secretary Babbitt's initiative had produced much organizational change but little else. Yet it was a necessary first step toward implementing an ecosystem management approach to Everglades restoration. In the new organizations that had been created—the South Florida Ecosystem Restoration Task Force, the Interagency Working Group, the Science Subgroup, and the Restudy team—the seeds of ecosystem management had been planted. Many of the attributes of ecosystem management were already visible and at play. The resource managers were adopting a systems perspective, formulating goals that would define success in the effort to restore ecological integrity, developing a science-based approach to decision making, and fostering interagency coordination. Although the South Florida Ecosystem Restoration Task Force was thus far a federal initiative, it would, once it received authorizing legislation, evolve to include representatives of state, local, and tribal governments. The organizational change provided a new institutional environment in which the idea of ecosystem restoration could grow and flourish.[56]

Endnotes

[1] House Committee on Interior and Insular Affairs Subcommittee on National Parks and Recreation, *Additions to the National Park System in the State of Florida: Hearings Before the Subcommittee on National Parks and Recreation of the Committee on Interior and Insular Affairs, House of Representatives*, 99th Cong., 1st and 2d sess., 1986, 140.

[2] Miller quoted in House Committee on Natural Resources Subcommittee on Oversight and Investigations and Subcommittee on National Parks, Forests and Public Lands and House Committee on Merchant Marine and Fisheries Subcommittee on Environment and Natural Resources, *Florida Everglades Ecosystem*, 103rd Cong., 1st sess., 1993, 14. On Everglades restoration as a test case for ecosystem management, see Congressional Research Service, "The Florida Everglades: An Ecosystem in Danger," as published in the hearing, 201.

[3] Byron Daynes, "Bill Clinton: Environmental President," in *The Environmental Presidency*, Dennis L. Soden, ed. (Albany: State University of New York, 1999), 259.

[4] Norman J. Vig, "Presidential Leadership and the Environment: From Reagan to Clinton," in Norman J. Vig and Michael E. Kraft, eds., *Environmental Policy in the 1990s*, 3rd ed. (Washington, D.C.: Congressional Quarterly, 1997), 104-105.

[5] Albert Gore, *Earth in the Balance: Ecology and the Human Spirit* (New York: Houghton Mifflin Company, 1992), 295-360.

[6] Browner had earlier served as a Senate environmental aide of Gore. Another one of Gore's environmental aides, Kathleen McGinty, was appointed head of the new Office of Environmental Policy. McGinty would become highly instrumental in Everglades restoration in Clinton's second term.

[7] Vig, "Presidential Leadership and the Environment: From Reagan to Clinton," 106; William Leary interview by Theodore Catton, 24 November 2004, Washington, D.C., 5; Stuart Stahl interview by Julian Pleasants, 22 February 2001, 25, Everglades Interview No. 3, Samuel Proctor Oral History Program, University of Florida, Gainesville, Florida [hereafter referred to as Stahl interview].

[8] Lettie McSpadden, "Environmental Policy in the Courts," in *Environmental Policy in the 1990s*, 176-177; Theodore Catton and Lisa Mighetto, *The Fish and Wildlife Job on the National Forests: A Century of Game and Fish Conservation, Habitat Protection, and Ecosystem Management* (Washington, D.C.: USDA Forest Service, 1998), 263-273.

[9] R. Edward Grumbine, "What is Ecosystem Management?" *Conservation Biology* 8 (March 1994): 27-38.

[10] Grumbine, "What is Ecosystem Management?" 31.

[11] W.B. Kessler, H. Salwasser, C. Cartwright, Jr., and J. Caplan as quoted in Grumbine, "What is Ecosystem Management?" 34.

[12] R.F. Noss and A. Cooperrider as quoted in Grumbine, "What is Ecosystem Management?" 34.

[13] Grumbine, "What is Ecosystem Management?" 31. In a follow-up article, Grumbine reported that many managers and academics found the ten themes to form a useful framework for understanding ecosystem

management. See R. Edward Grumbine, "Reflections on 'What is Ecosystem Management?'" *Conservation* Biology 11 (February 1997): 41-47.

[14] James W. Fourqurean and Michael B. Robblee, "Florida Bay: A History of Recent Ecological Changes," *Estuaries* 22 (June 1999): 349-350.

[15] Michael Collins interview by Theodore Catton, 13 July 2004, West Palm Beach, Florida, 1-2 [hereafter referred to as Collins interview].

[16] Collins interview, 1.

[17] Ben Iannotta, "Mystery of the Everglades," *New Scientist* (9 November 1996): 3535.

[18] Fourqurean and Robblee, "Florida Bay," 345.

[19] See Billy Causey interview by Theodore Catton, 20 January 2005, Marathon, Florida, 3-5 [hereafter referred to as Causey interview]; Collins interview, 2; Fourqurean and Robblee, "Florida Bay," 345; "Panel Blasts Florida Bay Neglect," *The Miami Herald*, 21 September 1993; "If Florida Bay Dies, An Industry Will, Too," *Tallahassee Democrat*, 17 February 1993.

[20] As quoted in "Ailing Florida Bay Endangers Coral," *The Miami Herald*, 11 August 1992.

[21] "Florida Bay: Catastrophe," *The Miami Herald*, 12 August 1992.

[22] As quoted in "Ailing Florida Bay Endangers Coral," *The Miami Herald*, 11 August 1992.

[23] Collins interview, 3.

[24] George M. Barley, Jr., Chairman, Advisory Council, to Honorable Jim Smith, Secretary of State, 27 August 1992, File Florida Bay, Box 157406, SFWMDAR.

[25] See "Ailing Florida Bay Endangers Coral," *The Miami Herald*, 11 August 1992.

[26] Collins interview, 1.

[27] As quoted in "Ailing Florida Bay Endangers Coral," *The Miami Herald*, 11 August 1992.

[28] Causey interview, 5-6.

[29] House Committee on Natural Resources Subcommittee on Oversight and Investigations, Committee on Agriculture Subcommittee on Specialty Crops and Natural Resources, and the Committee on Merchant and Marine Fisheries Subcommittee on Environment and Natural Resources, *Ecosystem Management: Joint Oversight Hearing on Ecosystem Management and a Report by the General Accounting Office, "Ecosystem Management—Additional Actions Needed to Adequately Test a Promising Approach,"* 103rd Cong., 2d sess., 1995, 99; Colonel Terrence C. "Rock" Salt interview by Theodore Catton, 19 January 2005, Miami, Florida, 5 [hereafter referred to as Salt interview—Catton].

[30] Richard Bonner interview by Theodore Catton, 13 May 2005, Jacksonville, Florida, 13.

[31] According to Joseph Browder, the plan was first published in 1992 and republished in 1993. Browder interview, 6. The 1993 document is in House Subcommittee on Oversight and Investigations et al., *Florida Everglades Ecosystem*, 309-347.

[32] Browder interview, 6.

[33] Everglades Coalition, "Past Conferences" <http://www.evergladescoalition.org/site/pastconference.html> (26 May 2005); Salt interview—Catton, 5.

[34] All quotations in Philip Elmer-Dewitt, "Facing a Deadline to Save the Everglades," *Time* 141 (21 June 1993): 57.

[35] George W. Griebenow, "A Team Called GREAT," *Water Spectrum* 9 (Winter 1976-1977): 19-20; U.S. Army Corps of Engineers, North Central Division, *Upper Mississippi River System Environmental Management Program, Sixth Annual Addendum* (Chicago: U.S. Army Corps of Engineers, North Central Division, 1991), 2-4; Upper Mississippi River Basin Coordinating Committee, *Upper Mississippi River Comprehensive Basin Study Main Report* (Chicago: Upper Mississippi River Basin Coordinating Committee, 1972), 1-3.

[36] Clinton had requested that EPA Administrator Carol Browner join his cabinet, as he wanted Congress to confer cabinet-level status on the agency.

[37] Thomas E. Lovejoy interview by Theodore Catton, 9 December 2004, Missoula, Montana [by telephone], 1 [hereafter referred to as Lovejoy interview].

[38] Lovejoy interview, 1.

[39] Causey interview, 7.

[40] As quoted in "Panel Blasts Florida Bay Neglect," *The Miami Herald*, 21 September 1993.

[41] "Testimony of George T. Frampton, Jr., Assistant Secretary for Fish and Wildlife and Parks," in House Committee on Natural Resources, et al., Subcommittee on Oversight and Investigations, *Ecosystem Management*, 100.

[42] Causey interview, 6-8; Salt interview—Catton, 6.

[43] Colonel Terry Rice interview by Brian Gridley, 8 March 2001, 29, Everglades Interview No. 4, Samuel Proctor Oral History Program, University of Florida, Gainesville, Florida [hereafter referred to as Rice interview].

[44] Salt interview—Catton, 1-2.

[45] Leary interview, 12; General Joe Ballard interview by Theodore Catton, 18 November 2004, Missoula, Montana [by telephone], 7-8.

[46] Salt interview—Catton, 5-6.

[47] Salt interview—Catton, 5-6.

[48] Quotations in Stuart Appelbaum interview by Brian Gridley, 22 February 2002, 18-19, Everglades Interview No. 11, Samuel Proctor Oral History Program, University of Florida, Gainesville, Florida [hereafter referred to as Appelbaum interview]; see also James Vearil, personal communication with the authors, 18 April 2006.

[49] Salt interview—Catton, 6. One year later, Colonel Terry Rice would make the same pitch to the Governor's Commission on a Sustainable South Florida. See Rice interview, 18.

[50] "Interagency Agreement on South Florida Ecosystem Restoration," 23 September 1993, Billy Causey's Task Force Files, Florida Keys National Marine Sanctuary administrative records [hereafter referred to as FKNMSAR].

[51] "Interagency Agreement on South Florida Ecosystem Restoration."

[52] Quotations in "Indians Sue U.S. Government," *Sun-Sentinel*, undated newspaper clipping in File Everglades Mediation Miccosukee, Box 19706, SFWMDAR; see also "Miccosukee Tribe Sues U.S. Agency Over Glades Plan," *The Herald*, 4 November 1994. The Miccosukee's accusation was nothing new; they had complained about being left out of financial discussions in 1993 as well. See Woodie Van Voorhees, Government & Public Affairs, to Til Creel, Irene Quincey, Steve Lamb, and Vince Katilus, 20 July 1993, File Everglades Mediation Miccosukee, Box 19706, SFWMDAR.

[53] Quoted in House Committee on Natural Resources, et al., Subcommittee on Oversight and Investigations, *Ecosystem Management*, 294.

[54] Richard G. Ring, Chairman, South Florida Ecosystem Restoration Working Group, to George Frampton, Chairman, South Florida Ecosystem Restoration Task Force, n.d., Billy Causey's Task Force Files, FKNMSAR; Stuart Appelbaum interview by Theodore Catton, 7 July 2004, Jacksonville, Florida, 1 [hereafter referred to as Appelbaum interview—Catton]; Salt interview—Catton, 6.

[55] Grumbine, "What is Ecosystem Management?" 31.

[56] Whitfield interview, 40; Water Resources Development Act of 12 October 1996 (110 Stat. 3658, 3771).

CHAPTER 14 **Searching for Consensus:** Sustainability and the Move Towards Everglades Restoration

As the South Florida Ecosystem Restoration Task Force coordinated federal activities regarding Everglades restoration, the Clinton administration, embracing the principle of sustainable development, made overtures to Florida's powerful sugar industry in order to gain its support of restoration efforts. These endeavors accorded with President Clinton's belief in the necessity of balance to resolve environmental disputes. Just as he demonstrated at his vaunted Forest Summit in Portland, Oregon, in April 1993, Clinton's aim in South Florida was to create common ground by persuading all sides to relinquish a little, end the fighting without declaring winners or losers, and move forward with a new consensus. Much to the discomfort of many environmentalists, this meant bringing Big Sugar into the circle.

President Clinton's environmental team had good reasons for wanting to work constructively with the sugar industry. Beyond the immediate goal of ending the litigation and clearing the way for cleanup to proceed, the Clinton team wanted to secure Big Sugar's philosophical and financial commitment to a long term ecosystem restoration plan. Such promises would ensure that growers made genuine progress in developing Best Management Practices (BMPs) aimed at reducing phosphorus levels in agricultural runoff; no one was in a better position than the industry itself to conduct research and development on the effects of farming on the natural ecosystem. Securing commitments would also ease the burden on federal and state coffers, and would create real, long range business incentives for the adoption of BMPs.[1]

According to Clinton administration officials, sugarcane growers had much to gain by supporting the restoration effort. They could improve their public image, deflect environmentalists' demands that

the sugar industry pay a far greater share of the cleanup, and place their business on an environmentally sustainable footing. This latter action was the Clinton administration's overarching goal, for the industry provided a livelihood to thousands of people in South Florida who would have little alternative employment if the industry collapsed. Many of the agricultural workers were underprivileged African Americans, Hispanics, and West Indian migrants who had worked their entire adult lives in the sugarcane fields and sugar mills. Despite frequent charges that the industry mistreated their workers, these rural inhabitants of the EAA were, for the most part, strongly attached to the region and supportive of the industry.[2]

As the Clinton team initiated settlement talks with sugar growers early in 1993, it sought to implement principles of "sustainable development." If ecosystem management was at the center of an intellectual ferment among scientists and resource managers, sustainable development was a concept that excited interest among economists and policy makers. Like ecosystem management, it predated the advent of the Clinton administration by a few years. It had first emerged as a concept for addressing disparities of wealth between developed and developing nations in the context of caring for the global environment. One of the first organizations to develop the idea was the United Nation's World Commission on Environment and Development, or Brundtland Commission, which first met in 1984. The Brundtland Commission defined sustainable development as "development that meets the needs of the present without compromising the ability of future generations to meet their own needs." The thrust of sustainable development, as explained in the Brundtland Commission's final report to the United Nations, *Our Common Future* (1987), was to meld economic and environmental concerns into a unified program. The environment could not be protected effectively without economic development, nor could economic development be sustained without environmental protection. Principles of sustainable development were outlined further at international conferences in New Delhi in 1990, Dublin in January 1992, and the Earth Summit in Rio de Janeiro in June 1992.[3]

Only after the Rio meeting did political leaders in the United States begin to suggest that sustainable development was a useful concept for domestic issues; President Clinton broadened the idea to include social justice perspectives. In June 1993, he formed the President's Council on Sustainable Development. This group was composed of 25 members drawn from government, industry, labor, and civil rights organizations. The council's guiding principle was to recognize the interdependence of economic prosperity, environmental protection, and social equity. Its mission was to explore "bold, new approaches to achieve our economic, environmental, and equity goals."[4] Sustainable solutions, like a three-legged stool, rested on the points of intersection between what was ecologically viable, what was economically feasible, and what was socially desirable. The council was to innovate ways to achieve "sustainable develop-

ment" through a *balance* of ecological, economic, and social values.[5] Clinton took a concept that had been steadily gaining ground in the international environmental community and made it central to his administration's domestic environmental policy. Sustainable development was an idea that would have great force in transforming the C&SF Project into the Comprehensive Everglades Restoration Plan.

The first milestone in the Clinton administration's efforts to co-opt Big Sugar was a much-ballyhooed "Statement of Principles," which Secretary of the Interior Bruce Babbitt, Florida Governor Lawton Chiles, and sugar industry leaders jointly announced on 13 July 1993 in a public ceremony held in the grand auditorium at the Department of the Interior in Washington, D.C. Largely orchestrated through the efforts of Gerald Cormick, the "leader of the school of alternative dispute resolution," the Statement of Principles represented give-and-take by the sugar industry, state, and federal agencies.[6] It called for a 90-day stay of Dexter Lehtinen's litigation; it provided an overview of a Technical Plan that was in development and would be finalized during the next 90 days as part of a final settlement agreement; and, most importantly, it spelled out financial commitments by the agricultural industry, the state, and the federal government.

A sugar cane field and canal in Moorehaven, Florida. (The Florida Memory Project, State Library and Archives of Florida)

Babbitt and Chiles hailed the accord as the closing chapter to a five-year court battle that had been costing valuable time and diverting money away from where it was most needed. "With this action," Babbitt related, "we expect to head off what could have been another decade of litigation and to immediately begin restoration."[7]

Industry representatives sounded the note on sustainable development. Nelson Fairbanks, president of U.S. Sugar Corporation, told the audience that he had "long believed that it was possible to save the Everglades while saving farm-related jobs," and this plan would do just that. "It asks farmers to pay a lot, much more than we wanted to pay," he said. "But it also lets us and our communities survive. That is what we have wanted all along." Robert Buker, senior vice president of U.S. Sugar and one of the chief negotiators for the industry, praised the Clinton administration for its role in the talks and called the breakthrough a "new paradigm" for resolving environmental disputes.[8] Alfonso Fanjul, president of Flo-Sun, said the Clinton administration had stood conventional wisdom on its head. "What's good for the environment can also be good for business," he said.[9]

The Statement of Principles began with a preamble asserting the parties' understanding of the problem. Nearly a century of human manipulation of the Everglades had made an attractive environment in South Florida that was now home to millions of people as well as a flourishing agricultural industry. "But in the last decade we have come to realize the tremendous cost this alteration of natural systems has exacted on the region," the statement read. "We pledge to inaugurate an unprecedented new partnership, joining the Federal and State governments with the agricultural industry of South Florida, to restore natural values to the Everglades while also maintaining agriculture as part of a robust regional economy." The parties further pledged to conduct future scientific research in a spirit of cooperation, and they expressed their hope that ecosystem restoration in South Florida would "become a national and international model for sustaining both the environment and the economy."[10]

Under the heading "Management Principles," the statement echoed the Settlement Agreement and Consent Decree in calling for the acquisition and establishment of flow-through filtration marshes, known as stormwater treatment areas (STAs), as the major component for cleanup of nutrients in the EAA. Water would pass through these marshes, allowing plants and other matter to cleanse the resource of phosphorus and other nutrients. The statement stipulated that parts of the Holey Land and Rotenberger tracts be used for these STAs, and it provided strong incentives for industry to implement BMPs. Unlike the earlier documents, however, it did not stipulate target levels for phosphorus outputs: these would be developed through subsequent research and calculations. In the following section on "Financial Principles," the statement described the respective commitments by the agricultural industry, the state, and the federal government in considerable detail. The agricultural industry agreed to pay up to $322 million over the next 20 years

for construction, research, monitoring, and operation and maintenance of the STAs. This constituted an impressive two-thirds of a $465 million plan. However, the contribution by Big Sugar was much less than these gross figures suggested. The state and federal governments would outspend the agricultural industry by more than two to one in the early going, and the agricultural industry's overall share, which would potentially escalate in the latter part of the 20-year period, would be substantially reduced through a credit system if target levels for phosphorus outputs were met according to schedule.[11]

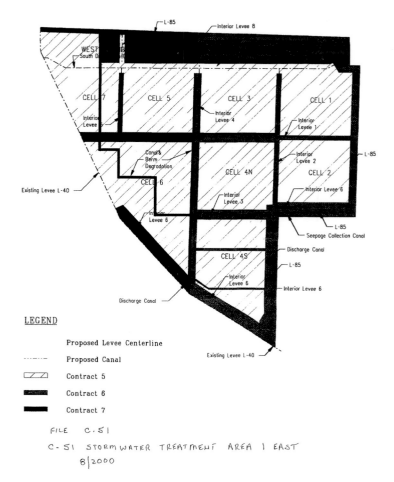

A diagram of STA-1 East. (U.S. Army Corps of Engineers, Jacksonville District)

Environmentalists almost unanimously denounced the Statement of Principles as vague, weak, and ingenuous. They did not like the provisions regarding the Rotenberger and Holey Land tracts, which they still wanted to use as buffer zones for the water conservation areas and Everglades National Park. They also wanted hard target levels of parts per billion, as they did not trust federal and state officials and industry representatives to calculate specific limits at a later time.[12] With regard to the financial commitments, environmentalists believed the sugarcane growers had obtained a sweetheart deal from the Clinton administration. They wanted the growers to give up more of their land for filtration marshes and to pay a greater share of the cleanup cost. The Statement of Principles was vague on how the money would be collected, they declared, as well as how the credits would be assessed. In addition, environmentalists pointed out, some large sugarcane growers were not parties to the agreement.[13]

STATEMENT OF PRINCIPLES

The Everglades is a wetland and wildlife resource unique in all the world. It has defined life in South Florida since humankind's introduction to the region. In acting to protect this important resource, we begin to define life for subsequent generations of Americans: what we choose to protect helps define us as a people.

In pursuit of human progress, South Florida has been ditched, diked, and drained for much of this century. By so doing, we have sought to provide a healthy, attractive living environment for millions of people safe from flooding and other natural forces; and to provide a base for a flourishing agricultural industry that provides important products, jobs, and income regionally and nationally.

But in the last decade we have come to realize the tremendous cost this alteration of natural systems has exacted on the region. This agreement will begin the renewal of the Everglades ecosystem, restoring natural flows of clean water. The result will benefit wildlife, urban drinking water supplies, and Florida Bay and other coastal waters and the life they sustain - - waters which are inextricably linked to the health of the Everglades themselves.

The Statement of Principles set out here is the basis on which the parties signing this agreement will seek a stay of pending litigation for 90 days, to reach a detailed settlement agreement resolving disputes that would otherwise continue for many years at enormous cost not just to the parties, but to the Everglades as well - - postponing the initiation of action to address critical threats to the system. Based on these principles, we will seek to include in the settlement discussions all parties to pending litigation who wish to contribute to the process.

We pledge more than a Plan; we pledge to provide the resources necessary for its successful implementation.

Moreover, we pledge to inaugurate an unprecedented new partnership, joining the Federal and State governments with the agricultural industry of South Florida, to restore natural values to the Everglades while also maintaining agriculture as part of a robust regional economy.

In addition, we will jointly conduct future scientific research on the ecological needs of the Everglades system and appropriate means to address those needs.

In so doing, we hope our efforts can become a national and international model for sustaining both the environment and the economy.

Management Principles

An End to Litigation.

In light of our commitment to implement these Principles, the parties to this Statement agree to join in motions to stay all Everglades litigation and administrative proceedings, including pending 298 District Administrative Litigation regarding Lake Okeechobee, for a period of 90 days, except for entry and access and the appeals pending before the Eleventh Circuit Court of Appeals.

This is necessary because, while this Statement signals a commitment to a process of mutual implementation of these Principles, it cannot and does not contain all provisions necessary for a comprehensive resolution of Everglades issues. We will use the 90-day period to resolve remaining issues and develop a complete settlement agreement.

A Commitment to Increasing Water Quantity to the Everglades.

A Technical Plan has been developed in intensive discussions over the past 120 days by experts from all sides. That Technical Plan addresses the improvement of water quality reaching the Everglades. It also commits to important steps in addressing pressing water quantity, sheet flow and other hydro-period restoration needs of the Everglades Ecosystem and of agricultural and other elements of South Florida's economy.

The first page of the Statement of Principles.

The Statement of Principles opened a fissure between environmentalists and the Clinton administration that would widen over the next three years. Environmentalists felt excluded by the mediation process, and although the Everglades Coalition and other environmental groups were invited to attend the ceremony at the Department of the Interior, the community was not mollified. EC members were quick to condemn the Statement of Principles as a sell-out to Big Sugar, whom they had been vilifying for years. Indeed, the fanfare surrounding its announcement, the photo opportunity for sugar industry moguls to share the podium with Secretary Babbitt and Governor Chiles, even the decision to unveil the Statement of Principles in the opulent auditorium of the Department of the Interior, all seemed calculated to offend in the view of some environmentalists. For Joe Browder, a consultant to the EC, the idea of inviting Big Sugar to proclaim its commitment to ecosystem restoration in the venerable old auditorium at Interior was no less than an act of defilement. "This [was] like pissing in the holy water," he would later comment.[14]

Browder had harsh words for Secretary Babbitt at the conclusion of the event. As the press conference was drawing to a close, Browder rose from his chair and said angrily, "It's an absolute betrayal, Bruce, and it won't stand." A few minutes later he buttonholed Babbitt on the floor of the auditorium. "This whole plan is bad science. I can't understand why you would agree to this." A *Miami Herald* reporter who was standing behind a pillar out of the secretary's view recorded the exchange. "Well that's my job, Joe, to find compromise," was Babbitt's exasperated reply.[15]

Dexter Lehtinen, who had spearheaded the earlier Settlement Agreement in 1991, blistered the federal government for giving away too much in the Statement of Principles. "It's vague and ambiguous on all the important points," he said. "It reminds me of Vietnam. You give up, declare victory, and go home."[16] A group called Clean Water Action immediately condemned the Statement of Principles as a taxpayer bailout of the sugar industry. Clean Water Action, more than any other group, appealed to people's pocket books. Florida taxpayers, particularly those in the SFWMD, would soon face a substantial hike in property taxes. Sylvia Kule, a member of Clean Water Action, promised to lead a bus load of senior citizens from Delray Beach to West Palm Beach to confront the governing board of the SFWMD when it met to approve the new $21 million ad valorem tax as called for in the Statement of Principles.[17]

Environmentalists were not the only ones who had problems with the Statement of Principles; the Miccosukee and Seminole Indians vilified the arrangement as well. Because no Miccosukee had participated in the negotiation of the principles, Billy Cypress, chairman of the tribe's business council, denounced them as sacrificing the Miccosukee's interests "on the alter [*sic*] of consensus." The document contained "shocking concessions to the special interests," Cypress continued, and he charged negotiators with deliberately preventing the Miccosukee and environmentalists from participating in the discussions because of their objections.[18] Lehtinen, speaking on behalf of the Miccosukee, went even further, claiming that the Statement of Principles would become known as "the Munich of the Everglades," where the federal government purchased "peace in our time with Big Sugar, leaving to others the difficult task of actually saving the Everglades."[19] The specific problems with the principles, according to Cypress and Lehtinen, who the tribe had hired as their attorney, was that they allowed delays in implementing water quality standards; they provided no "method or mechanism for achieving final [water quality] standards"; and they allowed Big Sugar to "pay less than the full cost of its own pollution." Instead, Cypress wanted to see the state and federal government adopt a final phosphorus standard of 10 parts per billion, achieved by 2002 in the water conservation areas, Big Cypress National Preserve, and Everglades National Park.[20] The Seminole were less strident in their comments, but still believed that the principles had several problems, including the possibility that they would "change the quantity of water flowing across the Big Cypress Reservation, . . . the timing and distribution of this water, and its quality."[21]

Although the environmental community and the Miccosukee had some legitimate complaints of the Statement of Principles, in many ways their opposition demonstrated a growing belief that a plan that made any kind of concessions to Big Sugar was wrong, regardless of the benefits it might produce. This belief stemmed from many factors, but the primary dynamic was the mutually beneficial relationship that the sugar industry had with the federal government. The industry profited from federal subsidies and price supports and returned the favors with large donations to politicians that looked out for sugar's interests. As one article reported, between 1979 and 1994, the sugar industry donated $12 million to both Democratic and Republican politicians, including more than $660,000 to sitting members of the House Committee on Agriculture between 1985 and 1990. In return, Florida's sugar industry alone had received more than $5 billion in government subsidies since the 1940s. These figures, coupled with the notion that the industry abused its labor force for large profits, made any kind of compromise with Big Sugar hard for many environmentalists to swallow.[22]

Therefore, it was not surprising that Jim Webb of the Wilderness Society was the only environmentalist who endorsed the Statement of Principles. Having earlier worked with Babbitt and officials in the Corps to get Congress to fund a restudy of the C&SF Project, he now accepted the compromise as a necessary step in moving the ecosystem restoration effort forward. Amid all the criticism from the environmental community, Webb's endorsement was a slender reed on which Babbitt and his team hoped to bring environmentalists back into the fold. But as some Democratic strategists soon observed, most environmentalists in Florida had nowhere else to go, as they would not vote Republican.[23]

Attention now turned to the state's Everglades Nutrient Removal Project, the prototype for the $465-million system of STAs mandated by the Statement of Principles and prescribed by the Settlement Agreement and Consent Decree prior to that. Begun in August 1991, the construction project was nearing completion. At a cost of $14 million, the constructed marsh occupied a 3,742-acre delta-shaped area situated on the border of the EAA and the Loxahatchee National Wildlife Refuge. Surrounded on all sides by earthen berms, the marsh was to be fed by a two-mile supply canal that would drain 35,000 acre-feet of water per year from the West Palm Beach Canal and farm seepage. The nutrient-laden water was to flow through a series of cells, each one filled with aquatic plants that would absorb phosphorus and "scrub" the water before it moved to the next cell. The first cell in the sequence, called the "buffer cell," was a 135-acre

area dense with cattails and algae that had a high capacity for taking phosphorus out of the water. As the cattails and algae died and decomposed, they would form a bottom layer of peat that would trap phosphorus permanently. From the buffer cell the water flowed through four massive cells, each covering several hundred acres and host to a different type of aquatic vegetation. Scientists hoped to compare the relative effectiveness of each type of aquatic plant for phosphorus removal. In addition, in Cells 1 and 2, separate 7.5-acre research cells would test different combinations of water depth, speed, and quantity. Engineers expected to apply the test results to the design of other, larger facilities that would be built at other locations around the EAA.[24]

Some scientists worried that the restoration effort relied too heavily on this single technological solution. The goal of the Everglades Nutrient Removal Project was to reduce phosphorus concentrations from 200 parts per billion (ppb) to 50. These results had been achieved from constructed wetlands before, but only where the wetlands were far larger in relation to the quantity of water flowing through them. Here the technology was being applied in an intensified form on an unprecedented scale, and it was being put forward as the primary solution to the problem. It remained to be seen whether the STAs could get phosphorus concentrations down to 50 ppb, and it was also unknown how effective the STAs would be on a long-term basis. Some skeptics complained that too much was riding on untried technology, that Babbitt and others were pushing "voodoo science." Other scientists shared these concerns, but emphasized that the Everglades cleanup could not wait for more answers. Richard Harvey of the Florida Department of Environmental Protection was one scientist who believed the gamble to be necessary. "Given a lot of time, waiting would be a valid argument," he told a reporter. "We're not willing to wait two to three years. We don't want the process to be slowed down."[25]

In November 1993, activation of this first experimental STA hit an unexpected snag. Water discharging from the STA into the Loxahatchee National Wildlife Refuge did not meet the 50 ppb standard. The EPA, citing authority under the Clean Water Act, decided that the SFWMD must obtain a federal permit to make further releases of this polluted water, a position consistent with the Settlement Agreement and Consent Decree of 1991-1992. It was also in step with plans in Congress to review the Clean Water Act in the upcoming session and extend its reach to farm-polluted water. But the requirement took SFWMD administrators by surprise. With water collecting in the STA and threatening to overtop the berms, the SFWMD resumed discharges into the Loxahatchee National Wildlife Refuge without a federal permit. Despite earlier threats, the EPA declined to levy a fine against the SFWMD. Nevertheless, the confrontation between EPA and the SFWMD alarmed farmers, who complained that they did not want to spend millions of dollars building filtration marshes only to have them commandeered by EPA. Moreover, they were

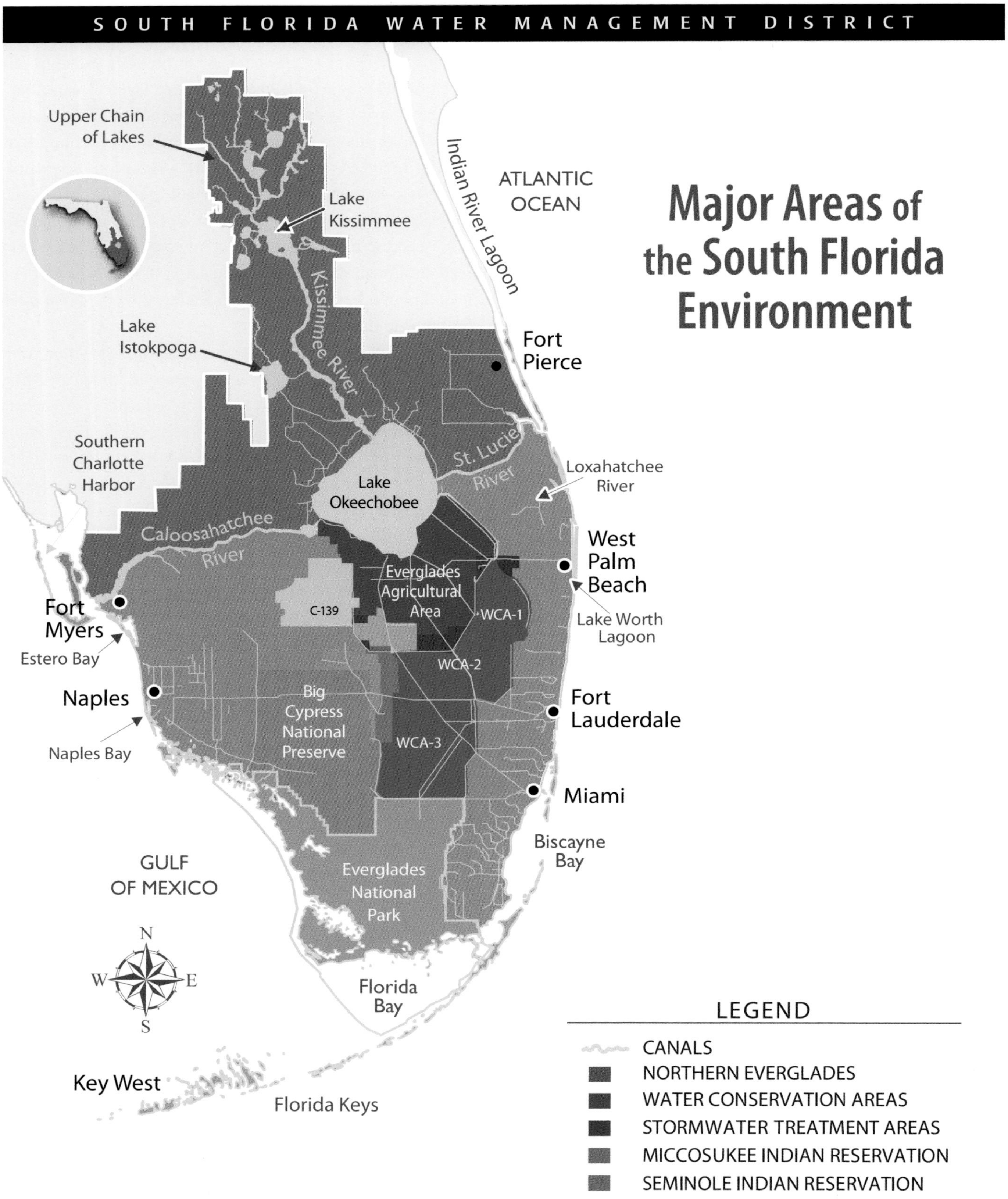
SOUTH FLORIDA WATER MANAGEMENT DISTRICT
Major Areas of the South Florida Environment
Upper Chain of Lakes
Lake Kissimmee
Indian River Lagoon
ATLANTIC OCEAN
Kissimmee River
Lake Istokpoga
Fort Pierce
Southern Charlotte Harbor
St. Lucie River
Loxahatchee River
Lake Okeechobee
Caloosahatchee River
West Palm Beach
Everglades Agricultural Area
C-139
WCA-1
Lake Worth Lagoon
Fort Myers
Estero Bay
WCA-2
Naples
Big Cypress National Preserve
Fort Lauderdale
WCA-3
Naples Bay
Miami
Biscayne Bay
GULF OF MEXICO
Everglades National Park
N
W
E
S
Florida Bay
Key West
Florida Keys
LEGEND
CANALS
NORTHERN EVERGLADES
WATER CONSERVATION AREAS
STORMWATER TREATMENT AREAS
MICCOSUKEE INDIAN RESERVATION
SEMINOLE INDIAN RESERVATION
ROTENBERGER AND HOLEY LAND WILDLIFE MANAGEMENT AREAS

Map of South Florida environmental features. (South Florida Water Management District)

concerned that the federal government would condemn a portion of their sugarcane fields for wetlands restoration without due compensation. In December, even as federal and state officials worked out their differences over the permit issue, agriculturists walked out of mediation talks, with representatives of the U.S. Sugar Corporation and the Sugar Cane Growers Cooperative of Florida refusing to sign the final version of the 1993 compromise. By early January, the litigation threatened to begin anew, as the administrative law judge set two hearings to schedule more than 150 depositions.[26]

Federal and state officials were stunned by the breakdown of negotiations. Their effort to forge consensus lay in tatters. Six months earlier they had alienated the Everglades Coalition; now at the end of 1993 they had lost the farmers as well. But the Clinton and Chiles administrations remained committed to working together on an Everglades restoration plan; there would be no more division between the federal government and the state. The Settlement Agreement and Consent Decree ensured against that. Moreover, the Clinton administration still had links to Alfonso Fanjul, Jr., president of Flo-Sun and a generous donor to the Florida Democratic Party. When the farmers broke off negotiations, Florida Crystals, Inc., a subsidiary of Flo-Sun (and the largest sugar producer in the EAA) kept the lines of communication open. In February 1994, Florida Crystals and federal negotiators quietly reached a separate agreement, whereby the corporation agreed to pay for nearly half of the construction costs of the STAs in exchange for not having to implement phosphorus standards until 2008. This arrangement infuriated nearly everyone: the Miccosukee Indians, environmentalists, and even Florida Crystals' counterparts, the U.S. Sugar Corporation and the Sugar Cane Growers Cooperative, whose strategy was now in disarray. Environmentalists and the Miccosukee took the issue to court, while Fanjul reaped the benefits: he was invited by Vice President Gore to attend an economic summit at the White House, and he hosted a tour of

An editorial cartoon showing the disgust that some felt with the continuing litigation over phosphorus cleanup.

a waste-to-energy facility on his sugar plantation by the President's Council on Sustainable Development.[27]

If the beginning of 1994 seemed to mark the nadir of the Clinton administration's effort to build consensus around South Florida's water management issues, it also galvanized public opinion for a renewal of that effort. On 3 March 1994, Governor Chiles announced that he was establishing the Governor's Commission for a Sustainable South Florida to solicit points of view and forge consensus in water matters. The commission would include 35 voting members appointed by the governor representing the business community, public interest and environmental organizations, county and city governments, and one representative each from the SFWMD, the South Florida Regional Planning Council, the Treasure Coast Regional Planning Council, the Florida Department of Environmental Protection, the Florida Game and Fresh Water Fish Commission, the Florida Senate, and the Florida House. In addition, it would include four non-voting federal officials representing the Corps, the Department of the Interior, EPA, and NOAA. The commission's primary charge was to "improve coordination among and within the private and public sectors regarding activities impacting the Everglades Ecosystem." Like the President's Council on Sustainable Development, it was to "recommend strategies for ensuring the South Florida economy is based on sustainable economic activities that can coexist with a healthy Everglades Ecosystem."[28]

Governor Chiles asked Estus Whitfield, the longtime advisor to Florida governors on environmental matters, to recommend a chairman for the commission. Whitfield suggested Richard Pettigrew, a former state legislator and speaker of the house. It was a fortunate choice. Pettigrew had the necessary prestige to make the commission visible to the public; he had experience at building consensus in the state legislature; and he had the right personality and temperament to control a large commission: patient, soft-spoken, empathetic, a skilled debater. In 1994, Pettigrew had been retired from state politics for some years and was practicing law in Miami, but he agreed to serve as chairman, holding the position until the commission completed its work in 1999.[29] Everglades hands who worked with the Governor's Commission universally praised his leadership: "a masterful job," "a fantastic job," "a master at bringing the interests together," "absolutely critical to the success."[30]

But this remarkable achievement still lay several years in the future. As the Governor's Commission began its work in the spring of 1994, federal and state legislators were working to enact two pieces of legislation—one federal and one state—that would further define the Everglades restoration process as a joint federal-state undertaking. The first of these, passed by Congress in March 1994, amended the Everglades National Park Protection and Expansion Act of 1989 to allow the secretary of the interior to take funds appropriated for flood control projects in the East Everglades and apply them for land acquisition in that area instead. The authorization paralleled another federal commitment to land acquisition in the Kissimmee River Valley and buttressed the state's ability to purchase land in the EAA. As such, it marked another step in the gradual transformation of the C&SF Project into the Comprehensive Everglades Restoration Plan. In a timely show of bipartisanship, Congressman Clay Shaw of Miami, a Republican, and Congressman Peter Deutsch of Broward County, a Democrat, co-sponsored the bill in the House, while Senator Bob Graham saw it through the Senate.[31]

State legislators, meanwhile, crafted a state law that went even further in solidifying federal support for Everglades restoration. The Everglades Forever Act, which Governor Chiles signed into law on 3 May 1994, codified construction projects and other cleanup efforts embodied in the Settlement Agreement (1991), Consent Decree (1992), and Statement of Principles (1993). The law described a treatment system, funding plan, regulatory program, research program, land use plan (including land acquisition in the East Everglades Area), and restoration schedule. The treatment system, which would be built by the SFWMD and known as the Everglades Construction Project, featured a combination of STAs and BMPs. In addition to the six STAs previously contemplated, the law required the

Corps to complete a seventh, STA 1-E as part of its work on a flood control project in the western C-51 basin.[32]

The funding plan called for a contribution of between $233 and $322 million by farmers (the same as in the Statement of Principles), and approximately $400 million by the state (a substantial increase over what had been outlined in the Statement of Principles). These amounts would be accompanied by an $87 million contribution by the federal government. The Florida Department of Environmental Protection and the EPA would regulate discharges by the STAs, and the SFWMD would supervise discharges by agricultural interests. The law mandated research to establish a scientific, numerically based standard for phosphorus levels and stipulated a default standard of 10 parts per billion if the Department of Environmental Protection did not set a standard by 2003. The restoration schedule called for the various STAs to become operational between 1997 and 2003, and all areas of the Everglades were to meet applicable water quality standards by December 31, 2006.[33]

In the spirit of achieving "balance," the Everglades Forever Act involved give and take by all sides. Governor Chiles could finally claim some success in bringing an end to the lawsuits and getting a restoration plan in place, while the Clinton administration had achieved its goal of establishing a long-range partnership between the federal government, the state, and the private sector. The sugar industry had held the line on its financial commitment, and it had obtained a reprieve of several years before more stringent guidelines on phosphorus levels would take effect. Environmentalists had won their point that the extensive acreage required for STAs should come out of agricultural lands, not the state-owned Rotenberger and Holey Land tracts.

Still, the environmental community, together with the Miccosukee Indians, believed that Big Sugar was the winner in this law, and that the environment and Florida taxpayers were the losers. The main problem, these groups contended, was that the Everglades Forever Act pushed back deadlines for agriculturists to meet water quality standards, essentially allowing the pollution of the Everglades to continue until 2006. These provisions convinced environmentalists and the Miccosukee that the Clinton and Chiles administrations, as well as state legislators, had sold out to the powerful sugar lobby, which, they said, had flooded the state capital with some 30 to 40 lobbyists.[34] They characterized the law as a disgraceful retreat from the Marjory Stoneman Douglas Everglades Protection Act of 1991. Indeed, the 1994 legislation began as an amendment to the earlier act and was only given a new title after Douglas, the grand dame of Everglades preservation, denounced the effort and demanded that her name be taken off the bill.[35] Although sugar interests contended with the notion that it had unduly influenced state politicians to pass the act, they did not disagree that the industry benefited from the law. According to Barbara Miedema, vice president of communications for the Sugar

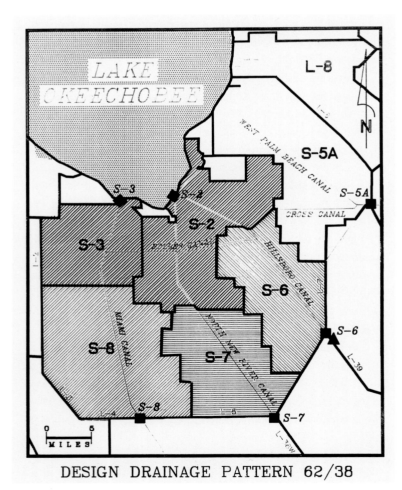

Drainage map of the Everglades Agricultural Area. (South Florida Water Management District)

Cane Growers Cooperative of Florida, the Everglades Forever Act provided "a far better, more comprehensive solution" than the Settlement Agreement or Consent Decree.[36]

The Miccosukee were especially vehement in their denunciations of the Everglades Forever Act, claiming that it merely codified the objectionable parts of the Statement of Principles and that it kowtowed to the sugar industry. In protest, the tribe took several actions. First, it, along with other entities, petitioned the Florida Department of Environmental Protection to implement a 10 ppb phosphorus standard immediately. When the department refused, the tribe filed a case in the federal district court, charging that the act changed Florida's water quality standards. At the same time, the Miccosukee—under the authority granted it by its water rights compact—began developing its own water quality standards, declaring that any water flowing onto reservation lands that exceeded the 10 ppb phosphorus limit would violate these standards. The tribe officially adopted these standards in December 1997, causing an uproar among the SFWMD and other agencies that would continue for the rest of the 1990s.[37]

Passage of the Everglades Forever Act was not the only setback for environmentalists and the Miccosukee in the spring of 1994. In its January meeting, the Everglades Coalition had made a strategic decision to endorse a petition drive to place a penny-a-pound pollution tax on sugar on the statewide election ballot.[38] The penny-a-pound tax was the brainchild of George Barley, an Orlando developer whom Joe Browder called "the strongest citizen Everglades leader in Florida."[39] As we have seen, Barley first got the attention of the environmental community for his efforts on behalf of Florida Bay. His success in bringing national attention to the degradation of Florida Bay was soon overshadowed, however, by his bold strategy to confront the Everglades polluters at the ballot box with the initiative for a tax on sugar. Barley's organization, the Save Our Everglades Committee, argued that Big Sugar was not only the major culprit in the decline of the Everglades, but that it was trying to pass the buck for cleaning up its own waste. This line of argument had broad public appeal, especially among the urban populace of South Florida who paid, according to Barley's organization, 111 times the amount that Big Sugar provided for water.[40]

The Everglades Coalition decided to get behind the Save Our Everglades campaign, a momentous decision as this was tantamount to the whole environmental community making a frontal attack on Big Sugar.[41] Anticipating a hard fight, coalition members concluded that they needed new leadership. They asked Joe Browder, the irascible critic of the Statement of Principles and no friend of the Clinton administration, to provide that direction. Browder agreed to serve, but wanted a co-chairman. The coalition elected Browder and Tom Martin as co-chairmen, while Theresa Woody of the Sierra Club was appointed grassroots coordinator.[42]

The penny-a-pound campaign soon acquired its own momentum. By the spring of 1994, more than half a million Florida voters had signed the petition, with 104-year-old Marjory Stoneman Douglas heading the list. The sugar industry fought back in the courts, complaining that the language in the petition unfairly passed judgment on the industry. In May, the Florida Supreme Court ruled that the initiative as written read too much like partisan rhetoric and could not go on the November ballot. With so many signatures already gathered, it was too late to reword the petition. The Save Our Everglades Committee, acknowledging that it had lost the battle but not the war, vowed to fight on and pursue a sugar tax by some other means, and the Everglades Coalition closed ranks behind it. The Supreme Court decision came less than three weeks after Governor Chiles signed into law the Everglades Forever Act. Locked in a struggle over the proposed sugar tax, the coalition had no choice but to place itself in opposition to the federal-state-agricultural partnership established under the Everglades Forever Act.[43]

The grassroots campaign to tax sugar—to make the "polluter pay"—had yet to reach full steam. That would happen in the context of presidential election year politics in 1996. In the meantime, environmentalists retreated into a skeptical funk as the state and federal governments moved ahead with the Everglades Construction Project—the name given to the system of STAs and other civil works mandated by the Everglades Forever Act.[44] As construction plans advanced through conceptual and preliminary design stages, the SFWMD acquired lands in portions of the EAA designed for STA 3 and 4. EPA granted a two-year extension of the SFWMD's operating permit for the Everglades Nutrient Removal Project, or STA 1 W. Sugar growers, playing their new role of public-spirited private enterprise, implemented BMPs, and the governor's office declared that the BMPs made a total phosphorus reduction of 44 percent compared to the baseline level of the previous decade.[45]

The federal and state governments proceeded as well with their respective efforts to build consensus for a comprehensive ecosystem restoration plan, aided by a workshop held in June 1994 for natural and social scientists funded by the U.S. Department of State's Man and the Biosphere Program. These scholars—both academicians and government personnel—discussed principles of ecosystem restoration and used South Florida as a case study of sustainability, primarily because "the Everglades and South Florida exemplify the complex set of issues that must be addressed to sustain human-dominated ecosystems." The group decided that federal and state forces needed to consider "urban, agricultural, and ecological systems" as they developed plans to maintain "fresh, flowing water" throughout the Everglades system. Based on a study of different hydrologic restoration scenarios, the group proclaimed the possibility of restoring the Everglades while continuing to meet urban and agricultural needs.[46]

At the same time, issues with Florida Bay continued to percolate. Scientists, including those from the NOAA, the SFWMD, and various universities, studied the issue in order to determine what was causing problems in the bay and how they could be resolved.

Different subgroups of the Interagency Working Group and the South Florida Ecosystem Restoration Task Force examined the issues as well; the Interagency Working Group on Florida Bay sponsored efforts to develop an interagency science plan for the bay. In 1994, the plan was completed, representing "the first interagency science plan for any South Florida subregion formulated under the aegis of the South Florida Ecosystem Task Force." Among other things, the plan called for trying to understand Florida Bay's condition prior to drainage and separating human-caused change from natural evolutions. It recommended the use of computer models to simulate how the bay would respond to change, and it posed a series of questions that needed answering.[47]

Despite these efforts, the end of 1994 saw little real progress, and the Florida Keys National Marine Sanctuary Advisory Council noted that no single issue was more important "to all of us, than getting restoration moving on Florida Bay." Likewise, George Frampton, assistant secretary of the interior who chaired the task force, emphasized the importance of getting general restoration efforts off the ground. "This is not rescuing an ecosystem at the last minute," he declared. "This is restoring something that has gone over the edge."[48]

Meanwhile, the Governor's Commission for a Sustainable South Florida led the state's effort to achieve consensus, meeting monthly and reporting to the governor on a quarterly basis. This body provided a crucial forum for representatives of the environmental community and the agricultural industry to go toe to toe and talk through their issues. In the early meetings, Chairman Richard Pettigrew enlisted the Florida Conflict Resolution Consortium to facilitate the process. With myriad issues to tackle, the commission had to decide whether to form committees or work through all the issues in a large group.

At first the members did not trust each other to divide into committees, but this soon changed. The commission met in a new location each month—Clewiston, Fort Myers, the Keys—and on the second day of these two-day meetings there was regularly a no-host event at which members had an opportunity to get to know one another as individual human beings, which built relationships of trust. This was essential to their mission of finding common ground. Gradually, commission members united behind a single vision: to put forward a plan for ecosystem restoration that would benefit all interests, be they agricultural, urban, recreational, or environmental.

Florida Bay. (South Florida Water Management District)

Indeed, Carol Rist, an environmentalist on the Governor's Commission, stated that a critical turning point for the commission came when agricultural and urban interests realized that they would not get federal money for reinventing the C&SF Project unless it was part of a program for restoring the Everglades. At that moment, Rist remembers, group members began to look for common ground with each other and with the environmental community.[49]

Meanwhile, the Corps proceeded with the reconnaissance phase of its restudy and, in 1995, presented a six-year plan for a feasibility study. An article in *Science* delineated the ultimate plan for restoration, stating that federal and state agencies wanted to "replumb the entire Florida Everglades ecosystem, including 14,000 square kilometers of wetlands and engineered waterways" at a cost of $2 billion, one-third of which was supposed to come from the federal government. The efforts would attempt to "take engineered swampland riddled with canals and levees and transform it into natural wetlands that flood and drain in rhythm with rainfall." Wetlands managers across the world were watching with interest, the article claimed, to see if the Florida plan would succeed, hoping to discover solutions for "their own ravaged regions." However, since nothing this complicated had ever been attempted, the restoration, which still did not have a "final blueprint," would have to operate on "a hefty dose of scientific uncertainty." In addition, politics threatened to capsize the undertaking, especially since it was unclear whether the "broad coalition of interests and money, from federal and state agencies to environmentalists and urban developers" could hold together over the life of the project. "We have the technical knowledge to do the restoration," John Ogden, a biologist for Everglades National Park stated, "but I worry about sustaining the political will."[50]

Indeed, despite state and federal efforts, environmentalists and sugarcane growers remained hostile. In January 1995, the EC announced that it would initiate a nationwide campaign against sugar price supports. If the growers refused to pay a fair amount to clean up their own waste, environmentalists reasoned, then the next step was to attack federal subsidies and allow market forces to drive some of the producers out of the EAA. Many now argued that sugarcane did not belong in the area at all: it was grown in the Caribbean at much less expense and without so much harm to the environment. Environmentalists soon found an unexpected ally in U.S. Senator Richard Lugar, a Republican from Indiana.[51] In the fall of 1995, Lugar was looking for voter support in Florida in his bid for the Republican Party presidential nomination. To court environmental interests, he proposed a federal tax on sugar, suggesting that the revenue be used to buy sugar plantations in the EAA for conversion into wetlands, thereby protecting the Everglades.[52]

Lugar's opponent in the Republican Party primaries, Senator Robert Dole, had a counter proposal. Under Dole's guidance, a section was inserted in the Federal Agricultural Improvement and Reform Act of 1996 appropriating $200 million (available to the sec-

retary of the interior on 1 July 1996) to acquire property in the Everglades ecosystem for restoration purposes and to "fund resource protection and resource maintenance activities in the Everglades ecosystem." Although this was one of the first federal appropriations specifically for Everglades restoration, it still upset some environmentalists because it did not force the sugar industry to contribute to the cost of these purchases.[53]

Not to be outdone, Vice President Albert Gore announced in February 1996 a comprehensive seven-year plan developed by the Clinton administration to restore the Everglades. This plan included both a slug of federal money to buy sugarcane plantations in the EAA and a penny-a-pound tax on Florida sugar. It proposed to double the federal government's current spending levels on Everglades protection to at least $500 million.[54] Sugar growers were not pleased; according to one account, Alfonso Fanjul called President Clinton after Gore unveiled the plan and "bitched" the President out. "He'd campaigned for Clinton, delivered a lot of votes," one lobbyist explained, "and here was Gore paying him back with a tax."[55]

In addition, the timing of Gore's announcement, coming on the heels of the two Republican proposals and on the eve of the state primaries in the 1996 presidential election campaign, gave some observers the impression that South Florida's environmental problems had ignited a bidding war. *Newsweek* saw the plan, which could ultimately total $1.5 billion, as "the high-water mark of reform," trumping Dole's "more modest plan to spend $200 million of taxpayer funds—not sugar money—to buy some of the sugar cane land for a water-restoration project."[56] *The Economist*, a conservative British magazine, described the administration's restoration plan under the jaundiced title, "The Florida Everglades, River of Money." This writer had no doubt that the federal largesse was aimed at capturing Florida's 25 electoral votes in the coming presidential election. "The federal cash has one source: election year politics," the article intoned.[57]

Election year politics continued to frame the issues. Buoyed by the administration's support for a penny-a-pound tax on sugar, the Save Our Everglades campaign secured enough signatures to get three proposed amendments to the Florida constitution on the November 1996 ballot. One would impose a penny-a-pound tax on sugar grown in the EAA, another would establish the principle that polluters were responsible for cleaning up their own waste, and the third would create a trust fund for Everglades restoration. The amendments were a bold and unusual step in two respects: they took the matter directly to a vote of the people, and they sought to hold one industry chiefly accountable for the pollution of the Everglades. After George Barley died in a 1995 plane crash on the way to an Everglades restoration meeting, his wife Mary headed the penny-a-pound campaign, using the financial backing of Paul Tudor Jones, the founder and chairman of the Tudor Group of Companies (a money management firm in Connecticut). Jones, a friend of the Barleys who had become interested in Florida's environmental health, pledged at George's funeral to pick up the environmental flag. With Jones' bankroll, Mary's citizen effort provoked a massive response by the sugar industry, which filed some 38 lawsuits challenging the amendments and spent around $35 million on advertising that opposed the penny-a-pound tax; environmental interests were only able to generate approximately $11 million for advertising. The advertising campaign reached a crescendo on Election Day, when the industry spent more than a million dollars to convince voters that the tax would ruin the industry and eliminate 40,000 jobs. Voters approved two amendments, but they rejected the crucial penny-a-pound tax.[58]

Environmentalists were not only stung by this second defeat of the tax initiative, some were embittered by what they viewed as a second betrayal by the Clinton administration. Once the Save Our Everglades campaign succeeded in getting the amendments on the ballot, the Clinton administration dropped its own proposal for a penny-a-pound tax on sugar. Ostensibly, the administration wanted to defer to Florida voters on this issue, a natural position, but environmentalists saw in this development the nefarious hand of Big Sugar. They were even more doubtful of the administration after it backpedaled from Gore's earlier pledge to take no less than 100,000 acres out of sugarcane production and rededicate the land for pollution abatement. Reportedly, the administration modified its position on this matter after another telephone call to the White House by Alfonso Fanjul.[59]

Embittered environmentalists claimed that the Clinton administration had politicized the planning process initiated by Secretary Babbitt in 1993 in order to win the state of Florida in the election of 1996. They accused Vice President Gore of grandstanding with the "Gore plan" while capitulating to the sugar interests, so that Clinton and Gore could win votes *and* maintain Big Sugar's political support.[60] In fact, Florida did swing narrowly into the Democratic column in President Clinton's re-election. It should be remembered that the Florida vote barely tipped to President George H. W. Bush in 1992, and would be so close in 2000 as to confound the national election until the U.S. Supreme Court decided the matter for President George W. Bush. It is impossible to draw a precise connection between Florida's crucial role in these national elections and the growing willingness by the White House and Congress to invest in Everglades restoration during the 1990s, but the connection cannot be ignored. As EPA administrator Carol Browner observed about the $200 million for Everglades restoration included in the 1996 farm bill, "suddenly, the political stars aligned."[61] The same thing would be said about CERP four years later.

But even with the Clinton administration's apparent backpedaling, there were glimmers of hope. The U.S. Army Corps of Engineers, for example, had completed the reconnaissance phase of its Restudy of the C&SF Project, declaring in November 1994 that "the

fundamental tenet of South Florida ecosystem restoration is that hydrologic restoration is a necessary starting point for ecological restoration." Using an environmental evaluation methodology that compared the hydrological effects of different restoration projects, the Corps determined that "the hydrologic function of the historic south Florida ecosystems can be recovered," and it recommended that it proceed with a feasibility study of the different restoration options that it could pursue.[62] Accordingly, the Clinton administration directed that the Corps complete, in the words of H. Martin Lancaster, Assistant Secretary of the Army (Civil Works), "a study to develop a comprehensive restoration plan for South Florida." This study, Lancaster explained, would try to "determine the feasibility of structural and/or operational modifications to the Central and Southern Florida Project to restore the Everglades and Florida Bay ecosystems."[63]

Editorial cartoon from the *St. Petersburg Times* about Everglades restoration. (Used by permission of the *St. Petersburg Times*)

Congress authorized the feasibility study in the Water Resources Development Act of 1996 (WRDA-96), drafted largely by Michael Davis, Deputy Assistant Secretary of the Army (Civil Works), directing that the Corps develop "a proposed comprehensive plan for the purpose of restoring, preserving, and protecting the South Florida ecosystem," including ways to protect water quality and to restore water to the Everglades, before 1 July 1999. The legislation stipulated that the Corps work with the South Florida Ecosystem Restoration Task Force (which it formally established) in this study, and it gave the Corps the authority to implement any restoration project "expeditiously" if it discovered that such an undertaking would "produce independent, immediate, and substantial restoration, preservation, and protection benefits."[64] To fund these efforts, the law appropriated $75 million, a large amount for projects that would normally fall under the umbrella of "continuing authorities." Such continuing authorities were usually capped at $5 million in order to preserve congressional control over them, meaning that it required, in the words of Davis, "some heavy lifting" on the part of the Corps before Congress would authorize the $75 million. The law also stipulated that non-federal interests share 50 percent in the cost of any restoration effort. Because of these features, and because of the relatively short time span for the study, Davis considered it a "watershed event" that "set the bar high" for future restoration endeavors.[65] The South Florida Ecosystem Restoration Task Force agreed, declaring in a 1998 biennial report that WRDA-96 was "an ambitious milestone in the goal of restoring a sustainable South Florida."[66]

By 1997, then, several pieces had fallen into place, expediting restoration of the Everglades ecosystem. Federal funding had been provided, both in the Federal Agricultural Improvement and Reform Act of 1996 and in WRDA-96. Congress had stipulated that the Corps complete a restoration study by 1999, and had also authorized it to begin restoration efforts that would have a significant effect on the ecosystem. These gains were achieved, in large part, because of the cooperation of federal, state, and non-government interests, largely through the workings of the South Florida Ecosystem Restoration Task Force and the Governor's Commission for a Sustainable South Florida.

But beneath this veneer of consensus, trouble still brewed, primarily between the sugar industry and environmentalists. Environmentalist criticism of the 1993 Statement of Principles and the Everglades Forever Act, which were supposed to end Dexter Lehtinen's lawsuit, upset sugar magnates who had compromised to get them enacted, and these hard feelings were intensified by the environmental community's efforts to enact the penny-a-pound sugar tax. Sugar forces, meanwhile, enraged environmentalists by filing new suits against water quality controls and by influencing politicians, including President Clinton, to weaken the industry's responsibility for cleanup efforts. Because of these conditions, restoration efforts would proceed with some difficulty, even though they now had a level of unprecedented federal support.

Endnotes

[1] G. H. Snyder and J. M. Davidson, "Everglades Agriculture: Past, Present, and Future," in *Everglades: The Ecosystem an Its Restoration*, Steven M. Davis and John C. Ogden, eds. (Delray Beach, Fla.: St. Lucie Press, 1994), 111.

[2] Snyder and Davidson, "Everglades Agriculture: Past, Present, and Future," 111. For more information on field workers (and for criticism of their treatment), see Wilkinson, *Big Sugar*.

[3] Lynn R. Martin and Eugene Z. Stakhiv, *Sustainable Development: Concepts, Goals and Relevance to the Civil Works Program*, IWR Report 99-PS-1 (Alexandria, Va.: Institute for Water Resources, Water Resources Support Center, U.S. Army Corps of Engineers, 1999), 3-5.

[4] "President's Council on Sustainable Development, Overview" <http://clinton2.nara.gov/PCSD/ Overview/index.html> (1 September 2005).

[5] F. Douglas Muschett, ed., *Principles of Sustainable Development* (Delray Beach, Fla.: St. Lucie Press, 1997): 53-57.

[6] Rodgers, "The Miccosukee Indians and Environmental Law," 10924.

[7] As cited in "Everglades Plan Comes Under Fire," *National Parks Magazine* 67 (September/October 1993): 10.

[8] Department of the Interior News Release, 13 July 1993, File Pro Ever Everglades Restoration History, Box 21213, SFWMDAR; *The Tampa Tribune*, 15 July 1993.

[9] *The Miami Herald*, 14 July 1993.

[10] Statement of Principles, 13 July 1993, File Pro Ever Everglades Restoration History, Box 21213, SFWMDAR.

[11] Statement of Principles.

[12] Environmental groups had practical as well as philosophical reasons to be mistrustful: in their public service role as watchdogs over government regulatory agencies, environmental groups needed hard target levels to help them monitor the government's job performance.

[13] Everglades Coalition to Bruce Babbitt, Secretary of the Interior, 30 July 1993, in House Subcommittee on Oversight and Investigations and House Subcommittee on National Parks, Forests and Public Lands, Committee on Natural Resources, and House Subcommittee on Environment and Natural Resources, Committee on Merchant Marine and Fisheries, *Florida Everglades Ecosystem*, 103rd Cong., 1st sess., 1993, 461-465; *Sun-Sentinel*, 13 July 1993; Mark Derr, "Redeeming the Everglades," *Audubon* 95 (September/October 1993): 130.

[14] Browder interview, 7.

[15] All quotations in *The Miami Herald*, 14 July 1993; see also Browder interview, 7; Grunwald, *The Swamp*, 298.

[16] *Sun-Sentinel*, 13 July 1993

[17] Browder interview, 7; *The Tampa Tribune*, 15 July 1993.

[18] Billy Cypress, Chairman, Miccosukee Tribe of Indians of Florida, to Honorable Harry Johnston, Member of Congress, 9 July 1993, File Everglades Mediation Miccosukee, Box 19706, SFWMDAR.

[19] As cited in Rodgers, "The Miccosukee Indians and Environmental Law," 10918.

[20] Billy Cypress, Chairman, and Dexter Lehtinen, General Counsel, "Position Statement Regarding Everglades 'Statement of Principles,'" 13 July 1993, File Everglades Mediation Miccosukee, Box 19706, SFWMDAR.

[21] "Seminole Tribe of Florida Comments and Concerns with the Everglades Mediation Technical Plan," 3, File Miccosukee WQ Standards, Box 19706, SFWMDAR.

[22] See Gary Barlow, "A Sweet Deal for the Sugar Industry," *In These Times* 20 (14 October 1996): 10. For an explanation of general federal policies towards the sugar industry, see Andrew Schmitz and Douglas Christian, "The Economics and Politics of U.S. Sugar Policy," in *The Economics and Politics of World Sugar Policies*, Stephen V. Marks and Keith E. Maskus, eds. (Ann Arbor: The University of Michigan Press, 1993), 49-78.

[23] Joe Browder, Tom Martin, and Theresa Woody to Everglades Coalition leadership, 7 April 1994, document provided by Joe Browder to author.

[24] David Kohn, "Polishing an Ecological Jewel," *Engineering News-Record* 231 (9 August 1993): 29.

[25] Kohn, "Polishing an Ecological Jewel," 29.

[26] *Sun-Sentinel*, 10 December 1993; *Palm Beach Post*, 10 December 1993; Malcolm Wade, Jr., interview by Julian Pleasants, 3 April 2001, 5,

Everglades Interview No. 5, Samuel Proctor Oral History Program, University of Florida, Gainesville, Florida [hereafter referred to as Wade interview]; "Discord Swamps Everglades Talks," *Engineering News-Record* 232 (3 January 1994): 9; "Everglades Case Heads Back to Court," *National Parks Magazine* 68 (March/April 1994): 11-12.

[27] Wade interview, 9; Levin, *Liquid Land*, 211; Joe Browder et al. to Everglades Coalition leadership, 7 April 1994; "Everglades Case Heads Back to Court," 11; Paul Roberts, "The Sweet Hereafter," *Harper's* 299 (November 1999): 62. Fanjul had Clinton's ear as well; it was alleged that he talked to the president in February 1996 about a proposed sugar tax, interrupting Clinton's break-up speech to Monica Lewinsky. Grunwald, *The Swamp*, 309.

[28] State of Florida, Office of the Governor, Executive Order Number 94-54, "Governor's Commission for a Sustainable South Florida," Governor's Commission for a Sustainable South Florida <http://www.state.fl.us/everglades/gcssf_eo.html> (8 September 2005).

[29] Whitfield interview, 39.

[30] Bonner interview, 4; Carol and Karsten Rist interview by Theodore Catton, 9 November 2004, 2 [hereafter referred to as Rist interview]; Causey interview, 9; Whitfield interview, 39.

[31] House, *Exchange of Certain Lands in California*, 103d Cong., 1st sess., 1993, H. Rept. 103-362; Act of 9 March 1994 (108 Stat. 98).

[32] Glynn D. Kay, "Outline of Florida Everglades Legislation," 20 April 1994, File PRO Ever Everglades Legislation, Box 21213, SFWMDAR.

[33] Kay, "Outline of Florida Everglades Legislation," 20 April 1994; "Governor's Office: Everglades Bill Summary," 3 May 1994, File PRO Everglades Forever Act 94, Box 15232, SFWMDAR. The full text of the bill with additional commentary is in House Committee on Natural Resources Subcommittee on National Parks, Forests, and Public Lands, *Oversight Hearing on the Land Use Policies of South Florida, With a Focus on Public Lands and What Impact These Policies are Having*, 103d Cong., 2d sess., 1994, 119-221.

[34] Save Our Everglades, "Everglades Forever Act," no date, and Florida Audubon Society, *The President's Letter* 19 (June 1994), File PRO Everglades Forever Act 94, Box 15232, SFWMDAR.

[35] Joe Browder, Tom Martin, Theresa Woody to Everglades Coalition leadership, 7 April 1994.

[36] Wedgworth and Miedema interview, 5.

[37] See Alfred R. Light, "Miccosukee Wars in the Everglades: Settlement, Litigation, and Regulation to Restore an Ecosystem," *St. Thomas Law Review* 13 (Spring 2001): 731-732; Miccosukee Environmental Protection Code, Subtitle B: Water Quality Standards for Surface Waters of the Miccosukee Tribe of Indians of Florida, copy in Miccosukee WQ Standards, Box 19706, SFWMDAR; Sam Poole for Frank Williamson, Jr., to Secretary Murley, Florida State Clearinghouse, 13 October 1997, ibid. The Seminole also adopted its own water quality standards, but since it set phosphorus levels at 50 ppb, it was not as controversial as the Miccosukee code. See Seminole Water Commission, Seminole Tribe of Florida, "Proposed Rules, Water Quality Protection and Restoration; Rules to Carry Out the Federal Clean Water Act and the Tribal Water Code including Water Quality Standards for the Big Cypress Reservation, File Indian Affairs," Seminole Water Quality Standard Research 94-98, Box 22792, SFWMDAR.

[38] The Everglades Coalition, "Past Conferences, 2005" <http://www.evergladescoalition.org/site/ pastconference.html> (13 September 2005).

[39] Joseph Browder, e-mail communication with author, 17 November 2004.

[40] Save Our Everglades, "Everglades Forever Act," no date, File PRO Everglades Forever Act 94, Box 15232, SFWMDAR.

[41] The Everglades Coalition, "Past Conferences, 2005" <http://www.evergladescoalition.org/site/ pastconference.html> (13 September 2005).

[42] Everglades Coalition Meeting Minutes, 13-16 January 1994, document provided by Joseph Browder to author.

[43] Save Our Everglades News Release, 26 May 1994, File PRO Everglades Forever Act '94, Box 15232, SFWMDAR, *Tallahassee Democrat*, 31 May 1994; Browder interview, 8.

[44] The Everglades Coalition invited Governor Chiles to deliver the keynote address at the coalition's annual meeting in January 1995, despite its condemnation of the Everglades Forever Act less than a year earlier.

[45] Office of the Governor, *Save Our Everglades: A Status Report by the Office of Governor Lawton Chiles* (n.p., 1994), 6-7.

[46] As reported in Victoria Myers, "The Everglades: Researchers Take a New Approach to an Old Problem," *Sea Frontiers* 40 (December 1994): 15. For more information on this study and on its findings, see Mark A. Harwell, "Ecosystem Management of South Florida," *BioScience* 47 (September 1997): 499-512. Harwell was the chair of the Man and the Biosphere Program Human-Dominated Systems Directorate.

[47] "Interagency Florida Bay Science Plan," March 1994, Billy Causey's Task Force Files, FKNMSAR.

[48] As quoted in Elizabeth Culotta, "Bringing Back the Everglades," *Science* 268 (23 June 1995): 1688.

[49] Rist interview, 2-3.

[50] All quotations in Culotta, "Bringing Back the Everglades," 1688.

[51] Levin, *Liquid Land*, 212.

[52] "The Florida Everglades, River of Money," *The Economist* 362 (30 March 1996): 32.

[53] Quotations in Federal Agricultural Improvement and Reform Act of 1996 (110 Stat. 888, 1022-1025); see also Kim O'Connell, "Gore Unveils Everglades Plan," *National Parks* 70 (May/June 1996): 13-14.

[54] Harvey Wasserman, "Cane Mutiny," *The Nation* 262 (11 March 1996): 6.

[55] As cited in Paul Roberts, "The Sweet Hereafter," *Harper's Magazine* 299 (November 1999): 62.

[56] Peter Katel, "Letting the Water Run into 'Big Sugar's' Bowl," *Newsweek* (4 March 1996): 56.

[57] "The Florida Everglades, River of Money," 32.

[58] Gail DeGeorge, "Big Sugar is Bitter over the Everglades," *Business Week* (4 November 1996): 192; Levin, *Liquid Land*, 216; Grunwald, *The Swamp*, 308-309; Reed interview, 7.

[59] Levin, *Liquid Land*, 214; Michael Grunwald, "When in Doubt, Blame Big Sugar," *The Washington Post*, 3 November 2004.

[60] Big Sugar's influence on the White House was not just direct; it also came through Florida politicians whose own re-election chances depended on Big Sugar's continued campaign funding support.

[61] Browner quoted in "The Florida Everglades: River of Money," 32.

[62] U.S. Army Corps of Engineers, Jacksonville District, *Central and Southern Florida Project, Reconnaissance Report, Comprehensive Review Study* (Jacksonville, Fla.: U.S. Army Corps of Engineers, Jacksonville District, 1994), EX-1—EX-4, 223-231.

[63] House Committee on Transportation and Infrastructure Subcommittee on Water Resources and Environment, *Water Resources Development Act of 1996: Hearings Before the Subcommittee on Water Resources and Environment, Committee on Transportation and Infrastructure, House of Representatives*, 104th Cong., 2d sess., 1996, copy at <http://web.lexisnexis.com.weblib.lib.umt.edu:2048 /congcomp> (3 January 2006).

[64] Water Resources Development Act of 1996 (110 Stat. 3658, 3767-3768).

[65] Michael Davis interview by Theodore Catton, 21 December 2004, 1-2.

[66] "Maintaining the Momentum: South Florida Ecosystem Restoration Task Force Biennial Report to the U.S. Congress, Florida Legislature, Seminole Tribe of Florida, and Miccosukee Tribe of Indians of Florida," Draft, December 1998, 1, File Administrative, 1998, FKNMSAR.

CHAPTER 15

Preserving Their Interests: The Seminole and Miccosukee Indians

As the clash between sugar interests and environmentalists threatened the consensus formula for Everglades restoration, the Seminole and Miccosukee Indians were waging battles that many perceived as equally divisive. These people had resided in the Everglades since the 1800s, obtaining state and federal reservations of land, making them especially concerned about water issues in the area. Because the Seminole did not have as much actual land in the Everglades, they generally used conciliatory approaches to promulgate their views. The Miccosukee Tribe, on the other hand, had a land base primarily in the Everglades, making them turn to litigation in the 1990s to protect their interests. Under the guidance of Dexter Lehtinen, the Miccosukee combated what they perceived as lax water quality standards through lawsuits and by setting their own guidelines. They justified their stance by declaring that the push towards ecosystem restoration had not sufficiently provided for their input, thereby threatening their interests.[1] The tribe's actions forced federal, state, and local interests to pay more attention to the Indians in restoration efforts, and, at least in the minds of the Miccosukee, furthered progress towards a restored Everglades.

As explained previously, by the 1960s, the Seminole and Miccosukee had several state and federal reservations in South Florida. In addition to these lands, the Miccosukee obtained a 40-year special use permit from the National Park Service in 1964 for a five and a half mile strip of land along the Tamiami Trail, comprising 333 acres and known as the Permit Area (this was later expanded to 667 acres in 1998). The permit allowed the Miccosukee to build offices, housing, and schools, but made such development subject to NPS approval. Moreover, both tribes obtained a license from the state for the use of a large chunk of land north of the Permit Area and east

of the reservation, although Conservation Area No. 3 flooded much of the northern part of this tract as well, and this license eventually transformed into a perpetual lease of the area.[2]

Because of the implementation of the C&SF Project in 1948, the Seminole and Miccosukee had little control over water policies that affected their land. The Corps of Engineers and the SFWMD made decisions about how much water went into Conservation Area No. 3 and the Everglades, as well as regulating the water in Lake Okeechobee.[3] These determinations affected not only the quantity of water flowing to Seminole and Miccosukee lands, but the quality as well. By the 1970s, the Seminole and the Miccosukee experienced the same ecological problems that the Everglades and the conservation areas faced as a consequence of water decisions and growth in South Florida. As Buffalo Tiger, a respected Miccosukee leader, related,

> As for Everglades' water, everything has changed. The water was very clean years ago. Miccosukees would swim in the Glades water and drink it. Today people are saying that the water is not clean. You can tell that is true because it is yellow-looking and does not look like water you would want to drink. You probably get sick from drinking it. That means that fish or alligators in the water are not healthy; white men did that, not Indians. Miccosukees were told that was what was going to happen many years ago, and now it has. We cannot just say that the water is no good or the land is no good and turn our back on that.[4]

Because of the conviction that the Indians had to do something to improve the quality of their water, even though the degradation was not of their own making, the Seminole and Miccosukee struggled to gain some measure of control over the water flowing over their reservations.

The Seminole's efforts began in 1973, when the tribe investigated whether or not the state of Florida ever compensated it for 16,000 acres of reservation land flooded by Conservation Area No. 3. The Seminole contacted Governor Reubin Askew, telling him that, in contrast with the state's practice regarding private land, the tribe had never obtained a fee for the easement. One reason for this was that the easement had actually been granted by the Board of Commissioners of State Institutions in August 1950, rather than by the Seminole themselves, since the Trustees of the Internal Improvement Fund held title to the land in trust for the Indians.[5] R. L. Clark, Jr., chairman of the FCD's governing board, explained that the Board of Commissioners of State Institutions had determined that the land flooded by Conservation Area No. 3 was worthless to the Seminole, and that its creation actually increased the value of other Seminole lands by making them "suitable for a higher and better use than previously existed." Besides, Clark noted, if compensation had been warranted, it would have gone to the state, not the tribe, since the state held title to the land.[6]

The controversy soon extended into litigation. Alleging that they wanted to regain control over land that was rightfully theirs, the Seminole filed a civil action in 1974 against the state of Florida to establish their rights to the acreage flooded by Conservation Area No. 3. Over the next 13 years this suit languished in the courts, with settlement talks proceeding intermittently, including proposals to exchange state land outside of reservation boundaries for the Seminole's 16,000 acres. The state especially wanted Seminole land known as the Rotenberger Tract in order to allow for additional impoundment as a part of Governor Bob Graham's Save Our Everglades program. Finally, in 1985, the Seminole declared that unless a reasonable settlement was negotiated, they would oppose construction of the Modified Hendry County Plan, a $20 million flood control project planned by the Corps of Engineers and to be built by the SF-WMD. This project was supposed to drain lands west of the EAA in order to provide acreage for citrus groves, placing the excess water in Conservation Area No. 3A. Because of this threat, the SFWMD and the state had more reason to resolve the litigation.[7]

In September 1986, the Seminole and the state finally reached a settlement. Under the agreement, the state would pay over $11 million to the Seminole for the Rotenberger Tract, the title and

Removal of invasive Melaleuca trees by the Miccosukee Tribe, behind a casino along Tamiami Trail. (South Florida Water Management District)

easement to other land flooded by Conservation Area No. 3, and for compensation for past projects conducted in Conservation Area No. 3A. The tribe, in turn, would withdraw its objections to the Modified Hendry County Plan. In addition, the settlement recognized that the Seminole wanted to develop a compact detailing its water rights and its responsibilities to preserve water quality.[8]

After state and tribal officials agreed to the settlement, the Seminole began work on the water rights compact. According to one source, the compact was "a way for the Tribe to integrate its own water use operations with most provisions of Florida water and environmental law."[9] It required the creation of a tribal water department and the development of a tribal water code, and it gave the Seminole responsibility for water management on their reservations. The tribe's legal representation claimed that the compact "recognize[d] the Tribe's sovereign power in the administration of reservation water resources," and that it allowed for "intergovernmental cooperation between sovereign governments," rather than "subordination of the Tribe's interests to the [SFWMD's]."[10] Upon the completion of the compact and its approval by the SFWMD, the entire settlement was forwarded to Congress for its authorization, provided in December 1987 under the Seminole Indian Land Claims Settlement Act. This law stipulated that the compact would have "the force and effect of Federal law for the purposes of enforcement of the rights and obligations of the tribe."[11]

Meanwhile, the Miccosukee made their own protests about the Modified Hendry County Plan. Fearing that the project would adversely impact water quality and vegetation in Conservation Area No. 3, the tribe filed an objection with the Corps, stating that the project had the potential of harming natural resources on tribal land. The SFWMD tried to assuage Miccosukee fears, stating that the project would have "no significant impact to the Indian land." Besides, the SFWMD continued, the plan was just a part of an "environmental enhancement" program that it was conducting for Conservation Area No. 3. Other components, according to the SFWMD, included restoring 30 square miles of the Everglades under the Holey Land project, using the Rotenberger Tract to prevent agricultural runoff from entering Conservation Area No. 3, and conducting the Shark River Slough Restoration to provide "major improvement to the natural flow system of the Everglades." Indeed, the SFWMD asserted that "the effects of the flood control portion of the Hendry County Project are inconsequential in comparison with the environmental benefits that will be associated with these three restoration projects."[12]

The Miccosukee were not so sure, and they worked with the SFWMD on a memorandum of agreement assuring that Miccosukee interests would not be harmed. This memorandum was concluded in May 1987. According to its provisions, the tribe agreed to withdraw its objection to the Modified Hendry County Plan and to develop a water rights compact and a tribal water resources department in

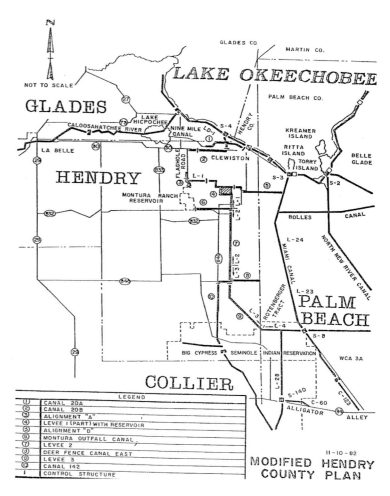

The Modified Hendry County Plan. (U.S. Army Corps of Engineers, Jacksonville District)

exchange for certain concessions from the SFWMD. These included the monitoring of discharges into Conservation Area No. 3A and on tribal land for pesticides, and the development of a monitoring program for water quality. As part of this plan, the SFWMD would make quarterly water quality reports to the tribe, and it would develop "nutrient standards" for 3A that would prevent "excessive nutrient enrichment." Likewise, if a proposed SFWMD project had the potential to impact "water quantity or quality" on the Miccosukee Reservation, the district would consult with the tribe.[13] The Miccosukee Business Council approved the memorandum of agreement on 11 May 1987, and began work on its compact soon after.[14]

In the late 1980s and early 1990s, the Seminole and Miccosukee became more concerned about water quality on their lands as the condition of Lake Okeechobee and the conservation areas worsened. In 1989, for example, the SFWMD issued an interim SWIM plan for Lake Okeechobee. This document noted that "in spite of intensified management efforts, . . . water quality conditions in Lake Okeechobee have not improved." Instead, they hit "the highest [phosphorus] levels yet recorded" in 1988. According to the report, the Surface Water Improvement and Management Act had mandated that phosphorus levels be lowered by 1992, and the SFWMD had developed a management strategy that emphasized controlling phosphorus inflows and implementing practices within sub-basins to reduce how much of the nutrient reached the lake, but, so far, no significant downturn in phosphorus had resulted. The SFWMD therefore proposed that "a more aggressive management approach" be used, part of which included continuing to divert nutrient-rich water from the EAA to the conservation areas—where it would affect Seminole and Miccosukee land.[15]

Because of these diversions, it was not surprising when the SWIM plan for the Everglades (defined as Conservation Area Nos. 1, 2A, 2B, 3A, 3B, the Arthur R. Marshall Loxahatchee National Wildlife Refuge, and the Everglades National Park) noted in 1992 that the discharge of water with high phosphorus levels into the conservation areas had "caused changes in existing vegetative species composition" that had the potential to "threaten fish and wildlife populations." Cattails and other non-native plants, such as melaleuca, had overrun sawgrass in several areas, the report declared. The diversion of water from the EAA to the conservation areas, it claimed, had brought an additional 45.4 metric tons of phosphorus per year. To combat these problems, the plan proposed, among other things, to create a Scientific Advisory Committee for the Everglades, with representation from various state, federal, and local agencies, including the Seminole and Miccosukee. The plan would develop strategies to protect and restore water quality in the Everglades. It also called for the implementation of STAs to cleanse the water, as well as monitoring programs and better regulation of landowner discharges.[16]

But the Miccosukee and Seminole believed more needed to be done to protect their interests. The Seminole were especially concerned about water quality on the Big Cypress and Brighton reservations. Under the authority of its water rights compact, the tribe had implemented a water quality monitoring program in 1989 that demonstrated that "the quality of water entering the [Big Cypress] reservation from upstream sources is severely degraded," exceeding phosphorus levels of 300 parts per billion (ppb).[17] The Seminole therefore received permission from the EPA to set its own water quality standards for its lands, and it began work on a water conservation plan for the Big Cypress Reservation, designed to "provide a comprehensive, fully integrated water management system" that could "support sustainable agriculture while contributing to the restoration of significant portions of the Everglades ecosystem."[18] As part of this plan, the tribe proposed to use surface water management structures to treat, control, and redirect water, and to implement BMPs in order to reduce nutrient levels, targeting 50 ppb as the accepted phosphorus level. In 1994, the Seminole completed the design phase of this plan, but more money was needed to implement it. In the meantime, the tribe discussed phosphorus levels in Lake Okeechobee with the SFWMD, including how they affected the Brighton Reservation (located to the northwest of the water body) since water from the lake was backpumped to the reservation.[19]

From 1993 to 1995, the SFWMD and the Seminole negotiated an agreement detailing water quality efforts for Big Cypress and Brighton. This accord attempted to define the sources of water for these reservations, as well as outline efforts to preserve water quality. The agreement delineated that the main source of water for the Big Cypress Reservation would be the Rotenberger Tract, and Lake Okeechobee would serve as a secondary source. The SFWMD would ensure that the design of STA-5 and -6 would, in the words of SFWMD General Counsel Barbara Markham, "effectuate deliveries from both primary and secondary sources," and it promised to conduct studies on water quality entering Big Cypress and Brighton reservations.[20] For its part, the tribe agreed to monitor the quality of water leaving the reservation, and it would also implement its Big Cypress water conservation plan as long as it could get the necessary funding.[21]

There were several objections to the agreement between the SFWMD and the Seminole. Both environmentalists and the Miccosukee were concerned that it required only that phosphorus concentrations not exceed 50 ppb in water both entering and leaving the reservation. Vida Verde, a non-profit organization dealing with social and environmental issues, stated that "water entering and leaving the Reservation should be below 50 ppb in phosphorus concentrations."[22] The association also objected to a provision in the agreement involving the Rotenberger Tract. The original design of STA-5, located in Hendry County and bordered by the L-3 canal on the west and the Rotenberger Tract on the east, contemplated taking water from C-139 and routing it through STA-5 to the Rotenberger Tract to allow for a more natural hydroperiod on the land. Water from C-139 could then be filtered to 50 ppb before entering Rotenberger, which would cleanse it further to 10 ppb before discharging it to Conservation Area No. 3A. Under the Seminole/

Aerial view of WCA 3. (South Florida Water Management District)

SFWMD agreement, however, the Seminole could take water from the Rotenberger Tract, use it on their citrus plantations and ranches, and then discharge it with a phosphorus level of 50 ppb. In Vida Verde's view, this constituted "receiving clean water, polluting it and releasing it downstream."[23]

Likewise, the Miccosukee objected to the plan, insisting that their own interests would be harmed if the Seminole did not mandate that water have a lower phosphorus level than 50 ppb. "Both the Seminole's [sic] and the District has [sic] been advised that it is the Tribe's intention to set a numeric standard for [total phosphorus] of approximately 10 ppb," Miccosukee Water Resources Director Truman E. "Gene" Duncan, Jr., explained. "If the District executes the Draft Agreement, this action will be interpreted as a willful anticipatory violation of the Tribe's water quality standards."[24] SFWMD officials did not see the situation in the same light, especially since the Everglades Forever Act did not set an immediate limit on phosphorus, pending more scientific research. "It is possible that phosphorus concentrations exceeding ten parts per billion (ppb) cause imbalances in Everglades' flora and fauna," Executive Director Samuel Poole III told Miccosukee Chairman Billy Cypress, "but we have not seen the scientific basis for this." He also reassured Cypress that the SFWMD had "no intention of violating any applicable water quality standards," including that of the Miccosukee.[25]

The problems that environmentalists and the Miccosukee articulated about the SFWMD/Seminole agreement spoke to the issues that both groups had with the Everglades Forever Act and its lack of a stringent phosphorus requirement. Whereas the Settlement Agreement had delineated 10 ppb as the appropriate criterion, the Everglades Forever Act had backed away from that figure, presumably at the bequest of the sugar industry and other agriculturists. Environmentalists and the Miccosukee used the scientific research of Ron Jones from Florida International University to show that any standard above 10 ppb would not effectively protect the Everglades. The Miccosukee were also concerned about the effects of the Everglades Construction Project—which would implement the structural features of the Everglades Forever Act, such as the STAs—on their water supply. The SFWMD insisted that STA-5 had to cleanse water only to 50 ppb, but the Miccosukee disagreed, stating that original conceptual design of STA-5 planned for a 10 ppb discharge.[26]

These differences convinced the Miccosukee that the state of Florida, through the SFWMD, had become misguided in its efforts to cleanse Everglades water, and that Miccosukee interests were at risk. The Miccosukee claimed that their reservation was being used as a "toilet" to collect phosphorus-laden farm water runoff. "Do you expect your neighbor to drink your garbage?" Miccosukee tribal chairman Billy Cypress asked. "We would not do that to anyone. We don't want it done to us."[27]

To protect its interests, the tribe decided to develop stringent water quality standards for its land, especially after the state of

Florida's Environmental Regulation Commission refused to look at the 10 ppb standard despite a petition from the Miccosukee and several environmental groups such as Friends of the Everglades. The Miccosukee thus requested authority from the EPA to set its own water quality standards, and the EPA granted permission in December 1994, giving the tribe "treatment as state" recognition under the Clean Water Act. Certain conditions existed, however. The standards had to apply only to those water resources that the Miccosukee actually held or that were held in trust for them by the federal government, and the tribe had to follow the same public participation regulations mandated by the EPA that states did. These included the publication of the proposed standards, public hearings on the criteria, and the opportunity for other groups to comment.[28]

Following these guidelines, the Miccosukee formulated their water quality standards. The stated purpose of these standards was not only to protect Miccosukee land, but also to ensure that water flowing into the conservation areas and Everglades National Park did not harm threatened and endangered species. Moreover, the tribe wanted to promote the social and economic well being of its members. Therefore, the tribe "vowed" that it would not "compromise" the health of its tribal members or of the Everglades ecosystem in setting its standards; instead, it would require that all water entering and leaving the reservation contain "a nutrient standard consistent with natural oligotrophic levels (including a total phosphorus limitation of 10 parts per billion of water)," oligotrophic meaning a low nutrient system. In enforcing these standards, the tribe insisted that it would not let "adjacent water users" utilize its water or its "vegetative communities . . . as a biological filter."[29]

When the Miccosukee held public hearings on these criteria in 1997, however, many objected to the stringent requirements. Although the Miccosukee considered them necessary in order to preserve the Everglades ecosystem, others disagreed, especially since the standards conflicted with the Everglades Forever Act, which had stipulated that 50 ppb would be used until the state developed a firm, numerically based criterion (which did not have to happen until 2003 and which did not have to be implemented until 2006). The Florida Sugar Cane League, for example, stated that the Miccosukee should wait to implement the 10 ppb standard until after the state had completed its scientific research on phosphorus loads.[30] Likewise, SFWMD official Frank Williamson, Jr., explained that the district's "most significant concern" with the Miccosukee's proposal was its "potential conflict" with both the Everglades Forever Act and the Settlement Agreement in *United States of America, et al. v. South Florida Water Management District, et al.* "If adopted," Williamson noted, the Miccosukee standards "would arguably establish a 10 ppb total phosphorus standard immediately." Williamson explained that this did not take into account "the needs and thresholds of the Everglades" and "the physical realities of water management within South Florida." In addition, he continued, the tribe was proposing only phosphorus limits "without providing a blueprint for improving water quality," and it had failed to produce any scientific analysis or data supporting its 10 ppb standard. The requirements in the Everglades Forever Act were in the general public's best interest, Williamson concluded, while the tribe's standards only benefited the Miccosukee.[31] In response, the Miccosukee declared that the law did not govern the tribe's federal reservation lands.[32]

Another problem that the SFWMD had with the Miccosukee's standards was that they did not conform to those developed by the Seminole Tribe. The Seminole's standards proposed phosphorus levels of only 50 ppb, keeping them in conformity with the Everglades Forever Act. The SFWMD foresaw difficulties if the Miccosukee adopted the 10 ppb rule, since this would mean that two different standards would exist for "the same water body," causing "unreasonable consequences" and "social and economic disruption."[33] In response to this concern, the EPA reminded the SFWMD that it had a dispute resolution mechanism in place that could "mediate disputes where the difference in water quality standards results in unreasonable consequences."[34]

Meanwhile, environmentalists applauded the Miccosukee's efforts. Joette Lorion, president of Friends of the Everglades and a Miccosukee consultant, stated that the tribe's action was necessary because "right now, state enforcement officers are like the Maytag repairman: They have nothing to do until Dec. 31, 2006." Charles Lee, senior vice president of the Florida Audubon Society, agreed, claiming that the sugar industry's "game plan" was to "prevent" the 10 ppb standard "from ever being set." The Miccosukee's water quality proposal, however, would make it more difficult for the industry to carry out its strategy. "We're tired of waiting," Miccosukee Water Resources Director Gene Duncan explained. "Broken promises—that's the history of the Indians and the Everglades."[35]

Accordingly, despite the concerns expressed by the SFWMD and others, the Miccosukee adopted its 10 ppb standard in December 1997 and submitted it to the EPA for approval.[36] The tribe also explained that it would determine whether water met the 10 ppb standard by measuring phosphorus content at five different locations: in the L-28 Interceptor Canal on the tribe's western boundary, in the L-28 Interceptor Canal at its dogleg (where water was discharged into the Gap Area), at a site in the C-60 Canal east of the S-140 pump station (measuring water emptying into the North Grass and South Grass areas), at the northeastern corner of the Alligator Alley Reservation in the North Grass region, and in the western portion of the Gap Area.[37]

Although the EPA usually had to approve water quality standards within 60 days, it took the agency two years before it issued a decision on the tribe's request. Some speculated that the reason for this was that the Miccosukee was the first entity—state or otherwise—to set a numeric criterion for phosphorus and it took considerable time

for EPA personnel to wade through the stacks of scientific literature on the subject. Finally, in May 1999, the EPA approved the Miccosukee's water quality standards, a significant victory for both the tribe and environmentalists. The EPA called the criteria "a significant step forward in protecting the health of the Everglades"; EPA Administrator Carol Browner, former secretary of Florida's Department of Environmental Protection, lauded the "tough standards," seeing them as a way to "protect and restore this national treasure [the Everglades] for future generations."[38] According to an article in *Time* magazine, the standards meant that "everyone" around the Miccosukee would have to meet the same criteria, even "sugar companies, which argue that they don't have the technology to comply."[39] The EPA agreed. Regional Administrator John Hankinson explained that the EPA's review provided "a strong foundation for developing future water quality standards and the technology necessary to meet those standards."[40] But the state continued its own scientific studies of phosphorus, unwilling to accept the 10 ppb without further review.

The S-140 pumping station. (South Florida Water Management District)

The Miccosukee also began pursuing means to end the Special Use Permit relationship with Everglades National Park for the 333 acres on the park's northern border. The catalyst for this action was Everglades National Park Superintendent Richard Ring's objections to the construction of houses in the Special Use area. In order to resolve the matter, the Miccosukee worked with Florida's congressional delegation—including Alcee Hastings and Carrie Meek—to pass legislation ending the Special Use relationship. In 1998, Congress enacted the Miccosukee Reserved Area (MRA) Act that terminated the Special Use Permit, expanded the area to approximately 660 acres, and granted the Miccosukee the right to govern the land "as though the MRA were a Federal Indian reservation."[41] After the passage of the act, the Miccosukee began developing water quality standards for that area as well, which, because of its location, affected Everglades National Park. The tribe essentially applied the same 10 ppb numeric criterion to the region as it did to its reserved lands, and the EPA approved this action in October 1999.[42]

Meanwhile, the Miccosukee employed another tactic in its fight to preserve the Everglades ecosystem: litigation. Throughout the 1990s, the tribe sued several federal and state agencies for many different reasons. In 1995, for example, the tribe filed a lawsuit, ultimately unsuccessful, against the U.S. Department of the Interior, the Corps of Engineers, and the SFWMD because of flooding on their land in 1994 and 1995 caused by Tropical Storm Gordon. The Miccosukee claimed that the Corps and the SFWMD did nothing to alleviate the flooding because of NPS opposition to receiving more water.[43] Other Miccosukee lawsuits included one in 1999 alleging that deviations from Conservation Area No. 3A's regulation schedule by the Corps and the SFWMD (done at the request of the National Park Service to preserve the endangered cape sable seaside sparrow) violated the Endangered Species Act by threatening the wood stork and the snail kite in 3A.[44]

Perhaps the most prominent Miccosukee litigation, however, dealt with water quality standards under the Everglades Forever Act. As explained above, in 1994, the tribe had joined other petitioners to request that the Florida Department of Environmental Protection establish a numeric water quality standard of 10 ppb. The department rejected the petition, but the state's Fourth District Court of Appeals reversed that decision (*Miccosukee Tribe v. Florida Department of Environmental Protection*), ruling that only Florida's Environmental Regulation Commission had the authority to either accept or reject the petition. The Environmental Regulation Commission decided that it would review the standards at some undesignated point, and the court subsequently found that this meant that the state was working as expeditiously as possible under provisions of the Everglades Forever Act. Only the Florida legislature, the court ruled, could hasten the timeline.[45]

At the same time, the Miccosukee sued the U.S. Department of the Interior, requesting that the U.S. District Court for the Southern District of Florida require the enforcement of the 1991 Settlement Agreement and 1992 Consent Decree. The genesis for this action was a 1994 settlement between the U.S. Department of the Interior and Flo-Sun Sugar Company, whereby the corporation agreed to pay $4 to $6 million a year in Everglades clean-up costs in exchange for the Interior Department not enforcing phosphorus standards until 2008. The Miccosukee objected to this arrangement, saying that it was opposed to "government attempts to substitute less stringent provisions of the Everglades Forever Act for those of the Settlement and Consent Decree."[46] The compromise between the Interior Department and Flo-Sun merely delayed the implementation of strict phosphorus standards to the detriment of the ecosystem. "Delay is the enemy of the Everglades," Cypress related. "The Miccosukee Tribe will not accept delay."[47] Dexter Lehtinen was even more forceful, claiming that the federal government, through its concurrence

with the Everglades Forever Act, had authorized not only the continued pollution of the Everglades, but had also "polluted the democratic process." Lehtinen vowed that "the Miccosukee Tribe will not allow their Everglades homeland to be sacrificed on the altar of political expediency."[48]

The Miccosukee continued its assault by filing an action against the EPA as well, charging that the Everglades Forever Act had changed Florida's water quality standard and that the EPA therefore had the responsibility to either approve or reject the changes, as stipulated by the Clean Water Act. According to one observer, the tribe claimed that, under the Everglades Forever Act, the state was allowing water with high levels of phosphorus to flow across South Florida, causing "an imbalance in the natural aquatic flora and fauna through 2006."[49] After representatives of the EPA testified that the presence of polluted waters did not necessarily mean that water quality standards had changed, the U.S. District Court for the Southern District of Florida rejected the tribe's claim. The tribe appealed

A Miccosukee Indian village. (The Florida Memory Project, State Library and Archives of Florida)

the ruling, leading the Eleventh Circuit Court of Appeals to remand the case back to the district court, instructing it to decide independently whether water quality standards had been altered. After its review, the district court ruled that a change had been made, and it instructed the EPA to take action. The EPA again stated that the act did not alter the standards, claiming, according to one legal scholar, that "it did not change any designated uses of downstream waters" and that "it did not change anti-degradation policy."[50]

In 1998, after conducting a judicial review of the EPA's decision under the Administrative Procedure Act, Judge Edward B. Davis of the district court overturned the EPA's decision as "arbitrary and capricious."[51] According to Davis, the Everglades Forever Act did not establish a legitimate compliance schedule as required by the Clean Water Act, and, because no numerical criterion had to be in place until 2006, the EPA was effectively allowing violations of state water quality standards by agricultural interests until that time. Therefore, Davis ordered the EPA to view the Everglades Forever Act as violating Florida's water quality standards. This seemed to be a significant ruling in favor of the Miccosukee, but the EPA stated that it would have to carefully analyze the decision before taking any action.[52] According to scholar William Rodgers, whatever the outcome, the case "had the collateral benefit of drawing EPA—the 'expert' water quality agency—into the South Florida water wars."[53]

Throughout the 1990s, then, the Miccosukee and the Seminole worked to protect the interests of their reservations and their interest in the Everglades—an area where they had resided for many decades. This fight focused on water quality, especially in relation to phosphorus concentrations. Although the Seminole generally used conciliatory methods to achieve their objectives—formulating water quality standards in conformance with the Everglades Forever Act, establishing a water conservation plan for Big Cypress Reservation in collaboration with several state and federal agencies—the Miccosukee took the opposite approach. Believing that the desire to achieve consensus was sacrificing its interests, the Miccosukee implemented water quality standards significantly more stringent than those set up by the Everglades Forever Act and sued federal and state entities over both water quality and quantity. Although these suits deepened conflicts with the SFWMD, the Corps, the EPA, the NPS—in short, almost every entity with a stake in water resource management—the Miccosukee regarded them as necessary to preserve the Everglades ecosystem. "The Everglades are dying," Buffalo Tiger declared. "The land cannot recover from this."[54] Besides, according to Gene Duncan, the "only time" the Miccosukee could "get anyone's attention is when we're in court."[55] If nothing else, the lawsuits focused attention in the late 1990s on the importance of lowering phosphorus concentrations to 10 ppb and made state and federal agencies take both tribes more seriously in water management decisions. As environmentalist Nathaniel Reed observed, the Seminole and the Miccosukee now had "a large say in how the Everglades is restored."[56]

Endnotes

[1] Billy Cypress, Chairman, Miccosukee Tribe of Indians of Florida, to Honorable Harry A. Johnston, Member of Congress, 9 July 1993, File Everglades Mediation Miccosukee, Box 19706, SFWMDAR.

[2] Robert H. Keller and Michael F. Turek, *American Indians & National Parks* (Tucson: The University of Arizona Press, 1998), 230; Parker Thomson to The Honorable Reubin O'D. Askew, Governor, 14 April 1977, File Conservation Area 3, Miccosukee Tribe, 1966-78, Re: Declaration of Trust, Box 02194, SFWMDAR. This license was later amended to become a perpetual lease.

[3] Harry A. Kersey, "Introduction," in Buffalo Tiger and Harry A. Kersey, Jr., *Buffalo Tiger: A Life in the Everglades* (Lincoln: University of Nebraska Press, 2002), 2.

[4] Buffalo Tiger and Kersey, *Buffalo Tiger*, 126.

[5] Joel Kuperberg to Hugh McMillan, 15 November 1973, File Re Seminole Tribe, Trustees Correspondence, State Lands Records Vault, Division of State Lands, Florida Department of Environmental Protection, Marjory Stoneman Douglas Building, Tallahassee, Florida.

[6] R. L. Clark, Jr., Chairman, to The Honorable Reubin O'D. Askew, 7 January 1974, File Re Seminole Tribe, Trustees Correspondence, State Lands Records Vault, Division of State Lands, Florida Department of Environmental Protection, Marjory Stoneman Douglas Building, Tallahassee, Florida.

[7] Kersey, "The East Big Cypress Case, 1948-1987," 466-474.

[8] Kersey, "The East Big Cypress Case," 474.

[9] Hobbs, Straus, Dean & Wilder to The Honorable Sidney R. Yates, Chairman, Subcommittee on Interior, House Committee on Appropriations, 9 March 1989, File Miccosukee Campground Project, Box 19707, SFWMDAR.

[10] Hobbs, Straus, Dean & Wilder to Ralph W. Tarr, Esq., Office of the Solicitor, 15 February 1989, File Miccosukee Campground Project, Box 19707, SFWMDAR.

[11] Quotation in Seminole Indian Land Claims Settlement Act of 1987 (101 Stat. 1556). For more information on the Seminole water rights compact, see Jim Shore and Jerry C. Straus, "The Seminole Water Rights Compact and the Seminole Indian Land Claims Settlement Act of 1987," *Journal of Land Use and Environmental Law* 6 (Winter 1990): 1-24.

[12] "Summary Information on the Modified Hendry County Plan," File Memorandum of Agreement, Miccosukee Tribe, Box 19707, SFWMDAR.

[13] "Memorandum of Agreement between the Miccosukee Tribe of Indians of Florida and the South Florida Water Management District," attachment to Miccosukee Tribe of Indians of Florida, Business Council Resolution No. MBC-27-87, File Memorandum of Agreement, Miccosukee Tribe, Box 19707, SFWMDAR.

[14] Miccosukee Tribe of Indians of Florida, Business Council Resolution No. MBC-38-88, File Indian Affairs Miccosukee Research 94, Box 22792, SFWMDAR.

[15] South Florida Water Management District, *Interim Surface Water Improvement and Management (SWIM) Plan for Lake Okeechobee, Part 1: Water Quality & Part VII: Public Information* (West Palm Beach, Fla.: South Florida Water Management District, 1989), Executive Summary—1-2, 7-8.

[16] South Florida Water Management District, *Surface Water Improvement and Management Plan for the Everglades: Planning Document* (West Palm Beach, Fla.: South Florida Water Management District, 1992), 5, 10-14, 36, 40.

[17] "Seminole Tribe Everglades Restoration Initiative: Water Conservation System Conceptual Plan, Briefing Paper for the U.S. Department of the Interior," 17 February 1995, File GOV 02-16-03 Federal Government 95 Interior Department, Indian Affairs—Seminoles, Conceptual Water Conservation System Design DOC, Box 15771, SFWMDAR.

[18] Seminole Tribe of Florida to the Honorable Ralph Regula, Chairman, and the Honorably Sidney Yates, Ranking Member, Subcommittee on the Interior, House Committee on Appropriations, 1 June 1995, File Seminole SWIM Big Cypress Plan, Box 19707, SFWMDAR.

[19] Woodie Van Voorhees, Government & Public Affairs, to Irene Quincey, et al., 2 March 1994, File Seminole SWIM Big Cypress Plan, Box 19707, SFWMDAR; Draft Agreement Between the South Florida Water Management District and the Seminole Tribe of Florida Providing for Water Quality, Water Supply and Flood Control Plans for the Big Cypress Seminole

Indian Reservation and the Brighton Seminole Indian Reservation, Implementing Sections V.C. and VI.D of the Water Rights Compact, 15 November 1993, ibid.

[20] Barbara A. Markham, General Counsel, to Governing Board Members, 30 June 1995, File Micco WQ Standards, Box 19706, SFWMDAR.

[21] Draft Agreement Between the South Florida Water Management District and the Seminole Tribe of Florida Providing for Water Quality, Water Supply and Flood Control Plans for the Big Cypress Seminole Indian Reservation and the Brighton Seminole Indian Reservation, Implementing Sections V.C. and VI.D of the Water Rights Compact, 26 June 1995, ibid.

[22] Phillip C. Garcia, President, and Shannon Larsen, Acting Secretary-Treasurer, to Valerie Boyd, et al., 8 December 1995, unlabeled file, Box 22792, SFWMDAR (emphasis in the original).

[23] Garcia and Larsen to Boyd, 8 December 1995.

[24] Truman E. Duncan, Jr., Water Resources Director, to Mr. Sam Poole, III, Executive Director, South Florida Water Management District, 25 September 1995, unlabeled file, Box 22792, SFWMDAR.

[25] Samuel E. Poole III, Executive Director, South Florida Water Management District, to Billy Cypress, Chairman, Miccosukee Tribe of Indians of Florida, 11 August 1995, File Micco WQ Standards, Box 19706, SFWMDAR.

[26] Duncan to Poole, 25 September 1995; see also Joette Lorion interview by Matthew Godfrey, 18 January 2006, 5 [hereafter referred to as Lorion interview].

[27] As quoted in "Indians Seek to Have Say in Everglades Suit," *The Miami Herald*, 28 March 1993.

[28] Elizabeth D. Ross to Distribution List, 4 September 1996, File Indian Affairs, Seminole Water Quality Standard Research 94-98, Box 22792, SFWMDAR; Lorion interview, 5; Rodgers, "The Miccosukee Indians and Environmental Law," 10926.

[29] "Miccosukee Environmental Protection Code Subtitle B: Water Quality Standards for Surface Waters of the Miccosukee Tribe of Indians of Florida" [hereafter referred to as Miccosukee Water Quality Standards], File Micco WQ Standards, Box 19706, SFWMDAR.

[30] "March 21, 1997 Summary of Comments Submitted by Interested Parties Concerning Micosukee Tribe's Proposed WQS," File Miccosukee WQ Standards, 1997-98, Box 19706, SFWMDAR.

[31] Frank Williamson, Jr., to Secretary Murley, Florida State Clearinghouse, Department of Community Affairs, 13 October 1997, File Micco WQ Standards, Box 19706, SFWMDAR.

[32] Miccosukee Tribe's Proposed Water Quality Standards (WQS)," File Miccosukee WQ Standards, 1997-98, Box 19706, SFWMDAR.

[33] Quotation in Williamson to Murley, 13 October 1997; see also Seminole Water Commission, Seminole Tribe of Florida, "Proposed Rules: Water Quality Protection and Restoration: Rules to Carry Out the Federal Clean Water Act and the Tribal Water Code, Including Water Quality Standards for the Big Cypress Reservation," File Indian Affairs, Seminole Water Quality Standard Research 94-98, Box 22792, SFWMDAR.

[34] Robert F. McGhee, Director, Water Management Division, United States Environmental Protection Agency, to Mr. Frank Williamson, Jr., Chairman, Governing Board, South Florida Water Management District, 17 November 1997, File Miccosukee WQ Standards, 1997-98, Box 19706, SFWMDAR.

[35] Lorion, Lee, and Duncan quotes all cited in Neil Santaniello, "Cleanup Pace Prompts Tribe to Take Hard Line," *Sun-Sentinel*, 15 October 1997.

[36] Miccosukee Water Quality Standards. The Miccosukee Reserved Area had previously been called the Permit Area, but under the Miccosukee Reserved Area Act of 1998, Congress recognized the 667 acres that the tribe had leased from the state as "Indian country." Rodgers, "The Miccosukee Indians and Environmental Law," 10926-10927.

[37] "Methodology for Determination of Compliance with the 10 Parts Per Billion Numeric Criterion for Total Phosphorus," 6 December 1998, Miccosukee WQ Standards, 1997-98, Box 19706, SFWMDAR.

[38] Quotations in "EPA Approves Tough Phosphorus Limit for Tribal Waters in Everglades," U.S. Environmental Protection Agency Press Release, 26 May 1999 <http://www.epa.gov/Region4/ oeapages/99press/052699. htm> (25 November 2003); see also Lorion interview, 5.

[39] Tim Padgett, "Last Stand," *Time* 154 (5 July 1999): 61.

[40] As quoted in Light, "Miccosukee Wars in the Everglades," 736.

[41] Act of 30 October 1998 (112 Stat. 2964).

[42] Lorion interview, 5-6.

[43] See Omnibus Order, *Miccosukee Tribe of Indians of Florida v. United States of America, et al.*, copy provided by James W. Vearil, Chief, Water Management Section, Met Section, Jacksonville District, U.S. Army Corps of Engineers; Rodgers, "The Miccosukee Indians and Environmental Law," 10920.

[44] See Dexter W. Lehtinen, Esquire, for the Miccosukee Tribe of Indians, to Honorable Secretary Bruce Babbitt, U.S. Department of the Interior, 16 March 1998, File Miccosukee WQ Standards, 1997-98, Box 19706, SFWMDAR.

[45] Keith W. Rizzardi, "Alligators and Litigators: A Recent History of Everglades Regulation and Litigation," *The Florida Bar Journal* 18 (March 2001): n.p. (copy provided by John D. Brady, principal assistant, Jacksonville District Office of Counsel, Jacksonville, Florida).

[46] Miccosukee Tribe of Indians of Florida, Press Release, 6 November 1995, File Everglades Mediation Miccosukee, Box 19706, SFWMDAR.

[47] As cited in Miccosukee Tribe of Indians of Florida, Press Release, 6 November 1995.

[48] As cited in Miccosukee Tribe of Indians of Florida, Press Release, 6 November 1995; see also "Tribe Sues U.S. Agency Over Cleanup," unidentified newspaper clipping in File Everglades Mediation Miccosukee, Box 19706, SFWMDAR.

[49] Quotation in Rodgers, "The Miccosukee Indians and Environmental Law," 10926; see also Light, "Miccosukee Wars in the Everglades," 731.

[50] Quotations in Rizzardi, "Alligators and Litigators"; see also Light, "Miccosukee Wars in the Everglades," 731-732.

[51] As cited in Rizzardi, "Alligators and Litigators."

[52] Rizzardi, "Alligators and Litigators"; Light, "Miccosukee Wars in the Everglades," 731-732.

[53] Rodgers, "The Miccosukee Indians and Environmental Law," 10926.

[54] Buffalo Tiger and Kersey, *Buffalo Tiger*, 3.

[55] Unidentified transcript, File Everglades Mediation Miccosukee, Box 19706, SFWMDAR.

[56] Reed interview, 36.

CHAPTER 16

A Laboratory for the Everglades: Kissimmee River Restoration in the 1990s

WHILE THE MICCOSUKEE FOCUSED on water quality problems, the Corps undertook a remarkable project in the 1990s—the restoration of the Kissimmee River. In the late 1980s, Kissimmee River restoration efforts had crystallized around the concept of watershed restoration, rather than focusing only on dechannelization of the river. The impetus was a growing mass of scientific literature informing the process of restoration, an accumulation of knowledge that increased in the 1990s and defined clearly what restoration meant. Kissimmee River restoration thus became the model for ecosystem management in South Florida, and it convinced federal and state agencies of three things: ecosystem restoration was possible; a clear definition of what was being restored was necessary; and reliance on scientific studies was crucial. The effort also provided many lessons to the Corps of Engineers of what was necessary to make restoration a success, but some still questioned whether the Corps was the proper agency to take the lead in such endeavors.

By the late 1980s, plans to restore the Kissimmee River were tenuous. The Corps could not recommend federal participation, both because it did not have congressional authorization and because dechannelizing the river did not have a positive benefit-cost ratio, at least in basic economic terms. Even when Congress appropriated money for restoration, the Reagan administration refused to allocate the funds. Meanwhile, agricultural interests and cattle ranchers in the Kissimmee basin continued to oppose restoration, concerned that it would infringe on the lands they used for their livelihood. SFWMD officials, however, became increasingly convinced that restoration was necessary, desirable, and possible, and the SFWMD proceeded with a demonstration project attempting to prove those points to detractors.

But questions arose over just what the term "restoration" meant. When Arthur Marshall, Johnny Jones, and other environmentalists had first called for the dechannelization of the Kissimmee in the 1970s, they did so because they believed that the river was facilitating the flow of nutrient-rich water to Lake Okeechobee, hastening that lake's demise. In the 1980s, however, studies showed that phosphorus was the limiting nutrient in the lake and that the Taylor Creek-Nubbin Slough area and the EAA were the greatest contributors of the mineral.[1] Because of these findings, the justification for restoring the Kissimmee River gradually began to turn towards reestablishing the ecological conditions of the Kissimmee River Basin to their pre-channelization state, rather than improving Lake Okeechobee's water quality.[2]

Throughout the 1980s, scientists such as Louis Toth, a biologist with the SFWMD, published studies of the conditions of the Kissimmee River and its floodplain before channelization. According to Toth, 94 percent of the floodplain was covered with water over 50 percent of the time, while seasonal wet-dry cycles occurred as well. These conditions produced "a mosaic of hundreds of distinct patches of intermingled vegetation," dominated by three community types: willow and buttonbush woody shrub wetlands; broadleaf marshes; and maidencane, beakrush, and mixed species wet prairies.[3] Toth continued that the basin housed approximately 35 species of fish, 16 species of waterfowl, 6 species of waterbirds, and numerous invertebrates. When channelization occurred, Toth explained, it drained over 30,000 acres of floodplain wetlands and disrupted the wet-dry cycles, causing water oxygen levels to drop. This caused an exodus of wading birds, a decline in waterfowl populations by 92 percent, a diminishment in fish populations, and exotic replacement of natural

Louis Toth, the guiding scientist of Kissimmee River restoration. (South Florida Water Management District)

vegetation. In order to restore natural conditions to the Kissimmee Basin, Toth argued for a holistic approach to restoration, taking into consideration all of these disparate factors.[4]

Toth's ideas were reinforced as the SFWMD monitored the effects of its demonstration project. In 1986, the SFWMD had completed construction of project features in Pool B, a 12-mile stretch of C-38 between S-65A and S-65B. Three weirs in the pool redirected water through seven miles of the old Kissimmee River bed, including three of its oxbows, allowing 1,300 acres of pasture to flood. By 1987, the SFWMD could already observe positive results. Marshland plants returned, as did clams and invertebrates. The rapidity of the change surprised even Toth, who stated that he did not "expect to see that much change in such a brief time in an area that hadn't been flooded in 20 years."[5] The SFWMD continued to monitor the project area until November 1988, claiming that it clearly showed that "restoration of wetland communities on the Kissimmee River floodplain is feasible" and that "restoration of ecological integrity of the river channel is possible." However, the SFWMD concluded, restoration could succeed only if a "holistic approach" was used to reestablish "both the form and function of the former ecosystem."[6]

The need for an integrated approach was emphasized at a Kissimmee River Restoration Symposium hosted by the SFWMD in

Results of the Kissimmee River Demonstration Project. (South Florida Water Management District)

Dr. Hsieh Wen Shen's Kissimmee River model. (South Florida Water Management District)

October 1988 in cooperation with the Florida Department of Environmental Regulation, the Florida Game and Fresh Water Fish Commission, the Office of the Governor, the Florida Dairy Farmers, Inc., and the Florida Sierra Club. Co-sponsored by several national environmental and engineering societies, the conference brought together scientists and engineers from a host of areas to examine data gained from the demonstration project, and to "provide a forum" for scientists and engineers to present their findings "as they relate to options for restoration."[7] Nearly 200 people attended, and 27 papers were presented on subjects ranging from vegetation to fish and wildlife to engineering concerns. According to M. Kent Loftin, the SFWMD representative in charge of the demonstration project, the symposium was "a major milestone—sort of a starting line for the final lap in a race to restore the Kissimmee River."[8]

By the end of the symposium, according to Toth and Nicholas Aumen, two scientists who attended the meetings, participants had decided that restoration efforts would have to revive the "physical, chemical, and biological characteristics, processes and interactions that governed the ecology and evolution of the historic ecosystem" in order to be successful.[9] Recognizing this, the conference developed a unified goal for Kissimmee River restoration: the reestablishment of the basin's "ecological integrity." Toth later elaborated that

this meant "reestablishing a river-floodplain ecosystem that is capable of supporting and maintaining a balanced, integrated, adaptive community of organisms having a species composition, diversity, and functional organization comparable to that of the natural habitat of the region."[10]

To achieve that goal, the symposium outlined specific guidelines and criteria. These included restoring both the "lateral connectivity" and the "longitudinal continuity" between the river and the floodplain, as well as reestablishing pre-channelization vegetation communities in order to recreate habitat. Such endeavors could be accomplished by allowing water flows in the river to approach pre-channelization levels in terms of duration and variability, and by restoring the hydroperiod of floodplain inundation. "The ecosystem restoration goal requires that all criteria [be] interdependent and mutually reinforcing," Toth and Aumen explained.[11]

With the development of these goals, guidelines, and criteria, the SFWMD now had a clear restoration objective in mind. In order to determine the best method to obtain that objective, the district, in 1986, had hired Dr. Hsieh Wen Shen, a sedimentation and environmental river mechanics expert at the University of California at Berkeley, a leading center for hydrology research, to examine different restoration proposals. Using computer models for guidance,

Shen constructed a 60- by 100-foot model of a square mile section of the river and used it to test various approaches, such as using weirs and plugs to redirect water into the old riverbed or backfilling C-38.[12]

In 1989, Shen provided SFWMD officials with the data he had gathered for the various proposals, allowing them to evaluate the information and decide which alternative they wanted to implement. He recommended backfilling the upper stretches of C-38 as the best plan, but he expressed some apprehension about the whole restoration process, given that it was another engineering manipulation of nature. "When you disturb a thing one way, and you disturb or distort it another way, there are always all types of dangers," Shen noted. "My concern is, if any, that in a few years from now, there may be other ecological concerns that we cannot foresee now."[13]

Despite Shen's trepidation, the SFWMD published a report in 1990 detailing what it perceived as the best plan in order to reestablish the "ecological integrity" of the Kissimmee basin. The district determined that a proposal designated as the Level II Backfilling Plan was the only strategy that had the potential to restore all aspects of the ecosystem. Under this plan, large sections of C-38 would be filled with dirt, thereby restoring 52 miles of the old river channel and flooding 24,000 acres of floodplain.[14] According to Jacksonville District Engineer Colonel Bruce A. Malson, it was "the most aggressive alternative," calling for the removal of four water control structures along C-38 in order to restore about 50 miles of the original river.[15] Despite strong opposition from agriculturists in the basin, Governor Bob Martinez endorsed the proposal, calling it "the only plan that will restore a functional, riverine, floodplain ecosystem with nearly the same biological characteristics as the preproject system," and he asked the Corps to use its Section 1135 authorization to help the state with the plan.[16]

Indeed, even though the Corps had recommended no federal involvement in 1987, SFWMD and Florida officials were optimistic that the Corps would now cooperate. For one thing, the presidential administration had changed from Ronald Reagan to George H. Bush. Although no strong supporter of environmental policies, Bush had made some campaign promises relative to the environment, including no net reduction of wetlands, and he had criticized Governor Michael Dukakis, his opponent in the 1988 presidential election, for ignoring environmental problems in Massachusetts. At the very least, the Bush administration was more open to environmental issues than the Reagan administration. Likewise, Lieutenant General Henry Hatch, who became Chief of Engineers of the Corps in 1988, wanted to "green" the Corps by becoming more responsive to environmental concerns. "We engineers must look at our work in a broad social and environmental context as well as in technical and short term economic terms," Hatch explained.[17] Encouraged by such comments, Senator Bob Graham contacted Hatch and asked him to support the Level II Backfilling Plan. Hatch was reluctant to request Section 1135 funding for the proposal because that sec-

tion authorized only environmental restoration projects that were consistent with authorized project purposes. The Level II Backfilling Plan, however, would effectively dismantle the Kissimmee River flood control project. Therefore, Hatch wondered whether the Corps could receive authorization and funding from Congress to dechannelize the Kissimmee, and the Jacksonville District prepared draft legislation for this purpose.[18]

With the Corps' support, and through the work of Graham, the Water Resources Development Act of 1990 (WRDA-90), passed in November 1990, contained a section specific to Kissimmee River restoration. Section 306 of the law also stipulated that environmental protection was a major Corps mission. The act therefore directed the Secretary of the Army to make a feasibility study of modifying the Kissimmee River flood control project in order to develop "a comprehensive plan for the environmental restoration" of the river.[19] In order to ensure that the Corps' ultimate proposal was in line with what the state wanted, the law required the Corps to base its study on the Level II Backfilling Plan developed by the SFWMD, and it set a deadline of 1 April 1992 for report submission to Congress.

According to one Corps official, the 16-month turnaround time for the study was a "wink of an eye" for the Corps since it usually took three to four years to complete a feasibility report. Yet the District would have to meet the deadline.[20] After the passage of WRDA-90, Graham told Corps leaders, in the words of one observer, that Congress would "hold [their] feet to the fire" and that there would be "hell to pay" if the deadline was missed.[21] This political pressure was stiff enough that the Corps made a policy decision to meet the 1 April deadline, and, in the Jacksonville District, District Engineer Colonel Bruce A. Malson pledged all of his efforts and resources to comply.[22]

To take charge of the feasibility study, Malson chose Stuart Appelbaum, Chief of the Jacksonville District's Flood Control and Floodplain Management Planning Division. Appelbaum, who had come to Jacksonville from the Baltimore District, had degrees in water resources engineering, but had no emotional connection with the Kissimmee flood control project, unlike several oldtimers within the District. "We [were] just looking to do whatever need[ed] to be done to solve problems," Appelbaum related. If that meant "removing the projects or structures that the Corps built in an earlier generation, so be it." The whole restoration philosophy was a "radical" concept for many in the Corps at the time, Appelbaum stated, but he embraced it as "the thrill of the project."[23]

Because of the strict deadline that the Corps faced, and because WRDA-90 specifically instructed the Corps to base its study on the Level II Backfilling Plan, Appelbaum put together a "game plan" that essentially involved putting the SFWMD report into "Corpsspeak," completing a project cost estimate and an EIS, and packaging them all together as one report.[24] A Special Review Conference, held in February 1991 to "resolve policy and procedural issues" on

Kissimmee River restoration and attended by the SFWMD, the Jacksonville District, the South Atlantic Division, Corps Headquarters, and the Assistant Secretary of the Army (Civil Works), essentially concurred with Appelbaum's plan, providing a Project Guidance Memorandum to govern the completion of the report.[25] The Jacksonville District then worked feverishly to produce the required document, holding weekly meetings with all of the chiefs of its various branches and using all of its resources. "I had carte-blanche [for] the people that worked for me [to do] whatever was needed," Appelbaum recollected. "Other projects or jobs suffered for resources because we got whatever we wanted."[26]

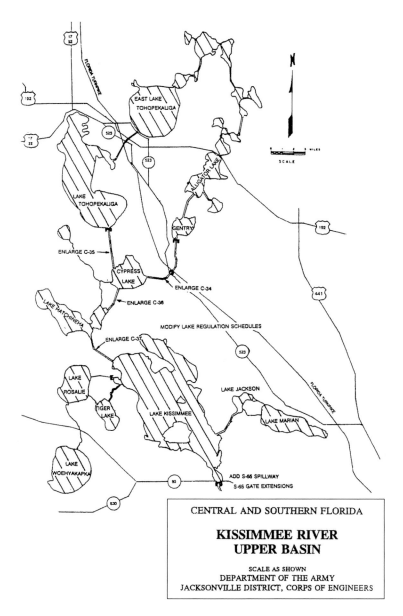

CENTRAL AND SOUTHERN FLORIDA

**KISSIMMEE RIVER
UPPER BASIN**

SCALE AS SHOWN
DEPARTMENT OF THE ARMY
JACKSONVILLE DISTRICT, CORPS OF ENGINEERS

Upper Kissimmee Basin Headwater Lakes. (U.S. Army Corps of Engineers, Jacksonville District)

At the same time that the Jacksonville District worked on the Level II Backfilling Plan, it also began a feasibility report on another aspect pertaining to Kissimmee River restoration: modification of the Upper Kissimmee Basin flood control project. The Upper Kissimmee Basin consisted of 15 lakes, including Kissimmee, Tohopekaliga, Hatchineha, Cypress, Gentry, and Alligator, and formed the headwaters of the Kissimmee-Okeechobee-Everglades ecosystem. After receiving permission in the 1954 Flood Control Act to implement a flood control plan in the upper basin, the Corps constructed eight regulatory structures in the headwater lakes. However, the plan created problems for the Lower Kissimmee Basin, as it caused the river to run dry 30 percent of the time.[27] As District Engineer Colonel Terrence "Rock" Salt (who replaced Malson in August 1991) explained, the Corps lowered water levels in the lakes before the rainy season began by discharging water to the Kissimmee, and then, once the rains came, it allowed the lakes to fill up with water, rather than releasing it to the Kissimmee. This meant that water flowed down the Kissimmee in the dry season, while the river was essentially dry during the rainy season, an unnatural effect. "In order to get the system to function more naturally," Salt related, "you have to have a way for the upper lakes to act the way they used to act so that the lower Kissimmee could act the way it used to act."[28]

Thus, the SFWMD had recommended in its 1990 report on Kissimmee restoration that work be done on the Upper Kissimmee chain of lakes in order to ensure more natural water flows in the river, and it suggested that this be done as Part I of the river's restoration. Because the Corps could perform headwaters revitalization and still maintain the flood control for which the project was authorized, the Corps undertook a study of how the upper lakes system could be modified under the authority of Section 1135 of WRDA-86. As the Board of Engineers for Rivers and Harbors explained, modifying the headwaters project was "a prerequisite for successful restoration of the lower Kissimmee River basin ecosystem."[29]

In August 1991, the Jacksonville District produced a draft feasibility report on Kissimmee River restoration and made it available for public comment. The report noted that, although channelization of the Kissimmee River had prevented flooding in the Lower Kissimmee Basin, it had also caused "long term degradation of the natural Lower Basin ecosystem," including a decline in fish and wildlife populations and a reduction in wetland acreage from 35,000 acres to 14,000 acres.[30] To correct these problems, the Corps had examined the SFWMD's Level II Backfilling Plan and had developed a Modified Level II Backfilling Plan. Under this new proposal, the Corps would

- backfill 29 miles of C-38;
- excavate 18 new river channel sections, totaling 11.6 miles;
- remove project structures in the area of backfilled reaches, including S-65B, S-65C, and S-65D;

- build a bypass spillway and channel at S-65 to serve as Lake Kissimmee's primary outlet; and
- modify the three weirs in Pool B from the demonstration project in order to "restore flows through oxbows and facilitate local flood plain inundation."[31]

To ensure the success of restoration, the report continued, the Corps would complete its Headwaters Revitalization Project, whereby it would modify some structures in the upper lakes to let water pass more easily, as well as acquire land so that it could better regulate discharges from the lakes. The Corps estimated that 68,395 acres of land would have to be obtained in either fee or easement for the entire project, necessitating the relocation of 356 residences. It outlined the total cost of the project as $422 million, with an additional $91 million for headwaters revitalization, and noted that negotiations were still underway as to how cost sharing would work. When completed (after an estimated 15 years), the project would produce a "restored ecosystem" consisting of "56 miles of restored river, 35,000 acres of restored wetlands, improved water quality, and improved conditions for numerous fish and wildlife species."[32]

In October 1991, the Corps held public hearings on the draft report. At these hearings, the Okeechobee County Commission went on record as opposing the project for various reasons, including the relocation plan and the belief that the project would harm existing environmental habitat.[33] Others expressed concerns about the differences between the Corps' plan and the SFWMD's. The SFWMD, for example, did not require the removal of any residences, but the Corps believed it was necessary to acquire land within the five-year floodplain so that the river could run its natural course. Price was another issue, as the Corps' estimated its project would cost significantly more than what the SFWMD had calculated. The reason for this discrepancy, the Corps stated, was that it had to pay a 25 percent contingency fee on any real estate costs, and because its "more refined" analysis had added items to the cost estimate, including containment levees and project monitoring.[34] Some, however, charged that the Corps was intentionally "jacking up the cost" in order "to kill the project."[35]

Other questions revolved around the effects of restoration on navigation and flood control. The Corps related that under its proposal, the river would be "at least 3 feet deep 90% of the time," allowing most crafts (aside from houseboats) to navigate the river. Likewise, even with the headwaters revitalization proposal, the Corps would continue to provide flood control in the upper basin "at the current authorized level." In the lower basin, flood control would occur through non-structural means, such as the purchase of lands in the five-year floodplain.[36] "Please remember that our findings to date are still tentative," Colonel Salt declared, "and do not necessarily represent the final results."[37]

After receiving comments from the general public and from other interested parties, the Jacksonville District produced its final feasibility report and EIS and transmitted it to the South Atlantic Division Engineer, who approved it in December 1991. This report explained that the major objections to the plan came from two sources: land owners opposed to relocation and recreational boaters concerned about navigation effects. Despite these concerns, the Jacksonville District still claimed that its recommended plan—consisting essentially of backfilling 29 miles of C-38 and excavating 11.6 miles of new river channel, among the other structural changes listed above—offered "the best solution for environmental restoration of the Kissimmee River."[38] Before submission to Congress, however, the report had to go to Corps Headquarters for review by the Washington Level Review Center (which, from 1989 to 1994, was responsible for technical and policy compliance reviews), the Board of Engineers for Rivers and Harbors, the Chief of Engineers, and the Secretary of the Army.

As these examinations were conducted, problems developed over cost sharing. In its initial draft feasibility report, the Corps had listed cost sharing requirements at 75 percent federal and 25 percent state, the normal division for fish and wildlife restoration projects. Before the draft was issued for public comment, a Corps review team composed of Headquarters and Division officials told Colonel Salt to leave the cost sharing requirement open. In the meantime, some environmentalists, including Theresa Woody of the Sierra Club, wondered if the federal cost could be lowered so that Congress would be more likely to approve the project. After receiving additional advice from Corps officials, Salt changed the cost share arrangement in the District's final report to 50/50.[39] This outraged Florida Governor Lawton Chiles. "It is not a showing of good faith to, at this critical point, back away and demand that the local sponsor should shoulder the cost of all lands, easements, rights-of-way, relocation, dredged material disposal areas, plus 50 percent of the construction costs," Chiles told Salt.[40] The Corps argued, however, that since the whole purpose of the proposal was ecosystem restoration, and since the state was basically requiring it to perform this action according to the state's plans, Florida should have to bear more of the costs.[41] "The feds and the state had made this problem," Salt explained, "and together they would fix it."[42]

Other disagreements arose over the number of residences that would have to be moved because of the relocation plan. The displacement proposal led residents of a subdivision and two trailer parks along the banks of C-38 to band together in November 1991 as ROAR—Realists Opposed to Alleged Restoration—in opposition to relocation.[43] After hearing numerous protests from concerned property owners, the SFWMD's governing board concluded in February 1992 that the Corps should not backfill any part of C-38 south of S-65D (essentially, seven miles of the canal in Pool E) and that it should not flood any homes south of U.S. Highway 98, which basically would allow the Kissimmee River Shores and Hidden Acres Estates communities to remain. If the Corps approved these arrangements, it would also mean that only 47 residences would have to be relocated, and that only 22 miles of C-38 would be refilled. At the same time, the SFWMD claimed, it would reduce any economic hardships that the restoration project might cause to local residents.[44]

In order to discuss these issues, Governor Chiles met with Assistant Secretary of the Army (Civil Works) Nancy P. Dorn in March 1992, the same month that President George Bush declared his support for restoration while on a campaign trip to Florida. In the meeting between Dorn and Chiles, the governor agreed that the backfilling of C-38 south of S-65D would be considered a "locally preferred option," meaning that if the state wanted that part of the project done, it would be responsible for it without federal participation. Chiles also accepted the modification of weirs in Pool B as a locally preferred option, and he agreed that the state would share construction costs on a 50/50 basis. In return, Dorn committed the federal government's 50/50 participation in land acquisition. This meant that the state would ultimately pay $286 million for the Level II Backfilling Plan—which now would fill only 22 miles of C-38—while the federal government would provide $139 million.[45]

With this arrangement in place, the report passed the Washington Level Review Center and went to the Board of Engineers for Rivers and Harbors for approval. According to Colonel Salt, members of the board were upset that they had had no input on the cost sharing formulation, which had already been worked out by the time they received the report, and characterized their review as merely "rubber stamping" a done deal.[46] In any case, the board explained in its official report that its review focused on "the proposed performance and effects of the recommended plan." After examining these aspects and investigating whether the plan conformed to guidelines outlined by the Water Resources Council, the board concurred with the recommendations in the Jacksonville District's final feasibility report and EIS. It noted, however, that Kissimmee River restoration was "unique" and "should not be viewed as precedent setting, or as a guideline for any future restoration projects," mainly because it involved "almost the total dismantling of a federally constructed flood control project" and because its "project formulation was constrained by congressional direction." In addition, the board emphasized that restoration would not succeed unless the Corps implemented the Section 1135 Headwaters Revitalization Project.[47] The feasibility report and EIS then went to Chief of Engineers Lieutenant General Henry Hatch, who approved it and transmitted it to the Secretary of

The channelized Kissimmee River. (South Florida Water Management District)

the Army. Thereafter, Assistant Secretary of the Army (Civil Works) Nancy Dorn sent the plan to Congress for its approval on 3 April 1992, requesting that it be included in the Water Resources Development Act of 1992 (WRDA-92).[48]

In the spring and summer of 1992, Congress considered the Corps' report. As it did so, supporters and detractors of the restoration project made their feelings known. The Florida Cattlemen's Association, for example, unanimously passed a resolution in June in opposition to the project, claiming that it would destroy "many thousands of acres" of wildlife habitat that had appeared since channelization had occurred. Claiming that the Corps had not adequately studied the environmental and economic impacts of restoration, the association registered its strong opposition to dechannelization until the plan went through "the same scrutiny and permitting procedures" as other "large project[s] in this area."[49] The association was not alone in its hostility, as ROAR continued to pepper the SFWMD with questions and concerns about the project, while the Florida Farm Bureau declared its opposition. At the same time, local ranchers made a video of boats floating down C-38 as "Let It Be" played in the background, and Okeechobee County issued a resolution of protest against the project.[50]

These objections were nothing new; almost since the time that environmentalists began advocating the restoration of the river, opponents had voiced their disapproval. Many critics used the destruction of existing environmental habitat as a reason to oppose the project. John B. Coffey, former chairman of the Glades County Commission and of the Kissimmee River Resource Planning and Management Committee, for example, stated that backfilling C-38 would cause "massive environmental disturbance." Instead, he advocated state purchase of "submerged lands of the old river."[51] Others believed that environmentalists were bullying the SFWMD, the Corps, and Congress to get what they wanted. "It seems that when Sierra Club or Audubon Society speaks, politicians wet their pants," W. H. Morse of the Kissimmee/Osceola County Chamber of Commerce declared before accusing environmental groups of misrepresenting the benefits of the project. Morse, as well as many residents around the Kissimmee River and Lake Okeechobee, wanted the Corps to leave the river alone, and he asked Congress not to be "conned" by environmentalist efforts.[52]

Other interests whole-heartedly supported restoration efforts. The Wilson Ornithological Society, for example, requested that the Corps "proceed with the Level II backfilling plan for the restoration of the flood plain marshes of the Kissimmee River" as

Lake Kissimmee and its connection to the Kissimmee River. (South Florida Water Management District)

soon as possible.[53] Meanwhile, the Everglades Coalition, led by Theresa Woody of the Sierra Club, advocated the restoration of the river through the Level II Backfilling Plan. Observing that restoration of the Kissimmee had been "a major environmental priority for the state of Florida since the mid-1970s," Woody related that the plan had "the broad bi-partisan support of the Florida Cabinet, state and national conservation organizations and the last four Governors of Florida," as well as that of President George Bush and the Corps of Engineers.[54] In another publication, Woody explained several ways in which the Level II Backfilling Plan was "a history-making project." For one thing, she said, it would be "the largest river restoration project ever undertaken"; for another, it would offer the Corps a way to "showcase" its "environmental enhancement" skills, thereby "expanding" its mission. Perhaps most importantly, Woody concluded, it would be a "pioneering work" in the field of ecosystem restoration, as it attempted "to heal a system that humans have torn apart."[55]

Heeding the arguments of the Wilson Ornithological Society and the Everglades Coalition, Congress included Kissimmee River restoration in WRDA-92, which became law on 31 October 1992. Not only did this act authorize "the ecosystem restoration of the Kissimmee River" (at a total cost of $426 million), it also instructed the Corps to carry out the Headwaters Revitalization Project (at a total cost of $92 million). The only caveat was that the Corps had to "ensure" that the project gave "the same level of flood protection as is provided by the current flood control project," but the Corps had already planned for that.[56]

With the passage of WRDA-92, the Corps began the "precedent-shattering" Kissimmee River restoration project.[57] In order to ensure the plan's success, the Corps tried to implement an effective partnership with the SFWMD, whose project it really was. In the early 1980s, SFWMD officials had evinced some dissatisfaction with the Corps, and its governing board had even discussed divorcing the district from its Corps relationship. The major problem was that the SFWMD focused less on flood control and more on water management after its reorganization in the 1970s, and several of its officials believed that the Corps did not share that vision. The Kissimmee River restoration project gave both agencies an opportunity to heal that relationship. "The Kissimmee, to me, was the beginning of the modern era [of working] with the Water Management District," Stuart Appelbaum would later recollect.[58]

One of the SFWMD's roles was to continue to supply necessary scientific information about the process of restoration. An area on which its researchers, including Louis Toth, focused was the importance of the Headwaters Revitalization Project. With his colleague Jayantha Obeyserka, former SFWMD scientist M. Kent Loftin, and William A. Perkins of Battelle Northwest, Toth published a paper in 1993 emphasizing that the "pulse-like" regulation schedule of the headwater lakes produced low dissolved oxygen levels in the river

and left it with either low or no flow for extended periods of time. This led to "repetitive fish kills" and "limited floodplain inundation." Therefore, "modified flow regulation [was] a key component" of the "plan to restore the ecological integrity" of the Kissimmee River.[59]

The Corps agreed. In 1992, it began work on an alternative regulation schedule for the Upper Kissimmee lakes, proposing to increase the permissible high levels in Lakes Kissimmee, Cypress, and Hatchineha from 52.5 feet to 54.0 feet during the dry season, and to raise Lake Kissimmee's level in the wet season to 52.5 feet (instead of 51.0). This would allow the Corps to elevate seasonal storage capacities by 100,000 acre feet, permitting it to more effectively "simulate the historic seasonal flow from Lake Kissimmee to the Lower Basin."[60] The Corps estimated that the revised schedule would also have positive environmental impacts on the upper basin, including increasing "the quantity and quality of the wetland habitat . . . to benefit fish and wildlife."[61] In addition, the Corps proposed several other measures for the headwaters, including the purchase of 16,000 acres of land bordering Lakes Hatchineha, Kissimmee, Cypress, and Tiger; the widening of C-36 (which connected Lake Hatchineha to Lake Cypress) and C-37 (which connected Lake Kissimmee to Lake Hatchineha) in order to "flatten the flood profile through the upper lakes and prevent excessive flooding"; and increasing S-65's outlet capacity in order to allow Lake Kissimmee to discharge more water.[62]

Meanwhile, the SFWMD created a Headwaters Revitalization Project 1135 Study Interagency Team, composed of representatives from the SFWMD, the Corps, the Florida Game and Fresh Water Fish Commission, and the U.S. Fish and Wildlife Service. This interagency group studied various plans, but ultimately concluded, through the use of computer simulation models, that the Corps' proposal provided the best opportunities for more natural releases from the headwaters.[63] At the same time, the team examined how the Corps' plan would enhance environmental features of the upper basin. Unfortunately, computer models showed that "the potential for environmental benefits within the upper chain of lakes is much lower than originally envisioned."[64] Instead, beneficial environmental impacts from headwaters revitalization would mainly be confined to the river itself. Fearing that headwaters revitalization would thus not meet necessary environmental criteria on its own, and in order to allow the Corps to spend money allocated specifically for headwaters revitalization on either the upper or lower basin plans, the team insisted that the two projects be combined into one.[65] Accordingly, a conference report on appropriations for fiscal year 1994 consolidated the two projects, and the Corps developed a Project Management Plan combining the two separate endeavors.[66]

As the headwaters revitalization program progressed, the Corps also began planning a test fill project for C-38 in order to determine the best way to plug the channel, including types of material to use and construction methods. It decided to conduct the test on a 1,000-foot stretch of Pool B, filling the canal with the dredged spoil taken from the area during the original construction of C-38. The test fill began in March 1994 and lasted until August.[67] In April, the Corps held an official groundbreaking ceremony near Lorida, located on U.S. Highway 98 between Okeechobee City and Sebring. State officials, federal authorities, and environmentalists all threw dirt into the canal, and newspapers heralded the ceremony as "the first actual step toward fixing South Florida's plumbing system."[68] Others tempered their enthusiasm. Richard Coleman, a member of the Sierra Club who had been pushing for restoration since the 1970s, characterized the groundbreaking as merely "another step in a long process," while Theresa Woody of the Sierra Club saw it as only "the first official bit of dirt to go back into the Kissimmee ditch."[69]

Even with the beginning of the test fill, much work remained. Since no one had ever restored a river on the scale of the Kissimmee plan, uncertainty existed as to how the ecosystem would react. However, several scientists in South Florida had been promoting the use of a strategy known as adaptive management as key to ecosystem restoration endeavors. Essentially "learning in the midst of doing," adaptive management was first used by the Northwest Power Planning Council in 1984 in its efforts to preserve salmon in the Pacific Northwest.[70] In the late 1980s and early 1990s, Florida scientists such as Buzz Holling, Carl Walters, and Lance Gunderson called for the use of the strategy in South Florida's ecosystem restoration endeavors. Holling and Walters, for example, organized

symposia and workshops with biologists and hydrologists to discuss how adaptive management could aid restoration efforts. The central concepts under adaptive management were that uncertainties were unavoidable in resource management. Instead of avoiding those uncertainties, adaptive management confronted them, using conceptual models to develop hypotheses and then testing them "in the real world." Based on the feedback from such experiments, strategies would be revised and altered. As one scholar explained it, the concept "aspires to create a new dialogue between humans and nature by treating policy as hypothesis and management as experiment, learning to live with and profit from the uncertainty and variability inherent in interacting ecological, economic and institutional systems."[71]

Because of the use of Dr. Hsieh Wen Shen's model to determine the best plan for Kissimmee River restoration, the waterway seemed like an ideal place to observe how adaptive management could aid in ecosystem restoration. Yet the strategy was not without its critics. For one thing, many correctly observed that its "trial-and-error" method was both slow and costly. Continuous testing and re-testing was frustrating, in large part because of the inability to cite accurate time and funding figures and to gage how those targets were being met. For another, adaptive management's embrace of uncertainty was unsettling to some. Many policy managers and scientists either preferred to "assume most uncertainty away" or to "seek spurious

Birds in a restored section of the Kissimmee River. (South Florida Water Management District)

Kissimmee River South plug, Phase IV-A. (South Florida Water Management District)

certitude," meaning "to break the problem or issue into trivial questions spawning answers and policy actions that are unambiguously 'correct,' but, in the end, are either irrelevant or pathologic."[72] Others merely did not want to admit that they did not have all of the answers. Just as importantly, adaptive management was essentially "experimentation that affects social arrangements and how people live their lives" since one of its purposes was to understand the impacts that human beings have on the natural world. Therefore, "those who have been users, owners, or governors of an ecosystem" often "resisted or sabotaged" the efforts.[73]

Regardless of the problems with adaptive management, many scientists and resource managers saw Kissimmee River restoration as an opportunity to integrate its concepts. In July 1991, for example, the SFWMD had established a scientific advisory panel of seven scientists to develop a comprehensive ecological evaluation plan. This commission recommended that any evaluation examine three specific things: whether the restored river and the floodplain met the necessary hydrologic objectives, whether specific biological and ecological characteristics were reestablished, and whether the SFWMD and the Corps had executed an adaptive management plan to track the first two objectives. It also suggested that the evaluation program consist of five stages: establishing reference conditions, ascertaining baseline conditions (or the current status of biological populations), assessing construction impacts, evaluating post-construction restoration, and "fine-tuning" restoration endeavors.[74]

The SFWMD took these suggestions to heart and formulated a restoration evaluation team in 1992. This team administered the evaluation program, which included the five stages recommended by the scientific advisory committee, as well as the extra step of developing conceptual models.[75] The SFWMD placed a high priority on evaluation; as Louis Toth observed, "Restoration evaluation will be the cornerstone of future environmental studies on the Kissimmee."[76] In the meantime, the Corps performed a construction evaluation of the test fill, publishing a report in 1995 on those efforts.[77]

Yet it took a few more years before construction actually began. For one thing, the Jacksonville District did not finalize its Section 1135 Project Modification Report on the Headwaters Revitalization Project until 1996 (after the District had completed an extreme drawdown of lake levels to clean up lake bottom and shorelines). Under that plan, the Corps proposed to modify regulation schedules, purchase 20,800 acres bordering the lake, and enlarge C-36 and C-37 in order to "restore the Kissimmee River and to expand the Upper Kissimmee Basin lake littoral zones."[78] For another, the Corps and the SFWMD monitored the results of the test fill for a two-year period. According to Jacksonville District Engineer Colonel Terry Rice, who replaced Salt as commander of the District in 1994, the agencies examined water quality improvement, fill stability, and vegetation reestablishment.[79] Finally, it took the SFWMD some time to acquire the necessary land along the river; by 1998, it had acquired most of the required tracts.[80] In 1999, all the pieces finally came together, resulting in the commencement of C-38 backfilling on 31 March. It did not take many years for positive results to appear; vegetation more characteristic of pre-channelized floodplain marshes soon returned, indicating that Kissimmee River restoration would succeed.[81]

According to Stuart Appelbaum, the Kissimmee River achievement was significant for the Corps in many ways. For one thing, it was one of the first times that the Corps had undone one of its projects solely for environmental reasons. This gave the Corps much more credibility in the eyes of environmental organizations, such as the Sierra Club, and it formulated better relations with environmental groups. Much of the credit for that goes to Colonels Bruce Malson, Rock Salt, and Terry Rice, who shepherded restoration plans through the Corps bureaucracy and were willing to make Kissimmee restoration a priority, not only because Congress had mandated it, but because they truly believed in the importance of ecosystem restoration.

Restoration efforts in the 1990s also created a successful partnership with the SFWMD that had been lacking in the 1980s, a time when the SFWMD, in its adjustment to its role as water manager rather than just flood control provider, occasionally clashed with the Corps. Because the state wanted Kissimmee restoration to succeed so badly, and because the Corps was the agency targeted by Congress for that effort, the two sides had to reach an agreement. Thus, both agencies exerted efforts to work together productively, including establishing a partnering charter with clear goals for each entity and conducting periodic partnering workshops.[82] But the Corps also worked with other local and federal groups, including environmental organizations such as the Everglades Coalition, in the Kissimmee project. The importance of outreach and partnerships was not lost on the Corps, which pledged to apply its experiences to the overall Everglades restoration.[83]

The Kissimmee River project also highlighted how important science was to ecosystem restoration. The work of Louis Toth and other scientists provided a blueprint for the SFWMD and the Corps to follow in its restoration efforts, and also created a process to evaluate how successful restoration would be. As Theresa Woody of the Sierra Club related, Toth was "our scientist, our guide. He believed. . . . He laid out the monitoring program that allowed us to build the science and lay down the baselines."[84] Likewise, the concept of adaptive management, which would take a prominent place in the

Kissimmee River oxbow near Phase IV-A. (South Florida Water Management District)

overall efforts to restore the Everglades, was introduced to many South Florida water planners through Kissimmee River restoration. Scientists regarded the restoration effort as a resounding success in the use of adaptive management in that it "achieved new ecological understanding and fundamental reorganization of large-scale water resource management approaches through iterative interaction of science and management, in a process that engaged stakeholders and generated social learning." Restoration efforts especially showed that in ecosystem restoration, resource managers needed to establish clear goals, expect surprises, be able to learn from mistakes, and keep communication lines open with all interested parties through public involvement.[85]

Clearly, efforts to restore the Kissimmee in the 1990s allowed the Corps, the SFWMD, and environmental groups to experiment with the best ways to conduct ecosystem management, and, if nothing else, it showed that if a clear definition of restoration was generated and then implemented, restoration efforts *would* work. Yet not all was rosy. For one thing, not everyone supported Kissimmee River restoration, including many residents living in the vicinity of the Kissimmee. This indicated that no matter what kind of consensus federal and state agencies reached on ecosystem restoration, some groups would be left on the sidelines. In addition, even though the Corps had effectively planned and begun Kissimmee restoration, some still wondered whether engineers were the best individuals to perform environmental restoration. Louis Toth, for example, claimed that many within the Corps still saw restoration as a "construction project" rather than an effort to reestablish an ecosystem. "These guys just don't get it," Toth declared in 2002. "I hate to say it, but [they] haven't learned anything about restoring an ecosystem." Toth's denunciations, although overstated, had a kernel of truth. To Toth, ecosystem restoration meant doing whatever was best for the environment, instead of "manipulating nature and managing different parts of the system for different things."[86]

In many ways, Kissimmee River restoration *was* just another human manipulation of nature—creating more storage in the headwaters so that the Corps could pattern water releases after pre-channelization tendencies, and forcing water back through old oxbows rather than through the C-38 channel. Was that true ecosystem restoration? If so, then the Corps was clearly the best agency to perform such work, given its experience on the Kissimmee River. Although a definitive answer to that question was elusive, it would continue to be debated throughout the 1990s in the context of Everglades restoration as a whole.

Endnotes

[1] See, for example, South Florida Water Management District, *Interim Surface Water Improvement and Management (SWIM) Plan for Lake Okeechobee, Part 1: Water Quality and Part VII: Public Information* (West Palm Beach, Fla.: South Florida Water Management District, 1989).

[2] Louis A. Toth and Nicholas G. Aumen, "Integration of Multiple Issues in Environmental Restoration and Resource Enhancement Projects in Southcentral Florida," in *Implementing Integrated Environmental Management*, John Cairns, Jr., Todd V. Crawford, and Hal Salwasser, eds. (Blacksburg, Va.: University Center for Environmental and Hazardous Material Studies, Virginia Polytechnic Institute and State University, 1994), 63.

[3] Louis A. Toth, "The Ecological Basis of the Kissimmee River Restoration Plan," *Florida Scientist* 56, no. 1 (1993): 29.

[4] Toth, "The Ecological Basis of the Kissimmee River Restoration Plan," 25-32; Joseph W. Koebel, Jr., "An Historical Perspective on the Kissimmee River Restoration Project," *Restoration Ecology* 3, no. 3 (1995): 149-152.

[5] Quotation in "Return of the River: Florida's Kissimmee May Meander Again," *The Wall Street Journal*, 8 September 1987; see also John Clark, Public Issue Specialist, to Kathy Cavanaugh, Public Issues Specialist, 14 April 1986, File PRO Kiss 380 Committee, Box 21213, SFWMDAR.

[6] Quotations in Louis A. Toth, *Environmental Responses to the Kissimmee River Demonstration Project*, Technical Publication 91-02 (West Palm Beach, Fla.: Environmental Sciences Division, Research and Evaluation Department, South Florida Water Management District, 1991), ii-iii; see also Toth, "The Ecological Basis of the Kissimmee River Restoration Plan," 39; "Kissimmee River Restoration/Closer Look," Fall 1988, File PRO Background Kissimmee River Restoration, Box 21213, SFWMDAR; Koebel, "An Historical Perspective on the Kissimmee River Restoration Project," 154-155.

[7] M. Kent Loftin, "Foreword," in *Kissimmee River Restoration Symposium: Proceedings*, M. Kent Loftin, Louis A. Toth, and Jayantha T. B. Obeysekera, eds. (West Palm Beach, Fla.: South Florida Water Management District, 1990), iii.

[8] Loftin, "Foreword," iii.

[9] Quotations in Toth and Aumen, "Integration of Multiple Issues in Environmental Restoration and Resource Enhancement Projects in Southcentral Florida," 64-65; see also Koebel, "An Historical Perspective on the Kissimmee River Restoration Project," 154-155.

[10] As quoted in Louis A. Toth, D. Albrey Arrington, and Glenn Begue, "Headwater Restoration and Reestablishment of Natural Flow Regimes: Kissimmee River of Florida," in *Watershed Restoration: Principles and Practices*, Jack E. Williams, Christopher A. Wood, and Michael P. Dombeck, eds. (Bethesda, Md.: American Fisheries Society, 1997), 432.

[11] Toth and Aumen, "Integration of Multiple Issues in Environmental Restoration and Resource Enhancement Projects in Southcentral Florida," 64-65.

[12] "Stakes High in Kissimmee Experiment," *The Miami Herald*, 4 August 1989; "Kissimmee River Restoration/Closer Look."

[13] As quoted in "Stakes High in Kissimmee Experiment," *The Miami Herald*, 4 August 1989; see also "Everglades: Render Back to Nature," *The Economist* 298 (9 December 1989): 50; "Kissimmee River Restoration/Closer Look."

[14] Major General C. E. Edgar III, Chairman, Board of Engineers for Rivers and Harbors, to Chief of Engineers, 12 March 1992, unlabeled black binder, Box 4386, JDAR.

[15] Quotations in Colonel Bruce A. Malson interview by Joseph E. Taylor, 15 August 1991, Jacksonville, Florida, 2, transcript in Library, Jacksonville District, U.S. Army Corps of Engineers, Jacksonville, Florida; see also Buker, *The Third E*, 88.

[16] Bob Martinez, Governor, to Honorable Robert W. Page, Assistant Secretary for Civil Works, Department of the Army, 28 March 1990, File PRO Kissimmee Restoration Plan Authorization, 1990, Box 21213, SFWMDAR.

[17] As quoted in "Corps of Engineers Focus on the Environment in the 1990's: A New Direction for a New Decade," U.S. Army Corps of Engineers, Jacksonville District, News Release, 18 January 1990, File PRO Kissimmee Restoration Plan Authorization, 1990, Box 21213, SFWMDAR.

[18] See Appelbaum interview, 8; "Kissimmee River Restoration/Section 1135 Project Briefing Paper," Binder Kissimmee River Section 1135 Study, Box 4383, JDAR; Lewis Hornung, Project Manager, Kissimmee River Restoration, "Presentation to the South Florida Water Management District," File PRO Kissimmee Restoration Plan Authorization, 1990, Box 21213, SFWMDAR.

[19] Water Resources Development Act of 1990 (104 Stat. 4604, 4624-4625).

[20] Appelbaum interview, 11.

[21] Appelbaum interview, 8.

[22] John Broaddus, CECW-ZE, to MG Kelly, 21 February 1991, Binder Kissimmee River Restoration General, Box 4384, JDAR.

[23] Appelbaum interview, 9-10.

[24] Appelbaum interview, 8-9.

[25] Quotation in Jimmy F. Bates, Chief, Policy and Planning Division, Directorate of Civil Works, Memorandum for Commander, South Atlantic Division, 8 March 1991, Binder Kissimmee River Restoration General, Box 4384, JDAR; see also Russell V. Reed, Study Manager, Memorandum for Record, 21 February 1991, ibid.; The Kissimmee River, Florida, Project," May 1999, 2, Kissimmee River Restoration Miscellaneous Facts & FDM-1 & FDM-2 Binder, Box 4382, JDAR.

[26] Appelbaum interview, 11, 13.

[27] "Kissimmee Restoration Flooded With Options," *The Palm Beach Post*, 26 June 1989.

[28] Colonel Terrence C. Salt interview by George E. Buker, 18 July 1994, Jacksonville, Florida, 22, transcript in Library, Jacksonville District, U.S. Army Corps of Engineers, Jacksonville, Florida [hereafter referred to as Salt interview—Buker).

[29] Quotation in Major General C. E. Edgar III, Chairman, Board of Engineers for Rivers and Harbors, 12 March 1992, in House, *Kissimmee River Restoration Study*, 102d Cong., 2d sess., 1992, H. Doc. 102-286; see also "Proposal for Project Modifications for Improvement of Environment Under Section 1135, Water Resources Development Act of 1986, Kissimmee River, Florida," December 1990, Binder Kissimmee River Restoration General, Box 4384, JDAR.

[30] "Summary of Corps Feasibility Report," 25 September 1991, 1-5, File DEIS, Box 1263, JDAR.

[31] "Summary of Corps Feasibility Report," 1-5.

[32] Quotations in "Summary of Corps Feasibility Report," 1-5; see also Salt interview—Buker, 22-23.

[33] Dave Johnson to Tom MacVicar, 4 October 1991, File PRO Kissimmee River Restoration, Box 21213, SFWMDAR.

[34] "Issues for Public Meeting," 30 September 1991, File DEIS, Box 1263, JDAR.

[35] "A New Kissimmee Curve," *The Palm Beach Post*, 10 September 1991.

[36] "Issues for Public Meeting," 30 September 1991.

[37] "Procedure for Public Meeting on Kissimmee River Restoration Feasibility Study," 8, File DEIS, Box 1263, JDAR.

[38] U.S. Army Corps of Engineers, Jacksonville District, *Central and Southern Florida Project, Final Integrated Feasibility Report and Environmental Impact Statement: Environmental Restoration, Kissimmee River, Florida* (Jacksonville, Fla.: U.S. Army Corps of Engineers, 1991), n.p. (quotation in syllabus section).

[39] Salt interview—Catton, 4; see also "Wildlife Agency Enters Kissimmee River Fray," *The Palm Beach Post*, 10 October 1991; Kent Loftin, "Restoring the Kissimmee River," *Geotimes* 35 (December 1991): 16.

[40] Lawton Chiles to Colonel Terrence Salt, Chief Engineer, Jacksonville District Corps of Engineers, 18 November 1991, File PRO Kissimmee River Restoration, Box 21213, SFWMDAR. SFWMD Executive Director Tilford Creel expressed these same views. See Creel to Salt, 19 November 1991, ibid.

[41] Appelbaum interview, 9.

[42] Salt interview—Catton, 4.

[43] The organization was originally known as Residents Opposing Alleged Restoration, but it changed its name in August 1992 because the original name indicated that it was only "a small group of homeowners who lived on the river and were just trying to save their homes." " 'Realists' Oppose Alleged River Restoration," *Okeechobee News*, 30 August 1992.

[44] See Kissimmee River Restoration Update, South Florida Water Management District, 22 October 1992, File PRO Kissimmee River Restoration, Box 21213, SFWMDAR; attachment to Tilford C. Creel, Executive Director, to Ms. Susan L. Marr, President, ROAR, 13 July 1992, ibid.; Louis A. Toth and Nicholas G. Aumen, "Integration of Multiple Issues in Environmental Restoration and Resource Enhancement Projects in Southcentral Florida," 68; Buker, *The Third E*, 88-89.

[45] See Kissimmee River Restoration Update, South Florida Water Management District, 22 October 1992; Lieutenant General H. J. Hatch, Chief of Engineers, to The Secretary of the Army, 17 March 1992, in House, *Kissimmee River Restoration Study*; Lawton Chiles to Ms. Nancy Dorn, Assistant Secretary of the Army for Civil Works, 11 March 1992, in ibid.; "U.S. Agrees to Pay Half of River Restoration," *Sun-Sentinel*, 6 March 1992; "Critics Vow To Keep Fighting Kissimmee River Restoration," *The Tampa Tribune*, 12 October 1992; "SFWMD Says Bush Administration Decision is Great News for Kissimmee Restoration," South Florida Water Management District Press Release, 5 March 1992, File PRO Kissimmee River Restoration, Box 21213, SFWMDAR.

[46] Salt interview—Catton, 4.

[47] Major General C. E. Edgar III, Chairman, Board of Engineers for Rivers and Harbors, 12 March 1992, in House, *Kissimmee River Restoration Study*.

[48] Nancy P. Dorn, Assistant Secretary of the Army (Civil Works), to Honorable Thomas S. Foley, Speaker of the House of Representatives, 3 April 1992, in House, *Kissimmee River Restoration Study*; Hatch to The Secretary of the Army, 17 March 1992.

[49] "Kissimmee River Restoration," attachment to Paul Genho, President, Florida Cattlemen's Association, to John Studt, Chief of Regulatory Branch, 25 June 1992, File 1110-2-1150a 1135 Kissimmee River Correspondence 1992, Box 1263, JDAR.

[50] Michael Grunwald, "An Environmental Reversal of Fortune: The Kissimmee's Revival Could Provide Lessons for Restoring the Everglades," *The Washington Post*, 26 June 2002; "ROAR Questions River 'Restoration' Plan," *Okeechobee News*, 5 April 1992; "County Resolution Protests Kissimmee River 'Restoration' Plans," *Okeechobee News*, 15 January 1992; "Critics Vow To Keep Fighting Kissimmee River Restoration," *The Tampa Tribune*, 12 October 1992.

[51] John B. Coffey, "Rerouting of Kissimmee River A Waste of Taxpayers' Money," *The Palm Beach Post*, 11 November 1991.

[52] Quotation in W. H. Morse, Chairman, Waterways Task Force, Kissimmee/Osceola County Chamber of Commerce, to Honorable Daniel Patrick Moynihan, 10 July 1990, File PRO Kissimmee Restoration Plan Authorization, 1990, Box 21213, SFWMDAR; see also Morse to The Honorable Henry J. Nowak, U.S. Representative, 1 October 1990, ibid.

[53] "Resolution by the Wilson Ornithological Society," attachment to Richard C. Banks, President, to Colonel Terrance Salt, District Engineer, 28 May 1992, File 1110-2-1150a 1135 Kissimmee River Correspondence to 1992, Box 1263, JDAR.

[54] Theresa Woody, Sierra Club, to The Honorable Robert A. Roe, 18 September 1992, File PRO Kissimmee River Restoration, Box 21213, SFWMDAR.

[55] Woody, "Grassroots in Action," 204-205.

[56] Water Resources Development Act of 1992 (106 Stat. 4797, 4802).

[57] Appelbaum interview, 9.

[58] Appelbaum interview, 11-12.

[59] Louis A. Toth, et al., "Flow Regulation and Restoration of Florida's Kissimmee River," *Regulated Rivers: Research & Management* 8 (1993): 155, 164.

[60] A. J. Salem, Chief, Planning Division, to Addresses on Attached List, 14 August 1992, File Kissimmee Headwaters Revitalization Section 1135, Box 1263, JDAR.

[61] Kissimmee River Headwaters Revitalization, Section 1135 Study, Team Meeting, 14 September 1993, Binder Kissimmee River Restoration Miscellaneous Facts & FDM-1 & FDM-2, Box 4382, JDAR.

[62] "Central and Southern Florida Project, Kissimmee River, Headwaters Revitalization Project: Draft Integrated Project Modification Report and Supplement to Final Environmental Impact Statement (EIS)," n.d., File Kissimmee Headwaters Revitalization Section 1135, Box 1263, JDAR.

[63] Michael A. Smith, CESAJ-PD-PF, Memorandum for Record, 30 November 1993, Binder Ukiss #3, Box 4386, JDAR; "Kissimmee 1135 Hydrologic Model Review, Scope of Work," Binder Joint Probability Coincident Frequency, Box 4382, JDAR; "Kissimmee 1135," 21 March 1994, ibid.

[64] Smith, Memorandum for Record, 30 November 1993.

[65] "State Plans Fight to Restore Money for Kissimmee River," *The Palm Beach Post*, 9 April 1993.

[66] "The Kissimmee River, Florida, Project," May 1999, 7, Kissimmee River Restoration Miscellaneous Facts & FDM-1 & FDM-2 Binder, Box 4382, JDAR.

[67] Koebel, "An Historical Perspective on the Kissimmee River Restoration Project," 157.

[68] "Fixing Nature's Filter," *The Palm Beach Post*, 2 May 1994.

[69] First quotation in "Restoring A River," *The Miami Herald*, 1 May 1994; second quotation in "Ground Broken, Returned to River," *The Orlando Sentinel*, 24 April 1994.

[70] National Research Council of the National Academies, *Science and the Greater Everglades Ecosystem Restoration* (Washington, D.C.: The National Academies Press, 2003), 63.

[71] Quotation in Steven Light, "Key Principles for Adaptive Management in the Context of Everglades History and Restoration," paper presented at Everglades Restoration Adaptive Management Strategy Workshop, October 2003, copy provided by James Vearil, Senior Project Manager, RECOVER Branch, Programs and Project Management Division, U.S. Army Corps of Engineers, Jacksonville District, Jacksonville, Florida; see also Lance Gunderson, "Resilience, Flexibility and Adaptive Management—Antidotes for Spurious Certitude?" *Conservation Ecology* 3, no. 1 (1999): 7, copy available at <http://www.consecol.org/vol3/iss1/art7> (28 April 2006); James Vearil, personal communication with authors, 19 April 2006.

[72] Gunderson, "Resilience, Flexibility and Adaptive Management."

[73] Kai N. Lee, "Appraising Adaptive Management," *Conservation Ecology* 3, no. 2 (1999): 3, copy at <http://www.consecol.org/vol3/iss2/art3> (28 April 2006).

[74] Koebel, "An Historical Perspective on the Kissimmee River Restoration Project," 157-158; see also "Restoration Evaluation Program," unlabeled black binder, Box 4386, JDAR.

[75] Tilford C. Creel, Executive Director, South Florida Water Management District, to Mr. David Ferrell, Field Supervisor, U.S. Fish and Wildlife Service, 25 November 1992, File 1110-2-1150a 1135 Kissimmee River Correspondence 1992, Box 1263, JDAR; "Restoration Evaluation Program."

[76] Toth, "The Ecological Basis of the Kissimmee River Restoration Plan," 47.

[77] For an example of the Corps' findings, see U.S. Army Corps of Engineers, Jacksonville District, *Central and Southern Florida Project, Kissimmee River Restoration Test Fill Project: Construction Evaluation Report, Preliminary Draft* (Jacksonville, Fla.: U.S. Army Corps of Engineers, Jacksonville District, 1995).

[78] Quotation in U.S. Army Corps of Engineers, Jacksonville District, *Kissimmee River, Florida, Headwaters Revitalization Project: Integrated Project Modification Report and Supplement to the Final Environmental Impact Statement* (Jacksonville, Fla.: U.S. Army Corps of Engineers, Jacksonville District, 1996), n.p. (included in syllabus); see also Buker, *The Third E*, 90-91; "The Kissimmee River, Florida, Project," May 1999, Kissimmee River Restoration Miscellaneous Facts & FDM-1 & FDM-2 Binder, Box 4382, JDAR.

[79] Rice interview, 27.

[80] South Florida Water Management District, *Save Our Rivers: 1998 Land Acquisition and Management Plan* (West Palm Beach, Fla.: South Florida Water Management District, 1998), 46-50.

[81] Grunwald, "An Environmental Reversal of Fortune."

[82] "Final Report for the Partnering Workshop, Kissimmee River Restoration Project, August 27-29, 1997," File Kissimmee Headwaters Revitalization Section 1135, Box 1263, JDAR.

[83] Kimberly Brooks-Hall interview by Theodore Catton, 7 July 2004, Jacksonville, Florida, 1.

[84] Theresa Woody interview by Theodore Catton, 18 January 2005, Naples, Florida, 7.

[85] "RECOVER Workshop Case Study: Kissimmee River Restoration," paper written by Steven Light and presented by Kent Loftin at the CERP Adaptive Management Workshop, June 2003, copy provided by James Vearil, Senior Project Manager, RECOVER Branch, Programs and Project Management Division, U.S. Army Corps of Engineers, Jacksonville District, Jacksonville, Florida.

[86] Both quotations cited in Grunwald, "An Environmental Reversal of Fortune."

CHAPTER 17   Conflicts and Difficulties: Water Distribution in the 1990s

As Kissimmee River restoration proceeded in the mid-1990s, no one doubted that a drastic overhaul of South Florida's water distribution system was necessary. The tricky part was to craft a plan that all sides—whether environmental, agricultural, or urban—could accept. The South Florida Ecosystem Restoration Task Force and the Governor's Commission for a Sustainable South Florida were attempting to build consensus for restoration, but water management in South Florida had a long history of contention that could not be put aside. Three plans that involved water distribution issues in the 1990s highlighted the difficulty in attaining consensus. The first dealt with Lake Okeechobee's regulation schedule, which influenced how much water interests downstream from the lake received, as well as the ecological quality of the surrounding area. The second consisted of an experimental Corps program (implemented as part of the Corps' Modified Water Deliveries Project) that analyzed how best to deliver water to Everglades National Park. This project became caught up in controversy over the flooding of agricultural areas and possible disruptions of the habitat of the Cape Sable seaside sparrow, which was first listed as an endangered species in 1967. The third, which involved the Modified Water Deliveries Project itself, emphasized land acquisition difficulties and their ultimate effects on interests such as the Miccosukee Indians. All of these examples saw hard-line stances exhibited by competing interests, indicating that the path to restoration would be rocky and difficult.

Lake Okeechobee Regulation

For many years, the major problem with Lake Okeechobee had been its phosphorus levels. Since the 1970s, federal, state, and local agencies had labored to diminish the mineral's concentration. The SWIM

plan developed for the lake in the late 1980s, for example, required an average annual phosphorus reduction of 40 percent. To achieve that goal, the SFWMD stipulated that no water could flow into the lake without a phosphorus standard of 0.18 milligrams per liter. The implementation of BMPs, as stipulated by the Everglades Forever Act, lowered levels as well, as did a dairy cow relocation effort. In 1996, the SFWMD reported that average annual amounts of phosphorus were half of what they were before the implementation of the SWIM plan, but further reductions were still necessary.[1]

Another issue with Lake Okeechobee was its regulation schedule. In the 1978, the Corps had modified its operational plan to allow lake levels between 15.5 and 17.5 feet (upwards from the existing 13 to 15 feet schedule). This provided more water for agricultural, urban, and environmental needs in times of drought, but it also meant that the Corps had to send larger slugs of water to the Caloosahatchee and St. Lucie estuaries during the rainy season in order to prevent flooding, even though residents of Martin County and other interested parties had protested such releases as far back as the 1950s. The releases to the areas, critics charged, disrupted salinity and deposited large amounts of sediment, killing fish and other life. Likewise, the higher stages caused problems in the lake's littoral zone because they pushed phosphorus concentrations to the shore and flooded areas used as habitat and feeding grounds for surrounding flora and fauna.[2]

In 1988, the Lake Okeechobee Littoral Zone Technical Group, composed of representatives from the SFWMD, the FWS, the Florida Game and Freshwater Fish Commission, Everglades National Park, the Florida Department of Natural Resources, the Florida Department of Environmental Regulation, and several universities

The Caloosahatchee estuary. (South Florida Water Management District)

issued a report on the lake's regulation schedule, claiming that it had "induced changes in the littoral zone" (meaning the water body's shore). These alterations included "loss of wading bird feeding habitat, decline in willow, and loss of moist-soil annual plant production." Because it considered the diminishment of the wading bird habitat critical, the group requested that an immediate lowering of the schedule occur "to improve fish and wildlife habitat." It recommended that the Corps implement a new plan fluctuating "between a high of 16 feet and a low of 14 feet," levels that would still allow the Corps to meet "the other demands placed on the lake."[3] In making this suggestion, the group was repeating only what other fish and wildlife experts had already proclaimed to be necessary. The Florida Game and Fresh Water Fish Commission, for example, had informed the SFWMD in April 1988 that "the existing schedule has reduced the probability of maintaining a productive littoral zone into the future."[4] Meanwhile, the SFWMD, in the midst of a three-year modeling study of lake levels, requested that the Corps use

a schedule, known as "Run 25," which would hold levels between 15.65 and 16.75 feet. According to the SFWMD, those levels would provide the most "benefit to the estuarine environment" while "having no negative impacts on the lake's water supply or littoral zone."[5] The Corps, however, made no changes to the schedule, in part because agricultural and urban interests demanded that higher levels in Lake Okeechobee were necessary.[6]

After environmental groups such as the Florida Lake Management Society protested the Corps' inaction, the SFWMD renewed its request in 1991, reiterating that Run 25 would "reduce damaging flows to the nearby St. Lucie Canal and Caloosahatchee River estuaries without sacrificing the flood control or water supply benefits derived from the lake."[7] The FWS, however, clamored for even lower levels, claiming that Run 25 would actually have little effect on the littoral zone. A group of 30 "environmental professionals" who examined the "overall ecological conditions of Lake Okeechobee" agreed with the FWS, insisting that lower levels would provide for "the sustained ecological health of Lake Okeechobee."[8] Accordingly, the SFMWD revisited Run 25 and developed a new schedule, known as Run 22, which would allow levels between 13.5 and 15.5 feet. The FWS applauded this proposal, as well as what it termed "the four step pulse release concept," whereby water would be discharged from the lake in a way that would mitigate damage to the St. Lucie and Caloosahatchee estuaries.[9] Other groups, such as the Okeechobee Waterway Association, the Glades County Board of County Commissioners, and the City of Belle Glade, disagreed with the Run 22 plan, stating that it would disrupt fishing, cause even more pollution in Lake Okeechobee, and jeopardize water supplies.[10]

In response to these Run 22 criticisms, scientists and environmentalists pointed to numerous studies indicating that high water levels in Lake Okeechobee caused increased phosphorus concentrations by "facilitat[ing] the movement of phosphorus laden water from the turbid center of the lake to the edge of the littoral zone."[11] Likewise, the studies showed that wading birds needed levels below 15 feet in the spring to feed and nest successfully, while marshlands also required lower levels for optimum health. Despite this evidence, and despite the SFWMD's suggestion that Run 22 be adopted, the Corps, hoping to achieve more of a middle ground between water supply and ecological protection, made a preliminary recommendation in 1994 to continue with Run 25 and to adopt it as its regular regulation schedule until the completion of the C&SF Restudy (which would analyze more thoroughly what levels were appropriate). Run 25, Corps officials claimed, would secure a sufficient water supply for downstream interests, including Everglades National Park and Florida Bay, while also causing some improvement in the littoral zone.[12] The Corps did not rule out the possibility of implementing Run 22 at some point, but explained that it would have to conduct an economic and environmental impact study before it could take such action.[13]

Predictably, environmentalists and fish and wildlife interests denounced the maintenance of Run 25. "Lake management decisions have placed other objectives above the lake's ecological health," the Florida Department of Environmental Protection declared, petitioning the Corps to stop "the sacrifice of the environment for other goals."[14] When the Sugar Cane Growers Cooperative of Florida supported the Corps' recommendation, saying that the Corps should not operate Lake Okeechobee "solely for in-lake environmental objectives," it seemed to prove that environmental surrenders were occurring.[15] Regardless of the criticism, the Corps did not budge on the issue, leading environmental interests to cry that urban and agricultural concerns had won the battle. Given the "environmental concerns" that the Corps' Restudy "was attempting to address," environmentalists were "surprised" that the Corps had seemingly paid no heed to their concerns, leaving them wary of what kind of a plan the Restudy would actually produce.[16]

Modified Water Deliveries and C-111 Projects

Along with the divisions evident in the debate over Lake Okeechobee regulation schedules, the Corps' Modified Water Deliveries and C-111 projects provoked controversy in the 1990s. In 1989, the Ev-

erglades National Park Protection and Expansion Act had authorized the Modified Water Deliveries Project. The law stated that in coordination with the Interior Department, which would purchase 107,600 acres of land, the Corps would modify structures of the C&SF Project in order to restore more natural water flows to Shark River Slough, as well as to provide flood mitigation for residents in the 8.5 Square Mile Area. The act required the Corps to complete a General Design Memorandum for the project, and in May 1993, the Secretary of the Army (Civil Works) approved the design memorandum, allowing for engineering design to begin. At the same time, the Corps worked on a General Reevaluation Report of the C-111 project, authorized by both the Flood Control Act of 1968 and by the Everglades National Park Expansion and Protection Act of 1989 to provide more water to Taylor Slough. Finalized in May 1994, the General Reevaluation Report proposed the construction of a water retention area on the eastern edge of Everglades National Park, as well as the development of a transition area to divide the park from agricultural lands. To accomplish these purposes, the reevaluation report recommended the purchase of the Frog Pond area.[17]

The Interior Department and the NPS agreed that acquisition of Frog Pond was essential, claiming that it would restore freshwater

The St. Lucie estuary and Indian River Lagoon. (South Florida Water Management District)

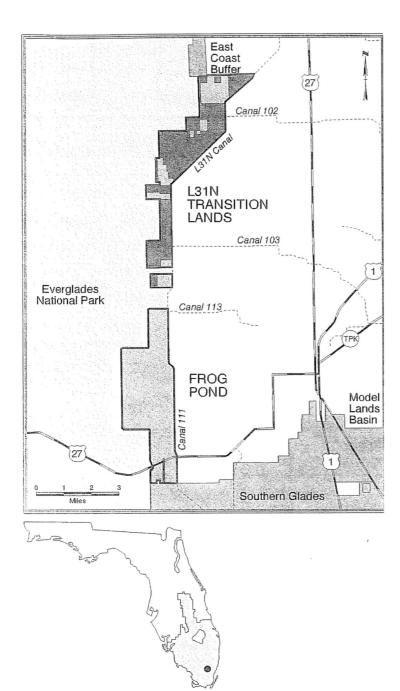

The Frog Pond area. (South Florida Water Management District, *Save Our Rivers: 1998 Land Acquisition and Management Plan* (1998)

flows to Florida Bay. As discussed earlier, the Frog Pond area had long been the subject of controversy, mainly because its agricultural production complicated an experimental delivery program that the Corps had undertaken in the 1980s to restore more natural water flows to Shark River and Taylor sloughs.[18] When the Corps had begun the experimental program in 1985, the Dade County Farm Bureau had sued because the plan "rerout[ed] . . . excess waters from federally-owned lands to privately-owned agricultural lands," causing flooding problems for farmers in the Frog Pond area (*Kendall, et*

al. v. Marsh, et al.).[19] In order to halt the litigation and to continue the experimental program, the Corps reached a settlement with the bureau and other agriculturists, whereby the farmers would drop their suit if the Corps reduced water levels in the L-31W canal (running north to south on the border of East Everglades), allowing for more drainage during wet periods. This arrangement lasted until the 1990s, when, as part of another iteration of the experimental program, Frog Pond lands were flooded again. In response, the South Dade Land Corporation sued the Corps in 1993, hoping to obtain an injunction against further flooding (*South Dade Land Corporation v. Sullivan*). The U.S. District Court for the Southern District of Florida, Miami Division, denied this injunction, stating that the Corps had adequately sought to mitigate any flooding, but the threat of litigation remained.[20]

Meanwhile, according to the Natural Resources Defense Council (NRDC), the drainage of Frog Pond lands had resulted in detrimental effects on both the Cape Sable seaside sparrow and on wetlands bordering the eastern portion of Everglades National Park. The council claimed that the experimental plan had drained the Rocky Glades (a wetlands area southeast of Shark River Slough), Taylor Slough, and the C-111 basin to such an extent that wet-season water levels were as much as two feet lower than before the experiment's beginning. This caused an invasion of exotic and woody plants that destroyed the sparrow's habitat and caused declines in fish populations. The experimental program was designed to reestablish more natural water conditions to Shark River and Taylor Slough, but instead, the NRDC claimed, it was causing unnatural circumstances that violated the Endangered Species Act.[21]

The NRDC was not alone in this belief; FWS and Everglades National Park officials also claimed that the experimental program adversely affected the Cape Sable seaside sparrow. As mentioned above, the experimental program had gone through several different iterations before 1996, the first five of which consisted of delivering water from Conservation Area No. 3 to the western and northeastern portions of Shark River Slough. When the Corps conducted a sixth iteration, which provided water to both Shark River Slough and Taylor Slough, FWS and Everglades National Park officials objected, citing the effects of the program on the sparrow and requiring that certain restrictions be implemented. In its preparation for the seventh iteration in 1996, the Corps asked that the restraints be removed, claiming that this phase, which proposed to establish a more natural hydroperiod in Northeast Shark River Slough and Taylor Slough, would not harm the sparrow. The FWS responded with a biological opinion stating that, although the program would improve long-term environmental conditions for the sparrow, it would have adverse effects in the short term because the sparrow needed a shorter hydroperiod to nest. To mitigate these effects, the FWS requested frequent consultations with the Corps, the SFWMD, and the park, and it also asked for the establishment of "a comprehensive monitoring and research program." Moreover, the FWS suggested that the Corps use the Modified Water Deliveries Project and the C&SF Restudy "to identify ways to redistribute regulatory releases from Water Conservation Area 3A more naturally over as large a geographic area as possible."[22]

Meanwhile, the NRDC and others wanted levels in the L-31W canal (as well as the L-31N and C-111) returned to "a minimum of initial design optimum settings," but the Corps refused, citing the need to look out for the interests of Frog Pond farmers. The NRDC argued that "concerns about flooding in the Frog Pond are increasingly irrelevant and unjustified," a view that seemed vindicated when the SFWMD purchased the Frog Pond area in 1995 for $12.5 million.[23] But the SFWMD allowed the corporation to lease the Frog Pond area for agricultural purposes, meaning that the Corps still faced the problem of coordinating the experimental program with the needs of both agriculture and the sparrow.

In February 1999, the FWS reiterated that the experimental program adversely affected the sparrow, and the Corps, with the approval of the Council on Environmental Quality, revised its experimental plan to retain more water in the conservation areas. It also pledged to reroute the water so that the point of discharge would occur south of where the sparrow lived. These proposals, however, led to two lawsuits filed by the NRDC and the Miccosukee Tribe of Indians of Florida. The NRDC action alleged violations of the Endangered Species Act and requested that the Corps follow a different water delivery plan, one that would carry water through the Northeast Shark River Slough and the 8.5 Square Mile Area to the potential detriment of agricultural and residential interests in western Dade County. The Miccosukee's lawsuit, meanwhile, attempted, in the words of one Jacksonville District legal advisor, "to preserve the status quo" and to prevent the impoundment of more water in the conservation areas.[24] The court eventually ruled in favor of the Corps in both actions, stating that the Jacksonville District was performing sufficient mitigation, but the dismissal did little to assuage the fears of the Miccosukee and some environmentalists that restoring water flows to Everglades National Park would harm other interests.[25]

The Frog Pond negotiations and the Cape Sable seaside sparrow situation created discontent exacerbated by battles in the 1990s over the 8.5 Square Mile Area. Under the Modified Water Deliveries Project, as authorized by the 1989 Everglades National Park Protection and Expansion Act, the Corps was supposed to construct flood mitigating works around the 8.5 Area to protect it from any flooding caused by water releases to Everglades National Park. Yet some, including NPS authorities, claimed that the easiest way to restore natural flows to Shark River Slough would be to purchase the entire 8.5 Square Mile Area and allow water to run across the area unhindered by any mitigation.[26] In recognition of these views, Congress amended the Everglades National Park Protection and Expansion Act in March 1994, specifically authorizing the secretary of the interior to use funds appropriated for the Modified Water Deliveries Project to acquire the 8.5 section, as long as the secretary did not contribute more than 25 percent of the total cost.[27]

But, as explained in an earlier chapter, property owners in the area had vehemently objected to any kind of relocation in the 1980s, and they did not retreat from that position in the 1990s. Instead, a 1992 Corps study declaring that acquisition was not necessary as long as flood mitigation structures were built fortified their stance. The Interior Department, however, disagreed with the Corps' conclusion and refused to release funding to construct the flood works (as the Everglades National Park Protection and Expansion Act had mandated), delaying the implementation of the Modified Waters Project. Not only did this mean that necessary water was not restored to the Everglades, it also caused more water to collect in Conservation Area No. 3A, leading to the drowning of tree islands and wildlife, much to the chagrin of the Miccosukee Indians. Miccosukee attorney Dexter Lehtinen thus blasted the Interior Department for its uncooperative attitude, accusing it of "scandalous" behavior, such as "selfish National Park Service obstructionism and outright misrepresentation" and the destruction of property rights.[28] The Interior Department denied that it had deliberately withheld funds, claiming instead that Hurricane Andrew (which hit Florida in August 1992) and its aftermath had "brought . . . things to a halt." Besides, Interior officials insisted, the Corps' plan would not give full flood protection to 8.5 Square Mile Area residents, especially if residential expansion in the area continued.[29] Regardless of who was right, the situation created a standstill, pitting the Interior Department against the Miccosukee, the Corps, and landowners.

In 1994, the state of Florida became involved in the controversy. That year, Governor Lawton Chiles issued an executive order requesting that the Corps hold any flood mitigation efforts for the 8.5 section "in abeyance" until more studies could be made.[30] To make those examinations, Chiles established a committee of Corps, SFWMD, Everglades National Park, and landowner representatives. When it issued its report in 1995, the committee took a middle-ground approach, stating that the best solution was to create a flow way buffer between the 8.5 Square Mile Area and the park. However, Jacksonville District Engineer Colonel Terry Rice dismissed the recommendation because the committee provided no information as to how to deal with the "increased budgets, more environmental problems, and congressional approvals" that would result.[31]

Because of the Corps' rejection, the SFWMD decided in 1996 to review the situation on its own, hoping to find a locally preferred option that it could implement. More studies just made the Miccosukee anxious, especially since flooding continued to be a problem on their lands, and the tribe appealed for speed in the SFWMD's review. Regardless, the district studied the issue for two years before finally deciding in November 1998 to support full acquisition of the 8.5 Square Mile Area. In the eyes of the Miccosukee, this was

C-111 project. (Everglades National Park)

the worst possible solution because it would lead only to protracted negotiations with landowners that would further delay implementation of the Modified Waters Project, meaning that they would obtain no relief. Some environmentalists, including Joette Lorion, president of Friends of the Everglades, supported the Miccosukee's stance, requesting that full acquisition be eliminated as an option. Likewise, landowners, such as Ibel Aguilera of the United Property Owners & Friends of the 8.5 Square Mile Area, Inc., contended that the SFWMD was employing coercive tactics to try to force property owners to sell their land, including sending letters that discussed acquisition as a foregone conclusion. Aguilera, spurred on by Representative John J. Duncan, Jr. (R-Tennessee), implied that such methods were more characteristic of the Cuban communist regime that she and many others in the 8.5 section had fled, rather than the United States.[32]

Interior Department officials, however, supported the SFWMD's decision, claiming that acquisition was necessary to restore "short

hydroperiod wetlands" that could support "habitat for populations of wading birds" on the eastern edge of the Everglades.[33] Moreover, environmentalists such as Mary Barley of the "penny-a-pound" campaign called for acquisition, believing that it would allow the establishment of a recharge area between Everglades National Park and the urban development of Dade County, thereby allowing more fresh water to flow to Florida Bay and the Florida Keys.[34] The different perspectives produced a stalemate, and by 1999, no progress had been made on any front, although the Corps had successfully completed two new control structures (S-355A and S-355B) to divert more water to Northeast Shark River Slough.[35]

In a similar way, differences between the Corps, the NPS, and other groups resulted in delays on the C-111 project, developed to restore "more natural quantity, quality, timing, and distribution of water deliveries to Taylor Slough and wetlands in the panhandle of Everglades National Park."[36] To accomplish this, the Corps and the SFWMD constructed S-332D, a new pump station along C-111 (which ran east to west from the park's eastern border), which would allow the Corps to increase water levels in the canal, thereby providing more water to the slough. But farmers located north of the pump objected to the higher canal levels, claiming that more water would damage their crops. When the NPS tried to purchase their lands, the agriculturists refused to sell, creating yet another standoff. By 1999, each side had hardened their stance, meaning that nothing more could be done on the C-111 work.[37]

According to many observers, one of the major reasons for the lack of progress in the Modified Water Deliveries and C-111 projects was the unwillingness of Everglades National Park officials to budge once they had declared their position. Critics claimed that NPS authorities saw the park's interests as supreme, and did not believe that any concessions could or should be made. This intransigence stemmed both from an overall insular mentality within the NPS, as well as from the attitude of specific leaders at Everglades National Park. For example, because officials believed that NPS personnel were the best authorities on park management, the NPS regarded each park, according to environmental scholar Richard West Sellars, as "a superintendent's realm," to be governed with as little outside interference as possible. The culture of the NPS, Sellars continued, produced "a strongly utilitarian and pragmatic managerial bent" that created "a management style" emphasizing "expediency and quick solutions" and disregarding "information gathering through long-term research."[38] Colonel Terry Rice put it more strongly. "The NPS culture is a selfish, self-centered culture," he declared. "They trust nobody."[39]

Many believed that Richard Ring, who was superintendent of Everglades National Park for most of the 1990s (and who also served for a period as chairman of the Task Force), epitomized these stubborn, self-centered characteristics. As superintendent of the park, critics charged, he clung to his beliefs about what was right or wrong and would not move from his position regardless of the situation. Although Ring held what he perceived to be the best interests of the park at heart, his uncompromising attitude irritated those who had to work with him. "I didn't know anybody that didn't like Dick Ring," Colonel Terrence "Rock" Salt, District Engineer of the Jacksonville District, explained, but "he was also the most frustrating guy to work with that you could find." The problem, in Salt's mind, was that Ring "had a tactical view of things and he would fight so hard for it that he would expend all the strategic capital he had built up."[40] Michael Collins, a member of the SFWMD's governing board, agreed. "Basically [Ring] believed that if you obstructed everything that came down the road you protected the park."[41]

But Ring was not alone in his actions. Michael Finley, superintendent of Everglades National Park in the mid-1980s, had resorted to litigation and polarized stances as well because he believed that it was the only way to get water management agencies to take the park seriously.[42] Environmentalist Joseph Browder of the Audubon Society applauded Ring and Finley, claiming that the superinten-

dents were only doing what was necessary to force the Corps to operate the C&SF Project to the benefit of fish and wildlife interests. In Browder's mind, the original project plan had required the Corps to provide flood control and water supply "in a way that does not damage the national interest lands protected by the NPS." Since the Corps disregarded this responsibility, Browder continued, Ring, Finley, and the entire environmental community had to expend efforts "to enforce the terms of the original agreement." "It's really that simple," he concluded.[43]

The Corps, however, claimed that the park's uncooperative attitude did nothing to help the cause of ecosystem protection; instead, it merely created an atmosphere of paranoia and polarization that frustrated any true restoration efforts. As proof, Corps personnel pointed to notes taken at park meetings entitled SWOT workshops, where the park's "strengths, weaknesses, opportunities and threats" (SWOT) were analyzed. These notes showed that park officials had listed "powerful organizations/interests with goals at odds with ours"—specifically the Corps, the SFWMD, the Miccosukee, developers, and agricultural interests—as threats.[44] As Rice, who by now was working as a consultant to the Miccosukee Indians, noted, SFWMD and Corps employees—"dedicated to Everglades restoration"—were "frustrated and beat down" by such statements and by the NPS's "culture of paranoia."[45]

Internal Corps emails corroborated Rice's statements. Richard Punnett, a modeler in the Jacksonville District's Hydrology and Hydraulics Branch, discussed the "real feeling of depression" that he experienced when considering the "delays, confusion, demands, red herrings, bad science, changing of positions" that the NPS used to "undercut" the restoration effort. Punnett claimed that Everglades National Park officials misrepresented scientific data and made "false claims" about restoration plans. Punnett and others were "doing [their] best to work through the gauntlet" of delays and obstructionist tactics employed by the NPS, but believed that the Service's "'red herrings' will never be an endangered species."[46]

Michael L. Choate, a hydrologist with the Jacksonville District, made even stronger statements, accusing Everglades National Park officials of conducting a "jihad" against "the infidels" (meaning the Corps, the SFWMD, and any other interest that disagreed with park stances). "We have wasted 8 years on [Modified Water Deliveries] and 6 years on C-111 arguing over L-28 (flood the Indians), 8.5 SMA (flood the homesteaders) and L-31N canal stages (flood the farmers)," Choate charged. Park officials had no problem with such delays, Choate declared, because "they have not lost until something is built." This attitude, Choate continued, meant that the NPS would obstruct any projects resulting from a comprehensive restoration plan, "resulting in time and cost escalations."[47] In the words of Hanley "Bo" Smith, chief of the Environment and Resources Branch, "it's hard to imagine worse interagency relationships than those" between the Corps, the FWS, and the NPS.[48]

Others within the Corps tried to be more philosophical about relationships with the NPS. James Vearil, a project manager for the Corps' Restoration Coordination and Verification team (RECOVER), for example, believed that the major reasons for the problems were "differences in our agencies missions, organizational structures, legal authorities, philosophy, customers, styles, rules and regulations, and organizational culture." Citing several publications on conflicts within water management (such as publications by William Lord and Peter Loucks), Vearil related that the conflicts between the Corps and the NPS lay in three arenas—cognitive (or technical issues), value, and interest—although, in Vearil's mind, the values and interests of both parties caused the majority of contention. "In order to adequately improve Corps/DOI communications," Vearil asserted, "our organizational leaders will also need to address the broader institutional/political/legal issues."[49] John Ogden, a former scientist with Everglades National Park who was now an ecologist for the SFWMD, agreed. "I still have hopes that we can forge a new, more productive relationship with the staff of Everglades National Park," he told Robert "Bob" Johnson, director of research at the park, but as long as Johnson and other park officials took "cheap shots" at restoration plans, the process would be difficult.[50]

To those dealing with the problems created by Lake Okeechobee regulation schedules, experimental water deliveries, the 8.5 Square Mile Area, and the C-111 project, Vearil's comments rang true. In all of these cases, the issues were not so much technical (although all sides used scientific information to support their positions) as they were questions of values and interests. What was important to one side was not necessarily important to the other, and if one refused to compromise, solutions were difficult to reach. Unfortunately, these problems portended difficulties for ecosystem restoration plans and projects. The whole basis for restoration, and the whole goal of the Task Force and the Governor's Commission for a Sustainable Florida, was to create consensus so that a restoration plan could move forward. But with sides clinging to their own values and interests, consensus was difficult to achieve. No one doubted that Lake Okeechobee had to be regulated in a different way in order to protect estuaries and to preserve water supply, just as no one questioned that restoring more natural flows to both Everglades National Park and Florida Bay were necessary. The problem was developing plans to which all sides could agree. If the Corps, the Interior Department, the SFWMD, agricultural interests, the Miccosukee Indians, environmentalists, and others could not agree on a Lake Okeechobee restoration schedule, or on how flows could best be restored to the park, how would they ever agree on an acceptable restoration plan in general?[51] That question plagued officials in the midst of developing a workable proposal and placed a pall of pessimism over the entire restoration effort. As Bradford Sewell, an attorney with the NRDC, observed, no "worse advertisements for Everglades restoration" could have appeared.[52]

Endnotes

[1] "Save Our Everglades: A Status Report by the Office of Governor Lawton Chiles," 30 June 1996, 5.

[2] See Herbert H. Zebuth, Environmental Coordinator, Florida Department of Environmental Regulation, to Ms. Sharon Trost, South Florida Water Management District, 31 March 1993, File Attachment 1, Box 782, JDAR; Joseph D. Carroll, Jr., Field Supervisor, Fish and Wildlife Service, to District Engineer, 4 June 1976, File 1520-03 (C&SF) Water Control Jan. 1976-Dec. 1976, Box 3579, JDAR.

[3] Lake Okeechobee Littoral Zone Technical Group, "Assessment of Emergency Conditions in Lake Okeechobee Littoral Zone: Recommendations for Interim Management," November 1988, File New EA Lake O, Box 782, JDAR.

[4] Colonel Robert M. Brantly, Executive Director, to Mr. John R. Wodraska, Executive Director, South Florida Water Management District, 25 April 1988, File New EA Lake O, Box 782, JDAR.

[5] John R. Wodraska, Executive Director, South Florida Water Management District, to Ms. Martha Musgrove, *The Miami Herald*, 31 May 1988, File Lake O (Background), Box 782, JDAR.

[6] Tilford A. Creel for Valerie Boyd, Governing Board Chairman, South Florida Water Management District, to Captain Ed Hansen, Marine Surveyor, 24 May 1993, File Attachment 1, Box 782, JDAR; Paul J. Trimble and Jorge A. Marban, "A Proposed Modification to Regulation of Lake Okeechobee," *Water Resources Bulletin* 25 (December 1989): 1249-1257.

[7] Colonel Terrence C. Salt, Corps of Engineers Commanding, Memorandum for CDR, South Atlantic Division, n.d., File Lake O (Background), Box 782, JDAR.

[8] Tilford C. Creel, Executive Director, South Florida Water Management District, to Colonel Terrence Salt, District Engineer, 12 December 1991, File Lake O (Reports), Box 782, JDAR.

[9] Quotation in Joseph D. Carroll, Acting Field Supervisor, to Colonel Terrence C. Salt, District Engineer, 14 May 1993, File Lake O Scoping Comments, Box 782, JDAR; see also A. J. Salem, Chief, Planning Division, Jacksonville District, to Addressees on Enclosed List, 8 March 1993, File Attachment 1, Box 782, JDAR.

[10] See Captain Ed Hansen, Past President, Okeechobee Waterway Association, to Ms. Valerie Boyd, Governing Board Chairman, South Florida Water Management District, 2 June 1993, File Attachment 1, Box 782, JDAR; Franklin D. Simmons, Chairman, Glades County Board of County Commissioners, to Mr. A. J. Salem, Chief, Planning Division, Environmental Studies Section, Jacksonville District, 23 March 1993, ibid.; Lomax Harrelle, City Manager, City of Belle Glade, to Colonel Terrence Salt, Department of the Army, 20 April 1993, ibid.

[11] Quotation in Zebuth to Trost, 31 March 1993; see also Colonel Robert M. Brantly, Executive Director, Florida Game and Fresh Water Fish Commission, to Mr. A. J. Salem, Chief, Planning Division, U.S. Army Corps of Engineers, 12 April 1993, ibid.; Karl E. Havens, "Water Levels and Total Phosphorus in Lake Okeechobee," *Journal of Lake and Reservoir Management* 13, no. 1 (1997): 16-25.

[12] Colonel Terrence C. Salt, U.S. Army, District Engineer, to Mr. Tilford C. Creel, Executive Director, South Florida Water Management District, 14 March 1994, File Lake O, Box 745, JDAR; Jacksonville District Corps of Engineers, "Public Meeting Notice on a Proposed Change in the Lake Okeechobee Regulation Schedule," 8 April 1994, File Lake O Scoping Comments, Box 782, JDAR.

[13] Richard E. Bonner, Deputy District Engineer for Project Management, to Mr. Tilford C. Creel, Executive Director, South Florida Water Management District, 9 August 1994, File Lake O EA Comments, 1994, Box 782, JDAR.

[14] "Department of Environmental Protection Staff Assessment, Lake Okeechobee Regulation Schedule Environmental Assessment," File Lake O EA Comments, 1994, Box 782, JDAR.

[15] Stephen A. Walker to Rea Boothby, Department of the Army, Jacksonville District, 10 May 1994, File Lake O EA Comments, 1994, Box 782, JDAR.

[16] Quotation in Thomas D. Martin, Co-Chair, Everglades Coalition, et al., to Col. Terrence Salt, U.S. Army Corps of Engineers, 9 May 1994, File Lake O EA Comments, 1994, Box 782, JDAR; see also Thomas D. Martin, Executive Director, and G. Thomas Bancroft, Senior Scientist, National Audubon Society, to Salt, 31 May 1994, ibid.

[17] U.S. Army Corps of Engineers, Jacksonville District, *Central and Southern Florida Project, Flood Control and Other Purposes, Canal 111 (C-111) South Dade County, Florida: Final Integrated General Reevaluation Report and Environmental Impact Statement* (Jacksonville, Fla.: U.S. Army Corps of Engineers, Jacksonville District, 1994), 3-4, 1-13; House Committee on Natural Resources Subcommittee on National Parks, Forests and Public Lands, *Issues Relating to the Everglades Ecosystem: Oversight Hearing before the Subcommittee on National Parks, Forests and Public Lands of the Committee on Natural Resources, House of Representatives, One Hundred Third Congress, Second Session, on the Land Use Policies of South Florida with a Focus on Public Lands and What Impact These Policies are Having,* 103d Cong., 2d sess., 1994, 5, 20 [hereafter referred to as Everglades Ecosystem Hearing].

[18] Supplemental Appropriations Act of 1984 (97 Stat. 1153, 1292-1293). The original legislation authorized the program for only two years, but Congress extended the deadline and finally, in the Everglades National Park Protection and Expansion Act, allowed the Corps to continue the program indefinitely.

[19] Quotation in *Dade County Farm Bureau v. John O. Marsh, Jr., et al.,* Case No. 83-1210, Memorandum in Support of Emergency Motion for a Temporary Restraining Order and Other Emergency Relief, 1-2, copy provided by James Vearil, Senior Project Manager, RECOVER Branch, Programs and Project Management Division, U.S. Army Corps of Engineers, Jacksonville District, Jacksonville, Florida; see also James Vearil, personal communication with authors, 26 May 2005; Brady interview, 1.

[20] *South Dade Land Corporation, et al. v. Gordon Sullivan, et al.* (853 F. Supp. 404), copy provided by James Vearil; Brady interview, 1.

[21] Sarah Chasis, Senior Attorney, Natural Resources Defense Council, to Mr. A. J. Salem, Chief, Planning Division, 12 January 1995, File Test 7, Box 15754, SFWMDAR.

[22] Noreen K. Clough, Regional Director, Fish and Wildlife Service, to Mr. A. J. Salem, Planning Division, District Corps of Engineers, 27 October 1995, File Test 7, Box 15754, SFWMDAR.

[23] Quotation in Chasis to Salem, 12 January 1995; see also "South Florida Ecosystem Restoration: FY 1996 Budget Cross-Cut, Summary of Federal Projects, Summary of State Projects," 15 May 1995, 13, Billy Causey Files, FKNMSAR.

[24] John D. Brady, Principal Assistant, Office of Counsel, Jacksonville District, U.S. Army Corps of Engineers, personal communication with the authors, 12 May 2005.

[25] Brady interview, 4.

[26] See House Committee on Resources Subcommittee on National Parks and Public Lands, *Issues Regarding Everglades National Park and Surrounding Areas Impacted by Management of the Everglades: Oversight Hearing Before the Subcommittee on National Parks and Public Lands of the Committee on Resources, House of Representatives,* 106th Cong., 1st sess., 1999, 2 [hereafter referred to as *Issues Regarding Everglades National Park*].

[27] Act of 9 March 1994 (108 Stat. 98).

[28] *Issues Regarding Everglades National Park,* 63-64; see also United States General Accounting Office, *South Florida Ecosystem Restoration: An Overall Strategic Plan and a Decision-Making Process Are Needed to Keep the Effort on Track,* GAO/RCED-99-121 (Washington, D.C.: U.S. General Accounting Office, 1999), 43.

[29] *Issues Regarding Everglades National Park,* 2-3, 8.

[30] "Executive Order: East Everglades/8.5 Square Mile Area," File PRO Florida Bay—General 94, Box 15232, SFWMDAR.

[31] *Issues Regarding Everglades National Park,* 8, 36.

[32] *Issues Regarding Everglades National Park,* 79-81, 99-100, 118-119.

[33] *Issues Regarding Everglades National Park,* 9, 31, 73.

[34] *Issues Regarding Everglades National Park,* 136-137.

[35] "Maintaining the Momentum: South Florida Ecosystem Restoration Task Force Biennial Report to the U.S. Congress, Florida Legislature, Seminole Tribe of Florida, and Miccosukee Tribe of Indians of Florida," March 1999 <http://www.sfrestore.org/documents/biennialreport/01.htm> (3 February 2006).

[36] "Maintaining the Momentum."

[37] U.S. General Accounting Office, *South Florida Ecosystem Restoration,* 44-45.

[38] Richard West Sellars, *Preserving Nature in the National Parks: A History* (New Haven, Conn.: Yale University Press, 1997), 283-284.

[39] Colonel Terry Rice interview by Theodore Catton, 6 December 2004, 5 [hereafter referred to as Rice interview—Catton].

[40] Salt interview—Catton, 10.

[41] Collins interview, 4.

[42] Finley interview, 1.

[43] Browder interview, 2.

[44] "Park Management Team SWOT Workshop Notes," 22 May 2000, copy provided by Colonel Terry Rice, Southeast Environmental Research Center, Florida International University, Miami, Florida.

[45] Terry Rice, Miccosukee Tribe of Indians, email to Mary Doyle, 21 June 2000, copy provided by James Vearil, Senior Project Manager, RECOVER Branch, Programs and Project Management Division, U.S. Army Corps of Engineers, Jacksonville District, Jacksonville, Florida.

[46] Richard Punnett email to James W. Vearil et al., 15 February 2001, copy provided by James Vearil, Senior Project Manager, RECOVER Branch, Programs and Project Management Division, U.S. Army Corps of Engineers, Jacksonville District, Jacksonville, Florida.

[47] Michael L. Choate email to Cheryl P. Ulrich, 26 January 2001, copy provided by Colonel Terry Rice, Southeast Environmental Research Center, Florida International University, Miami, Florida; see also Michael Grunwald, "An Environmental Reversal of Fortune," *The Washington Post,* 26 June 2002.

[48] Bo Smith email to Cheryl P. Ulrich, 31 January 2001, copy provided by James Vearil, Senior Project Manager, RECOVER Branch, Programs and Project Management Division, U.S. Army Corps of Engineers, Jacksonville District, Jacksonville, Florida.

[49] Jim Vearil email to Michael E. Magley, 30 March 2001, copy provided by James Vearil, Senior Project Manager, RECOVER Branch, Programs and Project Management Division, U.S. Army Corps of Engineers, Jacksonville District, Jacksonville, Florida.

[50] John Ogden email to Bob Johnson, Director, Everglades National Park Research Center, 23 September 2000, copy provided by James Vearil, Senior Project Manager, RECOVER Branch, Programs and Project Management Division, U.S. Army Corps of Engineers, Jacksonville District, Jacksonville, Florida.

[51] Grunwald, *The Swamp*, 315.

[52] Quotations in Michael Grunwald, "An Environmental Reversal of Fortune," *The Washington Post*, 26 June 2002.

CHAPTER 18

Getting the Water Right: The Restudy and Enactment of CERP, 1996–2000

In November 2000, the U.S. House of Representatives approved a bill authorizing the Comprehensive Everglades Restoration Plan (CERP), an ambitious project headed by the U.S. Army Corps of Engineers, to restore the South Florida ecosystem. Buoyed by the passage of the act, which President Bill Clinton signed on 11 December 2000, congressional leaders, Secretary of the Interior Bruce Babbitt, environmentalists, and even a man in a large green alligator costume (the mascot of the SFWMD) joined in a celebratory "love fest."[1] Sugar interests, environmental groups, state and county officials, Indian tribes, federal agencies, Republicans, and Democrats had all come together—through the Task Force, the Governor's Commission for a Sustainable Florida, and the Corps' Restudy team—in the latter part of the 1990s to develop a workable and agreeable restoration plan. Even though questions remained about the ultimate effects of CERP, it seemed that consensus had won the day. But it had been a long road to gain that consensus, fraught with pitfalls and obstacles.

As explained earlier, the Water Resources Development Act of 1996 (WRDA-96) directed the Corps to conduct a feasibility study on a comprehensive plan for Everglades restoration. At the same time, it mandated that the South Florida Ecosystem Restoration Task Force include non-federal interests. This stemmed from concerns expressed in 1994 and 1995 that non-federal groups—specifically the state of Florida and the Miccosukee and Seminole Indians—did not have enough input in the restoration effort. One of the problems was that the Federal Advisory Committee Act severely limited how non-federal organizations could participate on federal committees. Therefore, in the words of Colonel Terrence "Rock" Salt, now serving as the executive director of the Task Force, "Federal law does not provide a mechanism to build the sorts of relationships needed

in the Everglades restoration efforts without complications from the FACA."[2] Salt recommended that the Task Force support legislative proposals to exempt the Task Force from the Federal Advisory Committee Act, and that it then appoint the Miccosukee, the Seminole, the state, and the SFWMD as *ex officio* Task Force members.

The Task Force had the immediate opportunity to implement Salt's suggestions, as Congress was considering a bill to promote better cooperation between federal and non-federal interests. In March 1995, Congress passed the Unfunded Mandates Reform Act, stating that the Federal Advisory Committee Act would not apply to "meetings . . . held exclusively between Federal officials and elected officers of State, local, and tribal governments," where the conferences were "solely for the purpose of exchanging views, information, or advice relating to the management or implementation of Federal programs."[3] Accordingly, in June 1995, the Task Force expanded its membership to include the state of Florida, the Miccosukee Tribe of Indians, and the Seminole Tribe of Florida.[4]

To ensure that no difficulties would result from this arrangement, and to provide a congressional mandate for these groups' participation, Congress included a provision in WRDA-96 specifically delineating the membership of the Task Force. It stated that the committee would consist of the secretary of the Interior (who would chair the group), the secretaries of Army, Commerce, Agriculture, and Transportation, as well as the Attorney General, the administrator of the Environmental Protection Agency, and representatives from the state of Florida, the SFWMD, the Miccosukee, the Seminole, and a local government. The Task Force was instructed to consult with the Corps in the Corps' preparation of the Restudy; to coordinate different policies and plans for restoration; to facilitate coordination between the different agencies; and to manage the gathering of scientific data.[5] According to the Task Force's 1996 annual report, the inclusion of the non-federal groups allowed the organization to shift its focus "toward issue resolution," and it also "increas[ed] its emphasis on the urban and agricultural components" of restoration.[6]

With its official mandate, the Task Force delineated its goals for South Florida restoration. These included restoring the "diversity, interconnectedness and function of the region's predrainage landscape," as well as ensuring that the "working landscape" could sustain both "a healthy economy and a vibrant society while complementing the management of vital natural resources." The Task Force wanted to restore "estuarine and marine systems," and allow for "natural hydrologic functions in wetlands." It would strive to provide air "healthy to breathe," and it would educate South Floridians so that they could "understand and support the need to restore, preserve and protect the South Florida Ecosystem." To achieve these "desired future conditions," restoration had to consist of three components: getting the water right, restoring and enhancing the natural system, and transforming the built environment. It would also have to be based on sustainability, utilizing an "Ecosystem Approach." The Task Force committed to employ "sound science," as well as adaptive management, in its efforts, and it pledged to use "expanded Partnerships," including public involvement, to "integrate Restoration Planning."[7]

In its restoration activities, the Task Force had several good resources, including Colonel Salt, who, as executive director of the Task Force, assisted the secretary of the interior in the management of the group, and Colonel Terry Rice, who had replaced Salt as District Engineer of the Jacksonville District in 1994. Both believed that Floridians needed to coordinate human needs with environmental quality, and both were firmly committed to the restoration effort. During his 1991-1994 term as District Engineer of the Jacksonville District, environmentalists had embraced Salt as a "green" commander because of his work on both the Kissimmee River and the Restudy. According to Rice, Salt was "the bee's knees" in the eyes of many South Floridians.[8] Salt emphasized the same integrated system approach to South Florida's problems as he had used while dealing with salmon runs on the Columbia and Snake rivers as deputy commander of the Walla Walla District. The colonel had provided such strong leadership in environmental issues as District Engineer that restoration interests wrote letter after letter to Corps Headquarters in 1994, requesting that Salt be allowed to extend his three-year term.[9]

Because of Salt's popularity, Rice faced a daunting situation, but he committed to do his best to "continue what [Salt] had started." By the end of his tenure, restoration proponents were also asking for an extension of Rice's time.[10] Rice, who had a Ph.D. in water resources engineering, had spent most of his professional life in Africa, South America, Central America, and Europe, trying to solve water resource problems in developing countries. After becoming

immersed in Everglades issues, Rice discovered that "the problems we face in South Florida are really the same problems we face everywhere," namely, "how does man take care of himself and not destroy the environment in which he lives." He foresaw Everglades restoration as a prime opportunity to experiment with solutions to that fundamental question, and he regarded it as "one of our greatest chances on this earth" to develop "the model that we need to move forward into history."[11]

The Clinton Administration, especially Vice President Albert Gore, Jr., also provided strong support for Everglades restoration. Gore had been an early convert to the principles of ecosystem restoration and sustainability, and in the mid-1990s, he latched onto Everglades restoration as the crowning example of these concepts in action. He had strategic helpers in the Office of the Secretary of the Army, including Joseph Westphal and Michael Davis, Assistant Secretary and Deputy Assistant Secretary of the Army (Civil Works), respectively. Davis, for example, had studied wildlife ecology and wetlands science for 25 years, and had worked for the Council on Environmental Quality (CEQ) in the White House before becoming Deputy Assistant Secretary. Westphal, meanwhile, had been both a senior fellow at the Institute for Water Resources and the Senior Policy Advisor for Water at the EPA before his appointment as Assistant Secretary. Both held strong feelings in favor of Everglades restoration.[12]

Assistant Secretary of the Army (Civil Works) Joseph Westphal. (U.S. Army Corps of Engineers)

In Florida, the state government had restoration proponents as well. Governor Lawton Chiles, who died in office in 1998, had worked for years, both as a state senator and as governor, on environmental initiatives. Governor John Ellis "Jeb" Bush, who became governor in 1999, did not have as strong of an environmental record, but he readily embraced restoration as both necessary and desirable, telling a gathering of the Everglades Coalition in January 2000 that "there certainly should be no question about my personal commitment."[13]

Alarming statistics also fueled the desire for some kind of a comprehensive plan. According to a Task Force report, South Florida's population was expected to expand to as much as 12 million in "the next generation," fed by an estimated overall increase of 700 new residents every day. Such expansion, the Task Force declared, would continue to affect the natural resources of the area, which had already been "significantly disrupted" by "extensive drainage and flood control systems." It noted that half of the Everglades had disappeared due to drainage, wading bird populations had declined by 90 percent, exotic plants threatened to eliminate native vegetation, pollutants diminished water quality, and Florida Bay was "in a state of ecological collapse." All of these problems, the Task Force argued, revolved around the disruption of the "quantity, timing and distribution" in fresh water deliveries.[14]

An excess of rainfall in 1994 and 1995 emphasized the water distribution problems. The Corps' flood control system functioned adequately, but it created difficulties in other areas. In order to maintain acceptable levels in Lake Okeechobee, for example, the SFWMD had to flush hundreds of thousands of gallons of water down the St. Lucie and Caloosahatchee canals. This pounded the estuaries with slugs of fresh water and silt, disrupting their salinity balance, causing lesions on game fish, and killing seagrass. Likewise, pumping water from the EAA into the water conservation areas threatened deer populations and damaged tree islands, much to the dismay of the Miccosukee Indians.[15] Accordingly, the Miccosukee sued the United States in 1995, charging that the Corps had refused to send water to Everglades National Park because of objections from the NPS, causing undue damage to Miccosukee land in Conservation Area No. 3.[16] Facing such difficulties, the Task Force determined that restoration efforts should be centered on reestablishing the "historic hydrologic functions" of the Everglades, meaning balancing the "quantity, quality, timing, and distribution of fresh water" throughout the ecosystem.[17]

Another problem involved the resurrection of the idea to build a commercial airport near Everglades National Park, this time at a site formerly housing the Homestead Air Force Base. In August 1992, Hurricane Andrew, a Category Five storm, had pummeled South Florida, destroying the Homestead base, which lay between Everglades National Park and Biscayne National Park (established in 1968 to preserve Biscayne Bay). Several politicians, including then-

presidential candidate Bill Clinton and U.S. Senator Bob Graham, declared their support for redevelopment, and, soon after, Carlos Herrera, president of the Latin Builders Association, spearheaded a plan to construct a commercial airport on the former base's site. In January 1996, Dade County, in what many believed was a classic Miami backroom deal, voted to allow Herrera and his colleagues to begin plans for the development, and an EIS was concluded that year.[18]

However, the Everglades Coalition charged that the EIS was inadequate, stating that it did not take into account numerous factors. For one thing, there was no analysis of how runoff through the Military Canal (which removed water from the area) would affect Biscayne Bay, nor were there adequate considerations of the effects of noise on Everglades and Biscayne visitors. In addition, Task Force agencies had determined "that the Proposed Action likely conflicts with their restoration initiatives" and that it would not allow for adequate "protection of natural resources in Southern Dade County."[19] The coalition therefore requested that a supplemental EIS be performed, and Vice President Gore ordered that action in 1997.[20]

Meanwhile, the Natural Resources Defense Council, the Sierra Club, and the Friends of the Everglades contemplated filing a lawsuit. Yet not all environmentalists agreed with litigation, causing a rift within the environmental movement itself. The Audubon Society,

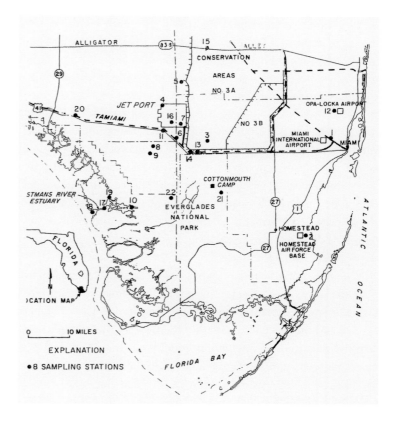

Map showing the location of the Homestead Air Force Base. (U.S. Geological Survey, "Preliminary Determinations of Hydrobiological Conditions in the Vicinity of the Proposed Jetport and Other Airports in South Florida" (1969).

for example, claimed that the Clinton administration's support of the airport, coupled with the fact that Senator Graham wanted it built to promote Homestead's economic development, made it useless to fight. Others, such as Joe Browder and Nathaniel Reed, however, were opposed to the airport and derided Audubon leaders for their stance.[21]

Such divisions generated by the Homestead airport, as well as the flooding problems in the water conservation areas, fragmented the foundation of consensus that the Task Force and the Governor's Commission were constructing. One of the ways that the groups hoped to repair the damage was through the inclusion of unbiased scientific studies in the restoration effort, something that the Task Force's Science Subgroup had recommended in 1996. The subgroup claimed that a South Florida Comprehensive Science Plan was essential, giving priority to studies detailing how the hydrologic system needed to be modified and operated to restore the ecosystem. Especially important was the delineation of pre-drainage hydrology and the use of adaptive management in the final comprehensive plan, including models, restoration support studies, and monitoring.[22]

Indeed, the strategy of adaptive management was a major focus for those developing the comprehensive plan. Proponents of adaptive management believed that it was a necessary component of CERP for several reasons. The uncertainty surrounding the effects of the restoration efforts and the size of the South Florida ecosystem were important factors, as was the fact that adaptive management recommended openness in the planning and implementation of components. Since water management decisions had generated so much mistrust over the years, such openness and commitment to conflict resolution was both desirable and necessary. Resource managers hoped that employing an adaptive management approach would "substantially improve the chance of success in achieving ecosystem goals" through several means. It would ensure a "proactive approach" in dealing with problems, while ensuring that "active collaboration" occurred between scientists, planners, and managers. It would provide a "formal mechanism to expedite and facilitate system-wide decision making, while also providing an "opportunity to develop best available science."[23]

In integrating adaptive management into the comprehensive plan, the Corps was merely riding a wave sweeping over the United States. Many land management agencies, including the U.S. Forest Service and the U.S. Department of the Interior, had implemented adaptive management into certain projects, such as the Forest Service's management plan of coastal forests in California, Oregon, and Washington in 1993, or the Interior Department's building of riparian habitat in the Grand Canyon in 1996. Yet South Florida was one of the first places to use adaptive management in a large ecosystem restoration project. As such, observers focused on whether the strategy would actually work, or whether it would merely result in a series of expensive and time-consuming disasters. At the same time,

scientists involved in the implementation of restoration endeavors charted the lessons whether the guiding principles of adaptive management proved true in South Florida. These included admitting that no easy answers and no experts existed; that new methodological approaches and new scientific methods were needed to solve ecological problems; that uncertainty should be embraced and acknowledged rather than minimized; that the ultimate goal of management was resilience, meaning "enhancing the ability of a system to persist and function in the face of extreme disturbance"; that surprises were the rule, not the exception; that the devil was in the dynamics of an ecosystem's function, not just the details; that human and ecological systems were always changing and uncertain; and that management cannot be separated from the scientific process.[24] But the relative newness of adaptive management, coupled with the fact that the strategy could potentially result in expensive experiments that wasted time and money without doing anything to restore the ecosystem, meant that criticisms and second-guessing were sure to arise in the process, as they certainly did.

Yet in dealing with scientific issues and with adaptive management, restoration proponents already had a good base. In 1989, the SFWMD and Everglades National Park had co-sponsored a symposium at Key Largo to examine what scientists knew about the Everglades ecosystem. Several hundred scientists attended, continuing the discussions after the conference in six adaptive environmental assessment workshops and in informal interactions and contacts. The result of these efforts was the publication in 1994 of *Everglades: The Ecosystem and Its Restoration*, edited by Steven M. Davis of the SFWMD and John C. Ogden of Everglades National Park. This volume, which demonstrated the interdisciplinary nature of ecology and ecosystem studies, included contributions by 57 scientists from various universities and organizations, including the National Audubon Society, the SFWMD, the National Marine Fisheries Service, the Bureau of Land Management, Everglades National Park, the University of Florida, and the University of Miami, among others. Each chapter was peer-reviewed by at least three outside evaluators, and the final product covered a host of subjects, including agriculture, wetland protection, fresh water flows to Florida Bay, climate, fire, hydrology, vegetation, fish and wildlife, phosphorus, and ecosystem restoration. Contributors discussed these topics in both their historic and contemporary contexts. The main focus of the work, however, was on "the interrelated roles of ecosystem size, disturbance patterns, and hydrology as determinants of large-scale ecosystem restoration," and its final chapter, co-authored by Davis and Ogden, synthesized the hypotheses and conclusions of the contributors into a set of problems that ecosystem restoration would have to solve.[25]

According to Ogden and Davis, "the reduction in ecosystem size and compartmentalization of the remaining system are trends that must be reversed in any Everglades restoration initiative." Likewise, hydrologic and fire fluctuations needed to approach natural characteristics. Moreover, water delivery should occur according to historical rainfall patterns in an attempt to reestablish natural hydrology. Such a plan, they concluded, should also have components allowing for natural "volumes and distributions" and "depth patterns in time and space." Finally, water delivery needed to "mimic extended periods of flooding" in Everglades marshes. These objectives would succeed, Davis and Ogden insisted, only if they occurred under a "regional ecosystem-level planning process."[26] The Corps agreed, and used the publication as "a primary source for the basic hypotheses and technical understandings of the Everglades system."[27]

Using the blueprint that Davis and Ogden's book provided, the Corps began working in earnest on a comprehensive plan. Instrumental in the program's development was the Governor's Commission for a Sustainable South Florida. In the eyes of many, the commission played a major role in the creation of CERP because it brought together a host of divergent interests—environmentalists, dairy farmers, sugar growers, vegetable producers, state agencies, county governments, and federal groups—and persuaded them to agree to a central plan. The task of building consensus was not easy; Richard Pettigrew, chairman of the commission, noted that when the commission first came together in 1994, distrust abounded among the different groups. In the words of Robert Dawson, former Assistant Secretary of the Army (Civil Works) during the Reagan administration and current lobbyist for the sugar industry, a lot of "scar tissue" had developed around water management issues through the years, influencing how interests perceived each other.[28] To eradicate this baggage, Pettigrew forced the different representatives to mingle socially, planning evening "happy hours" for this purpose. "The happy hours were critical," Pettigrew later recollected, because they made commission members "g[e]t to know each other as people."[29]

As Pettigrew slowly built relationships of trust between the different interests, the sides compromised and negotiated in a productive manner. Their efforts were accelerated after Rice met with them in June 1995 and declared that if the Governor's Commission could outline a comprehensive restoration program that met the approval of all its members, it would serve as the template for the Corps' restudy. Although commission members expressed skepticism with Rice's pledge, they spent the next year developing a plan, using staffing and other support from the Corps. According to Rice, the negotiations were at times "excruciating" because of the manifold interests that had to be satisfied. "Every issue that came up [was] a major discussion," the colonel noted, leading to the creation of numerous working groups to sort out the problems.[30] Finally, the commission constructed a plan unanimously approved by the whole group, and it submitted the proposal to the Corps in 1996.

After receiving the plan, the Jacksonville District's interdisciplinary Restudy team molded and formatted it into a feasibility report. In essence, the Corps' team had to take the recommendations made by the Task Force and the Governor's Commission and formulate them into a clear and workable restoration plan (accepted by all interested parties) that the Corps could implement in South Florida. Stuart Appelbaum, of Kissimmee River restoration fame, headed this effort, using the same novel approaches to the comprehensive plan as he did in the preparation of the Kissimmee feasibility study. Patterning the organization of his team after Skunk Works, the nickname for Lockheed Martin's Advanced Development Program that built fighter planes, Appelbaum placed the different members—ecologists, biologists, engineers, economists, hydrologists, planners, public relations personnel, real estate specialists, and so forth—in the same area to facilitate communication. He also integrated similar personnel from federal, local, state, and tribal agencies into the process, making the Restudy team as inclusive as possible. Shaking things up was the only approach to take, Appelbaum insisted, because "the Corps was perceived as the bad guys." If the agency presented the "traditional" Corps arrogance, Appelbaum believed, it would not be able to generate the necessary cooperation that it needed.[31]

With Appelbaum's fresh approach—designated the "Something Tells Me We're Not in Kansas Anymore, Toto" method[32]—the Corps

Richard Pettigrew when he was Speaker of the House in Florida's state legislature. (The Florida Memory Project, State Library and Archives of Florida)

analyzed the conceptual plan submitted by the Governor's Commission. This report recommended that numerous water projects be undertaken to "achieve a healthy ecosystem" capable of supplying "vital water resources" for "a sustainable South Florida." It grouped these proposals under 13 thematic concepts:

- Regional Storage Within the Everglades Headwaters and Adjacent Areas
- Lake Okeechobee Operational Plan
- Everglades Agricultural Area Storage
- Water Preserve Areas
- Natural Areas Continuity
- Water Supply and Flood Protection for Urban and Agricultural Areas
- Adequate Water Quality for Ecosystem Functioning
- Increase Spatial Extent and Quality of Wetlands Beyond the Everglades
- Invasive Plant Control
- Aquifer Storage and Recovery
- Protection and Restoration of Coastal, Estuarine, and Marine Ecosystems
- Conservation of Soil
- Operation, Management, and Implementation of the C&SF Project Modifications and Related Lands.[33]

The Corps took the recommended projects and ran them through a model-based screening process, using the resulting data to determine, in Appelbaum's words, "what ideas made the most sense."[34] The Internet facilitated the compilation and sharing of this data, allowing team members to post and download documents and reams of data.

By the middle of 1997, the team had targeted six options as feasible. After making additional analyses, the group made one proposal the preferred choice and formulated an initial draft plan for that option, releasing it to the public in June 1998. After receiving comments on the plan, the Corps finished its draft feasibility report in October 1998, "a record pace given the scale and complexity of this work," and Vice President Gore, who happened to be at the West Palm Beach airport the day of its release, officially submitted it to the public.[35]

Equating the importance of ecosystem restoration in South Florida to "the first pyramid, the first dam, the first skyscraper, [and] the first trip to the moon," the report outlined the pre-drainage conditions of the Everglades and the environmental effects of the C&SF Project. Although the project adequately fulfilled its designated responsibilities of flood control and water supply, "it significantly changed the way water moved and paused in the Everglades." Therefore, "a rethinking" of the project was "in order." The major problem was that project works and operation allowed as much as 1.7 billion gallons of water to flow to either the Atlantic Ocean or the Gulf of Mexico per day, reducing the amount of the resource "needed for the ecosystem and regional water supplies." These flows

also damaged estuaries, causing imbalances in salt and fresh water and killing seagrass, fish, and animal life. Other areas, such as Florida Bay and Everglades National Park, had too little water, causing another kind of ecological damage. Moreover, Lake Okeechobee was "often managed as if it were a reservoir," diminishing the quality of its water and the water it emitted to the rest of South Florida. "These conditions," the document concluded, "seriously threaten the natural and human environment of south Florida."[36]

The Corps proposed 12 solutions. To conserve some of the water flowing to the ocean, and to reduce freshwater discharges to the Caloosahatchee and St. Lucie rivers, the Corps would construct surface water storage reservoirs, water preserve areas (between urban areas and the eastern portion of the Everglades), and more than 300 underground units used for aquifer storage and recovery. Aquifer storage and recovery, or ASR, was a process by which excess surface and ground water was injected underground into the Floridan Aquifer, where it was stored in a "freshwater bubble" until it was needed. Then, the water was extracted, requiring only disinfection before being placed in water distribution systems.[37]

In addition to ASR and above-ground reservoirs, the Corps also proposed the use of limestone quarries in northern Miami-Dade County for water storage. In addition, it would change the existing "rainfall-driven operational plan" so that it could "mimic nature" in its water deliveries to Everglades National Park and the water conservation areas. The removal of approximately 500 miles of canals and levees and the reconstruction of 20 miles of the Tamiami Trail into bridges would also allow for more natural sheet flow while improving water deliveries to Florida Bay and Biscayne Bay. In order to improve water quality, the Corps would manage Lake Okeechobee "as an ecological resource," and it would construct over 30,000 acres of additional stormwater treatment areas (STAs) to cleanse urban and agricultural runoff. Implementing these proposals, the Corps stated, would provide "a comprehensive solution for ecosystem restoration" while still maintaining "the same level of flood protection, if not more, for south Florida."[38]

The admission of the ecological damage caused by the C&SF Project, and, especially, the proposals to correct such problems (including the removal of canals and levees), constituted a major concession for the Corps. According to Michael Davis, the Deputy Assistant Secretary of the Army (Civil Works), "we have people who worked on the original draining of the Everglades and who now, at the end of their careers, are seeing our work turn around 180 degrees." Yet Davis did not perceive the Corps' proposals as anything out of the ordinary. "Our traditional mission was flood control and navigation," he explained, "but really it's always been about problem solving."[39] In addition, the 12 plans of action were really just more engineering solutions, something that Appelbaum considered entirely appropriate. "You have to understand the system has been irrevocably altered," he related. "50 percent of the spatial

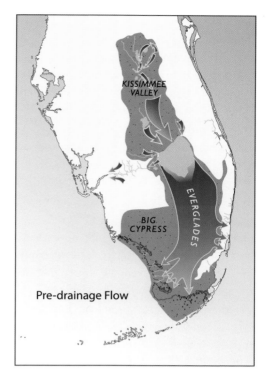

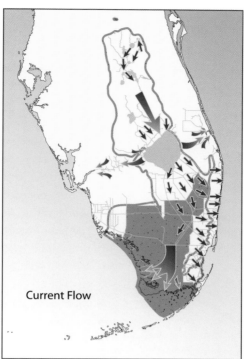

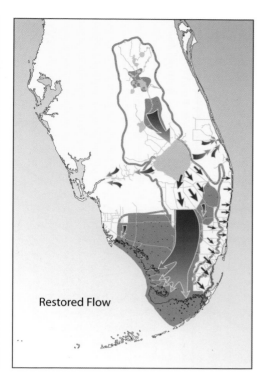

Historic water flow in the Everglades. (South Florida Water Management District)

Current water flow in the Everglades. (South Florida Water Management District)

Future water flow in the Everglades (with implementation of CERP). (South Florida Water Management District)

extent of the original [Everglades] is gone, so the patient is always going to have to be on a respirator."[40]

But the plan dissatisfied many, something that the Corps readily admitted. Criticisms ranged from those who claimed that the restoration projects were a waste of money to those who believed that the Corps had not gone far enough. Some expressed skepticism about the ASRs, calling them a new, untested method of water storage. The cost of the plan—which the Corps estimated at $7.8 billion—gave others pause. The Corps agreed that the price tag was high, constituting a "major investment," but it claimed that it would be worth it: "the overall beneficial effects of the recommended plan are expected to far outweigh its adverse effects."[41]

Perhaps the most stinging and surprising criticisms came from NPS officials and environmentalists who had worked with the Corps' Restudy team.[42] They claimed that the proposal placed agricultural and urban interests above environmental concerns and did not significantly enhance water supplies to Florida Bay, Biscayne Bay, Northeast Shark River Slough, and Taylor Slough. According to Everglades National Park scientists, the plan increased flows to the park from 60 percent of predrainage levels to only 70 percent. Yet, the scientists argued, the plan provided immediate opportunities for developers, agriculture, and property owners to extend their holdings.[43] "It doesn't take $8 billion to get restoration benefits—it takes $1 billion," Brad Sewell of the Natural Resources Defense Council explained. "Our concern is that most of the money will be spent on water-supply projects rather than restoration projects."[44]

What environmental critics really wanted was the implementation of a modeling scenario developed by Everglades National Park and the U.S. Fish and Wildlife Service, called D13R1-4. This model proposed that water which normally ran to tide in the C-51 basin in Palm Beach County and in the C-14/C-13 basins in Broward County be cleansed of impurities and stored in Conservation Area No. 2A for future release to the park. Such a plan could produce an additional 245,000 acre feet of water for the Everglades. The Corps agreed to study the plan, even though it appeared it would adversely affect the conservation areas.[45]

Based on the input that it received at public meetings, as well as comments from other federal and state agencies, the Corps revised its draft report and issued the final version in April 1999. This report, which totaled over 4,000 pages, represented the efforts of over 150 people representing 30 different federal, state, and local agencies, including American Indian tribes, all of which had a slightly different perspective on water management. Some were from environmental organizations, others represented agricultural interests, while still others were from urban areas. According to Appelbaum, the final report dealt with essentially one question: how to enlarge the "water pie" so that "everybody can get a bigger slice."[46] The report itself explained that the Restudy team had focused on certain guiding principles in its preparation of the plan, including restoring the South Florida ecosystem without ignoring other "water needs"; using an "inclusive and open process" to engage "all stakeholders"; partnering with all interested federal,

Figure III-1. Schematic representation of the conceptual design changes between Alternative D and Alternative D13R within the Water Conservation Areas.

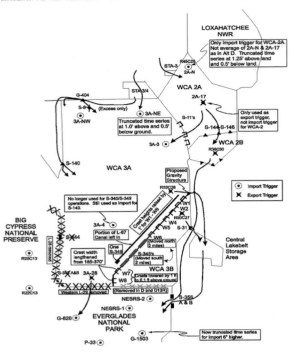

The D13R1 plan. (U.S. Army Corps of Engineers, Jacksonville District, *Central and Southern Florida Project Comprehensive Review Study: Final Integrated Feasibility Report and Programmatic Environmental Impact Statement* (Jacksonville, Fla.: U.S. Army Corps of Engineers, Jacksonville District, 1999).

state, local, and tribal agencies; using the best available science; and advocating adaptive management techniques in order to make the program as flexible and successful as possible.[47]

In essence, the final comprehensive plan was little different from the draft proposal, including the ultimate cost. As one summary put it, the final version attempted to conserve 1.7 billion gallons of water a day through the construction of above- and below-ground reservoirs. It also proposed that 240 miles of levees and canals be removed, that 35,600 acres be used for the creation of additional STAs, and that over 200,000 acres of land be acquired for both the STAs and for the reservoirs.[48]

However, there were some changes. For one, the Corps proposed in the final report to implement a series of pilot projects in order to investigate whether certain proposals were really feasible. These included "wastewater reuse, seepage management, Lake Belt technology, and three aquifer storage and recovery projects" in the vicinity of Lake Okeechobee, the C-43 basin, and Site 1. It also outlined that a Florida Keys Water Quality Protection Program was "critical for restoration of the South Florida ecosystem," and it proposed that water deliveries to Everglades National Park and the conservation areas be based on a rain-driven regulation schedule rather than a calendar-based one. Because of the uncertainty of how successful some of the proposals would be, the report called for the implementation of an

Adaptive Assessment Program to monitor the accomplishments of the different phases of the plan. "Adaptive assessment provides an organized process for confronting and reducing the levels of uncertainty" resulting from insufficient information, the document noted. Although listing the D13R1-4 scenario as a tentatively promising plan, the report did not fully commit to the proposal, indicating only that more studies would be conducted in order to assess its impacts on the conservation areas. The Corps did promise, however, that it would "provide for an improved capability for delivery of additional water to Everglades National Park and Biscayne Bay by capturing additional runoff from urban areas."[49]

Perhaps because the final proposal differed in only minor ways from the draft, criticism of the plan continued, largely at the hands of environmentalists. The Sierra Club, for example, argued that the ASRs would merely worsen urban sprawl in South Florida by providing more water for growth, while six ecologists, including Stuart Pimm from the University of Tennessee and Gordon Orians from the University of Washington, claimed that the lack of a firm commitment to provide 245,000 additional acre feet of water to Everglades National Park proved that the plan was based upon science that was faulty at best and manipulated at worst. Citing "deep, systematic problems," they composed a letter to Secretary of the Interior Bruce Babbitt, asking for an independent scientific review of the proposal.[50] Pimm also criticized the plan for its structural aspects. "We should just take out the damn dikes, for God's sake, and leave the area alone," he proclaimed, fearing that the Corps' plan would just "maintain a managed, fragmented structure instead of restoring the natural system."[51] Likewise, Orians asserted that the Corps had allowed flood control and urban and agricultural water supply concerns to supersede ecological needs.

Meanwhile, the U.S. General Accounting Office (GAO) reviewed the final plan and concluded that the restoration process still needed a "strategic plan that clearly lays out how the initiative will be accomplished and includes quantifiable goals and performance measures." The main problem, according to the GAO, was that the Task Force had delineated three specific goals for restoration—getting the water right, restoring and enhancing the natural system, and transforming the built environment—but the Corps' plan focused only on the first, ignoring the other two. The GAO also expressed concern that the comprehensive plan included no clear way to resolve conflicts. Using the C-111 and Modified Water Deliveries Project as examples, the GAO declared that turf wars were sure to develop between the Corps and the Interior Department if a resolution mechanism was not included, resulting in delays and increased costs. "Without some means to resolve agencies' disagreements and conflicts in a timely manner," the GAO concluded, "problems . . . could continue to hinder the initiative."[52]

Michael Davis disagreed with the GAO's assessment, presenting to Senate subcommittees a joint statement with the Interior Department indicating "the close working relationship between the Army and [Interior] on all levels on Everglades issues." He defended the Corps' focus on water distribution by saying that getting the water right was an integral part of restoring the natural system and transforming the built environment, and he emphasized that the Corps was attempting to develop "an overarching strategic framework that ties all the pieces together." "Waiting until we complete a detailed strategic plan would not be prudent in light of the declining health of this ecosystem," Davis argued. Although "in certain cases decision making could have been more efficient," Davis claimed that the Restudy team, the Task Force, and the Governor's Commission had been "efficient and successful in resolving issues."[53]

Davis may have overstated his case, as disagreements still existed about the plan. Yet many environmentalists decided to support the project despite their reservations, believing it was the best proposal they could get. Charles Lee of the Florida Audubon Society, for

example, backed the plan even though he preferred the removal of "every man-made barrier in the Everglades."[54] As Tom Adams, lobbyist for the National Audubon Society related, "First, you get the deal—and then [you] decide if it's a good deal."[55]

There were also encouraging signs that the sugar industry would not hinder the plan's land acquisition proposals. For one thing, the industry had proved somewhat cooperative in the late 1990s in the purchase of what was known as the Talisman property. This land consisted of approximately 52,000 acres south of Lake Okeechobee and immediately north of the Holey Land Wildlife Management Area. It was divided into a central 32,000-acre parcel and smaller tracts totaling 20,000 acres. Owned by the Talisman Sugar Corporation, a subsidiary of the St. Joe Paper Company (which had purchased the acreage in 1971), these tracts had been used for sugar cultivation, but had experienced decreased productivity in recent years because of heavy soil subsidence. Therefore, Talisman Sugar wanted to sell the

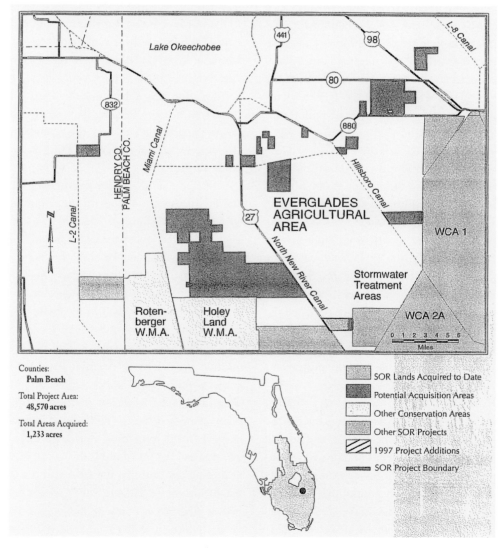

The Talisman tracts (shaded in dark gray). (South Florida Water Management District, *Save Our Rivers: 1998 Land Acquisition and Management Plan* (1998)

land. Other sugar companies were interested, but environmentalists had different ideas, advocating the purchase of the property, which it termed "the golden fleece," so that it could be used as part of a large storage reservoir between the Miami and North New River canals in Palm Beach County.[56]

After the 1996 Farm Bill provided $200 million for the acquisition of lands for conservation purposes, the Interior Department began looking in earnest at the Talisman property, as well as at other EAA lands necessary for the storage reservoir. Because the department proposed taking as much as 150,000 to 200,000 acres in the EAA, the sugar industry was initially uncooperative, claiming that no scientific evidence indicated that such a large reservoir was necessary. The industry even filed a lawsuit against the federal government over Talisman. Facing this standoff, some hydrologists and biologists suggested that the government start with a 60,000-acre reservoir and then add additional space later if necessary. Subsequently, the sugar industry softened its stance and proposed that the two sides engage in a land swap. The Interior Department agreed, and William Leary (senior counselor to the assistant secretary of the interior for fish, wildlife, and parks), John Hankinson (a regional administrator of the EPA), Barry Roth (an Interior Department solicitor), and Buff Boland (a private consultant experienced in land purchases for conservation) began negotiations between the department, Florida's governor's office, the SFWMD, the Nature Conservancy, representatives of the St. Joe Paper Company, and a joint venture of sugar growers.

In 1998, an agreement was reached, and Vice President Gore announced its terms on a trip to South Florida. The sugar industry agreed not to obstruct the acquisition of the central 32,000 acres of the Talisman property and to provide additional EAA acreage to the federal government for the reservoir in exchange for some of the scattered pieces of land owned by St. Joe, as well as the right to farm the Talisman property for at least five years. This would enable the federal government to acquire a little over 60,000 acres, 50,400 of which would be used for the reservoir, while the remaining 10,700 acres would serve as filtering marshes. The Task Force regarded the settlement as "a testimony to how cooperation among government agencies and stakeholders can help to accomplish South Florida ecosystem restoration and sustainability." Restoration proponents hoped that this was an accurate observation.[57]

Meanwhile, the final feasibility report made its way through the higher echelons of the Corps. The usual practice was to have the Chief of Engineers review the report, comment on it, and send it to Congress. However, tension between Chief of Engineers Lieutenant General Joe Ballard and Deputy Assistant Secretary of the Army (Civil Works) Michael Davis, exacerbated by continuing denunciations of the report by environmentalists and Everglades National Park officials, complicated this arrangement. Because of Davis's connections to the Clinton Administration, Ballard claimed that Davis and Joseph Westphal, Assistant Secretary of the Army (Civil

Works), were actually trying to use Corps projects to garner more support for Vice President Gore in the 2000 presidential elections. At the very least, Ballard stated, Davis was acting as the "stovepipe" for Kathleen McGinty, chief of the CEQ and a strong advocate for Everglades restoration. Therefore, Ballard averred, Davis pushed the comprehensive plan incessantly because of Gore's public commitment to it. Such actions infuriated Ballard, not because he disagreed with ecosystem restoration (he actually saw it as a prime market for the Corps), but because he believed that Davis "had an overbearing personality" without "a good understanding of the issues."[58]

The problems between the two reached a head in 1999 over General Ballard's Chief of Engineers report, which Davis supposedly rewrote—expanding it from two to twenty-five pages—to include a Corps commitment to provide an additional 245,000 acre feet of water to Everglades National Park. This assurance caused a storm of controversy. According to Ballard, different interests who had participated on the Restudy team were "surprised" by the recommendation, believing "they had been back-doored" by the Corps.[59] In the words of Stuart Appelbaum, "some of the stakeholders viewed it as a [firm] commitment to deliver the water through the D13R4" scenario.[60] Especially livid were the Miccosukee Indians, who worried that the D13R1-4 plan would cause even more flooding in the conservation areas. Therefore, the Miccosukee filed a lawsuit against the Corps in 1999, alleging that the pledge to provide 245,000 acre feet to the park violated both the National Environmental Policy Act (because it had not undergone the EIS process) and the Corps' restudy authority.[61]

In spite of the controversy, Vice President Gore presented the final feasibility report on the comprehensive plan to Congress in July 1999, calling it "the single largest ecosystem restoration ever attempted anywhere in the world."[62] At the same time, Davis composed a bill authorizing the plan, entitled "Restoring the Everglades, An American Legacy Act" (REAL), for inclusion in the Water Resources Development Act of 2000. The purpose of the bill, according to Davis, was to obtain congressional authorization of the Comprehensive Everglades Restoration Plan (CERP) "as the conceptual road map for restoring the Everglades."[63] It would also include provisions to begin four of the six recommended pilot projects—the Caloosahatchee River (C-43) Basin ASR, the Lake Belt In-Ground Reservoir Technology project, the L-31 Seepage Management Project, and the Wastewater Reuse Technology Project—while providing a programmatic authority for the Corps. In addition, the act would authorize 10 initial construction projects, including reservoirs in the EAA and the C-44 Basin, as well as the establishment of STAs in various locations. The bill proposed the use of adaptive assessment and monitoring to analyze the success of the efforts, and it encouraged the completion of the Modified Water Deliveries Project. Davis pointed out that the "bulk" of the comprehensive plan would be authorized in future water resources

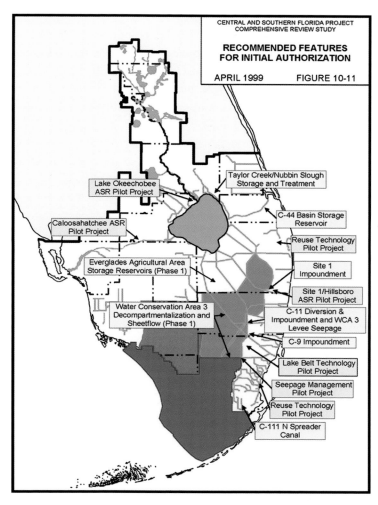

Taylor Creek/Nubbin Slough
Storage and Treatment

Lake Okeechobee
ASR Pilot Project

Caloosahatchee ASR
Pilot Project

C-44 Basin Storage
Reservoir

Reuse Technology
Pilot Project

Everglades Agricultural Area
Storage Reservoirs (Phase 1)

Site 1
Impoundment

Site 1/Hillsboro
ASR Pilot Project

Water Conservation Area 3
Decompartmentalization and
Sheetflow (Phase 1)

C-11 Diversion &
Impoundment and WCA 3
Levee Seepage

C-9 Impoundment

Lake Belt Technology
Pilot Project

Seepage Management
Pilot Project

Reuse Technology
Pilot Project

C-111 N Spreader
Canal

Pilot projects and other recommended features in CERP. (U.S. Army Corps of Engineers, Jacksonville District, *Central and Southern Florida Project Comprehensive Review Study: Final Integrated Feasibility Report and Programmatic Environmental Impact Statement* (Jacksonville, Fla.: U.S. Army Corps of Engineers, Jacksonville District, 1999)

development acts, and the Corps was therefore requesting only a little over $1 billion for the initial appropriation (only half of which the federal government would have to provide, as project costs would be shared 50/50 with the state).[64]

After its introduction into Congress, the bill went to the Senate Committee on Environment and Public Works, where, proponents believed, danger lurked. Senator John Chafee, a Republican from Rhode Island who was a firm supporter of the restoration effort, had just passed away, replaced by Senator Robert C. Smith of New Hampshire, a conservative who had once declared that the Republican party was too liberal for his taste. According to Davis, Smith's chairmanship worried many environmentalists, who believed that Smith's conservatism would cause him to shoot the restoration effort out of the water. Yet to Davis's surprise, Smith declared at the first hearing on the comprehensive plan—located in a packed, hot auditorium at the Naples Golf Club in Naples, Florida—that "you will not find daylight between John Chafee and Bob Smith on the

support for the Everglades."[65] Smith pledged his full commitment to the comprehensive plan, and to those who questioned the large cost of the project, he replied that it would cost Americans only the price of "a can of Coke a year," or "about 50 cents per person, per year."[66] Because of Smith's support, Davis considered him as an instrumental force in the authorization of CERP. Having the complete backing of the Clinton Administration did not hurt either; Davis related that Gore made Everglades restoration a high priority, regarding it as an unprecedented and important program that required the full attention of senior staff.[67]

Another important facet to the actual authorization of the program was the fact that representatives from such divergent groups as the sugar industry and the National Audubon Society joined forces in support of the bill. Tom Adams of Audubon and Robert Dawson, a representative of sugar and urban interests, for example, patrolled the halls of Congress together, drumming up support. The culmination of the consensus-based effort that the Task Force and the Governor's Commission had promoted since the mid-1990s, the unified support showed that interest groups were committed enough to Everglades restoration to support CERP, even though they might not agree to the full plan and despite their ideological differences with others who backed it.[68]

Yet as the bill made its way through Congress, criticisms continued to resound. Dexter Lehtinen, attorney for the Miccosukee Indians, testified before Smith's Committee on Environment and Public Works that the "parochial attitudes" of the restoration effort discriminated against the Miccosukee and caused continuing and accelerated damage to Miccosukee land in the conservation areas. Lehtinen compared restoration to the Vietnam War, saying that everyone believed that things were "fine" in the Everglades, even as officials of the Florida Game and Fresh Water Fish Commission reported that Conservation Area No. 3A had experienced more ecological deterioration in the past five years than in the preceding 40. Lehtinen charged that the restoration effort was in "chaos" and needed to be exposed, much like the emperor who had no clothes.[69] Sugar interests, represented by Malcolm "Bubba" Wade, senior vice president of the U.S. Sugar Corporation, as well as other agricultural interests and the Board of Commissioners for Miami-Dade County, also called for Congress to place assurances in the authorizing legislation that agricultural and urban interests would not have their water supplies reallocated for environmental restoration "without replacement water being available on comparable terms."[70]

In contrast, environmentalists and Everglades National Park authorities continued to claim that CERP offered no real solutions to the park's water problems. Joe Browder was especially adamant in his opposition, as was the Natural Resources Defense Council and scientists such as Stuart Pimm, all of whom wanted a guarantee in the legislation that the park would have the extra 245,000 acre feet of water it needed. Yet, in part because of the Miccosukee lawsuit, the Clinton administration emphasized that the Corps would merely study the proposal; the legislation would contain no such guarantee. Environmentalists such as Browder continued to push for the commitment, but leaders of the Audubon Society, concerned that such efforts would submarine the bill, exerted pressure on the agitators to hold their tongues, believing, in the words of Audubon President John Flicker that "this isn't perfect, but it's more good than bad."[71]

Hearing such complaints, Senator Max Baucus (D-Montana), one of the members of the Committee on Environment and Public Works, began to question the entire restoration plan. "I have this funny feeling that I might be buying something that sounds good," Baucus stated, "but on down the road, I am going to leave to my successors here a huge, huge problem." Baucus was especially concerned because no one seemed to be able to definitively list the problems, the solutions, and the issues that the restoration program would address. "Nobody," he concluded, "has really provided a compelling case that this plan is going to work."[72] In response, Senators Smith and Graham jokingly threatened to leave Baucus in the Everglades as alligator fodder.[73] With such pressure, whether serious or not, Baucus eventually muted his criticisms, although affirming that the plan was an imperfect proposal.[74]

By the summer of 2000, the complaints about CERP threatened to shatter the consensus surrounding the plan. As they had with Baucus, Senators Smith and Graham, together with Florida's other senator, Connie Mack, coerced the major interests to concede to the plan, offering some compromises along the way. To satisfy agricultural and urban interests, for example, the bill specifically declared that no water could be taken from agriculture, urban areas, Everglades National Park, fish and wildlife, or the Miccosukee and Seminole Indians unless other comparable water sources were available. To reassure the Miccosukee and other critics, it stipulated that no conveyance of 245,000 acre feet of water to Everglades National Park would be made until the Corps had conducted a feasibility study and until Congress had specifically authorized such an action. To appease environmental concerns, the bill declared that its major purpose was ecosystem restoration. Senator Smith also ensured that a report of the Senate Committee on Environment and Public Works stated that "the water necessary for restoration, currently estimated at 80 percent of the water generated by the Plan, will be reserved or allocated for the benefit of the natural system."[75] Since this pledge was not proposed in the law itself (as were the urban and agricultural assurances), some hard-line, such

Senator Bob Graham and President Bill Clinton. (The Florida Memory Project, State Library and Archives of Florida)

as the Friends of the Everglades, as well as Everglades National Park scientists, were not happy, and they denounced CERP as "a plan for ecological inaction."[76]

Despite these few voices in the wilderness, most groups, including environmentalists, rejoiced when Congress passed the Water Resources Development Act of 2000 in November 2000 (signed by President Clinton in December), complete with the CERP provisions.[77] As passed, the act authorized CERP "as a framework for modifications and operational changes to the Central and Southern Florida Project" necessary to "restore, preserve, and protect the South Florida ecosystem while providing for other water-related needs of the region, including water supply and flood protection." The law required that the $7.8 billion cost of the project be shared 50/50 with the state of Florida, and that the state be responsible for only 50 percent of the project's operation and maintenance (estimated at $172 million annually). In accordance with the recommendations of the GAO, it required the Secretary of the Army and Florida Governor Jeb Bush to establish "an agreement for resolving disputes" within 180 days, and it stipulated that "an independent scientific review panel" would oversee "the Plan's progress toward achieving the natural system restoration goals." So that Congress could also track progress, it mandated that the Corps and the Interior Department submit a joint report "on the implementation of the Plan" at least every five years.[78]

With the passage of the act, the Corps, the Task Force, the Governor's Commission, and all interests that had worked on CERP celebrated a great victory, and reactions were effusive. "The Everglades legislation is the most important piece of environmental legislation" of its generation, Secretary of the Interior Bruce Babbitt remarked. "It's going to open an entirely new chapter in conservation history." Governor Jeb Bush agreed, calling the measure the first step in "the restoration of a treasure for our country."[79] In the words of Senator Graham, the signing of the bill was "a signal day for the movement around the world to try to repair damaged environmental systems."[80] *Audubon* magazine was no less enthusiastic, stating that "on an overcast, bone-chilling December morning in Washington, D.C., President Bill Clinton launched the restoration of the Everglades and quietly made conservation history." The culmination of more than 50 years of effort, *Audubon* continued, CERP held the promise of a new day for the South Florida ecosystem.[81]

In many ways, the stars seemed to align to ensure the passage of CERP. Even though it carried the huge price tag of $7.8 billion—making it the "most expensive environmental project ever," according to *Audubon*—it came at a time when the economy of the United States was surging and when the federal government had money to spend, largely because of the Clinton administration's efforts to balance the federal budget.[82] In addition, the work that environmentalists, state officials, and federal agencies had done since the 1970s to restore the Kissimmee River, to preserve Lake Okeechobee, and

to reestablish more natural water flows to Everglades National Park had made Everglades restoration a non-partisan issue, one that both Republicans and Democrats supported. Florida had gone through a series of Republican and Democratic governors in the 1970s, 1980s, and 1990s, but each, beginning with Reubin O'D. Askew, had supported Everglades measures. Governor Jeb Bush, a Republican who assumed Florida's governorship in 1999, promised to keep that bipartisanship alive, "pressur[ing] and cajol[ing]" the state legislature "to pick up half the tab for restoration."[83] According to Charles Lee, senior vice-president of the Florida Audubon Society, "you cannot overstate [Bush's] importance" to the passage of CERP.[84] Likewise, the Clinton Administration, and especially Vice President Gore and Michael Davis, had embraced Everglades restoration as one of its primary environmental focuses, providing crucial leadership and

Editorial cartoon from the *St. Petersburg Times* showing the skepticism some expressed about CERP. (Used by permission of the *St. Petersburg Times*)

backing. The fact that Florida's electoral votes played such an important role in the 2000 presidential election did not hurt either, nor did having two environmentally aware District Engineers—Rock Salt and Terry Rice—as commanders of the Jacksonville District between 1991 and 1997.

Because federal, state, local, and tribal interests had accomplished something that no one else had been able to do—the inauguration of a massive ecosystem restoration effort that would extensively modify a U.S. Army Corps of Engineers project—observers from across the United States looked to this success as a template for other environmentally troubled areas. Dan McGuinness, director of the National Audubon Society's Upper Mississippi River campaign, for example, noted that "the prescription for restoring the [Upper Mississippi] means an Everglades-scale congressional appropriation and the same kind of national focus."[85] William C. Baker, president of the Chesapeake Bay Foundation, agreed, relating that it would cost at least $8.5 billion over the next 10 years to restore the bay. The funding of CERP left him hopeful that, in the words of one article, "the environment's big win in the Everglades can be repeated."[86]

But had the environment won in the Everglades, or was CERP just another way to provide more water for the continued growth of agriculture and population in South Florida? What would actually be restored by the various projects proposed under CERP? Would vegetation and fish and wildlife populations really return? Did CERP do enough to revive Lake Okeechobee, and would it really provide for cleaner water in South Florida? These questions tempered the enthusiasm occasioned by CERP's authorization, as did the fact that the Homestead airport proposal was still alive and kicking, despite nearly 40 years of environmentalist efforts to keep airports from infringing on Everglades National Park. In many ways, these questions were nothing that CERP proponents had not foreseen and were a major reason why adaptive management was implemented as an overall strategy. "It would be almost arrogant to get up and say we've got a plan, we're just going to go down this track for the next thirty-plus years and we've got all the answers," Stuart Appelbaum explained. Uncertainty was a part of ecosystem restoration because "we don't have 100 years of experience" in that field. "In fact," Appelbaum continued, "nobody's ever done this before at this scale."[87] The ambiguity unnerved some observers, especially because technology and nature had collided frequently throughout the 1900s in South Florida, leaving only ecological devastation in its wake. Whether or not additional technologies could heal that damage remained to be seen.

Endnotes

[1] Mary Doyle, "Implementing Everglades Restoration," *Journal of Land Use and Environmental Law* 17 (Fall 2001): 59-60.

[2] Col. Terrence Salt, Executive Director, to Mr. George Frampton, Chairman, South Florida Ecosystem Restoration Task Force, 7 February 1995, Billy Causey's Files, FKNMSAR. EPA Assistant Administrator Robert Perciasepe had also called for the expansion of the Task Force's membership in 1994, specifying the inclusion of the Florida Department of Environmental Protection, the Florida Department of Community Affairs, the SFWMD, and the Governor's Commission on a Sustainable South Florida. Perciasepe to Frampton, 7 September 1994, Billy Causey's Task Force Files, FKNMSAR.

[3] Unfunded Mandates Reform Act of 1995 (109 Stat. 48, 65-66).

[4] Rock Salt, Executive Director, to Task Force Members, South Florida Ecosystem Restoration Task Force, 26 June 1995, Billy Causey's Task Force File, FKNMSAR.

[5] Water Resources Development Act of 1996 (110 Stat. 3658, 3772-3773); "1996 Annual Report of the South Florida Ecosystem Restoration Working Group," 12 December 1996, 5, Administrative Files, 1998, FKNMSAR.

[6] South Florida Ecosystem Restoration Task Force, "Annual Report 1996," copy in Everglades Digital Library <http://everglades.fiu.edu/taskforce/ar1996/index.html> (19 October 2004).

[7] Quotations in "The Science Plan" (excerpt from Cross Cut Budget FY 99," available at Everglades Information Network, South Florida Ecosystem Restoration Task Force Collection <http://everglades.fiu.edu/taskforce/index.html> (1 February 2006); see also U.S. General Accounting Office, *South Florida Ecosystem Restoration: An Overall Strategic Plan and a Decision-Making Process Are Needed to Keep the Effort on Track*, GAO/RCED-99-121 (Washington, D.C.: U.S. General Accounting Office), 6.

[8] Rice interview, 11.

[9] Salt interview—Catton, 5; "Update Regarding Colonel Rice," File S. Fla. Water Mgt. Dist., Box 14, S1824, Executive Office of the Governor Subject Files, 1991-1996, FSA.

[10] Rice interview, 11.

[11] Rice interview, 10-11.

[12] General Joe N. Ballard interview with Theodore Catton, 18 November 2004, Washington, D.C., 6 [hereafter referred to as Ballard interview].

[13] Cyril.T. Zaneski, "The Players: The Politicians," *Audubon* 103 (July/August 2001): 74.

[14] Quotations in "South Florida Ecosystem Restoration: FY 1996 Budget Cross-Cut Summary of Federal Projects, Summary of State of Florida Projects," 19 May 1995, 1, Billy Causey Files, FKNMSAR; see also "A Last-Ditch Attempt to Save the Everglades," *ENR* 240 (8 June 1998): 36.

[15] "Testimony of Nathaniel P. Reed Before the Water Management District Review Commission," 26 September 1995, File GOV 16 Water Management District Review Commission, 1995, Box 15771, SFWMDAR; Michael Grunwald, "Water World," *The New Republic* 230 (1 March 2004): 23; Melanie Steinkamp to High Water File, 28 August 1996, File 1996 High Water Review Panel, FWSVBAR.

[16] See *Miccosukee Tribe of Indians of Florida v. United States of America, et al.*, Case No. 95-0532-CIV-DAVIS, Omnibus Order, 31 July 1997, copy provided by James W. Vearil, Chief, Water Management Section, Met Section, Jacksonville District, U.S. Army Corps of Engineers.

[17] "South Florida Ecosystem Initiative," 9 December 1994, Billy Causey's Task Force Files, FKNMSAR.

[18] Grunwald, *The Swamp*, 309.

[19] Bradford H. Sewell to William J. Perry, Secretary of Defense, 28 October 1996, Administrative Files, 1996, FKNMSAR.

[20] Grunwald, *The Swamp*, 313.

[21] Grunwald, *The Swamp*, 310, 335.

[22] "South Florida Ecosystem Restoration, Scientific Information Needs: A Science Subgroup Report to the Working Group of the South Florida Ecosystem Restoration Task Force," 1996, iii-iv, copy at Everglades Information Network, South Florida Ecosystem Restoration Task Force Collection <http://everglades.fiu.edu/taskforce/scineeds/ index.html> (1 February 2006).

[23] RECOVER, "Comprehensive Everglades Restoration Plan Adaptive Management Strategy," April 2006, copy provided by James Vearil, Senior Project Manager, RECOVER Branch, Programs and Project Management Division, U.S. Army Corps of Engineers, Jacksonville District, Jacksonville, Florida.

[24] Steven Light, "Key Principles for Adaptive Management in the Context of Everglades History and Restoration," paper presented at Everglades Restoration Adaptive Management Strategy Workshop, October 2003, copy provided by James Vearil, Senior Project Manager, RECOVER Branch, Programs and Project Management Division, U.S. Army Corps of Engineers, Jacksonville District, Jacksonville, Florida.

[25] Quotation in Steven M. Davis and John C. Ogden, "Introduction," in *Everglades: The Ecosystem and Its Restoration*, Steven M. Davis and John C. Ogden, eds. (Delray Beach, Fla.: St. Lucie Press, 1994), 4; see also John

Ogden interview by Brian Gridley, 10 April 2001, 4, 6, 16, Everglades Interview No. 7, Samuel Proctor Oral History Program, University of Florida, Gainesville, Florida [hereafter referred to as Ogden interview].

[26] Steven M. Davis and John C. Ogden, "Toward Ecosystem Restoration," in *Everglades: The Ecosystem and Its Restoration*, 792-794.

[27] "Responses by Dr. Joseph Westphal to Additional Questions from Senator Graham," in Senate Committee on Environment and Public Works Subcommittee on Transportation and Infrastructure, *U.S. Army Corps of Engineers' Budget for Fiscal Year 2001: Hearing Before the Subcommittee on Transportation and Infrastructure of the Committee on Environment and Public Works, United States Senate*, 106th Cong., 2d sess., 2000, 68.

[28] As cited in Cyril T. Zaneski, "Anatomy of a Deal," *Audubon* 103 (July/August 2001): 50.

[29] As cited in Zaneski, "Anatomy of a Deal," 52.

[30] Quotations in Rice interview, 17-18; see also "Central and Southern Florida Project Comprehensive Review Study," 3 January 1996, File General Information, Interagency MTGS/GRPS, Box 15079, SFWMDAR.

[31] Appelbaum interview, 39.

[32] Appelbaum interview, 19.

[33] "The Conceptual Plan of the Governor's Commission for a Sustainable South Florida" <http://fcn.state.fl.us/ everglades/gcssf/concept/conc_tc.html> (3 February 2006).

[34] Appelbaum interview, 28.

[35] Quotation in U.S. Army Corps of Engineers, Jacksonville District, "Overview: Central and Southern Florida Project Comprehensive Review Study," October 1998, 22, copy in File Restudy Feasibility, Box 1365, JDAR; see also Appelbaum interview, 29-30; Buker, *The Third E*, 114-115.

[36] U.S. Army Corps of Engineers, Jacksonville District, "Overview," 9-11.

[37] U.S. Army Corps of Engineers, Jacksonville District, "Overview," 16-17; CERP Aquifer Storage and Recovery Program <http://www.evergladesplan.org/facts_info/sywtkma_asr.cfm> (25 May 2006).

[38] U.S. Army Corps of Engineers, Jacksonville District, "Overview," 16-17.

[39] As cited in David Helvarg, "Destruction to Reconstruction: Restoring the Everglades," *National Parks Magazine* 72 (March/April 1998): 24.

[40] As cited in Helvarg, "Destruction to Reconstruction," 24.

[41] U.S. Army Corps of Engineers, Jacksonville District, "Overview," 20-21.

[42] Appelbaum interview, 31.

[43] As quoted in Grunwald, *The Swamp*, 320-321.

[44] As quoted in Zaneski, "Anatomy of a Deal," 53.

[45] See South Florida Restoration Office, U.S. Fish and Wildlife Service, "Final Fish and Wildlife Coordination Act Report, March 1999," 1-5, copy in Library, Jacksonville District, U.S. Army Corps of Engineers, Jacksonville, Florida; Appelbaum interview, 31-32.

[46] Appelbaum interview, 30.

[47] U.S. Army Corps of Engineers, Jacksonville District, *Central and Southern Florida Project Comprehensive Review Study: Final Integrated Feasibility Report and Programmatic Environmental Impact Statement* (Jacksonville, Fla.: U.S. Army Corps of Engineers, Jacksonville District, 1999), ii [hereafter referred to as *Final Integrated Feasibility Report*].

[48] See Tom Ichniowski, "Everglades Plan Sent to Hill," *Engineering News-Record* 243 (12 July 1999): 12.

[49] *Final Integrated Feasibility Report*, ix-x, 9-30—9-33, 9-38—9-39, 9-54—9-55.

[50] Quotation in "Saving the Everglades: Water In, Water Out," *The Economist* (6 February 1999): 30; see also Mark Alpert, "Replumbing the Ever-

glades," *Scientific American* 281 (August 1999): 16; Grunwald, *The Swamp*, 324.

[51] As cited in Alpert, "Replumbing the Everglades," 16.

[52] Quotations in U.S. General Accounting Office, *South Florida Ecosystem Restoration*, 17, 19; see also Andrew Wright and Tom Ichniowski, "GAO Hits Everglades Program," *Engineering-News Record* 242 (10 May 1999): 13.

[53] Quotations in Senate Committee on Energy and Natural Resources Subcommittee on National Parks, Historic Preservation, and Recreation and the Committee on Appropriations Subcommittee on Interior and Related Agencies, *South Florida Ecosystem Restoration: Joint Hearing Before the Subcommittee on National Parks, Historic Preservation, and Recreation of the Committee on Energy and Natural Resources and the Subcommittee on Interior and Related Agencies of the Committee on Appropriations, United States Senate*, 106th Cong., 1st sess., 1999, 26; see also Wright and Ichniowski, "GAO Hits Everglades Program," 13.

[54] All quotations in Alpert, "Replumbing the Everglades," 16.

[55] As quoted in Zaneski, "Anatomy of a Deal," 53.

[56] Quotation in Leary interview, 7; see also "Questions & Answers: The Talisman Property," File Everglades, Box 13, S1824, Executive Office of the Governor Subject Files, 1991-1996, FSA; "Babbitt Enters Talisman Land Discussions," *The Palm Beach Post*, 13 October 1995; *Final Integrated Feasibility Report*, F-34—F-35.

[57] Quotation in South Florida Ecosystem Restoration Task Force, "Maintaining the Momentum"; see also Michael Davis interview by Theodore Catton, 21 December 2004, 2 [hereafter referred to as Davis interview]; Wade interview, 32-33; Leary interview, 8. Nathaniel Reed pointed to the fact that in 2000, the sugar industry was trying to negotiate an additional 10-year delay for the completion of the reservoir as evidence that the industry would "do everything in its power quietly . . . to prevent Talisman from becoming a reservoir." Reed interview, 2.

[58] Ballard interview, 6; see also Davis interview, 3-4.

[59] Ballard interview, 8.

[60] Appelbaum interview, 32.

[61] Appelbaum interview, 33; see also John D. Brady, personal communication with the authors, 12 May 2005. This lawsuit was eventually dismissed by mutual agreement, but it had the effect of removing any references to the Chief of Engineers report from the eventual authorizing legislation.

[62] Quotation in Ichniowski, "Everglades Plan Sent to Hill," 12; see also Appelbaum interview, 30.

[63] As quoted in Quotations in Senate Subcommittee on Transportation and Infrastructure and the Committee on Environment and Public Works, *Everglades Restoration: Hearings Before the Subcommittee on Transportation and Infrastructure and the Committee on Environment and Public Works, United States Senate*, 106th Cong., 2d sess., 2000, 24 [hereafter referred to as *Everglades Restoration*].

[64] *Everglades Restoration*, 3, 24-25; see also Senate, *Restoring the Everglades, An American Legacy Act*, 106th Cong., 2d sess., 2000, S. Rept. 106-363, 7-13.

[65] As cited in *Everglades Restoration*, 3; see also Davis interview, 5.

[66] *Everglades Restoration*, 162.

[67] Davis interview, 5.

[68] Doyle, "Implementing Everglades Restoration," 62.

[69] *Everglades Restoration*, 46.

[70] *Everglades Restoration*, 121, 320, 348.

[71] As quoted in Grunwald, *The Swamp*, 341.

[72] *Everglades Restoration*, 219.

[73] *Everglades Restoration*, 224-225.

[74] Grunwald, *The Swamp*, 346.

75 Senate, *Water Resources Development Act of 2000*, 106th Cong., 2d sess., 2000, S. Rept. 106-362, available at <web.lexis-nexis.com> (22 February 2006).

76 As quoted in Grunwald, *The Swamp*, 344 (see also p. 343).

77 Davis interview, 4; "Clinton Signs Bill Endorsing $7.8-Billion Everglades Job," *Engineering News-Record* 245 (18 December 2000): 9.

78 Water Resources Development Act of 2000 (114 Stat. 2572, 2681, 2684-2686, 2690-2692).

79 Both quotations in Zaneski, "Anatomy of a Deal," 48-50.

80 As quoted in Grunwald, *The Swamp*, 352.

81 Zaneski, "Anatomy of a Deal," 48-50.

82 Zaneski, "Anatomy of a Deal," 51.

83 As quoted in Zaneski, "The Players," 74.

84 As quoted in Zaneski, "The Players," 74.

85 As quoted in Jon R. Luoma, "Blueprint for the Future," *Audubon* 103 (July/August 2001), copy available at <http://magazine.audubon.org/features0107/nation/nation0107.html> (7 February 2006).

86 As quoted in Luoma, "Blueprint for the Future."

87 Appelbaum interview, 35.

ALTHOUGH A FEELING OF EXUBERANCE followed the authorization of CERP in the Water Resources Development Act of 2000, a long road lay ahead for Everglades restoration. To expedite efforts, the Jacksonville District morphed its Restudy team into the Restoration Coordination and Verification (RECOVER) branch, with Stuart Appelbaum remaining as chief, and the Clinton Administration created an Office of Everglades Restoration within the Department of the Interior, headed by Michael Davis (although Secretary of the Interior Gale Norton abolished it in November 2001). Environmentalists gained another victory when the Clinton administration issued its decision against the Homestead airport on 16 January 2001. The major problem was ensuring that the fragile coalition that had coalesced around the development of CERP held together. As Appelbaum related in 2002, "Keeping the focus, keeping the camaraderie, the partnerships, is . . . challenging," especially with all of the different interest groups that had staked out a claim to water.[1]

One of the major interests, of course, was the U.S. Army Corps of Engineers, builders of the C&SF Project. As authorized by Congress, two of the purposes of that project, as stated by Chief of Engineers Lieutenant General R. A. Wheeler in 1948, were to provide "a high degree of flood protection" in South Florida and to control water levels in the region "for agricultural use of lands . . . and for maintenance of municipal water supplies." The Corps would fulfill these objectives by removing "excess waters in wet seasons" and storing them for dry season use.[2] As designed, the C&SF Project fulfilled these responsibilities admirably. It allowed for the creation of the Everglades Agricultural Area, a 700,000-acre region that became a bastion for sugar and vegetable growing, generating millions of dollars in revenue, while also providing the necessary mechanisms to allow South Florida to achieve phenomenal growth in the last half of the twentieth century, going from a population of about 500,000 in 1950 to six million in 2000. Moreover, the project successfully impeded saltwater intrusion, a large problem in South Florida urban areas at the time that the C&SF Project was proposed.

Its flood control functions also worked remarkably well. Hurricane Frances, a Category Four hurricane that dumped 13 inches of water on Florida in September 2004, for example, did not cause any major flooding within the borders of the C&SF Project, largely because of the Corps' flood control works. The Corps and the SFWMD redirected much of the water into Lake Okeechobee, storm basins, estuaries, and marshes, providing protection against significant flood damage. Compared with the devastation wreaked by hurricanes in 1926, 1928, and 1947, the contrast was striking. As Richard Bonner, Deputy District Engineer for the Jacksonville District, related, "There was a time when thousands of people were killed because we couldn't provide protection." Now, the C&SF Project safeguarded both lives and property.[3]

Another purpose of the project was "the preservation of fish and wildlife resources," something that the Corps considered an "important feature" of its C&SF plan. But the C&SF Project did not fulfill this objective as well as the others, in part because the original plan was vague as to exactly what "benefits" would accrue and how the project would allow for them.[4] Some advantages did result from the creation of water conservation areas, but in general, the C&SF Project damaged South Florida's ecosystem by disrupting hydroperiods and patterns of water flow in order to provide flood control and water supply. This damaged vegetation and fish and wildlife habitat, leading to startling decreases in certain populations. By 2000, 50 percent of the historic Everglades had disappeared, used instead for agriculture or urban growth, leading to a 90 percent reduction in the number of wading birds and the listing of 68 South Florida animal and plant species as either threatened or endangered.[5] According to many environmentalists, this destruction largely occurred because the Corps reneged on its promise to provide sufficient water to Everglades National Park. The Corps countered that the problems resulted more from innocent ignorance rather than malicious intent.

Yet some concerns, such as the NPS and the FWS, were convinced that the Corps elevated flood control and urban and rural water supply above the needs of fish and wildlife, and they objected accordingly. The NPS, for example, continually asked for more water for the park and eventually got Congress to pass different pieces of legislation mandating this practice. The FWS, meanwhile, requested that the Corps operate the water conservation areas to benefit fish and wildlife, which occurred as long as no flood control or water

supply problems resulted. In cases of high water, however, the Corps pumped water from the EAA into the conservation areas, drowning deer and vegetation. The lack of water to Everglades National Park, coupled with the management of the conservation areas, created an early adversarial relationship between the Corps, the NPS, and the FWS that never really improved. By the 1980s, superintendents of the park felt that they had to take drastic measures, including litigation, to get the Corps and the SFWMD to respond to their concerns. The FWS took a less contentious approach, but still frequently commented that Corps plans ignored fish and wildlife.

From a legal standpoint, the Corps considered fish and wildlife preservation to be on the same level as flood control and water supply since Congress did not give a higher priority to any of these authorized purposes. Yet from a practical standpoint, at least in the early years of the C&SF Project, the Corps seemed to place more emphasis on flood control and water supply. The primary reason for this was that, until the 1970s, state of Florida officials and many

The Everglades landscape. (National Park Service)

Floridians themselves were more concerned about flood protection and water supply than they were about preserving flora and fauna. The Corps, which in those decades still focused primarily on flood control and navigation itself, responded accordingly. E. Manning Seltzer, general counsel for the Department of the Army, for example, opined in the 1960s that "the project is primarily for flood control, . . . water conservation and storage." Because House Document 643 referred to Everglades National Park, Seltzer claimed that "it was the congressional intent that the planning and operation of the project be carried out in such a way that the national Park would not suffer damage." The park would receive "additional benefits" (undefined by Seltzer) as long as these benefits were "consistent with the other purposes of the project." Otherwise, the park would lose out. This mantra seemed to guide the Corps throughout the second half of the twentieth century, although by the 1980s and 1990s, fish and wildlife concerns were considered more equally in policy decisions.[6]

In its attempts to balance various purposes and interests in South Florida water management decisions, the Corps had precedents to follow. As early as the Progressive Era in American history, federal agencies advocated the creation of "multiple-use stewardship of the nation's water resources." They thus developed projects that would "support navigation, irrigation, hydroelectric power and also prevent wasteful flooding by the operation of regulated reservoirs." Prominent examples included the Tennessee Valley Authority, created in the 1930s, and the Pick-Sloan Plan of the 1940s that enabled management of the Missouri River.[7] In addition, the Corps had developed other large-scale water management plans, including the Mississippi River and Tributaries Project in 1928, and had also established environmental restoration projects for rivers and other waterbodies, including the Upper Mississippi River Environmental Management Program.

Yet the development of the C&SF Project was different in several ways. For one thing, the water in question originated and remained in Florida, meaning that state and local entities would not have to worry about sharing the resource with other areas. Perhaps because of this, the state of Florida took on much of the responsibility for developing programs and water management plans through the SFWMD and other state agencies. Indeed, for much of the 1970s and 1980s, major water management actions in South Florida were proposed and developed by the state, not by the federal government. This was not unusual per se, as Louisiana, for example, floated several ideas in the late twentieth century about restoring the Atchafalaya Basin, but the innovations proposed by Florida officials and the strong support over time for environmental measures were remarkable. The SWIM plans for Florida waterbodies, for example, and their level of funding represented a significant effort on Florida's part to try to resolve water quality problems at a time when the federal government seemed lax on such issues. Likewise, the

fact that most elements of Governor Graham's Save Our Everglades program continued into the twenty-first century was noteworthy. All of these efforts, however, had to be coordinated with federal and local agencies, meaning that a comprehensive restoration plan was only feasible after commissions and committees such as the South Florida Ecosystem Restoration Task Force and the Governor's Commission for a Sustainable South Florida had brought all interests to the table.

And there were numerous interests to please. In the 1960s, environmental groups began to support NPS and FWS stances, riding the wave of a strengthening environmental movement in the United States to stake out their claim in South Florida water management. Groups such as the National Audubon Society (and its Florida chapter), the Friends of the Everglades, the Sierra Club, and others made their presence known with their protests over the proposed Everglades Jetport. They used their success in stopping this project—revolving around appeals to the Nixon administration—as a blueprint for future protests about Kissimmee River channelization, Lake Okeechobee water quality, water deliveries to Everglades National Park, and the preservation of environmentally endangered lands. Led by such luminaries as Marjory Stoneman Douglas, Arthur Marshall, Johnny Jones, and Nathaniel Reed, environmentalists effectively informed the Corps in the 1970s that they were major players at the table, and they continued that oversight role through the subsequent decades.

The sugar industry was also becoming more powerful in the 1960s as well, and it began exerting its own political pressure on water management issues. Because of the need for drainage to keep lands dry for farming, and because agricultural runoff contained pollutants adversely affecting South Florida's water quality, environmentalists and sugar quickly developed a mutual antagonism that colored each side's approaches to water management. Dealing with water quality in Lake Okeechobee and in the Everglades in general in the 1980s and 1990s made the differences especially pronounced, culminating in the 1988 lawsuit over water quality and the penny-a-pound tax campaign. The animosity between the two groups remained a serious threat throughout the development of CERP, which is why Tom Adams and Robert Dawson walking the halls of Congress together was such a striking symbol.

Likewise, conflicts existed between those advocating the protection of the Everglades and property owners, especially in southwest Florida and in the East Everglades region. These Floridians—who, in many instances, were the ones actually residing in the remaining Everglades—claimed that government officials needed to consider property, hunting, fishing, and airboating rights as supreme to environmental preservation. But restoration of the ecosystem clashed with these ideas, especially after the state of Florida made land acquisition a priority in the 1980s. The SFWMD, the Corps, and property owners around the Kissimmee River were able to compromise in order to allow Kissimmee restoration to proceed, but when Everglades National Park clamored for the removal of property owners from the Frog Pond and the 8.5 Square Mile Area in order to improve water deliveries, only messy controversy resulted. The discord in that matter delayed the implementation of the Modified Water Deliveries Project and eventually cast a shadow over the entire development of CERP. The purchase of the Talisman property from sugar interests indicated that settlements were possible, but it was clear that protracted negotiations would be necessary.

The Miccosukee and Seminole Indians, meanwhile, claimed that no federal, state, or local agencies adequately protected their interests in water management. Holding land in the water conservation areas, these tribes had a large stake in seeing the resolution of water quality and water supply issues. When solutions were proposed that contradicted Seminole and Miccosukee ideas of what the Everglades needed (such as in the Settlement Agreement stemming

Dade County urban and agricultural areas. (South Florida Water Management District)

from the 1988 lawsuit and the Everglades Forever Act), the Miccosukee resorted to litigation to ensure that their lands were protected. Both tribes also developed their own water quality standards to govern the resource on reservation lands. Some observers regarded such tactics as divisive, but the Miccosukee did not care about that perception, placing the health of the Everglades ecosystem, as they saw it, above all other concerns.

With all of these different interests, it truly was amazing that architects of CERP such as Richard Pettigrew, Colonel Rock Salt, Colonel Terry Rice, Stuart Appelbaum, Senator Bob Graham, and others were able to forge a restoration plan that all sides could accept, or at least that was reasonable enough to prevent outright opposition. Many heralded CERP as a model for other regional ecosystems to follow, but there were unique South Florida characteristics that influenced the development of the plan and that ultimately allowed the necessary consensus to occur.

For one thing, had the state of Florida not experienced an influx of politicians and officials who truly cared about the environment, it is doubtful that restoration efforts ever would have amounted to anything. Several were critical to transforming Florida from a state primarily concerned about flood control and drainage to one intent on restoring ecosystem quality. Fundamental in that transformation was Governor Reubin Askew's administration, which was the first to admit that significant environmental problems existed in South Florida. Listening to the concerns of Arthur Marshall, Askew oversaw the passage of several significant acts in 1972, such as the Florida Environmental Land and Water Management Act, which proposed measures to protect environmentally endangered lands.

Yet Askew's successor, Bob Graham, deserves the lion's share of the credit for making ecosystem restoration a critical component of state planning. Graham, who had supported environmental measures while a state senator, became convinced while governor that Florida had to provide more leadership in environmental issues, in part because the federal government under the Reagan administration had little concern for such matters. Graham therefore inaugurated the "Save Our Everglades" program, which focused attention on measures to save the Kissimmee-Okeechobee-Everglades watershed. After becoming one of Florida's U.S. senators, Graham continued his leadership on a national level. As many South Floridians have observed, it is impossible to overestimate his influence on Everglades restoration.[8]

Another important factor was the increasing environmental awareness of the Corps of Engineers, especially the Jacksonville District. In large part because of issues such as the Everglades Jetport and the Cross-Florida Barge Canal, the Corps began to take environmental concerns into consideration in its project planning. The Jacksonville District had several District Engineers that moved this process along, including Colonel Donald A. Wisdom, Colonel Rock Salt, and Colonel Terry Rice. Each of these individuals attempted to implement a more environmentally friendly attitude in the District. The Jacksonville District was not alone in this process, as other Corps districts also evinced more concern for the environment in the last half of the twentieth century. Some St. Paul District personnel, for example, were called "ecofreaks" as early as the late 1970s because of environmental enhancement efforts that they promoted.[9] Yet the level of support that Salt and Rice in particular evidenced for environmental restoration was remarkable, indicating that the Jacksonville District cared just as deeply about environmental quality as it did about engineering proficiency.

In addition, the establishment of an Environmental Advisory Board in 1970, which provided recommendations and aid to Corps leadership on environmental issues, facilitated the development of environmental concern among higher Corps authorities. The need to consider ecosystem health in Corps projects convinced Corps officials in the 1980s to accept environmental restoration, and Congress included Section 1135 in the Water Resources Development Act of 1986, authorizing the Corps to perform that function on existing projects. Accordingly, the Jacksonville District began looking in earnest at dechannelizing the Kissimmee River (as requested by environmentalists and Florida authorities). When Salt and Rice assumed district leadership in the 1990s, they furthered the acceptance of ecosystem restoration, relying on individuals such as Stuart Appelbaum to change the District's usual methods of operation.

Also affecting the Corps' attitude was the degree to which Everglades issues played out on a national stage and became embroiled in federal politics. Environmental organizations discussed matters such as water supply to Everglades National Park and the Everglades jetport in their national publications, bringing unwanted publicity to the Corps, while presidential administrations, including those of Richard Nixon and Bill Clinton, grasped on to Everglades issues, hoping to use them to make their environmental mark. This presidential support was not surprising, considering how crucial Florida was in several presidential elections, including those in 1996 and 2000, but the fact that presidents saw the environment—rather than economic or social issues—as the way into Floridians' hearts is significant.

Water issues in South Florida also forced the Corps (and Floridians in general) to rethink whether technological and engineering solutions to environmental problems were really the right directions to take, paving the way for the more open process in the Jacksonville District's Restudy project. Indeed, the extent of the environmental damage that canals and levees had produced in South Florida indicated that other methods needed to be considered. Ultimately, the Corps still relied on structural solutions in its CERP proposal, but it expressed a willingness to consider non-structural answers throughout the process. The problem was that so much development had occurred in South Florida that it was impossible to reverse the entire

S-9 pumping station. (The Florida Memory Project, State Library and Archives of Florida)

C&SF Project. Yet the Jacksonville District's work on the Kissimmee River indicated that the Corps was no longer as wedded to technology and engineering as the ultimate answers.

In fact, ecological and biological studies became almost as significant in Corps plans for restoration as engineering reports, highlighting the growing importance and influence of science in water issues, as well as showcasing the growth of ecology in the latter half of the twentieth century and the increasing sophistication of scientific measuring techniques. Spurred on by Arthur Marshall in the late 1960s, the Everglades, Lake Okeechobee, and the Kissimmee River became enmeshed in an avalanche of scientific studies, many of which embraced the principles of systems ecology, whereby mathematical and computer models were increasingly used to analyze complex ecosystems. By the 1990s, seminal publications such as *Everglades: The Ecosystem and Its Restoration* (1994) provided a blueprint for restoration efforts, while Louis Toth's work on the Kissimmee River informed much of the plans to reestablish that waterway. The implementation of Kissimmee River restoration efforts as well as CERP also provided a platform on which to test the strategy of adaptive management, a set of ideas and methods that embraced scientific uncertainty and used the "real world" as a testing ground for the management and restoration of ecosystems. Although the jury was still out in 2000 on whether or not adaptive management was a proper tool for South Florida, the uncertainty surrounding restoration efforts and the disparate interests with a stake in restoration had necessitated its implementation.

Nearly 60 years ago, Marjory Stoneman Douglas noted that the people of South Florida faced an "eleventh hour" in regard to the "dying Everglades." She hoped that the U.S. Army Corps of Engineers' C&SF Project would save the region, but she understood that politics would influence its implementation. "How far [the Corps] will go with the great plan for the whole Everglades," she continued, "will depend entirely on the co-operation of the people of the Everglades and their willingness, at last, to do something intelligent for themselves."[10] Douglas's statement was prophetic. For many years, the people of South Florida generally regarded flood control and water supply as more important than fish and wildlife issues, and, in the eyes of many observers, the Corps' C&SF Project accordingly shunted ecological concerns aside.

In the late 1990s, the people of South Florida once again faced an "eleventh hour," leading the Corps to develop another plan. Because of the numerous interest groups that had emerged between 1948 and 2000 to influence water management decisions, environmental restoration now took its place besides flood control and water supply as a primary purpose of the C&SF Project. Whether this primacy would continue was largely in the hands of these conflicting interests and the fragile consensus they had built. Whatever happened, the process of implementing and operating the C&SF Project, with all of its attendant controversies and difficulties, had left an indelible mark on the South Florida landscape, for good and for bad. CERP now had a chance to make its own impression, but, much like its C&SF parent, its form and depth depended on the communication, compromises, and collaboration that disparate stakeholders were willing to make.

Endnotes

[1] Quotation in Appelbaum interview, 39; see also Grunwald, *The Swamp*, 354.

[2] House, *Comprehensive Report on Central and Southern Florida for Flood Control and Other Purposes*, 80th Cong., 2d sess., 1948, H. Doc. 643, Serial 11243, 2.

[3] Quotations in Michael Grunwald, "This Time, Man Defeated Nature: Florida's Flood-Control System Kept Frances from Swamping Plains," *The Washington Post*, 9 September 2004.

[4] House, *Comprehensive Report on Central and Southern Florida*, 2.

[5] U.S. Army Corps of Engineers, "Overview: Central and Southern Florida Project Comprehensive Review Study, October 1998," 11, copy in File Restudy Feasibility, Box 1365, JDAR.

[6] Both quotations as cited in James W. Vearil personal communication with Matthew Godfrey, 16 May 2006.

[7] John E. Thorson, *River of Promise, River of Peril: The Politics of Managing the Missouri River* (Lawrence: The University Press of Kansas, 1994), 59-61.

[8] See, for example, Reed interview, 23.

[9] Ben A. Wopat interview by Matthew Pearcy, St. Paul, Minnesota, 25 April 2002, 19, Oral History File, St. Paul District, U.S. Army Corps of Engineers, St. Paul, Minnesota.

[10] Douglas, *The Everglades*, 385.

Administering the Cure: Implementing the CERP Program, 2001–2010

THE PASSAGE OF CERP seemed to herald a new environmental age for South Florida—one in which repairing damage to the ecosystem would be a priority to stakeholders in water management. Yet ten years later, few CERP projects were constructed and the Corps had yet to execute most of the plan. Implementing CERP proved to be just as difficult and time-consuming as its development. In the years between 2000 and 2010, criticisms of CERP mounted, heightened by a belief that the federal government was not as committed to restoration as proponents had hoped. The reasons for the lack of progress stemmed mainly from factors that no one foresaw in 2000, including the need to develop project processes, an economic recession that siphoned money from the restoration effort, a presidential administration caught up in high-profile distractions that detracted focus from CERP, and the escalation of concerns over water quality. Whether the Corps and other stakeholders could overcome these issues and actually make CERP effective remained to be seen, but the slow progress in implementation frustrated both stakeholders and Corps personnel. As the ecosystem continued to decline, some wondered whether CERP could ever deliver the healing it promised.

Initially, CERP implementation rode the momentum of the Water Resources Development Act of 2000 (WRDA-2000). Even before WRDA-2000 passed, the Corps had initiated either project management plans or the formation of project delivery teams for 12 CERP projects, and it planned on starting the process for other projects in the spring of 2001.[1] However, implementation soon slowed, in large part because of project requirements contained in WRDA-2000. Deciding how to best meet these stipulations took the Corps and its partners a few years, as it involved complexities that no side really

understood. "Between 2000 and 2004 the [Corps] and the SFWMD largely focused on developing a complex coordinating structure for planning and implementing CERP projects," a report stated, leading to delays in actual CERP construction.[2] As Stuart Appelbaum of the Jacksonville District explained, "[T]he whole nature of the relationship and . . . setting up the process was much more complex [and] cumbersome than anybody would have imagined in 2000." According to Appelbaum, "we just basically underestimated how difficult it was going to be to set this up."[3]

For one thing, WRDA-2000 directed the Corps to develop programmatic regulations for CERP. These regulations would provide "processes and procedures" to "guide the Army Corps of Engineers in the implementation of the Comprehensive Everglades Restoration Plan." The Corps first issued draft regulations on 2 August 2002, and received 820 comments on them. Wading through the comments and adjusting the regulations accordingly took time, involving a process of "reconcil[ing] different points of view and . . . find[ing] consensus solutions to common concerns." One item of debate was how to know when "restoration" had occurred. The Corps proposed using the level of performance delineated in CERP's Final Integrated Feasibility Report and Programmatic Environmental Impact Statement of April 1999, but others wanted success defined "in terms of hydrologic and ecologic targets." Still others saw restoration simply as "getting the water right," whatever that might mean. Taking into account these comments, the Corps ultimately defined restoration "as the recovery and protection of the South Florida ecosystem so that it once again achieves and sustains the essential hydrological and biological characteristics that defined this ecosystem in an undisturbed condition." Having attempted to resolve such debates (but

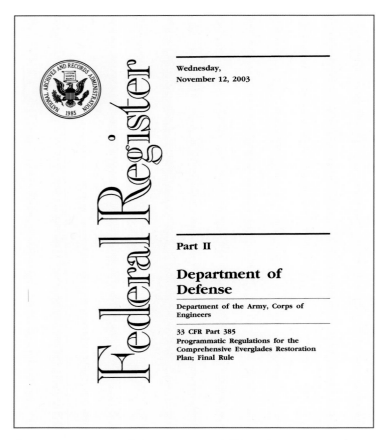

The programmatic regulations governing CERP implementation, published on 12 November 2003.

recognizing that in some cases it was not possible to please all sides), the Corps published the final regulations on 12 November 2003.[4]

The concern over defining "restoration" highlighted the fact that some environmental organizations believed that CERP—"with its aquifer storage wells, mining pits, and rebuilt canals"—was more "a water delivery system for cities and suburbs" than a restoration program.[5] According to some environmentalists, the Corps and the SFWMD needed reminders that, as Robert Johnson of the NPS's South Florida Natural Resources Center said, they "didn't get funding for the Central and Southern Florida Restudy to provide cheaper water to the South Florida residents or to greatly enhance flood protection."[6] Others claimed that merely having the Corps in charge of restoration meant that the environment would never get its due. As Joe Browder declared, putting the Corps as the lead on restoration was like placing the "Bureau of Reclamation in charge of protecting the Grand Canyon." The Corps relied on "expensive, unproven, high-technology efforts"—namely, the aquifer storage and recovery proposals—rather than "biologically friendly, natural-system based restoration," environmentalists argued. Until the Corps' approach changed, they asserted, the plan would ultimately benefit urban interests more than the ecosystem.[7]

These criticisms were similar to what the Corps had heard during the development of CERP, but in the 2000s, such perspectives

contributed to CERP delays. As one report explained, "The greatest challenge in the project planning process has been developing technically sound project plans that are acceptable to the many agencies and stakeholders involved."[8] According to Carol Wehle, who became executive director of the SFWMD in 2005, the situation was exacerbated by the disbanding after CERP's passage of the Governor's Commission for a Sustainable South Florida. Had "the core group that crafted Everglades restoration" continued to meet on a regular basis, Wehle asserted, it would have "facilitat[ed] communication, [kept] consensus and provid[ed] a stronger lobby . . . for the issues that we needed Congress to address."[9]

With differences of opinion persisting about the Corps' environmental commitment , it became important for the Jacksonville District to highlight how each proposed CERP project met the ecological needs of South Florida. One way of doing this was through the development of project implementation reports, another requirement stipulated in WRDA-2000. According to the act, the Corps had to submit a project implementation report to the House Committee on Transportation and Infrastructure and the Senate Committee on Environment and Public Works for every project authorized under CERP before construction could begin. These reports had to include a justification that the project would result in "environmental benefits" for South Florida and that the project was "cost-effective." Project implementation reports also had to comply with NEPA requirements, and they had to include feasibility and cost effectiveness reports. In addition, projects had to have assurances that the water "generated by the [project] will be made available for the restoration of the natural system." A "binding agreement" between the president of the United States and the governor of Florida supplemented these assurances, specifically ensuring that the state would make no "consumptive use" of water "until such time as sufficient reservations of water for the restoration of the natural system are made under State law."[10]

The Jacksonville District took the preparation of project implementation reports very seriously. According to Dennis Duke, former Jacksonville District project manager for CERP, the District asked itself several questions when putting the reports together: "Is this the right plan? Is it still cost effective? Does it still produce the benefits? Is it justified? Will it be justified if nothing else is built?"[11] Answering these questions took more time than anyone had realized during the CERP planning process, frustrating the different stakeholders in South Florida, including the Jacksonville District. As one report concluded, "PIR completion . . . represents a major hurdle in the implementation process for all CERP projects."[12]

Indeed, taking a CERP project from planning to construction was even more convoluted than a traditional Corps project. Corps regulations required very specific processes in traditional civil works projects, including the completion of a reconnaissance study, a feasibility and cost-sharing agreement, a design agreement, and a project cooperation agreement, leading some to assert that the Corps was "not an institution . . . built for speed and flexibility."[13] CERP projects were even more complicated, requiring a design agreement, a master program management plan, a project management plan, a project implementation report, a report containing detailed designs, plans, and specifications, and a project cooperation agreement. Moreover, after the Jacksonville District prepared each of these reports, both Corps leadership and the SFWMD (as local sponsor) had to provide their approval before the District could proceed to the next stage.[14] "You've got to be very nimble to do Everglades restoration because the circumstances are constantly changing," Robert Johnson of the NPS asserted, and the Corps' "standardized, . . . very rigid process" was not conducive to that.[15] In the words of Paul Souza of the U.S. Fish and Wildlife Service, the Corps' planning process was not "designed [with] such a multifaceted restoration project in mind."[16]

To streamline procedures, the Corps and the SFWMD tried to develop a plan explicating how the two sides would work together. According to Stuart Appelbaum, such an agreement would provide "a legal mechanism" outlining each organization's responsibilities for each project. "We didn't want to go back and have to re-negotiate every one of these terms every time for every project," Appelbaum stated.[17] A master agreement would serve as "an umbrella framework,"[18] streamlining how the Corps and the SFWMD would work together by covering such elements as how the SFWMD would receive credits for work previously done, how lands, easements, and rights-of-way would be provided, and how the sides would deal with historic properties. Begun in 2002, the master agreement took seven years to complete because of the complexities of the issues. When it was finally executed in 2009, some considered that to be "one of the major accomplishments over the last decade."[19]

Signing of the CERP Master Agreement on 13 August 2009. (South Florida Water Management District)

Additional delays in CERP implementation resulted from a lack of federal funding and project authorization. Under CERP, the State of Florida and the federal government were to share project costs 50/50. The Corps estimated that the federal government would need to appropriate over $1.6 billion between 2001 and 2010 (and nearly $4 billion over 30 years) to meet CERP requirements as delineated in WRDA-2000. Congress would also need to authorize 26 additional CERP plan components at a cost of $6.2 billion in water resource development acts passed every two years (from 2002 to 2014).[20] Yet Congress passed no additional water resource development acts until 2007, causing projects to "stack up in a pipeline." By 2007, the Bush administration had spent only $358 million on CERP projects, while the State of Florida had expended $2 billion.[21] In terms of federal financial support, the future did not look bright either, as one 2006 report noted that planned federal expenditures for the next five years were only 21 percent of the amount the federal government was supposed to spend.[22]

One reason for the lack of funding was the terrorist attack of 11 September 2001 that destroyed the World Trade Center towers in New York City. After this strike, President George W. Bush focused his efforts on what he called the Global War on Terror, which included an invasion of Iraq, beginning in 2003. These national defense endeavors siphoned money from other federal programs, leaving little for CERP appropriations. When Hurricane Katrina hit Louisiana and Mississippi in 2005, virtually destroying New Orleans in the process, it too diverted restoration funds. Another reason for insufficient appropriations was the Bush administration's lack of commitment to Everglades restoration. Vice President Al Gore had championed the Everglades during the Clinton administration, but Bush had no similar proponent. Everglades restoration was "a really high priority in the 1990s," Shannon Estenoz of the SFWMD recalled. "We were not a high priority in the next decade."[23] According to Thomas Van Lent of the Everglades Foundation, the fault did not lie entirely with the Bush administration. In Van Lent's view, the Corps did not do a good enough job of explaining in terms that non-engineers could understand what CERP would accomplish and how individual projects would result in environmental benefits. The Corps did not "relate [CERP] in a way that the average constituent" could grasp, Van Lent said, reducing the amount of support for restoration.[24] Whatever the reasons, Congress appropriated few funds to CERP for much of the 2000s.

Because of funding and project planning factors, as a 2005 article in *The Economist* explained, "[f]ive years [after the passage of CERP], not a single CERP project has been built."[25] The Committee on Scientific Review of Everglades Restoration Progress, which was appointed in 2004 by the Corps and the U.S. Department of the Interior to provide biennial reviews of CERP, painted a bleak picture of restoration progress in its 2006 report. While noting that "much good science has been developed to support the restoration efforts" and that "progress has been made in CERP program support," the committee explained that the ten CERP components scheduled in the original plan for completion by 2005 were delayed, as were the six pilot projects that should have been finished in 2004.[26] The U.S. Government Accountability Office agreed with the committee's assessment in 2007, stating that CERP projects "are significantly behind their original implementation schedule," some by "as many as 6 years."[27]

Delays also plagued non-CERP projects essential to restoration. The Modified Water Deliveries Project (Mod Waters), for example, was a "poster child" for problems.[28] Characterized by the Miccosukee as a project "seriously delayed by a lack of leadership and constantly changing plans,"[29] Mod Waters, according to the Interior Department's Office of Inspector General, was eight years behind schedule by the middle of the decade and its costs had ballooned from approximately $80 million to nearly $400 million.[30] Disagreements between the Corps and other stakeholders contributed to the problems. For example, the Corps' proposed acquisition of 2,100 acres in the 8.5 Square Mile Area and 77 residential tracts led to litigation from landowners. Not only did residents claim that the Corps could not exercise eminent domain, they also argued that the Corps' flood control plan for the area—known as Alternative 6D and consisting of "a perimeter levee, seepage canal, pump station, and storm

water drainage for flood protection in the 8.5 SMA"—would protect only between a third and a half of the residents, rather than everyone living in the area. Congress, they declared, had not authorized the Corps "to implement a plan that does not protect the entire 8.5 SMA from flooding." The U.S. District Court initially ruled against the Corps in 2002, but Congress eventually authorized Alternative 6D in a joint resolution dated 20 February 2003.[31]

Other issues included whether or not the quality of water flowing into Everglades National Park once Mod Waters was constructed would meet the necessary standards.[32] In response to this concern, Congress included a provision in the Interior Department Appropriation Act for 2004 making funding for the Mod Waters project contingent on whether the quality of water going into Loxahatchee National Wildlife Refuge and Everglades National Park met "applicable State water quality standards and numeric criteria adopted for phosphorus."[33] Trying to resolve these water quality and land acquisition issues seriously hindered Mod Waters' implementation, causing in part its delays.

Even water quality seemed to suffer setbacks in the 2000s. Although some observers lauded the SFWMD's efforts to improve water quality, pointing to the 44,000 acres of stormwater treatment areas built under its Everglades Construction Project (which resulted in a reduction in phosphorus concentrations in agricultural

Vegetable farming in the Everglades Agricultural Area. (South Florida Water Management District)

C-111 Spreader Canal emptying into Florida Bay. (South Florida Water Management District)

runoff from 147 ppb to 41 ppb),[34] others claimed the state did not take water quality seriously. Environmentalists and the Miccosukee Indians pointed specifically to actions taken by the Florida legislature. In 2003, the legislature passed an act extending the deadline for meeting the phosphorus guidelines delineated in the Everglades Forever Act from 2006 to 2016. This extension applied to all areas in the Everglades except for the Arthur R. Marshall Loxahatchee National Wildlife Refuge and Everglades National Park.[35] Meanwhile, a state commission in charge of setting the phosphorus standard had decided in July 2003 to "allow the measurement of pollution levels to be averaged over time and at many different measuring stations." This had the potential of allowing phosphorus levels in one location "to drastically exceed the standard as long as other areas remain below it."[36] Claiming that these developments were "an absolute betrayal," in the words of Charles Lee of the National Audubon Society, many expressed disappointment with the state, convinced that the sugar industry was behind its actions, something that state officials denied.[37] Yet to individuals such as Mary Munson, Sun Coast regional director of the National Park Conservation Association, "Big Sugar's political influence and backroom deals overwhelmed good sense."[38]

With criticism of its water quality efforts mounting, and with federal funding lagging, State of Florida officials decided to take restoration matters into their own hands, thereby hoping to demonstrate the state's commitment to the environment. At the same time, the SFWMD hoped to break out of the Corps' plodding pace on restoration that had contributed to implementation delays. Such delays meant that "CERP was $1 billion over budget and two years behind schedule" by 2004, according to one report. Thus, in 2004, Governor Jeb Bush implemented what became known as Acceler8, a program developed by SFWMD Executive Director Henry Dean. Under this program, Bush allocated $1.5 billion of state money to "jump-start" the design and construction of eight different CERP projects.[39] These included:

- Work on water preserve areas (specifically Site 1, C-9, and C-11 impoundments, Acme Basin B, and seepage management in Water Conservation Areas 3A and 3B);
- Expansion of the EAA Stormwater Treatment Area (Compartments B and C);
- Construction of the EAA Reservoir (A-1) and Bolles and Cross canals;

- Hydrologic restoration of Southern Golden Gate Estates (Picayune Strand);
- Implementation of Biscayne Bay Coastal Wetlands Phase I;
- Construction of C-111 Spreader Canal;
- Construction of C-43 (Caloosahatchee River) West Storage Reservoir; and
- Construction of C-44 (St. Lucie Canal) Reservoir and Stormwater Treatment Area.[40]
- In the words of one SFWMD report, Acceler8 was supposed to "achieve 70 percent of the restoration plan's goals by 2011—five years ahead of the current schedule—while maintaining CERP's momentum."[41]

According to the scientific review committee, however, the eight projects under Acceler8 focused on preventing the flushing of water to the St. Lucie and Caloosahatchee estuaries and mainly benefited Lake Okeechobee, the northern estuaries, the Ten Thousand Islands National Wildlife Refuge, and Biscayne Bay.[42] Yet the program did ignite work on several projects that were passed on to the Corps once the Corps had authorization and funding to work on them (see below). By 2006, the SFWMD had completed full design work on the EAA reservoir, while also beginning some construction work on the Picayune Strand Restoration Project and the C-44 Reservoir and Stormwater Treatment Area. Over the next couple of years, it also started construction on the C-111 Spreader Canal (which did have positive benefits for Everglades National Park) and on the Biscayne Bay Coastal Wetlands project. Such progress led Ken Ammon, the SFWMD's deputy executive director over Everglades Restoration and Capital Projects, to call Acceler8 "a huge jumpstart."[43]

Yet that jumpstart came at a cost, including straining the already tense relationship between the Corps and the SFWMD. The SFWMD claimed the Corps moved too slowly, while the Corps, according to SFWMD Executive Director Carol Wehle, found it "unsettling . . . to have a local sponsor be able to step up and more quickly plan, design and permit, and construct projects."[44] According to Shannon Estenoz, Acceler8 worsened an anti-federal culture among the SFWMD that was difficult to eradicate, especially since it "damaged a lot of the [person-to-person] relationships that existed" between the Corps and the SFWMD.[45] Even as late as 2009, when Colonel Alfred Pantano became district engineer of the Jacksonville District,

Plugging the Prairie Canal, part of the Golden Gate Estates, Picayune Strand Restoration Project, under the Acceler8 program. (South Florida Water Management District)

tension remained. The SFWMD "had some very serious issues and concerns with us," Pantano stated.[46]

Meanwhile, some environmentalists did not like the idea of the State of Florida leading the CERP effort, fearing that the state's major goals lay in water supply for agricultural and urban interests, and that the state would not be subject to the same environmental assurance requirements under WRDA-2000 that the Corps had to meet. The SFWMD's development of a water supply plan for South Florida in 2005 seemed to confirm these fears. The plan "call[ed] for the forecast water needs of cities over the next 50 years to be met by 2010; for farmers' needs to be met by 2015; and for basic environmental standards to be met only after 2020," seemingly prioritizing water use for cities and agriculture above the environment.[47]

These criticisms highlighted the difficulty the Corps had in maintaining the consensus it had built around CERP. According to Pantano, CERP was based on a "fragile coalition of the willing" that could easily be broken. Indeed, Pantano continued, relationships were key to the success of CERP, but they could be damaged. When that occurred, the entirety of CERP was endangered.[48] Such disruption was evidenced in discussions over different CERP projects. As Thomas Van Lent of the Everglades Foundation explained, the Corps "had consensus of restoration of the Everglades as a concept," but not consensus on the many different aspects of CERP. In the eyes of Van Lent and other environmentalists, many parts of CERP "still remained very controversial."[49] According to Stuart Appelbaum, CERP issues were "more contentious" after the passage of the legislation "because they're real. You can't just paper them over; you've got to solve the issue."[50] Robert Johnson of the NPS agreed. "When you actually start[ed] implementing [projects] on the ground," he asserted, "not everybody loved [CERP] as much as they thought they did." The Corps and its partners, said Johnson, had to work to resolve "tradeoffs and conflicts."[51]

One example of the disagreements between stakeholders over specific CERP projects came with discussions of decompartmentalization, defined as the removal of "canals, levees and other barriers that impede the natural sheetflow of water into and through the historic Everglades."[52] Under CERP, the Corps planned to "reestablish the ecological and hydrologic connection of Water Conservation Area (WCA)-3A with WCA-3B and Everglades National Park."[53] Realizing the complexity of making such connections—and the scientific uncertainty surrounding it—the Corps proceeded slowly. Before moving forward with project planning, for example, the Jacksonville District wanted to see how Mod Waters worked and then, under the adaptive management principle, adjust how decompartmentalization proceeded. This cautious approach maddened environmentalists who saw decompartmentalization as the key to Everglades restoration. "We've got to get this decomp project moving along," Van Lent asserted. "That's a pretty uniformly held view from just about every major environmental group."[54] The NPS agreed. "If

we're going to restore the river of grass," Superintendent Dan Kimball declared, "we need to start looking at how we bring water down" into Everglades National Park.[55] Even though some Corps officials agreed that it was "crucial to get storage north of the conservation areas so that we can begin decompartmentalizing some of the central Everglades,"[56] some environmentalists did not think that the two sides were "speaking the same language" over how to proceed with decompartmentalization.[57]

Environmentalists' frustration stemmed in part from the fact that the Corps was unable to make any real progress on projects that would move water south of Lake Okeechobee into the Everglades. Jacksonville District officials recognized the need to stop flushing water to the St. Lucie and Caloosahatchee estuaries, but adequate conveyance and reservoir systems did not exist to enable the Corps to move the water south. Even if the Corps could get authorization to proceed on developing those systems, other issues would prevent it from moving water south. These included issues with endangered species, specifically the Cape Sable seaside sparrow and the need to prevent flooding of its habitat in the conservation areas (see below). They also included water quality problems (see below), as the technology had not yet been implemented to effectively treat large amounts of water moving south. Faced with these issues, decompartmentalization could not get off the ground, and the stakeholders could not agree on how to proceed. Water therefore continued to inundate the estuaries, but, as Jacksonville District Colonel Alfred Pantano explained, the Corps was unable to "fix" the problem.[58]

As mentioned above, the endangered species issue proved especially problematic to decompartmentalization. The Miccosukee Indians, for example, accused the Corps of "discriminatory water management" in its efforts to protect the endangered Cape Sable seaside sparrow. The sparrow had habitat in Everglades National Park, Big Cypress National Preserve, and the Southern Glades Wildlife and Environmental Area (Units 1 and 2), and the timing and amount of water flowing into this habitat was critical for its survival. Recognizing this, the Corps sometimes closed the S-12 gate so that water would not run from WCA 3A into the park and flood the sparrow's habitat. The tribe claimed that this action stacked water on tribal lands and degraded tree islands in the conservation area. The tribe also contended that an excess of water in WCA 3A adversely affected the snail kite, another endangered species, causing an alarming drop in its numbers. The U.S. Fish and Wildlife Service disagreed, stating that the decline of the snail kite had occurred because of long-term drought, not because of the shutting of the S-12 gate. The Corps, meanwhile, was caught in the middle, as the tribe clamored for "an end" to "more than a decade of discriminatory water management actions" and the FWS pleaded for sparrow protection.[59]

With cracks appearing in the coalition and no federal CERP project under construction, many observers predicted the demise of CERP. As early as 2006, the *St. Petersburg Times* declared that CERP was "on its last legs," while *Southeast Construction* asked "Is the Everglades Restoration Dying?" *The New York Times*, meanwhile, proclaimed that "the rescue of the Florida Everglades . . . is faltering."[60] Cynicism and frustration had replaced the hope and optimism of 2000, while the South Florida ecosystem continued to waver on the brink of total collapse. According to the CERP scientific review committee, "the delays afflicting CERP and foundational non-CERP projects are creating increased concern that the Everglades ecosystem may suffer irreparable losses before major restoration actions are taken to reverse the ecosystem decline."[61] Journalist Michael Grunwald put it more bluntly. Lake Okeechobee was fast "becoming a dead zone," he declared, while the St. Lucie and Caloosahatchee estuaries were experiencing "a massive increase in toxic 'red tides'" as well as "dramatic die-offs of manatees, dolphins and oysters."[62]

But the Everglades were not dead yet, and neither was CERP. After being elected governor of Florida in 2006, Charlie Crist overhauled the governing board of the SFWMD, helping to reduce some

Ft. Lauderdale area looking into WCA-2B. (South Florida Water Management District)

of the anti-federal sentiment that had simmered since Acceler8. The new SFWMD board promptly shifted policy by declaring that municipalities needed to have additional sources of water, other than what the SFWMD was already providing, if they wanted to continue to grow. "[T]hat was a major wakeup call" for the municipalities, stated Stuart Appelbaum, forcing them to start exploring wastewater reuse and conservation techniques.[63]

The passage of a water resources development act in 2007 (over President Bush's veto) was another important milestone. This law contained authorizations for the expenditure of $1.82 billion on three CERP projects: the Picayune Strand restoration project in Collier County, the Indian River Lagoon-South project, and the Site 1 Impoundment project. Not only did WRDA-2007 enable these projects to go forward but, as Stuart Appelbaum explained, it had a drastic psychological effect on restoration because "it got the authorization ball rolling again."[64] According to Jacksonville District Engineer Colonel Alfred Pantano, it showed "meaningful" progress on CERP— "actually doing something, as opposed to fighting and talking about it."[65]

The Picayune Strand restoration project, formerly known as the Southern Golden Gate Estates Ecosystem Restoration Plan, dealt with land included in the proposed Southern Golden Gate Estates development of the 1960s. Located between Alligator Alley and the Tamiami Trail, the project involved 55,000 acres and was called the "missing piece" in the restoration of southwestern Florida. According to project plans, the Corps and the SFWMD would plug 49 miles of canals and redirect the fresh flow of water across drained wetlands. The project also involved the removal of 227 miles of roads. Planners hoped that such actions would restore fish and wildlife habitat (including 82 square miles of Florida panther habitat), protect against saltwater intrusion by recharging the aquifer, and re-establish wetlands in the area. The Corps had finished the project implementation report and environmental impact statement for the project in 2004. The SFWMD got a jump on the project under Acceler8 by completing most of the design work by 2007 and by plugging the Prairie Canal, which allowed for increased water flow into Fakahatchee Strand State Preserve. When WRDA-2007 authorized the project, the Corps began in earnest with design and construction.[66]

Meanwhile, the Indian River Lagoon-South project, located in Martin and St. Lucie counties on Florida's east coast, was one on which the Corps and the SFWMD had worked since 1996. Characterized as "a poster child for environmental restoration in the long-term Everglades plan,"[67] its goal was to reverse the "damaging effects of pollution and unnaturally large fresh-water discharges" into Indian River Lagoon (described as "the most biologically diverse estuarine system in all of North America") and St. Lucie Estuary.[68] The project had several components, including construction of four above-ground water storage reservoirs and four stormwater treatment areas, enabling the capture of water from the C-23, C-24,

Florida Bay mangrove islands near Fakahatchee Pass. (South Florida Water Management District)

and C-25 north and southfork basins, the return of that water into a reservoir demarcated as C-44 for water supply and environmental benefits, and the reduction of phosphorus and nitrogen going into the St. Lucie Estuary. The Corps had completed the project implementation report in 2004, enabling Congress to authorize it in WRDA-2007. Before Congress gave its authorization, the SFWMD had built a test cell for the C-44 reservoir and had finished designing the project under the Acceler8 program.[69]

The Site 1 Impoundment project involved capturing excess water discharged to the Atlantic Intracoastal Waterway by the Hillsboro Canal in southern Palm Beach County and delivering that water into the Hillsboro Canal Basin, thereby alleviating demand on water from the Arthur R. Marshall Loxahatchee National Wildlife Refuge and from Lake Okeechobee. To store this water, the Corps would construct a 1,660-acre impoundment. According to Paul Souza of the U.S. Fish and Wildlife Service, this would "allow water levels in Loxahatchee to be more natural" by providing more "flexibility" in the management of water supply.[70] The Corps completed the project implementation report in October 2006, allowing for its inclusion in WRDA-2007. Prior to that authorization, the SFWMD completed approximately 30 percent of the design work for the project under Acceler8, and upon the passage of WRDA-2007, the Corps completed the design work and broke ground on construction.[71] With the

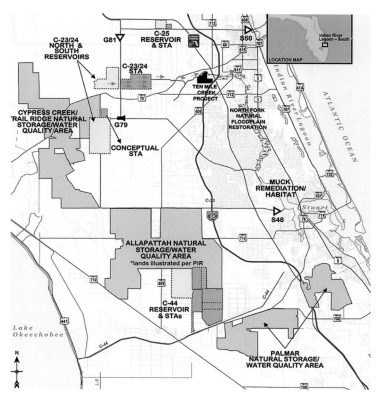

Indian River Lagoon-South project map. (U.S. Army Corps of Engineers, Jacksonville District)

successful authorizations and the beginning of construction of these projects, proponents of CERP finally began to feel that some movement was occurring. As Paul Souza, field supervisor for the U.S. Fish and Wildlife Service's South Florida Ecological Services office, declared, work on these projects constituted "an important historical time in the history of Everglades restoration."[72]

This feeling was heightened in June 2008 when Governor Crist—"against a backdrop of water, grass and birds"—announced that the State of Florida was planning to purchase 187,000 acres of U.S. Sugar lands in the EAA for $1.75 billion. Listening to the announcement, "dozens of advocates gathered in small groups, gasping with awe," one newspaper reported, "as if at a wedding for a couple they never thought would fall in love."[73] Characterized by Crist as "as monumental as the creation of the nation's first national park,"[74] the plan called for the state to allow farming for six years while it negotiated land swaps with other property holders in the EAA. After six years, the SFWMD would use the acreage to restore natural water flow and to store approximately a million acre feet of water. SFWMD officials hoped to have all details of the purchase finalized by 30 November 2008.[75]

Lt. Gen. Robert Van Antwerp, chief of engineers of the U.S. Army Corps of Engineers, speaks to the SFWMD Governing Board prior to the signing of the Site 1 Impoundment Project Partnership Agreement.
(South Florida Water Management District)

After the initial enthusiasm for the purchase passed, criticisms emerged. Some questioned whether the SFWMD would really use the lands to benefit the ecosystem, or whether it would just use them to further South Florida's development. Others saw the purchase as a bailout of U.S. Sugar, which was struggling economically. Still others believed that the buyout would have no impact on Everglades National Park unless the Mod Waters project (including modifications to Tamiami Trail to allow water to pass under the road) was completed. "If the federal government does not release the water down south," Kirk Fordham of the Everglades Foundation stated, "then the State of Florida will have built the world's largest swimming pool with absolutely no purpose at all." Perhaps most disconcerting, however, was the high price tag for the purchase. Some, including former Jacksonville District Engineer Colonel Terry Rice, who continued to advise the Miccosukee Indians (which sued the SFWMD over the purchase), feared that the expenditure of $1.75 billion by the state would divert resources from other restoration projects, causing delays. Others wondered whether the high price would prevent the purchase from ever happening. As Damien Cave wrote in *The New York Times*, the "eye-popping price" might mean "that the purchase will be too expensive and too unwieldy to reach its full potential."[76]

These criticisms proved prophetic. With Florida's property values shrinking (thereby reducing the ad valorem taxes that funded land purchases), the state was forced to downsize the purchase to 181,000 acres for $1.34 billion in November 2008. In the first months of 2009, it reduced it again to 72,800 acres for $536 million. As the economic recession deepened and property values continued to fall, the state had to reconfigure the deal in August 2010, reducing it to 26,800 acres for $197 million. In the meantime, other problems with the deal surfaced, including reports that some of the land was not suitable for restoration and that appraisals showed that the state was paying too much per acre for it. In addition, over a dozen projects were delayed because of the state's efforts to retain money for the purchase. The most controversial of these was the construction of the A-1 Reservoir in West Palm Beach County, one of the projects jumpstarted under the Acceler8 program. Although some entities, such as the Natural Resource Defense Council, opposed construction of the A-1 Reservoir because of fears it would destroy wildlife habitat,[77] the Miccosukee pressed for its construction, claiming that it would significantly improve water quality. On 15 May 2008, the SFWMD halted construction on the reservoir, claiming it "could not afford both" the reservoir and the U.S. Sugar land purchase. This stoppage led to the Miccosukee Indians successfully petitioning the U.S. District Court for the Southern District of Florida to issue an order on 31 March 2010 requiring the SFWMD to build the reservoir.[78]

Despite the controversy over A-1 Reservoir, the SFWMD completed the purchase of 27,000 acres of U.S. Sugar lands in October 2010.[79] Although not nearly on the level envisioned by Crist in 2008, the purchase sparked optimism in some observers. As Shannon

Construction of the A-1 Reservoir. (South Florida Water Management District)

Estenoz of the SFWMD explained, "If you would have told me 15 years ago that U.S. Sugar was going to put itself up for sale to the government for Everglades restoration, I would have laughed you out of the room. But," she continued, "it happened," proving that CERP itself—which at times seemed impossible—could succeed as well.[80] Some environmentalists were also pleased, believing, in the words of Charles Lee of Audubon of Florida, that "out of all the possible lands that they could have targeted, they got the best ones."[81] Nathaniel Reed, vice chairman of the Everglades Foundation, agreed. "The U.S. Sugar lands," he claimed, "are the most important lands that could possibly be acquired for the restoration of the Everglades."[82] At the same time, critics lamented the shrinking size of the land purchase, wondering if the acreage would really be enough to make any kind of difference.[83]

Indeed, as a whole, concerted progress on CERP was still lacking, the sugar purchase notwithstanding. A 2008 report by the scientific review committee emphasized that CERP was still "bogged down in budgeting, planning, and procedural matters and is making only scant progress toward achieving restoration goals." "Meanwhile," it continued, "the ecosystems that the CERP is intended to save are in peril."[84] Newspaper articles trumpeted this "harsh review,"

quoting William L. Graf, chairman of the committee, as saying that the nation was "in danger of losing" the Everglades "for our kids and their kids."[85] But restoration leaders pointed to signs that CERP was slowly gaining momentum, such as the development of an integrated water delivery schedule (completed in 2008). This schedule was "a strategy that coordinates and prioritizes projects for planning and construction based on project benefits, costs and funding," providing much-needed guidance for how to proceed with CERP projects once funding and authorization was available.[86]

Meanwhile, the nation's economic recession actually proved in some ways to be a boon. With the United States dealing with high unemployment, President Barack Obama signed into law a stimulus bill on 13 February 2009 intended to create new jobs and to "spur economic activity" by both businesses and consumers. One key component of the act—entitled the American Recovery and Reinvestment Act of 2009 (ARRA)—allocated $2 billion to the Corps for civil works construction projects. From this pot, the Corps designated more than $80 million to the Site 1 Impoundment project and to the Picayune Strand Restoration project (over $40 million each), enabling construction to begin on both in 2010. Other CERP projects also received ARRA funding, including the L-31N Seepage

Management pilot project, a melaleuca eradication proposal, and an adaptive assessment and monitoring program. In addition, the Kissimmee River restoration project received funding. As one report indicated, "The infusion of federal dollars [for restoration programs]—more than $200 million in fiscal year 2009—is the largest amount in any single year since Congress passed CERP in 2000."[87] According to Shannon Estenoz, the stimulus funding was a "huge shot in the arm" for CERP.[88]

In addition to funding, CERP got a boost with the appointment of two leaders to prominent positions in the Army. In 2009, Jo-Ellen Darcy, a former senior staffer in the U.S. Senate who had worked Everglades issues for many years, became assistant secretary of the Army (civil works). Terrence "Rock" Salt, former commander of the Jacksonville District and former head of the South Florida Ecosystem Restoration Task Force (his tenure as chairman had ended in 2003), became her deputy that same year. Both Darcy and Salt brought wisdom gained from years of working on restoration issues, and both provided credibility with environmental groups and with the Corps. Jacksonville District Commander Colonel Alfred Pantano explained that Darcy had "a very informed perspective . . . and appreciation" for CERP, while Salt, who had worked as the director of the Department of the Interior's restoration initiatives since 2003, brought the potential to ease the sometimes contentious relationship between Everglades National Park and the Corps. Their appointments brought optimism to all sides.[89]

Yet these glimmers of hope were dimmed by continuing issues with water quality. Because of the changes the State of Florida had made to the Everglades Forever Act in 2003, including the extension of the time limit for phosphorus reduction compliance, the Miccosukee Tribe sued the U.S. Environmental Protection Agency, claiming that the alterations violated water quality standards in the Clean Water Act but that the EPA had refused to take any action.[90] On 29 July 2008, the U.S. District Court agreed that the amendments to the act "change[d] Florida's previous water quality standards" and ordered the EPA "to comply with its duty under the Clean Water Act to approve or disapprove those change in a manner consistent" with the court's findings and conclusions.[91] When the EPA did not act on the court's order, the Miccosukee petitioned the court to find the agency in contempt. On 14 April 2010, the court directed the EPA to issue an amended determination no later than 3 September 2010 that would "specifically direct the State of Florida to correct the deficiencies in the Amended EFA and the Phosphorus Rule that have been invalidated." The court also insisted that the EPA "notify the State of Florida that it is out-of-compliance with the narrative and nutrient standards for the Everglades Protection Area" and that it direct the state "to measure on a yearly basis the cumulative impacts and effects of phosphorus intrusion beyond the 10 ppb standard."[92] The EPA complied with this order on 3 September 2010, requiring the state to reduce total phosphorus levels of water flowing into the Everglades

Protection Area so that stormwater treatment area discharge would "not cause an exceedance of the long-term criterion of 10 ppb." The EPA also told the state to expand stormwater treatment areas in the EAA by 42,000 acres. According to the determination, these actions would "ensure" that the quality of water flowing from the EAA and the C-139 Basin (located southwest of Lake Okeechobee and northwest of the Everglades Protection Area) would meet "a scientifically sound Water Quality Based Effluent Limit (WQBEL) in permits in the shortest time possible."[93]

Meanwhile, in November 2008 and June 2009, total phosphorus levels in the Arthur R. Marshall Loxahatchee National Wildlife Refuge exceeded the Consent Decree's long-term levels. The Miccosukee Tribe therefore filed motions under existing litigation asking the court to do three things: make a formal declaration of Consent Decree violations; declare that the state had breached its commitment to the Special Master in the case to restore water quality; and order the state to complete construction of the A-1 reservoir in the EAA (see above discussion). The court complied with this request on 31 March 2010, ordering the Special Master to conduct evidentiary hearings on the issues.[94]

The results of this litigation—which, in the eyes of the Miccosukee, was necessary to get the State of Florida to fulfill its water quality commitments—had the potential to drastically affect the state's ability to fund CERP projects. Would the state be "focused on water quality to the exclusion of everything else?" Stuart Appelbaum wondered.[95] Shannon Estenoz believed that was the case. The state "has been plopped on a water quality path for the foreseeable future," she declared, and it no longer had the flexibility to pursue CERP projects according to the priorities it and the Corps had set.[96] In the words of the NPS, "existing shortfalls in State funding will make compliance with these water quality requirements very difficult, and severely limit the SFWMD's ability to continue their cost sharing agreements for the CERP."[97]

Carol Wehle, executive director of the SFWMD, was even blunter. Litigation over water quality, she said, "could bring [CERP] to its knees." For one, she explained, the SFWMD did not "have the financial wherewithal to do those water-quality projects and be the fiscal partner" in CERP. For another, she wondered whether it was even technically feasible to construct "a large-scale facility" that could consistently cleanse water to 10 ppb "through rainy seasons and drought seasons." The SFWMD would make its best effort, but if it could only consistently clean water to 12 ppb, she asked hypothetically, "are you going to starve the Everglades of water forever?"[98] Jacksonville District Commander Alfred Pantano sympathized with these difficulties, but he claimed that CERP could continue unhindered until 2014 without any additional outlays from the state because of credits the SFWMD had accrued. Yet he still considered water quality "the single most critical item . . . that somehow needs to be addressed."[99] Dan Kimball, superintendent of Everglades National

Park, agreed. "Ultimately, we want the quality, quantity, and timing of distribution of water to be right."[100]

Aside from the funding issues, the problems over water quality had the capacity to halt progress on CERP in other ways. As Ken Ammon of the SFWMD explained, if the 10 ppb standard had to be met before water could be pumped into the Everglades, it could take "a decade or decades to get to that level." In the meantime, areas of the Everglades in desperate need of water would continue to suffer, and "a large part of the Everglades [could] be lost." Ammon did not dispute that maintaining water quality was important, pointing to the fact that by the middle of 2011, the SFWMD would have nearly 60,000 acres of stormwater treatment areas in operation. In his mind, however, it was preferable to send water into the Everglades to stop soil oxidation, even if that water had phosphorus levels exceeding 10 ppb.[101] But until the state met the 10 ppb standard, litigation would continue because the Miccosukee had pledged not to "ignore the water quality problems that exist in the Everglades."[102]

Because of the need for more stormwater treatment areas, the increased water storage proposed by CERP would not be forthcoming in the near future. Although the SFWMD announced at the end of 2010 that it would restart construction on the A-1 Reservoir, the structure was now planned to be a flow-equalization basin (or shallow reservoir) that would more effectively move water through stormwater treatment areas, rather than the originally proposed 190,000 acre-feet reservoir. In addition, the SFWMD would have to use the U.S. Sugar land for stormwater treatment instead of for storage. The NPS held out hope that the state would continue to acquire parcels of U.S. Sugar land so that the SFWMD could eventually store an additional 500,000 acre-feet of water, but the economic woes of the state and the need to focus on water quality made that possibility dubious.[103]

Questions also remained about aquifer storage and recovery (ASR). In the original CERP program, ASRs were promoted as an alternative to reservoirs, but there were questions about whether ASRs would actually work. The CERP program proposed several ASR pilot projects to gather scientific information, but they had made only modest progress in the 2000s. At least one had pumped water contaminated with arsenic back to the surface. In 2008, the Corps and the SFWMD issued an interim report on the ASR program, stating that pilot projects had been constructed along the Kissimmee River

The Arthur R. Marshall Loxahatchee National Wildlife Refuge. (Everglades National Park)

and the Hillsboro Canal to allow for "cycle testing" over the next several years, but other pilots (such as the Lake Okeechobee ASR pilot) were still in process. "To date, no 'fatal flaws' have been uncovered that might hinder the implementation of CERP ASR," the report concluded. "It is believed that ASR will work almost anywhere in south Florida on some scale and with some degree of efficiency."[104] Others were more lukewarm about ASR prospects. As Colonel Pantano explained, the Corps had not "dismissed" ASRs, but he did not believe they would be "a practical solution any time soon."[105] If they were used, Ken Ammon of the SFWMD asserted, it would be on a "much smaller scale than we had originally envisioned."[106]

With storage possibilities diminishing, and with water quality dominating the discussion, the Committee on Scientific Review of Everglades Restoration Progress noted again in 2010 the slow progress of CERP and the continuing decline of the health of the overall South Florida ecosystem. Yet the report saw some positive trends as well, including the stability of or increase in populations of wading birds, Cape Sable seaside sparrows, and panthers, as well as successful control of some exotic plant species. In addition, water conservation efforts, coupled with a slower-than-predicted popula-

tion growth, had resulted in "substantially lower" water demands by urban populations, increasing the amount of water available for the environment. Moreover, with three CERP projects under construction, the committee could report "tangible progress" on CERP implementation. "After years of delay," the committee concluded, "it is critically important to maintain this momentum to minimize further degradation of the system."[107]

Some believed that issuing plaudits of what CERP had accomplished was just grasping at straws to try to find anything positive in the plodding pace of CERP. Indeed, many actively involved in South Florida restoration were insistent that more needed to be done. Even though projects such as Picayune Strand, Site 1, and Indian River Lagoon-South were proceeding, some observers argued, they were mere showcase projects that would not have any real meaningful impact on CERP. Picayune Strand, Colonel Alfred Pantano explained, "is a good project. But in the big scheme of Everglades restoration, you would never do that project first." The projects were authorized and funded, Pantano continued, in part because they were non-controversial. "We can't continue to do these compartmentalized projects, which on their own may function, but don't achieve the

The Hillsboro ASR pilot project. (South Florida Water Management District)

end," Pantano explained. "We've got to get to the heart of the matter, which is the central part of the Everglades, so we can convey the water south where it needs to go."[108]

One project that had the potential to impact the Everglades positively was the construction of a one-mile bridge over Tamiami Trail, part of the Mod Waters project designed to accelerate the flow of water into Everglades National Park. Several interest groups saw the Tamiami Trail bridge as "an important first step in reconnecting the natural flow path of the Everglades and putting water where it belongs."[109] Since its construction, Tamiami Trail had served as a virtual dam to the southward flow of water into Shark River Slough. Allowing water to flow again would provide benefits to Everglades National Park—as long as the Corps could actually get meaningful amounts of water to the trail. Regardless, some claimed that the Tamiami Trail project would also benefit Miccosukee Indian lands by alleviating the backup of water in WCA 3A.[110]

The Miccosukee, however, had a different perspective, calling the project the "White Elephant Bridge" and the "Bridge to Nowhere." It accused the Department of the Interior and the Corps with developing the plan for its construction "behind closed doors" and without Miccosukee involvement, even though it had the potential to affect Miccosukee land.[111] Indeed, the Miccosukee declared that the project would "destroy ancient sacred archeological sites and artifacts of the Tribe."[112] Instead of elevating Tamiami Trail, the tribe proposed that the Corps "clean out the sediment and vegetation built up downstream of the existing culverts and structures" which, tribal officials believed, would enable necessary southward flow of water to proceed.[113] The Corps and the Interior Department, however, continued to move forward with the bridge, causing the tribe to accuse them of pursuing "an unconstitutional path that has resulted in an ad-hoc process in which the Tribe was not allowed to participate in a fulsome manner."[114] In Pantano's view, "the tribe ultimately supports what we're trying to achieve" with the bridge, but it was using litigation to address concerns about water quality, seepage, and especially the prospect of flooding tribal lands to clean the water.[115]

Although the Miccosukee objected to the Tamiami Trail bridge, others saw it as a concrete indication that Mod Waters and Everglades restoration was proceeding. But clearly much more needed to be accomplished in the years after 2010 to alleviate continuing damage to the South Florida ecosystem. Some, such as the Miccosukee, saw water quality as the most important issue. Colonel Pantano of the Corps did not disagree, declaring that effectively resolving the federal government's role in water quality would be a key factor in the future, as would deciding how to clean up Lake Okeechobee and its "legacy source of phosphorus" from years of agricultural runoff. In fact, the Miccosukee characterized Lake Okeechobee water quality as "the elephant in the restoration room," stating that it was essential that the EPA more vigorously implement National Pollutant Discharge Elimination System permits and enforcement for the lake.

The Tamiami Trail. (South Florida Water Management District)

Otherwise, tribal representatives declared, "there is no assurance that water quality will ever be met in Lake Okeechobee or in its discharges to the Everglades and the estuaries."[116]

Just getting more water directed south to the Everglades was also important. But in order to do this, the Corps, the SFWMD, and the Interior Department had to solve seepage issues so that landowners surrounding Everglades National Park would be protected from flooding. As Everglades National Park Superintendent Dan Kimball explained, seepage management presented some "challenging" problems, especially because seepage projects as proposed in the CERP plan were "no longer feasible." Until the issue was solved, and until adequate storage and conveyance systems could be built south of Lake Okeechobee, no more water could be moved into the Everglades, and Lake Okeechobee would continue in a never-ending cycle of being either too low (in drought years) or too high (causing destruction to the Caloosahatchee and St. Lucie estuaries).[117]

Controlling invasive species was another important component of Everglades restoration that needed more attention. Although popular television programs focused on the presence of pythons in South Florida, other non-native species were also problematic, including lygodium (or climbing fern), melaleuca, Brazilian pepper, Australian pine, and exotic fish. The Corps had taken various measures to control these species, including partnering with the University of Auburn to use black Labradors to sniff out pythons. The Corps also worked on a melaleauca eradication program as part of CERP (experiencing some success when it released a moth that successfully helped eliminate the exotic vegetation), but according to Pantano, the Corps did not have the money to battle all of South Florida's invasive species.[118] One solution, according to Lieutenant Colonel Michael Kinard, deputy district engineer of the Jacksonville District, was to include provisions for dealing with exotic species in management plans so that funding was available when invasives inevitably appeared in a project area. "We're trying to implement" this measure "in current and future projects," Kinard explained, but "that's an area [where] we still need to do a better job."[119] According to Paul Souza of the U.S. Fish and Wildlife Service, managing the invasive species problem was "a very key part of Everglades restoration" because one of CERP's goals was to improve conditions for natural flora and fauna in South Florida.[120]

With all of these issues, the key, according to Robert Johnson of the NPS, was patience. "I don't think any of us expected that we would go for almost ten years before we'd see a groundbreaking in CERP," Johnson asserted, but that was the nature of environmental restoration. "You dig the hole one shovel load at a time," he said, and "you're not going to fill it back very quickly."[121] But that patience was sometimes hard to maintain, and Stuart Appelbaum's "biggest concern" was whether the interests that cooperated to get CERP passed would stick with "multi-generational implementation." CERP was "more complex and more difficult than anybody thought," Appelbaum said, but if stakeholders would stay the course, Appelbaum was "very optimistic that we'll achieve the result."[122] Carol Wehle shared that optimism, recognizing the "passion" of those working on CERP. "They truly believe in Everglades restoration," Wehle said. "And they care about it so much they will not allow it to fail."[123]

Others were more concerned, seeing the construction of compartmentalized and out-of-sequence projects, the litigation over water quality, and the lack of adequate progress in moving water south from Lake Okeechobee as signs that CERP was in serious jeopardy. The fact that the relationships conceived in the program's development were falling apart, and no entity had enabled all sides to come together, exacerbated the problems. "Looking back, I don't know how we ever got CERP authorized," Colonel Pantano declared. "What happened back then, we need it now, desperately." But instead of cooperation, Pantano saw litigation and tensions between the SFWMD and the Corps. "We're at such a critical point with this [program]," Pantano explained. "We need people to step up and lead, to lay the sword on the table."[124]

Part of the responsibility, in Pantano's view, lay with the South Florida Ecosystem Task Force which, he and others believed, had failed to effectively coordinate the interests of different parties. Composed of high-ranking federal, state, local, and tribal representatives, the task force had shown little ability to get the major stakeholders to meet at the negotiating table, giving way instead to an atmosphere of litigation and intransigence. Rather than providing the necessary coordination, compromise, and leadership, Pantano asserted, "the cynicism and sarcasm about [the task force] has become so great that people go there with an attitude that nothing is going to be accomplished." Pantano hoped that the organization would take stronger action in confronting the "difficult and controversial and litigious" issues surrounding CERP. "All the good work [the task force] did leading up to 2000," he declared, was close to disappearing.[125]

As hope in CERP's ability to restore the South Florida ecosystem ebbed and flowed, one Corps project stood as "a sign" that restoration could really produce what it promised: the Kissimmee River restoration.[126] In 2001, the Corps had finished Phase I of the restoration, with Phase II completion following in 2009, resulting in "continuous water flow" to 19 miles of the river. Phase III, which consisted of backfilling C-38 and restoring water flow to another 8 miles of the river, was in progress, with an estimated completion date of 2015. When Phase III was done, the Corps declared, "more than 40 square miles of river-floodplain ecosystem will be restored, including almost 20,000 acres of wetlands and 44 miles of historic river channel."[127]

Even with only the first two phases complete, the ecological improvements in the Kissimmee River Basin were remarkable. A 2010 Jacksonville District publication detailed some of these advances: thriving wetland plants, decreases in organic deposits on the river bottom, increases in dissolved oxygen, the return of largemouth bass and sunfishes, and the increase and restoration of several bird species, including white ibis, great egret, snowy egret, little blue heron, ducks, and black-necked stilts.[128] These improvements were noted in a 25 January 2009 article in the St. Petersburg Times by Jeff Klinkenberg, who described a trip along the Kissimmee River with guide Paul Gray, a biologist with the National Audubon Society. As they traversed portions of the restored river, the two saw ducks, a great blue heron, sandhill cranes, and glossy ibis, all indications that restoration was working. Gray told Klinkenberg about the flooding of four miles of the old floodplain in August 2008 after Tropical Storm Fay hit Florida. "It was the most exciting thing I've seen," Gray remarked, because, as Klinkenberg noted, it indicated that "the canal remembers how to be a river." As the pair took in the scene, it seemed almost idyllic:

A marsh wren hollered. A Northern harrier, a beautiful hawk, soared inches above the millet. We saw a pair of wood storks. We heard a limpkin squawking. A raccoon and her young swam across the creek in front of us. An otter poked its head up, glanced our way and vanished as quickly. Alligators tried to get warm on the bank.

Seeing this tumult of life, Klinkenberg remembered something his friend Richard Coleman once told him: "A river is a cauldron of birth, death and diversity."[129] With restoration proceeding, C-38 was again becoming that cauldron. It was again becoming a river.

Success in Kissimmee restoration not only provided hope, but served as a good example of what could happen if patience with CERP were maintained. Indeed, the years between 2000 and 2010 provided several important lessons learned for the Corps and other stakeholders in water management. The first—as indicated by the Kissimmee restoration—was merely that reversing the ecological damage in South Florida was possible, if not extremely difficult to achieve. "Does ecological restoration work?" Appelbaum asked. "I think we've answered that on the Kissimmee."[130] Yet these years also

Kissimmee River restoration. (South Florida Water Management District)

showed the need for the Corps to be flexible in its project processes. Ecological restoration required dexterity, an ability to change directions if conditions warranted it, and a need to implement projects quickly before more damage occurred. The Corps' planning processes did not necessarily provide that flexibility—a situation that not only made Everglades restoration difficult but that also had repercussions for a Corps that was trying to become a major player in ecosystem restoration.

The efforts to implement CERP also indicated that firm commitments to restoration were necessary from both the federal government and the state of Florida. CERP could limp along if one party evinced a stronger commitment than the other, but difficulties invariably resulted, including inequities in project spending and hard feelings between agencies. Finally, the years since 2000 had shown that maintaining the fragile coalition of interest groups that had developed CERP was as difficult as originally expected. Invariably, some groups wanted CERP and the Corps to address their own issues, regardless of whether solving those issues advanced CERP in a systematic way. For CERP to succeed, those groups had to be willing to compromise and even sacrifice some of their own interests for the good of the program. If groups were unwilling to do that, stakeholders would be unable to maintain the "laser beam" focus that Paul Souza believed was essential for CERP to succeed. "Our success is solely founded upon our ability as partners to set differences aside [and] focus on our common vision," Souza stated, but the willingness of interest groups to do this remained an issue.[131]

Even though these lessons seemed clear to observers in 2010, exactly how the Corps would use them in the upcoming years remained to be seen. By the end of 2010, Everglades restoration stood at "a crucial crossroads," teetering on a line separating failure and success.[132] To some watching this balancing act, it was only a matter of time before it all came crashing down. "If [the stakeholders] don't get our act together" soon, Colonel Pantano observed in 2011, "we're going to have a massive crater in the Everglades restoration program, and we're going to come to a screeching halt."[133]

Events early in 2011 prompted further concern. Florida's state legislature passed a budget "requiring deep budget cuts for the state's five water districts," including the SFWMD, leaving some wondering whether CERP projects would end up on the cutting-room floor. "I think it's going to shut Everglades restoration down," Miami attorney Eric Buermann predicted. "I don't think you can accomplish anything with that level of funding."[134] Meanwhile, U.S. District Judge Alan Gold issued an order in the water quality litigation requiring the EPA to take responsibility for improving water quality in South Florida, since the state seemed incapable of doing so. "Protection of the Everglades requires a major commitment which cannot be simply pushed aside in the face of financial hardships, political opposition, or other excuses," Gold warned. "These obstacles will always

exist, but the Everglades will not—especially if the protracted pace of preservation efforts continues at the current pace."[135]

Such bleak developments and alarming pronouncements recalled Marjory Stoneman Douglas's declaration in 1948 that the Everglades were in their "eleventh hour."[136] In the sixty years that followed, the ecosystem had faced drainage, drought, and destruction, leaving it on its deathbed. CERP had promised healing, but the doctors could not agree on how to administer the medicine, nor could they find the funding to implement the necessary cures. Meanwhile, life continued to ebb from the land, although a heartbeat remained. CERP proceeded against the backdrop of an ecosystem in danger, at the confluence of numerous and disparate competing interests. As urgency mounted, the bonds of cooperation strained. At stake was no less than "the greatest concentration of living things on this continent."[137]

Endnotes

[1] Dennis Duke, "Comprehensive Everglades Restoration Program," Powerpoint presentation, File TP 200-1F CERP Program Files (2000), CERP Code: 01.09.02, South Florida Ecosystem Restoration Task Force, Box 1, Location 4078, JDAR. Before the passage of CERP, the Corps had executed a master design agreement with the SFWMD that covered pre-construction engineering and design work on six pilot projects and 56 other components, spanning 38 years at an estimated $712 million, to be cost-shared between the Corps and the SFWMD. U.S. Army Corps of Engineers and South Florida Water Management District, "The Comprehensive Everglades Restoration Plan: Implementing CERP," Powerpoint presentation, File LOW Kick-off Meeting, Box 2, Lake Okeechobee Watershed, JDAR; South Florida Water Management District, *2001 Executive Summary: Everglades Consolidated Report, January 1, 2001* (West Palm Beach, Fla: South Florida Water Management District, 2001), 2, 24.

[2] Committee on Independent Scientific Review of Everglades Restoration Progress (CISREP), Water Science and Technology Review Board, Board on Environmental Studies and Toxicology, Division on Earth and Life Studies, *Progress Toward Restoring the Everglades: The First Biennial Review – 2006* (Washington, D.C.: The National Academies Press, 2007), 8-9.

[3] Stuart Appelbaum interview by Joshua Pollarine, 27 September 2010, Jacksonville, Florida, 4 (hereafter referred to as Appelbaum II interview). A 2002 article in *National Parks* noted that "the first step" in CERP was the completion of "a cumbersome planning process." Phyllis McIntosh, "Reviving the Everglades," *National Parks* 76 (February 2002): 30.

[4] All quotations in this paragraph are from "Programmatic Regulations for the Comprehensive Everglades Restoration Plan," *Federal Register* 68, 12 November 2003, 64200-64206.

[5] Kim Todd, "Who Gets the Water?" *Sierra* 88 (January 2003): 10.

[6] Robert Johnson interview by Joshua Pollarine, 18 October 2010, Homestead, Florida, 6 (hereafter referred to as Johnson II interview).

[7] All quotations in Jack E. Davis, *An Everglades Providence: Marjory Stoneman Douglas and the American Environmental Century* (Athens: The University of Georgia Press, 2009), 599-600. Critics still claimed that the Corps was relying too much in its planning on ASRs, which were still considered expensive and untested. See National Research Council et al.,

Aquifer Storage and Recovery in the Comprehensive Everglades Restoration Plan: A Critique of the Pilot Projects and Related Plans for ASR in the Lake Okeechobee and Western Hillsboro Areas (Washington, D.C.: National Academy Press, 2001), 1.

[8] CISREP et al., *Progress Toward Restoring the Everglades: The Second Biennial Review – 2008* (Washington, D.C.: The National Academies Press, 2008), 7.

[9] Carol Wehle telephone interview by Nicolai Kryloff, 17 March 2011, transcript, 4.

[10] Water Resources Development Act of 2000 (114 Stat. 2572 at 2681-2683, 2686-2690).

[11] Dennis Duke interview by Joshua Pollarine, 29 September 2010, Jacksonville, Florida, 9.

[12] CISREP et al., *Progress Toward Restoring the Everglades: The Second Biennial Review – 2008*, 82.

[13] Shannon Estenoz telephone interview by Joshua Pollarine, 20 October 2010, transcript, 7; see also Ken Ammon telephone interview by Joshua Pollarine, 4 March 2011, transcript, 6.

[14] "The Comprehensive Everglades Restoration Plan: Implementing CERP," Powerpoint presentation.

[15] Quotations in Johnson II interview, 11; see also Colonel Alfred Pantano telephone interview by Matthew Godfrey, 13 January 2011, transcript, 7.

[16] Paul Souza telephone interview by Joshua Pollarine, 8 March 2011, transcript, 15.

[17] Appelbaum II interview, 5.

[18] Estenoz interview, 12.

[19] Quotation in Brooks Moore telephone interview by Joshua Pollarine, 3 March 2011, transcript, 3; see also "Master Agreement Between the Department of the Army and South Florida Water Management District for Cooperation in Constructing and Operating, Maintaining, Repairing, Replacing, and Rehabilitating Projects Authorized to be Undertaken Pursuant to the Comprehensive Everglades Restoration Plan," 13 August 2009 <http://www.evergladesplan.org/pm/program_docs/master_agreement _cerp.aspx> (22 November 2010). According to Estenoz, the master agreement "was languishing for six and a half years" before Terrence "Rock" Salt became deputy assistant secretary of the army (civil works) and took action.

[20] See "The Comprehensive Everglades Restoration Plan: Implementing CERP," Powerpoint presentation; Appelbaum II interview, 4; "Overall CERP Schedule and Funding: Projected Combined Funding Requirement," in "Comprehensive Everglades Restoration Plan – CERP," Powerpoint presentation, File TP 200-1F CERP Program Files (2001), CERP Code 01.09.02, South Florida Ecosystem Restoration Task Force, Box 1, Location 4078, JDAR.

[21] Quotations in Ammon interview, 5; see also CISREP et al., *Progress Toward Restoring the Everglades: The Second Biennial Review – 2008*, 84; Abby Goodnough, "Vast Effort to Save Everglades Falters as U.S. Funds Dwindle," *The New York Times*, 2 November 2007. The State of Florida had overpaid its required costs by 2007.

[22] CISREP et al., *Progress Toward Restoring the Everglades: The First Biennial Review – 2006*, 10.

[23] Quotation in Estenoz interview, 3; see also Goodnough, "Vast Effort to Save Everglades Falters as U.S. Funds Dwindle."

[24] Thomas Van Lent interview by Joshua Pollarine, 19 October 2010, Palmetto Bay, Florida, 4.

[25] "Water, Bird and Man – The Everglades," *The Economist* 371 (8 October 2005): 32-34.

[26] CISREP et al., *Progress Toward Restoring the Everglades: The First Biennial Review – 2006*, 1-2, 9. The six pilot projects included the Lake Okeechobee, Caloosahatchee River (C-43), and Site 1 Impoundment aquifer storage and recovery projects, as well as the L-31N Seepage Management project, the Wastewater Reuse Technology project, and the Lake Belt In-Ground Reservoir Technology project.

[27] U.S. Government Accountability Office, *South Florida Ecosystem: Restoration is Moving Forward but Is Facing Significant Delays, Implementation Challenges, and Rising Costs*, Report GAO-07-520 (Washington, D.C.: Government Accountability Office, 2007), 6-7.

[28] Dan Kimball telephone interview by Joshua Pollarine, 1 March 2011, transcript, 3-4.

[29] Dexter W. Lehtinen, "The Myth of Everglades Restoration: An Additional View of the Miccosukee Tribe of Indians of Florida," 6 (documents pertaining to water quality litigation in possession of authors).

[30] U.S. Department of the Interior, Office of Inspector General, *Modified Water Deliveries to Everglades National Park: Audit Report*, Report No. C-IN-MOA-0006-2005 (Washington, D.C.: Office of Inspector General, 2006), i.

[31] Quotations in Pervase A. Sheikh, "Everglades Restoration: Modified Water Deliveries Project," 5, CRS Report for Congress, 17 March 2005, Order Code RS21331; see also Consolidated Appropriations Resolution, 2003 (117 Stat. 11).

[32] "Mod Waters & CSOP: 'It's Still the Park's way or the highway'" (documents pertaining to water quality litigation in possession of authors).

[33] Department of the Interior and Related Agencies Appropriations Act, 2004 (117 Stat 1241, 1251).

[34] CISREP et al., *Progress Toward Restoring the Everglades: The First Biennial Review – 2006*, 6, 151-156.

[35] U.S. Department of the Interior, National Park Service, Everglades National Park, "South Florida Natural Resources Center Water Quality Program," document provided by South Florida Natural Resources Center, National Park Service, Homestead, Florida.

[36] Ryan Dougherty, "Sugar Deal Sours 'Glades Restoration," *National Parks* 77 (September/October 2003): 10-11.

[37] As quoted in Michael Grunwald, "Sugar Plum: Jeb v. the Everglades," *The New Republic* 228 (12 May 2003): 16.

[38] As quoted in Dougherty, "Sugar Deal Sours 'Glades Restoration," 10.

[39] "Water, Bird and Man – The Everglades"; Nathaniel Reed telephone interview by Nicolai Kryloff, 11 March 2011, transcript, 4 [hereafter referred to as Reed telephone interview]. As Stuart Appelbaum remembered, the state "got pretty frustrated" and decided to just "get it done" themselves. Appelbaum II interview, 10.

[40] Joni Warner email to Nicolai Kryloff, 28 March 2011, personal communication in possession of the authors.

[41] Quotation in South Florida Water Management District, *2006 South Florida Environmental Report: Executive Summary* (West Palm Beach, Fla.: South Florida Water Management District, 2006), 26; see also Ammon interview, 9-10.

[42] CISREP et al., *Progress Toward Restoring the Everglades: The First Biennial Review – 2006*, 9, 140.

[43] Quotation in Ammon interview, 12 (see also pp. 9-11); see also CISREP et al., *Progress Toward Restoring the Everglades: The Second Biennial Review – 2008*, 82-84. Even though Acceler8 did not have a huge impact on Everglades National Park, the NPS still liked the program because, as Superintendent Dan Kimball explained, "it is really important to show progress on the ground." Kimball interview, 17-18.

[44] Wehle interview, 9.

45 Estenoz interview, 4.

46 Colonel Alfred Pantano telephone interview by Matthew Godfrey, 13 January 2011, transcript, 7.

47 "Water, Bird and Man – The Everglades."

48 Pantano interview, 3, 5-6.

49 Van Lent interview, 3.

50 Appelbaum II interview, 11.

51 Johnson II interview, 7-8.

52 "CERP Project: Water Conservation Area 3, Decompartmentalization & Sheet Flow Enhancement (Decomp)" <http://www.evergladesplan.org/pm/projects/proj_12_wca3_1.aspx> (13 November 2010).

53 U.S. Army Corps of Engineers, Jacksonville District, "Project Status Report: CERP WCA 3 DECOMP" <http://www.evergladesplan.org/pm/projects/project_docs/status/proj_12_current.pdf> (13 November 2010).

54 Van Lent interview, 8.

55 Kimball interview, 11.

56 Kim Taplin telephone interview by Nicolai Kryloff, 25 February 2011, transcript, 15.

57 Van Lent interview, 8.

58 Pantano interview, 21-23 (quotation on p. 23).

59 Quotations in Dexter W. Lehtinen to Susan Conner, Planning Division, U.S. Army Corps of Engineers, Jacksonville District, 1 February 2010 (documents pertaining to water quality litigation in possession of authors); see also U.S. Fish and Wildlife Service South Florida Ecological Services Office, "Species Conservation Guidelines, South Florida: Cape Sable seaside sparrow," Draft, 8 October 2003 <http://www.fws.gov/verobeach/images/pdflibrary/Cape_Sable_Seaside-Sparrow_Conservation_Guidelines.pdf> (30 March 2011); Paul Souza telephone interview by Joshua Pollarine, 8 March 2011, transcript, 6.

60 Craig Pittman, "A Fearful Bog," St. Petersburg Times, 12 March 2006; Debra Wood, "Life Support: Is the Everglades Restoration Dying?" Southeast Construction 7 (1 November 2007): 28; Abby Goodnough, "Vast Effort to Save Everglades Falters as U.S. Funds Dwindle," The New York Times, 2 November 2007.

61 CISREP et al., Progress Toward Restoring the Everglades: The Second Biennial Review – 2008, 88.

62 Michael Grunwald, "Everglades," Smithsonian 36 (March 2006): 57; see also Grunwald, The Swamp, 362-363.

63 Appelbaum II interview, 23-24.

64 Quotation in Armistead, "Cash and Commitment Rescue Everglades Replumbing"; see also Duke interview, 12.

65 Pantano interview, 3.

66 U.S. Army Corps of Engineers, Jacksonville District, Reviving the Picayune Strand (Jacksonville, Fla.: U.S. Army Corps of Engineers, Jacksonville District, 2010), 1-4; "Comprehensive Everglades Restoration Plan, Picayune Strand Restoration (formerly Southern Golden Gate Estates Ecosystem Restoration Plan), Final Integrated Project Implementation Report and Environmental Impact Statement" < http://www.evergladesplan.org/pm/projects/ docs_30_sgge_pir_final.aspx> (20 November 2010); "CERP Project: Picayune Strand Restoration" <http://www.evergladesplan.org/pm/projects/proj_30_sgge.aspx> (20 November 2010); Appelbaum II interview, 14; Thomas F. Armistead, "Feds Award First Everglades Contract," Engineering News-Record 253 (23 November 2009): 1; Taplin interview, 4; Souza interview, 13, 17. Lieutenant Colonel Michael Kinard, deputy district engineer of the Jacksonville District, noted that just the work that the SFWMD had done under Acceler8 had resulted in progress. "Getting to see [the area] after just seven years' worth of growth is just amazing," he asserted. "Once you return nature back and just let the water take its natural course it's amazing how fast that it starts to rebound." Kinard telephone interview by Joshua Pollarine, 7 March 2011, transcript, 16.

67 As quoted in Libby Wells, "Last Public Hearing on Indian River Lagoon Restoration," The Palm Beach Post, 12 January 2004, copy in File: IRL-S (04), CERP Code 07.98, Box 1, Location 4413, JDAR.

68 First quotation in U.S. Army Corps of Engineers, Jacksonville District, Indian River Lagoon – South: Facts & Information (Jacksonville, Fla.: U.S. Army Corps of Engineers, Jacksonville District, 2010), 1; second quotation in "CERP Project: Indian River Lagoon – South" <http://www.evergladesplan.org/pm/projects/ proj_07_irl_south.aspx> (20 November 2010).

69 "CERP Project: Indian River Lagoon – South" <http://www.evergladesplan.org/pm/projects/ proj_07_irl_south.aspx> (20 November 2010); U.S. Army Corps of Engineers and South Florida Water Management District, "Final Integrated Project Implementation Report and Environmental Impact Statement, Indian River Lagoon – South" <http://www.evergladesplan.org/pm/studies/irl_south_pir.aspx> (20 November 2010); John Paul Woodley, Jr., Assistant Secretary of the Army (Civil Works), "Record of Decision: Central and Southern Florida Project, Indian River Lagoon—South," 25 January 2006, File: IRL-S (06), CERP Code: 07.02.01, Box 1, Location 4413, JDAR; Appelbaum II interview, 16.

70 Souza interview, 19.

71 U.S. Army Corps of Engineers, Jacksonville District, Fran Reich Preserve/Site 1 Impoundment (Jacksonville, Fla.: U.S. Army Corps of Engineers, Jacksonville District, 2010), 1; "Central and Southern Florida Project, Comprehensive Everglades Restoration Plan, Site 1 Impoundment Project, Final Integrated Project Implementation Report and Environmental Impact Statement" <http://www.evergladesplan.org/pm/projects/docs_40_site_1_pir.aspx> (20 November 2010); "CERP Project: Site 1 Impoundment" <http://www.evergladesplan.org/pm/ projects/proj_40_site_1_impoundment.aspx> (20 November 2010); Appelbaum II interview, 15.

72 Souza interview, 12.

73 Quotations in Mary Williams Walsh, "Helping the Everglades, or Big Sugar?" The New York Times, 14 September 2008; see also Kris Hundley, "Deal Would Be the End for Florida Sugar Giant," St. Petersburg Times, 24 June 2008.

74 As quoted in Walsh, "Helping the Everglades, or Big Sugar?" The New York Times, 14 September 2008.

75 Quotation in Damien Cave, "Florida Buying Big Sugar Tract for Everglades," The New York Times, 25 June 2008; see also Erik Stokstad, "Big Land Purchase Triggers Review of Plans to Restore Everglades," Science 321 (4 July 2008): 22. One report indicated that in addition to the U.S. Sugar lands, the State of Florida needed an additional 40,000 acres owned by the Okeelanta Corporation to allow it to restore natural flow to the EAA. Walsh, "Helping the Everglades, or Big Sugar?" The New York Times, 14 September 2008.

76 All quotations in Damien Cave, "Possible Flaws in State Plan to Rescue the Everglades," The New York Times, 2 July 2008; see also Walsh, "Helping the Everglades, or Big Sugar?" The New York Times, 14 September 2008.

77 Bradford Sewell, Senior Attorney, NRDC, to The Honorable Jeb Bush et al., 13 December 2006 (documents pertaining to water quality litigation in possession of authors). According to former Assistant Secretary of the Interior Nathaniel Reed, who, in 2011, was serving as vice chairman of the Everglades Foundation, environmental groups also opposed the reservoir because the SFWMD "refused to guarantee that the water stored in that reservoir [would] be used for restoration of the Everglades." In Reed's words, after "it became increasingly clear that the water stored in that reservoir could be used to irrigate sugarcane land . . . the entire environmental community turned against" the reservoir. Reed telephone interview, 5.

78 Quotations in Don Van Natta Jr. and Damien Cave, "Deal to Save Ev-erglades May Help Sugar Firm," *The New York Times*, 8 March 2010; see also *United States of America and Miccosukee Tribe of Indians v. South Florida Water Management District*, Case No. 88-1886-CIV-MORENO, Miccosukee Tribe of Indians' Emergency Motion for Injunctive Relief and Supporting Memorandum of Law, 1-2; *United States of America and Miccosukee Tribe of Indians v. South Florida Water Management District*, Case No. 88-1886-CIV-MORENO, Miccosukee Tribe of Indians of Florida's Closing Brief on the July 26, 2010 Evidentiary Hearing, 6-7.

79 Don Van Natta Jr. and Damien Cave, "Deal to Save Everglades May Help Sugar Firm," *The New York Times*, 8 March 2010; Simone Baribeau, "Florida Weighs Swapping Cash for Debt in Scaled Back Everglades Pur-chase," <http://www.bloomberg.com/news> (10 August 2010); Adam Play-ford, "Everglades Restoration Buy of U.S. Sugar Land May Be Sharply Down-sized," *The Palm Beach Post*, 4 August 2010. The Miccosukee Tribe disagreed with this reasoning, stating that the U.S. Sugar lands were not "instrumental to remedying violations of the Consent Decree." *United States of America and Miccosukee Tribe of Indians v. South Florida Water Management District*, Case No. 88-1886-CIV-MORENO, Miccosukee Tribe of Indians' Emergen-cy Motion for Injunctive Relief and Supporting Memorandum of Law, 6-7.

80 Estenoz interview, 11.

81 As quoted in Adam Playford, "Everglades Restoration Buy of U.S. Sugar Land May Be Sharply Downsized," *The Palm Beach Post*, 4 August 2010.

82 Reed telephone interview, 17. According to Superintendent Dan Kimball of Everglades National Park, the NPS was "very supportive" of the acquisition because of the "strategic location" of the lands. "When you have an opportunity like that from a willing seller with a strategic location . . . you should take it," he explained. Kimball interview, 16.

83 For example, see Damien Cave, "In the Everglades, the Miracle that Wasn't," <http://green.blogs.nytimes.com/2010/08/13/the-miracle-that-wasnt-everglades-restoration/> (11 August 2010).

84 CISREP, et al., *Progress Toward Restoring the Everglades: The Second Biennial Review – 2008*, 1-12.

85 As quoted in Damien Cave, "Harsh Review of Restoration in Ever-glades," *The New York Times*, 30 September 2008; see also Craig Pittman, "Red Tape Snarls Glades Rescue," *St. Petersburg Times*, 30 September 2008; "Biennial Report Finds Little Progress, Many Hindrances," *Engineering News-Record* 261 (13 October 2008): 16.

86 "Biennial Report Finds Little Progress, Many Hindrances," 16; U.S. Army Corps of Engineers, Jacksonville District, *Integrated Delivery Sched-ule: Facts & Information* (Jacksonville, Fla.: U.S. Army Corps of Engineers, 2009), 1-2; Pittman, "Red Tape Snarls Glades Rescue," *St. Petersburg Times*, 30 September 2008.

87 Quotations in "Everglades Projects Receive $103 Million in Recov-ery Act Funds," *Everglades Report* (May/June 2009): 1, 3; see also "American Recovery and Reinvestment Act of 2009, Army Corps of Engineers, Civil Works Construction" <http://www.usace.army.mil/recovery/Documents/ComprehensiveConstruction.pdf> (8 December 2010); U.S. Army Corps of Engineers, "Stimulus Money for US Army Corps of Engineers" <http://www.usace.army.mil/recovery/Pages/CWMoney.aspx> (8 December 2010).

88 Estenoz interview, 12-13. Appropriations in the fiscal year 2010 budget for CERP projects continued the momentum. U.S. Army Corps of Engineers, Jacksonville District, *Comprehensive Everglades Restoration Plan, Central and Southern Florida Project: Report to Congress* (Jacksonville, Fla.: U.S. Army Corps of Engineers, 2010), viii.

89 Quotation in Pantano interview, 4; see also "Everglades Leader Ap-pointed to Pentagon Post," *Everglades Restoration* (March/April 2009): 1, 4; Estenoz interview, 12.

90 See *Miccosukee Tribe of Indians of Florida v. United States of America, the Environmental Protection Agency, Michael O. Leavitt, and Jimmy Palmer*, Complaint, U.S. District Court for the Southern District of Florida.

91 *Miccosukee Tribe of Indians of Florida v. United States of America, et al.*, Lead Case No. 04-21448-CIV-GOLD/McALILEY, U.S. District Court, Southern District of Florida, Order Granting Plaintiffs' Motions [DE 357]; [DE 364] in Part; Granting Equitable Relief; Requiring Parties to Take Ac-tion by Dates Certain, 14 April 2010, 2, 44-45.

92 *Miccosukee Tribe of Indians of Florida v. United States of America, et al.*, Lead Case No. 04-21448-CIV-GOLD/McALILEY, U.S. District Court, Southern District of Florida, Order Granting Summary Judgment; Closing Case, 29 July 2008, 98-99.

93 Quotations in United States Environmental Protection Agency, Amended Determination, i-v, (documents pertaining to water quality litiga-tion in possession of authors); Sylvia R. Pelizza, Refuge Manager, Arthur R. Marshall Loxahatchee National Wildlife Refuge, and Dan B. Kimball, Su-perintendent, Everglades and Dry Tortugas National Parks, Briefing Memo-randum, Current Water Quality Litigation (Moreno and Gold cases), 7 May 2010, document provided by Alice Clarke, Everglades National Park.

94 Quotation in *United States of America v. South Florida Water Man-agement District, et al.*, Case No. 88-1886-CIV-MORENO, U.S. District Court for the Southern District of Florida, Miami Division, Order Granting Motion to Adopt the Special Master's Report, Motion Seeking Declaration of Violations, and Motion for Declaration of Breach of Commitments, 31 March 2010, 19; see also *United States of America v. South Florida Water Management District, et al.*, Case No. 88-1886-CIV-MORENO, U.S. District Court for the Southern District of Florida, Miami Division, Order Requir-ing Special Master to Hold a Hearing on the Issue of Remedies and Submit a Report to the Court; Pelizza and Kimball, Briefing Memorandum, Current Water Quality Litigation (Moreno and Gold cases), 7 May 2010; Pelizza and Kimball, Briefing Memorandum, Everglades Consent Decree Compliance, 7 May 2010, document provided by Alice Clarke, Everglades National Park.

95 Appelbaum II interview, 18.

96 Estenoz interview, 12.

97 "Department of Interior Vision and Strategic Plan for Achieving Ev-erglades Restoration," 24 September 2010, 5, document provided by Robert Johnson, Director, South Florida Natural Resources Center, National Park Service, Homestead, Florida.

98 Wehle interview, 13-14.

99 Pantano interview, 9-10, 17-19. Wehle said that the SFWMD had al-ready purchased nearly 65 percent "of all lands necessary for CERP projects" and 100 percent of the land needed "for the first suite of projects." Wehle interview, 2. Exacerbating the water quality issue was a new challenge result-ing from converting lands that previously had been used for agriculture to reservoirs. Because of high levels of fertilization and crop protection chemi-cals used on the land, much of the soil was contaminated to levels that would violate water quality. Cleansing the soil took money, and it was unclear who would be responsible for these costs. See Pantano interview, 14-15; Taplin interview, 18-19; Ammon interview, 7.

100 Kimball interview, 11.

101 Ammon interview, 15-20.

102 Lehtinen, "The Myth of Everglades Restoration," 2.

103 Quotations in "Department of the Interior Vision and Strategic Plan for Achieving Everglades Restoration, Sept 24, 2010," 6-7, document pro-vided by Robert Johnson, Director, South Florida Natural Resources Center,

National Park Service, Homestead, Florida; see also Debra Wood, "Everglades Restoration: What's Going On?" *Southeast Construction* 6 (1 April 2006): 74; Pantano interview, 18; Taplin interview, 16-17.

[104] Quotations in South Florida Water Management and U.S. Army Corps of Engineers, *Aquifer Storage & Recovery Program: Interim Report 2008* (West Palm Beach, Fla: South Florida Water Management District, 2008), v-vi; see also Taplin interview, 8-9.

[105] Pantano interview, 11.

[106] Ammon interview, 13.

[107] CISREP et al., *Progress Toward Restoring the Everglades: The Third Biennial Review, 2010*, 5-6.

[108] Pantano personal communication with Nicolai Kryloff, 2 June 2011.

[109] Quotations in Souza interview, 12; see also Reed telephone interview, 6; Kimball interview 8.

[110] See, for example, Souza interview, 12-13.

[111] Lehtinen, "The Myth of Everglades Restoration," 6-7.

[112] Steve Terry, Truman Eugene Duncan, Jr., Rory Feeny, Colonel (Ret'd) Terry L. Rice, and Ronald D. Jones memorandum to Chairman Colley Billie, 11 January 2011 (documents pertaining to water quality litigation in possession of authors).

[113] Lehtinen, "The Myth of Everglades Restoration," 6-7.

[114] Terry et al. memorandum to Chairman Colley Billie, 11 January 2011 (documents pertaining to water quality litigation in possession of authors).

[115] Pantano personal communication with Nicolai Kryloff, 2 June 2011.

[116] Lehtinen, "The Myth of Everglades Restoration," 4.

[117] Quotations in Kimball interview, 20, 34; see also Pantano interview, 20-23; Sean Smith telephone interview by Joshua Pollarine, 24 February 2011, transcript, 36.

[118] Quotations in Pantano interview, 28-30; see also Taplin interview, 10-12; Smith interview, 30-31; Ammon interview, 29-32.

[119] Kinard interview, 9. Paul Souza agreed with Kinard, stating that "the best way to deal with invasive species is to keep them from becoming problems in the first place." Souza interview, 28. Dan Kimball lauded the work that the Corps had done on invasive species, saying that it was "the first time I've really seen them focused on it as part of CERP." Kimball interview, 24.

[120] Souza interview, 22-23.

[121] Johnson II interview, 22-23.

[122] Appelbaum II interview, 28-29.

[123] Wehle interview, 21.

[124] Pantano personal communication with Nicolai Kryloff, 2 June 2011.

[125] Pantano personal communication with Nicolai Kryloff, 2 June 2011.

[126] Pantano interview, 15.

[127] U.S. Army Corps of Engineers, Jacksonville District, *Kissimmee River Restoration Progress: Facts & Information* (Jacksonville, Fla.: U.S. Army Corps of Engineers, Jacksonville District, 2010), 1.

[128] U.S. Army Corps of Engineers, Jacksonville District, *Kissimmee River Restoration Progress*, 1.

[129] Paul Klinkenberg, "Regarding a River's Reincarnation," *St. Petersburg Times*, 25 January 2009.

[130] Appelbaum II interview, 22-23.

[131] Souza interview, 35.

[132] "Department of Interior Vision and Strategic Plan for Achieving Everglades Restoration," 24 September 2010, 1.

[133] Pantano personal communication with Nicolai Kryloff, 2 June 2011.

[134] As quoted in Curtis Morgan, "Critics Fear Everglades Projects Won't Escape Budget Ax," *The Miami Herald*, 4 June 2011.

[135] Quotation in *Miccosukee Tribe of Indians of Florida and Friends of the Everglades v. United States of America*, Case No. 04-21448-CIV-GOLD, U.S. District Court for the Southern District of Florida, Omnibus Order, 26 April 2011, 38.

[136] Douglas, *The Everglades*, 385.

[137] U.S. Army Corps of Engineers, "Overview: Central and Southern Florida Project Comprehensive Review Study, October 1998," 6.

The authors would like to acknowledge the gracious help and support of several individuals in the preparation of this history. Martin Reuss, one of the most noted scholars of water history in the United States, served as the original technical advisor on this project and provided helpful direction and comments throughout the process. He also opened research and interview doors, and was a good sounding board for our ideas about the direction of the history. When Marty retired, Matthew Pearcy became the technical advisor and provided valuable help and encouragement. Daniel Hayes, formerly with the Jacksonville District's West Palm Beach office, also provided valuable assistance in making contacts with individuals and in facilitating the research process. We are grateful for their help.

In the course of preparing this history, the authors and their research team made several trips to Florida, consulting numerous repositories, including records of the Jacksonville District, U.S. Army Corps of Engineers; the South Florida Water Management District; the U.S. Fish and Wildlife Service; Everglades National Park; Florida Keys National Marine Sanctuary; and the Florida State Archives. They also consulted manuscript collections at the University of Florida, Florida State University, and the University of Miami, including those of Arthur Marshall, Marjory Stoneman Douglas, and Spessard Holland. The authors researched in National Archives records for the Corps, the National Park Service, the U.S. Fish and Wildlife Service, and the Environmental Protection Agency. In addition, the authors used transcripts of additional interviews done by the Samuel Proctor Oral History Program at the University of Florida.

Several individuals were instrumental to this research effort, including Colonel Alfred Pantano, Oriana Armstrong, James Vearil, and Jean Pavlov of the Jacksonville District; Lori Ojala, Earl Johnson, and Ann Gangitano of the South Florida Water Management District; Joseph Carroll of the U.S. Fish and Wildlife Service; Nancy Russell of Everglades National Park and the NPS's staff at the South Florida Natural Resources Center; Billy Causey of the Florida Keys National Marine Sanctuary; Julian Pleasants, director of the Samuel Proctor Oral History Program; and Andre Wilkerson of the Federal Records Center in Atlanta, Georgia.

In addition to documentary research, we conducted more than fifty oral history interviews. We thank those who took time out of their busy schedules to speak with us: Kenneth Ammon, Stuart Appelbaum, General Joe Ballard, Ernie Barnett, Frank Bernardino, Richard Bonner, John Brady, Kim Brooks-Hall, Joseph Browder, Eric Bush, Billy Causey, Michael Collins, Wesley Bowman "Bo" Crum, Michael Davis, Gene Duncan, Dennis Duke, Lester Edelman, Shannon Estenoz, Michael Finley, Susan Gray, Daniel Hayes, Eric Hughes, Robert Johnson, Haynes Johnson, Drew Kendall, Dan Kimball, Lieutenant Colonel Michael Kinard, William Leary, Joette Lorion, Thomas Lovejoy, John "Jack" Maloy, Barbara Miedema, Gerald Miller, Brooks Moore, Colonel Alfred Pantano, Garth Redfield, Nathaniel Reed, Colonel Terrence Rice, Karsten and Carol Rist, Colonel Terrence "Rock" Salt, W. Ray Scott, Deborah Scerno, David Sikkema, Paul Souza, Randy Smith, Rick Smith, Sean Smith, Kimberley Taplin, Tom Teets, Thomas Van Lent, James Vearil, Herschel Vinyard, George Wedgworth, Carol Wehle, General Arthur Williams, and Theresa Woody. Nanciann Regalado of the Jacksonville District's Public Relations Office and Oriana Armstrong, Jacksonville District Librarian & Regional Knowledge Coordinator, facilitated many of our interviews with District personnel, and Carol and Karsten Rist provided us with a list of Everglades Coalition members.

Many individuals also provided us with copies of documents from their own research files, including Joseph Browder, Terrence Rice, Joette Lorion, and James Vearil. We thank them for their assistance in gathering information. Thanks also to Susan Bennett, Olga Cruz, Patrick Lynch, and Randy Smith of the South Florida Water Management District for their assistance in sharing digital photographs and other indispensible visual elements.

Several reviewers commented on the draft manuscript and provided invaluable suggestions. We would especially like to thank Lisa Mighetto, Martin Reuss, James Vearil, Susan Sylvester, Matthew Pearcy, Steven Robinson, and Kent Sieg for their comments. Within HRA, David Strohmaier, Nicolai Kryloff, and Joshua Pollarine provided both helpful reviews and invaluable research assistance. Despite their help, however, the conclusions and assessments in this report remain the sole responsibility of the authors.

The research for this book encompassed several different sources. Of most significance were manuscript collections in a variety of repositories. Especially important were the records of the South Florida Water Management District, located in the district's headquarters in West Palm Beach, Florida. These records not only detail SFWMD management of water in South Florida, but also contain information about other interested parties in the state, including the Seminole and Miccosukee Indians, county and city governments, and federal agencies. In a similar way, the records of the Jacksonville District of the U.S. Army Corps of Engineers, both in Jacksonville and in the Federal Records Center in Atlanta, Georgia, contain an abundance of information on not only the C&SF Project and CERP, but on different perspectives of other agencies and the general public. With the increased trend of digital documentation in more recent years, inter-agency documents accessed via the world wide web proved invaluable to researching the recent history discussed in this work. Public access to this digital record, as well as access provided by the Army Corps to the Jacksonville District's digital archive, allowed the authors to review material not available in hard copy.

The authors also researched in the files of Everglades National Park, the Florida Keys National Marine Sanctuary (which provided many documents on the South Florida Ecosystem Restoration Task Force and Florida Bay problems), the U.S. Fish and Wildlife Service, and the Office of History, U.S. Army Corps of Engineers. Other valuable sources included records of the National Park Service in the National Archives and various governors' papers in the Florida State Archives (which also houses an extensive photograph collection). Supplementing these record collections were perusals of private manuscript collections at various Florida universities, including the papers of Spessard Holland, Claude Pepper, Bob Graham, Arthur Marshall, and Marjory Stoneman Douglas, among others.

In order to gain more insight into the environmental perspective, the authors used publications from several national periodicals, including *National Parks Magazine*, *Audubon*, *Time*, *Newsweek*, *Reader's Digest*, *U.S. News & World Report*, and *Sports Illustrated*. Scientific information was mainly gleaned from reports and publications in the libraries of the SFWMD and the Jacksonville District, although periodicals such as *Science* and *Engineering News-Record* were helpful as well.

A host of books also exist on South Florida water management. Particularly insightful were Nelson Manfred Blake's *Land Into Water—Water Into Land*, David McCally's *The Everglades: An Environmental History*, Michael Grunwald's *The Swamp*, and Ted Levin's *Liquid Land: A Journey Through the Florida Everglades*. All of these sources provide a good overview of the problems that the Everglades have faced historically, as well as some of the proposed solutions.

Particularly important to this project were numerous oral histories, which provided first-hand accounts of many of the issues discussed. The authors and their research team performed more fifty oral histories themselves, but also relied heavily on interviews conducted by the Samuel Proctor Oral History Program at the University of Florida. Also important were oral histories conducted with District Engineers of the Jacksonville District, available in that agency's library. These not only provided useful insights, but also added some rhetorical flair to the manuscript.

A complete bibliography of sources used and consulted follows.

Manuscripts

Administrative Records of the Florida Keys National Marine Sanctuary, National Oceanic and Atmospheric Administration. Florida Keys National Marine Sanctuary Headquarters, Marathon, Florida.

Administrative Records of the Jacksonville District, U.S. Army Corps of Engineers. Jacksonville District, Jacksonville, Florida.

Administrative Records of the South Florida Natural Resources Center, National Park Service. Homestead, Florida.

Administrative Records of the South Florida Water Management District. West Palm Beach, Florida.

Administrative Records of the Vero Beach Office, U.S. Fish and Wildlife Service. Vero Beach Office. Vero Beach, Florida.

Arthur R. Marshall Papers. Manuscript Series 73. Special and Area Studies Collections, George A. Smathers Library (East), University of Florida, Gainesville, Florida.

Central Records of Everglades National Park. Everglades National Park Archives, Everglades National Park, Homestead, Florida.

Claude Pepper Collection. Claude Pepper Library, Florida State University, Tallahassee, Florida.

D. Robert "Bob" Graham Papers. Manuscript Series 148. Special and Area Studies Collections, George A. Smathers Library (East), University of Florida, Gainesville, Florida.

Florida Department of Agriculture and Consumer Services. Surface Water Improvement and Management Plan Files. Series 1497. Florida State Archives, Tallahassee, Florida.

Florida Executive Office of the Governor. Brian Ballard, Director of Operations, Subject Files, 1988. Series 1331. Florida State Archives, Tallahassee, Florida.

Florida Executive Office of the Governor. Press Secretary Subject Files 1979-86. Series 172. Florida State Archives, Tallahassee, Florida.

Florida Executive Office of the Governor. Subject Files, 1987-1988. Series 1401. Florida State Archives, Tallahassee, Florida.

Florida Executive Office of the Governor. Subject Files, 1991-1996. Series 1824. Florida State Archives, Tallahasseee, Florida.

Florida Game and Fresh Water Fish Commission. Everglades Conservation Files, 1958-1982. Series 1719. Florida State Archives, Tallahassee, Florida.

Florida Game and Fresh Water Fish Commission. Lake and Stream Restoration Project Files, 1950-1986. Series 1570. Florida State Archives, Tallahassee, Florida.

Florida Governor's Office. Jay Landers Subject Files. Series 949. Florida State Archives, Tallahassee, Florida.

Florida State Board of Conservation. Water Resources Subject Files, 1961-1968. Series 1160. Florida State Archives, Tallahassee, Florida.

Florida State Land Records. State Lands Records Vault, Division of State Lands, Florida Department of Environmental Protection, Marjory Stoneman Douglas Building, Tallahassee, Florida.

John Henry Davis Papers. Manuscript Series 23. Special and Area Studies Collections, George A. Smathers Library (East), University of Florida, Gainesville, Florida.

Marjory Stoneman Douglas Papers. Manuscript Collection 60. Archives and Special Collections, Otto G. Richter Library, University of Miami, Miami, Florida.

Records of the Environmental Protection Agency. Record Group 412. National Archives and Records Administration Southeast Region, Atlanta, Georgia.

Records of Everglades National Park. Documents provided by Alice Clarke. Homestead, Florida.

Records of the U.S. Fish and Wildlife Service. Record Group 22. National Archives and Records Administration Southeast Region, Atlanta, Georgia.

Records of the National Park Service. Record Group 79. National Archives and Records Administration II, College Park, Maryland.

Records of the Office of the Chief of Engineers. Record Group 77. Federal Records Center, Atlanta, Georgia.

Records of the Office of History, Headquarters, U.S. Army Corps of Engineers. Office of History, Headquarters, U.S. Army Corps of Engineers, Alexandria, Virginia.

Robert N. "Bert" Dosh Papers. Manuscript Series 25. Special and Area Studies Collections, George A. Smathers Library (East), University of Florida, Gainesville, Florida.

Spessard L. Holland Papers. Manuscript Series 55. Special and Area Studies Collections, George A. Smathers Library (East), University of Florida, Gainesville, Florida.

Trustees Correspondence. State Lands Records Vault, Division of State Lands, Florida Department of Environmental Protection, Marjory Stoneman Douglas Building, Tallahassee, Florida.

Water Quality Litigation Documents. In possession of authors.

Interviews

Adams, Colonel James W. R. Interview by George E. Buker. 18 November 1981. Transcript in Library, Jacksonville District, U.S. Army Corps of Engineers, Jacksonville, Florida.

Ammon, Ken. Interview by Joshua Pollarine. 4 March 2011.

Appelbaum, Stuart. Interview by Brian Gridley. 22 February 2002. Everglades Interview No. 11. Samuel Proctor Oral History Program, University of Florida, Gainesville, Florida.

_____. Interview by Joshua Pollarine. 27 September 2010.

_____. Interview by Theodore Catton. 7 July 2004.

Ballard, General Joe. Interview by Theodore Catton. 18 November 2004.

Barnett, Ernest "Ernie." Interview by Matthew Godfrey. 23 September 2004.

Bonner, Richard. Interview by Theodore Catton. 13 May 2005.

Brady, John D. Interview by Theodore Catton. 12 May 2005.

Brooks-Hall, Kimberly. Interview by Theodore Catton. 7 July 2004.

Browder, Joseph. Interview by Theodore Catton. 17 November 2004.

Causey, Billy. Interview by Theodore Catton. 20 January 2005.

Collins, Michael. Interview by Theodore Catton. 13 July 2004.

Davis, Michael. Interview by Theodore Catton. 21 December 2004.

Devereaux, Colonel Alfred B. Interview by George E. Buker. 23 April 1984. Transcript in Library, Jacksonville District, U.S. Army Corps of Engineers, Jacksonville, Florida.

Duke, Dennis. Interview by Joshua Pollarine. 29 September 2010.

Edelman, Les. Interview by Theodore Catton. 17 November 2004.

Ehrlichman, John D. "Presidential Assistant with a Bias for Parks." An oral history conducted in 1991 by William Duddleson. In *Saving Point Reyes National Seashore, 1969-1970: An Oral History of Citizen Action in Conservation*. Regional Oral History Office, The Bancroft Library, University of California, Berkeley, 1993.

Estenoz, Shannon. Interview by Joshua Pollarine. 20 October 2010.

Finley, Michael. Interview by Matthew Godfrey. 20 October 2004.

Herndon, Colonel Robert L. Interview by Joseph E. Taylor. 26 May 1989. Transcript in Library, Jacksonville District, U.S. Army Corps of Engineers, Jacksonville, Florida.

Johnson, Robert. Interview by Joshua Pollarine. 18 October 2010.

Johnson, Robert. Interview by Theodore Catton. 16 July 2004.

Jones, John C. Interview by Brian Gridley. 23 May 2001. Everglades Interview No. 9. Samuel Proctor Oral History Program, University of Florida, Gainesville, Florida.

Kimball, Dan. Interview by Joshua Pollarine. 1 March 2011.

Kinard, Lieutenant Colonel Michael. Interview by Joshua Pollarine. 7 March 2011.

Leary, William. Interview by Theodore Catton. 24 November 2004.

Lorion, Joette. Interview by Matthew Godfrey. 18 January 2006.

Lovejoy, Thomas E. Interview by Theodore Catton. 9 December 2004.

Maloy, John "Jack." Interview by Matthew Godfrey. 14 July 2004.

Malson, Colonel Bruce A. Interview by Joseph E. Taylor. 15 August 1991. Transcript in Library, Jacksonville District, U.S. Army Corps of Engineers, Jacksonville, Florida.

Moore, Brooks. Interview by Joshua Pollarine. 3 March 2011.

Myers, Colonel Charles T. III. Interview by George E. Buker. 30 December 1987. Transcript in Library, Jacksonville District, U.S. Army Corps of Engineers, Jacksonville, Florida.

Ogden, John. Interview by Brian Gridley. 10 April 2001. Everglades Interview No. 7. Samuel Proctor Oral History Program, University of Florida, Gainesville, Florida.

Pantano, Colonel Alfred. Interview by Matthew Godfrey. 13 January 2011.

Reed, Nathaniel. Interview by Julian Pleasants. 18 December 2000. Everglades Interview No. 2. Samuel Proctor Oral History Program, University of Florida, Gainesville, Florida.

————. Interview by Nicolai Kryloff. 11 March 2011.

Rice, Colonel Terry. Interview by Brian Gridley. 8 March 2001. Everglades Interview No. 4. Samuel Proctor Oral History Program, University of Florida, Gainesville, Florida.

————. Interview by Theodore Catton. 6 December 2004.

Rist, Carol and Karsten. Interview by Theodore Catton. 9 November 2004.

Salt, Colonel Terrence C. "Rock." Interview by George E. Buker. 18 July 1994. Transcript in Library, Jacksonville District, U.S. Army Corps of Engineers, Jacksonville, Florida.

————. Interview by Theodore Catton. 19 January 2005.

Souza, Paul. Interview by Joshua Pollarine. 8 March 2011.

Smith, Sean. Interview by Joshua Pollarine. 24 February 2011.

Stahl, Stuart. Interview by Julian Pleasants. 22 February 2001. Everglades Interview No. 3. Samuel Proctor Oral History Program, University of Florida, Gainesville, Florida.

Taplin, Kim. Interview by Nicolai Kryloff. 25 February 2011.

U.S. Army Corps of Engineers. Office of the Chief of Engineers. *Engineering Memoirs: Major General William E. Potter*. EP 870-1-12. Washington, D.C.: U.S. Army Corps of Engineers, 1983.

————. *Water Resources, People and Issues: An Interview with William Gianelli*. EP 870-1-24. Washington, D.C.: U.S. Army Corps of Engineers, 1985.

Van Lent, Thomas. Interview by Joshua Pollarine. 19 October 2010.

Vinyard, Herschel. Interview by Nicolai Kryloff. 4 August 2011.

Wade, Malcolm, Jr. Interview by Julian Pleasants. 3 April 2001. Everglades Interview No. 5. Samuel Proctor Oral History Program, University of Florida, Gainesville, Florida.

Waterston, Lieutenant Colonel Robert J. III. Interview by George E. Buker. 12 April 1984. Transcript in Library, Jacksonville District, U.S. Army Corps of Engineers, Jacksonville, Florida.

Wedgworth, George and Barbara Miedema. Interview by Matthew Godfrey. 9 July 2004.

Wehle, Carol. Interview by Nicolai Kryloff. 17 March 2011.

Whitfield, Estus. Interview by Brian Gridley. 15 May 2001. Everglades Interview No. 8. Samuel Proctor Oral History Program, University of Florida, Gainesville, Florida.

Wisdom, Colonel Donald A. Interview by George E. Buker. 22 and 23 December 1978. Transcript in Library, Jacksonville District, U.S. Army Corps of Engineers, Jacksonville, Florida.

Woody, Theresa. Interview by Theodore Catton. 18 January 2005.

Wopat, Ben A. Interview by Matthew Pearcy. 25 April 2002. Transcript in Oral History File, St. Paul District, U.S. Army Corps of Engineers, St. Paul, Minnesota.

Personal Communications

Brady, John D. Personal communication with Theodore Catton, 12 May 2005.

Browder, Joseph. Personal communication with Theodore Catton, 17 November 2004.

James, Thomas. Personal communication with Theodore Catton, 1 April 2005.

Vearil, James. Personal communication with Theodore Catton, 31 March 2005, 26 May 2005.

————. Personal communication with Matthew Godfrey, 16 May 2006.

Warner, Joni. Personal communication with Nicolai Kryloff, 28 March 2011.

Internet Sites

"American Recovery and Reinvestment Act of 2009, Army Corps of Engineers, Civil Works Construction" < http://www.usace.army.mil/recovery/Documents/ComprehensiveConstruction.pdf> (8 December 2010).

"CERP Project: Indian River Lagoon—South" <http://www.evergladesplan.org/pm/projects/ proj_07_irl_south.aspx> (20 November 2010).

"CERP Project: Picayune Strand Restoration" <http://www.evergladesplan.org/pm/projects/ proj_30_sgge.aspx> (20 November 2010) .

"CERP Project: Site 1 Impoundment" <http://www.evergladesplan.org/pm/projects/proj_40_site_1_impoundment.aspx> (20 November 2010).

"CERP Project: Water Conservation Area 3, Decompartmentalization & Sheet Flow Enhancement (Decomp)" <http://www.evergladesplan.org/pm/projects/proj_12_wca3_1.aspx> (13 November 2010).

"Central and Southern Florida Project, Comprehensive Everglades Restoration Plan, Site 1 Impoundment Project, Final Integrated Project Implementation Report and Environmental Impact Statement" <http://www.evergladesplan.org/pm/projects/docs_40_site_1 _pir.aspx> (20 November 2010).

"Comprehensive Everglades Restoration Plan, Picayune Strand Restoration (formerly Southern Golden Gate Estates Ecosystem Restoration Plan), Final Integrated Project Implementation Report and Environmental Impact Statement" < http://www.evergladesplan.org/pm/projects/docs_30_sgge_pir_final.aspx> (20 November 2010).

"The Conceptual Plan of the Governor's Commission for a Sustainable South Florida" <http://fcn.state.fl.us/ everglades/gcssf/concept/conc_tc.html> (3 February 2006).

Everglades Coalition. "Letters and Position Statements" <http://www.evergladescoalition.org/Issues.htm> (23 September 2010).

Everglades Coalition. "Past Conferences" <http://www.evergladescoalition.org/site/ pastconference.html> (26 May 2005).

Everglades Information Network. South Florida Ecosystem Restoration Task Force Collection <http://everglades.fiu.edu/taskforce/index.html> (1 February 2006).

"Everglades Litigation and Restoration" <http://exchange.law.miami.edu/everglades> (29 August 2005).

Florida Audubon. "Lake Okeechobee 2007 Report" <http://fl.audubon.org/PDFs/pubs_policydocs-LakeOReport_1-07.pdf> (23 September 2010).

————. "Biennial Everglades Summary Report," Biennial Reports, 2006-2008 <http://fl.audubon.org/pubs_EvergladesReport.html> (23 September 2010).

Florida Defenders of the Environment. "Restoring the Ocklawaha River Ecosystem" <http://www.fladefenders.org/publications/restoring3.html> (18 January 2005).

Florida Department of Environmental Protection. "History of Florida's Conservation Efforts" <http://www.dep.state.fl.us/lands/acquisition/P2000/BACKGRND.htm> (20 December 2005).

Florida Fish and Wildlife Conservation Commission. "Rotenberger Wildlife Management Area" <http://www.floridaconservation.org/recreation/rotenberger/history.asp> (March 22, 2005).

Florida Supreme Court Briefs & Opinions. College of Law Library, Florida State University, Tallahassee, Florida <http://www.law.fsu.edu/library/flsupct> (15 February 2006).

Governor's Commission for a Sustainable South Florida <http://www.state.fl.us/ everglades/gcssf_eo.html> (8 September 2005).

Library of Congress. "Reclaiming the Everglades: South Florida's Natural History, 1884-1934" <http://memory.loc.gov/ammem/award98/fmuhtml/everhome.html> (1 December 2004).

"Marjorie Harris Carr Cross Florida Greenway—History" <http://www.dep.state.fl.us/gwt/cfg/ history.htm> (18 January 2005).

"Master Agreement Between the Department of the Army and South Florida Water Management District for Cooperation in Constructing and Operating, Maintaining, Repairing, Replacing, and Rehabilitating Projects Authorized to be Undertaken Pursuant to the Comprehensive Everglades Restoration Plan." <http://www.evergladesplan.org/pm/program_docs/ master_agreement _cerp.aspx> (22 November 2010).

National Research Council, Committee on Independent Scientific Review of Everglades Restoration Progress. "Progress Toward Restoring the Everglades: The First Biennial Review, 2006" <http://books.nap.edu/catalog.php?record_id=11754> (23 September 2010).

_____. "Progress Toward Restoring the Everglades: The Second Biennial Review, 2008" <http://www.nap.edu/catalog.php?record_id=12469> (23 September 2010).

_____. "Progress Toward Restoring the Everglades: The Third Biennial Review, 2010" <http://www.nap.edu/catalog.php?record_id=12988> (23 September 2010).

"President's Council on Sustainable Development, Overview" <http://clinton2.nara.gov/ PCSD/Overview/index.html> (1 September 2005).

Ronald Reagan Presidential Library Archives <http://www.reagan.utexas.edu/archives> (20 December 2005).

South Florida Ecosystem Restoration Task Force <http://www.sfrestore.org> (15 February 2006).

_____. "FY 2001-2002 Strategic Plan and Biennial Report" <http://www.sfrestore.org/documents/work_products.html> (23 September 2010).

_____. "FY 2002-2004 Strategic Plan and Biennial Report," Biennial Reports, 2002-2008 <http://www.sfrestore.org/documents/work_products.html> (23 September 2010).

_____. "Integrated Financial Plan," Biennial Reports, 2003-2007 <http://www.sfrestore.org/documents/work_products.html> (23 September 2010).

_____. "Land Acquisition Strategy," Annual Reports, 2004-2009 <http://www.sfrestore.org/documents/work_products.html> (23 September 2010).

South Florida Ecosystem Restoration Task Force, Science Subgroup. "South Florida Ecosystem Restoration: Scientific Information Needs" <http://everglades.fiu.edu/taskforce/ scineeds/sub2.pdf> (30 July 2004).

Southwest Florida Regional Planning Council. "Demographics" <http://www.swfrpc.org> (16 December 2005).

South Florida Water Management District. "South Florida Environmental Report," Annual Reports, 2005-2010 <http://my.sfwmd.gov/sfer> (23 September 2010).

United Nations Environmental Programme, Division of Technology, Industry, and Economics, International Environmental Technology Centre. "Planning and Management of Lakes and Reservoirs: An Integrated Approach to Eutrophication" <http://www.unep.or.jp/publications/techpublications/techpub-11/1-4-2.asp> (5 April 2005).

U.S. Army Corps of Engineers. "Stimulus Money for US Army Corps of Engineers" <http://www.usace.army.mil/recovery/Pages/CWMoney.aspx> (8 December 2010).

U.S. Army Corps of Engineers, Jacksonville District. "Project Status Report: CERP WCA 3 DECOMP" <http://www.evergladesplan.org/pm/projects/project_docs/status/proj_12_current.pdf> (13 November 2010).

U.S. Army Corps of Engineers and South Florida Water Management District. "Final Integrated Project Implementation Report and Environmental Impact Statement, Indian River Lagoon—South" <http://www.evergladesplan.org/pm/studies/irl_south_pir.aspx> (20 November 2010).

U.S. Fish and Wildlife Service South Florida Ecological Services Office. "Species Conservation Guidelines, South Florida: Cape Sable seaside sparrow." Draft, 8 October 2003 <http://www.fws.gov/verobeach/images/ pdflibrary/Cape_Sable_Seaside-Sparrow_Conservation_Guidelines.pdf> (30 March 2011).

Newspapers

The Christian Science Monitor.
The Daily Business Review.
The Evening Star (Washington, D.C.).
Evening Times.
The Florida Times-Union.
Fort Lauderdale News and Sun-Sentinel.
Fort Myers News-Press.
Miami Daily News.
The Miami Herald.
Miami New Times.
Naples Daily News.
New York Daily News.
New York Times.
The News Tribune.
Okeechobee News.
The Orlando Sentinel.
Orlando Sentinel-Star.
The Palm Beach Post.
The Palm Beach Times.
Pensacola (Fla) News-Journal.
The Post (West Palm Beach).
Sarasota Herald-Tribune.
St. Petersburg Times.
The South Dade News Leader.
Stuart (Fla.) News.
Sun-Sentinel.
Tallahassee Democrat.
The Tampa Tribune.
The Tampa Tribune-Times.
The Times-Union and Journal.
The Wall Street Journal.
The Washington Post.

Government Publications

Abberger, Will and Estus Whitfield. *Save Our Everglades: The Kissimmee River.* Tallahassee, Fla.: State of Florida, Office of the Governor, Office of Planning Budgeting, Natural Resources, Unit, 1985.

Catton, Theodore and Lisa Mighetto. *The Fish and Wildlife Job on the National Forests: A Century of Game and Fish Conservation, Habitat*

Protection, and Ecosystem Management. Washington, D.C.: USDA Forest Service, 1998.

Committee on Independent Scientific Review of Everglades Restoration Progress, Water Science and Technology Review Board, Board on Environmental Studies and Toxicology, Division on Earth and Life Studies. *Progress Toward Restoring the Everglades: The First Biennial Review—2006*. Washington, D.C.: The National Academies Press, 2007.

_____. *Progress Toward Restoring the Everglades: The Second Biennial Review—2008*. Washington, D.C.: The National Academies Press, 2008.

_____. *Progress Toward Restoring the Everglades: The Third Biennial Review, 2010*. Washington, D.C.: The National Academies Press, 2010.

Coordinating Council on the Restoration of the Kissimmee River Valley and Taylor Creek-Nubbin Slough Basin. *Legislative Summary: First Annual Report to the Florida Legislature*. Tallahassee, Fla.: Coordinating Council on the Restoration of the Kissimmee River Valley and Taylor Creek-Nubbin Slough Basin, 1977.

Florida Department of Administration, Division of State Planning. *Final Report on the Special Project to Prevent Eutrophication of Lake Okeechobee*. Tallahassee: Florida Department of Administration, Division of State Planning, 1976.

Florida Department of Environmental Regulation. *Report of Investigations in the Kissimmee River-Lake Okeechobee Watershed*. Tallahassee: State of Florida Department of Environmental Regulation, 1976.

Florida Office of the Governor. *Save Our Everglades: A Status Report by the Office of Governor Lawton Chiles*. N.p., 1994.

Hartwell, J. H., H. Klein, and B. F. Joyner. *Preliminary Evaluation of Hydrologic Situation in Everglades National Park, Florida*. Miami, Fla.: United States Department of the Interior, Geological Survey, Water Resources Division, 1963.

Land Acquisition Selection Committee for the Board of Trustees of the Internal Improvement Trust Fund. *Florida Statewide Land Acquisition Plan*. Tallahassee, Fla.: Board of Trustees of the Internal Improvement Trust Fund, 1986.

Leach, S. D., Howard Klein, and E. R. Hampton. *Hydrologic Effects of Water Control and Management of Southeastern Florida*. Report of Investigations No. 60. Tallahassee, Fla.: State of Florida, Bureau of Geology, 1972.

Loftin, M. Kent, Louis A. Toth, and Jayantha T. B. Obeysekera, eds. *Kissimmee River Restoration Symposium: Proceedings*. West Palm Beach, Fla.: South Florida Water Management District, 1990.

Martin, Lynn R. and Eugene Z. Stakhiv. *Sustainable Development: Concepts, Goals and Relevance to the Civil Works Program*. IWR Report 99-PS-1. Alexandria, Va.: Institute for Water Resources, Water Resources Support Center, U.S. Army Corps of Engineers, 1999.

National Research Council et al. *Aquifer Storage and Recovery in the Comprehensive Everglades Restoration Plan: A Critique of the Pilot Projects and Related Plans for ASR in the Lake Okeechobee and Western Hillsboro Areas*. Washington, D.C.: National Academy Press, 2001.

Sheikh, Pervase A. "Everglades Restoration: Modified Water Deliveries Project." CRS Report for Congress. 17 March 2005. Order Code RS21331.

South Florida Water Management District. *2001 Executive Summary: Everglades Consolidated Report, January 1, 2001*. West Palm Beach, Fla: South Florida Water Management District, 2001.

_____. *2006 South Florida Environmental Report: Executive Summary*. West Palm Beach, Fla.: South Florida Water Management District, 2006.

_____. *Interim Surface Water Improvement and Management (SWIM) Plan for Lake Okeechobee, Part 1: Water Quality & Part VII: Public In-*

formation. West Palm Beach, Fla.: South Florida Water Management District, 1989.

_____. *Save Our Rivers: 1998 Land Acquisition and Management Plan*. West Palm Beach, Fla.: South Florida Water Management District, 1998.

_____. *Surface Water Improvement and Management Plan for the Everglades: Planning Document*. West Palm Beach, Fla.: South Florida Water Management District, 1992.

South Florida Water Management and U.S. Army Corps of Engineers. *Aquifer Storage & Recovery Program: Interim Report 2008*. West Palm Beach, Fla: South Florida Water Management District, 2008.

Toth, Louis A. *Environmental Responses to the Kissimmee River Demonstration Project*. Technical Publication 91-02. West Palm Beach, Fla.: Environmental Sciences Division, Research and Evaluation Department, South Florida Water Management District, 1991.

Upper Mississippi River Basin Coordinating Committee. *Upper Mississippi River Comprehensive Basin Study Main Report*. Chicago: Upper Mississippi River Basin Coordinating Committee, 1972.

U.S. Army Corps of Engineers. Jacksonville District. *Central and Southern Florida Project, Flood Control and Other Purposes, Canal 111 (C-111) South Dade County, Florida: Final Integrated General Reevaluation Report and Environmental Impact Statement*. Jacksonville, Fla.: U.S. Army Corps of Engineers, Jacksonville District, 1994.

_____. *Central and Southern Florida Project for Flood Control and Other Purposes: Rule Curves and Key Operating Criteria, Master Regulation Manual, Volume 2, Part 1*. Jacksonville, Fla.: Department of the Army, Jacksonville District, Corps of Engineers, 1978.

_____. *Central and Southern Florida, Kissimmee River: Executive Summary*. Jacksonville, Fla.: U.S. Army Corps of Engineers, 1984.

_____. *Central and Southern Florida: Kissimmee River, Florida*. Jacksonville, Fla.: U.S. Army Corps of Engineers, 1985.

_____. *Central and Southern Florida Project Comprehensive Review Study: Final Integrated Feasibility Report and Programmatic Environmental Impact Statement*. Jacksonville, Fla.: U.S. Army Corps of Engineers, Jacksonville District, 1999.

_____. *Central and Southern Florida Project, Final Integrated Feasibility Report and Environmental Impact Statement: Environmental Restoration, Kissimmee River, Florida*. Jacksonville, Fla.: U.S. Army Corps of Engineers, 1991.

_____. *Central and Southern Florida Project, Kissimmee River Restoration Test Fill Project: Construction Evaluation Report, Preliminary Draft*. Jacksonville, Fla.: U.S. Army Corps of Engineers, Jacksonville District, 1995.

_____. *Central and Southern Florida Project, Plan of Survey: Everglades National Park Water Requirements*. Jacksonville, Fla.: U.S. Army Engineer District, Jacksonville, Corps of Engineers, 1964.

_____. *Central and Southern Florida Project, Reconnaissance Report, Comprehensive Review Study*. Jacksonville, Fla.: U.S. Army Corps of Engineers, Jacksonville District, 1994.

_____. *Central and Southern Florida Project: Special Report on Local Cooperation in the Part of the Project Authorized by the Flood Control Act of 1954*. Jacksonville, Fla.: Corps of Engineers, U.S. Army, 1956.

_____. *Comprehensive Everglades Restoration Plan, Central and Southern Florida Project: Report to Congress*. Jacksonville, Fla.: U.S. Army Corps of Engineers, 2010.

_____. *Cross Florida Barge Canal Restudy Report: Final Summary*. Jacksonville, Fla.: Department of the Army, Jacksonville District, Corps of Engineers, 1977.

_____. *Fran Reich Preserve/Site 1 Impoundment*. Jacksonville, Fla.: U.S. Army Corps of Engineers, Jacksonville District, 2010.

_____. *Indian River Lagoon—South: Facts & Information*. Jacksonville, Fla.: U.S. Army Corps of Engineers, Jacksonville District, 2010.

_____. *Integrated Delivery Schedule: Facts & Information*. Jacksonville, Fla.: U.S. Army Corps of Engineers, 2009.

_____. *Kissimmee River, Florida, Headwaters Revitalization Project: Integrated Project Modification Report and Supplement to the Final Environmental Impact Statement*. Jacksonville, Fla.: U.S. Army Corps of Engineers, Jacksonville District, 1996.

_____. *Kissimmee River Restoration Progress: Facts & Information*. Jacksonville, Fla.: U.S. Army Corps of Engineers, Jacksonville District, 2010.

_____. *Kissimmee River Study Including Taylor Creek—Nubbin Slough Basins*. Jacksonville, Fla.: U.S. Army Corps of Engineers, 1980.

_____. *Reviving the Picayune Strand*. Jacksonville, Fla.: U.S. Army Corps of Engineers, Jacksonville District, 2010.

_____. *Survey-Review Report on Central and Southern Florida Project: Water Resources for Central and Southern Florida, Main Report*. Jacksonville, Fla.: Department of the Army, Jacksonville District, Corps of Engineers, 1968.

U.S. Army Corps of Engineers. North Central Division. *Upper Mississippi River System Environmental Management Program, Sixth Annual Addendum*. Chicago: U.S. Army Corps of Engineers, North Central Division, 1991.

U.S. Army Corps of Engineers and South Florida Water Management District. *2006: An Annual Update, Comprehensive Everglades Restoration Plan*. 2006.

_____. *Central and Southern Florida Project, Comprehensive Everglades Restoration Plan, Project Management Plan: Florida Bay & Florida Keys Feasibility Study*. 2002.

_____. *Central and Southern Florida Project, Project Management Plan: Southwest Florida Feasibility Study*. 2002.

U.S. Army Corps of Engineers and U.S. Department of the Interior. *Central and Southern Florida Project: Comprehensive Everglades Restoration Plan 2005 Report to Congress*. Washington, D.C.: U.S. Army Corps of Engineers and U.S. Department of the Interior, 2005.

U.S. Department of the Interior and Luna B. Leopold. *Environmental Impact of the Big Cypress Swamp Jetport*. Washington, D.C.: United States Department of the Interior, 1969.

U.S. Department of the Interior. Bureau of Sport Fisheries and Wildlife, Region 4. *A Fish and Wildlife Report for Inclusion in the Corps of Engineers' General Design Memorandum, Part 1: Agricultural and Conservation Areas, Supplement 27—Plan of Regulation for Conservation Area #2, Central and Southern Florida Flood Control Project*. Vero Beach, Fla.: Branch of River Basins, 1958.

U.S. Department of the Interior. Fish and Wildlife Service, Region 4. "A Preliminary Evaluation Report of the Effects on Fish and Wildlife Resources on the Everglade Drainage and Flood Control Project, Palm Beach, Broward, and Dade Counties, Florida." October 1947. Copy in Library, Jacksonville District, U.S. Army Corps of Engineers, Jacksonville, Florida.

U.S. Department of the Interior. Office of Inspector General. *Modified Water Deliveries to Everglades National Park: Audit Report*. Report No. C-IN-MOA-0006-2005. Washington, D.C.: Office of Inspector General, 2006.

U.S. Engineer Office. *Definite Project Report on Cross-Florida Barge Canal*. Jacksonville, Fla.: U.S. Engineer Office, 1943.

U.S. General Accounting Office. *South Florida Ecosystem Restoration: An Overall Strategic Plan and a Decision-Making Process Are Needed to Keep the Effort on Track*. GAO/RCED-99-121. Washington, D.C.: U.S. General Accounting Office, 1999.

U.S. Government Accountability Office. *Report to the Committee on Transportation and Infrastructure, House of Representatives: South Florida Ecosystem: Restoration is Moving Forward but is Facing Significant Delays, Implementation Challenges, and Rising Costs*. Washington, D.C.: U.S. Government Accountability Office, 2007.

U.S. Government Accountability Office. *South Florida Ecosystem: Restoration is Moving Forward but Is Facing Significant Delays, Implementation Challenges, and Rising Costs*. GAO-07-520. Washington, D.C.: U.S. Government Accountability Office, 2007.

U.S. House. *Caloosahatchee River and Lake Okeechobee Drainage Areas, Florida*. 70th Cong., 1st sess., 1928. H. Doc. 215. Serial 8900

_____. *Caloosahatchee River and Lake Okeechobee Drainage Areas, Florida (Side Channels)*. 79th Cong., 2d sess., 1947. H. Doc. 736. Serial 11059.

_____. *Comprehensive Report on Central and Southern Florida for Flood Control and Other Purposes*. 80th Cong., 2d sess., 1948. H. Doc. 643. Serial 11243.

_____. *Establishing the Big Cypress National Preserve in the State of Florida, and for Other Purposes*. 93d Cong., 1st sess., 1973. H. Rept. 93-502. Serial 13020-5.

_____. *Exchange of Certain Lands in California*. 103d Cong., 1st sess., 1993. H. Rept. 103-362.

_____. *Kissimmee River Restoration Study*. 102d Cong., 2d sess., 1992. H. Doc. 102-286.

_____. *Water Resources for Central and Southern Florida*. 90th Cong., 2d sess., 1968. H. Doc. 369.

U.S. House. Committee on Interior and Insular Affairs Subcommittee on National Parks and Public Lands. *Everglades National Park Protection and Expansion Act of 1989: Hearing before the House Subcommittee on National Parks and Public Lands of the Committee on Interior and Insular Affairs*. 101st Cong., 1st sess., 1989.

U.S. House. Committee on Interior and Insular Affairs Subcommittee on National Parks and Recreation. *Additions to the National Park System in the State of Florida: Hearings Before the Subcommittee on National Parks and Recreation of the Committee on Interior and Insular Affairs, House of Representatives*. 99th Cong., 1st and 2d sessions., 1985 and 1986.

U.S. House. Committee on Natural Resources Subcommittee on National Parks, Forests and Public Lands. *Issues Relating to the Everglades Ecosystem: Oversight Hearing before the Subcommittee on National Parks, Forests and Public Lands of the Committee on Natural Resources, House of Representatives, One Hundred Third Congress, Second Session, on the Land Use Policies of South Florida with a Focus on Public Lands and What Impact These Policies are Having*. 103d Cong., 2d sess., 1994.

_____. *Oversight Hearing on the Land Use Policies of South Florida with a Focus on Public Lands and what Impact these Policies are Having*. 103d Cong., 2d sess., 1994.

U.S. House. Committee on Natural Resources, Subcommittee on Oversight and Investigations and Subcommittee on National Parks, Forests and Public Lands and Committee on Merchant Marine and Fisheries, Subcommittee on Environment and Natural Resources. *Florida Everglades Ecosystem*. 103rd Cong., 1st sess., 1993.

U.S. House. Committee on Natural Resources, Subcommittee on Oversight and Investigations, Committee on Agriculture, Subcommittee on Specialty Crops and Natural Resources, and the Committee on Merchant and Marine Fisheries, Subcommittee on Environment and Natural Resources. *Ecosystem Management: Joint Oversight Hearing on Ecosystem Management and a Report by the General Accounting Office, "Ecosystem Management—Additional Actions Needed to Adequately Test a Promising Approach."* 103rd Cong., 2d sess., 1995.

U.S. House. Committee on Public Works. *Central and Southern Florida Flood Control Project.* Report prepared by the Library of Congress. 84th Congress, 2d session, 1956. Committee Print 23.

U.S. House. Committee on Resources. *Everglades National Park Expansion.* 108th Congress, 2d sess., 2004. H. Rpt. 108-516.

U.S. House. Committee on Resources Subcommittee on National Parks and Public Lands. *Issues Regarding Everglades National Park and Surrounding Areas Impacted by Management of the Everglades: Oversight Hearing Before the Subcommittee on National Parks and Public Lands of the Committee on Resources, House of Representatives.* 106th Cong., 1st sess., 1999.

U.S. House. Committee on Transportation and Infrastructure. *Modified Water Deliveries to Everglades National Park Tamiami Trail Modifications Final Limited Reevaluation Report and Environmental Assessment, Communication from the Assistant Secretary of the Army (Civil Works), the Department of the Army.* 111th Cong., 1st sess. 2009. H. Doc. 111-11.

————. Committee on Transportation and Infrastructure, Subcommittee on Water Resources and Environment. *Comprehensive Everglades Restoration Plan—The First Major Projects: Hearing Before the Subcommittee on Water Resources and Environment, House of Representatives.* 108th Cong., 2d sess., 2004.

————. *Water Resources Development Act of 1996: Hearings Before the Subcommittee on Water Resources and Environment, Committee on Transportation and Infrastructure, House of Representatives.* 104th Cong., 2d sess., 1996.

U.S. Senate. *Caloosahatchee River and Lake Okeechobee Drainage Areas, Florida.* 70th Cong., 2d sess., 1929. S. Doc. 213. Serial 9000.

————. *Caloosahatchee River and Lake Okeechobee Drainage Areas, Fla.* 71st Cong., 2d sess., 1930. S. Doc. 115. Serial 9219.

————. *Establishing the Big Cypress National Preserve, Florida.* 93d Cong., 2d sess., 1974. S. Rept. 93-1128. Serial 13057-7.

————. *Everglades of Florida.* 62d Cong., 1st sess., 1911. S. Doc. 89. Serial 6108.

————. *Public Works for Water, Pollution Control, and Power Development and Atomic Energy Commission Appropriation Bill, 1970.* 91st Cong., 1st sess., 1969. S. Rept. 91-528. Serial 12834-4.

————. *Restoring the Everglades, An American Legacy Act.* 106th Cong., 2d sess., 2000. S. Rept. 106-363.

————. *River Basin Monetary Authorizations and Miscellaneous Civil Works Amendments.* 91st Cong., 2d sess., 1970. S. Rept. 91-895. Serial 12881-3.

————. *Water Resources Development Act of 2000.* 106th Cong., 2d sess., 2000. S. Rept. 106-362.

U.S. Senate. Committee on Appropriations, Subcommittee on Public Works. *Water Supply for Central and Southern Florida and Everglades National Park: Meeting Arranged by Subcommittee of the Committee on Appropriations, United States Senate.* 91st Cong., 2d sess., 1970.

U.S. Senate. Committee on Commerce. *Rivers and Harbors: Hearings Before the Committee on Commerce, United States Senate, Parts 1-3.* 71st Cong., 2d sess., 1930.

U.S. Senate. Committee on Energy and Natural Resources. *Everglades National Park.* 108th Cong., 2d sess., 2004. S. Rpt. 108-298.

U.S. Senate. Committee on Energy and Natural Resources Subcommittee on National Parks, Historic Preservation, and Recreation and the Committee on Appropriations Subcommittee on Interior and Related Agencies. *South Florida Ecosystem Restoration: Joint Hearing Before the Subcommittee on National Parks, Historic Preservation, and Recreation of the Committee on Energy and Natural Resources and the Subcommittee on Interior and Related Agencies of the Committee on Appropriations, United States Senate.* 106th Cong., 1st sess., 1999.

U.S. Senate. Committee on Energy and Natural Resources, Subcommittee on Public Lands, National Parks and Forests. *El Malpais National Monument and Big Cypress National Preserve: Hearing Before the Subcommittee on Public Lands, National Parks and Forests of the Committee on Energy and Natural Resources, United States Senate.* 100th Cong., 1st sess., 1987.

U.S. Senate. Committee on Energy and Natural Resources, Subcommittee on Public Lands, National Parks and Forests. *Everglades National Park Protection and Expansion Act of 1989: Hearing before the Subcommittee on Public Lands, National Parks and Forests of the Committee on Energy and Natural Resources.* 101st Cong., 1st sess., 1989.

U.S. Senate. Committee on Energy and Natural Resources, Subcommittee on Public Lands, Reserved Water and Resource Conservation. *Additions to the Big Cypress National Preserve; Establishing the San Pedro Riparian National Conservation Area; Designating the Horsepasture River as a Component of the National Wild and Scenic Rivers System; and Amending FLPMA: Hearing Before the Subcommittee on Public Lands, Reserved Water and Resource Conservation of the Committee on Energy and Natural Resources, United States Senate.* 99th Cong., 2d sess., 1986.

U.S. Senate. Committee on Environment and Public Works. *Implementing the Comprehensive Everglades Restoration Plan (CERP): Hearing Before the Committee on Environment and Public Works, Senate.* 107th Cong., 2d sess., 2002.

U.S. Senate. Committee on Environment and Public Works Subcommittee on Transportation and Infrastructure. *U.S. Army Corps of Engineers' Budget for Fiscal Year 2001: Hearing Before the Subcommittee on Transportation and Infrastructure of the Committee on Environment and Public Works, United States Senate.* 106th Cong., 2d sess., 2000.

U.S. Senate. Committee on Foreign Relations. Subcommittee on International Operations and Organizations, Democracy, and Human Rights. *Everglades: Protecting Natural treasures Through International Organizations: Hearing Before the Subcommittee on International Operations and Organizations, Democracy, and Human Rights, Senate.* 110th Cong., 1st sess., 2007.

U.S. Senate. Committee on Interior and Insular Affairs. *Everglades National Park: Hearings Before the Committee on Interior and Insular Affairs, United States Senate, Ninety-First Congress, First Session, on the Water Supply, the Environmental, and Jet Airport Problems of Everglades National Park.* 91st Cong., 1st sess., 1969.

U.S. Senate. Committee on Interior and Insular Affairs, Subcommittee on Parks and Recreation. *Everglades-Big Cypress National Recreation Area: Hearing Before the Subcommittee on Parks and Recreation of the Committee on Interior and Insular Affairs, United States Senate.* 92d Cong., 1st sess., 1971.

————. *Everglades-Big Cypress National Recreation Area: Hearings Before the Subcommittee on Parks and Recreation of the Committee on Interior and Insular Affairs, United States Senate, Part 2.* 92d Cong., 2d sess., 1972.

U.S. Senate. Committee on Public Works, Subcommittee on Flood Control—Rivers and Harbors. *Central and Southern Florida Flood Control*

Project: Hearing Before the Subcommittee on Flood Control—Rivers and Harbors of the Committee on Public Works, United States Senate. 91st Cong., 2d sess., 1970.

U.S. Senate. Subcommittee of the Committee on Public Works. *Rivers and Harbors—Flood Control Emergency Act: Hearings Before a Subcommittee of the Committee on Public Works, United States Senate.* 80th Cong., 2d sess., 1948.

Statutes at Large

Act of 1 March 1929 (45 Stat. 1443).
Act of 30 May 1934 (48 Stat. 816).
Act of 6 December 1944 (58 Stat. 794).
Act of 30 June 1948 (62 Stat. 1171).
Act of 10 October 1949 (63 Stat. 733).
Act of 3 September 1954 (68 Stat. 1248).
Act of 15 October 1966 (80 Stat. 931).
Act of 13 August 1968 (82 Stat. 731).
The National Environmental Policy Act of 1969 (83 Stat. 852).
Act of 19 June 1970 (84 Stat. 310).
Act of 11 October 1974 (88 Stat. 1258).
Act of 30 November 1983 (97 Stat. 1153).
Supplemental Appropriations Act of 1984 (97 Stat. 1153).
Act of 17 November 1986 (100 Stat. 4082).
Seminole Indian Land Claims Settlement Act of 1987 (101 Stat. 1556).
Act of 29 April 1988 (102 Stat. 443).
Act of 18 November 1988 (102 Stat. 4571).
Everglades National Park Protection and Expansion Act of 1989 (103 Stat. 1946).
Water Resources Development Act of 1990 (104 Stat. 4604).
Water Resources Development Act of 1992 (106 Stat. 4797).
Act of 9 March 1994 (108 Stat. 98).
Unfunded Mandates Reform Act of 1995 (109 Stat. 48).
Federal Agricultural Improvement and Reform Act of 1996 (110 Stat. 888).
Water Resources Development Act of 1996 (110 Stat. 3658).
Act of 30 October 1998 (112 Stat. 2964).
Water Resources Development Act of 2000 (114 Stat. 2572).
Department of the Interior and Related Agencies Appropriations Act, 2004 (117 Stat 1241).

Legal Cases

Miccosukee Tribe of Indians of Florida v. United States of America, et al. Lead Case No. 04-21448-CIV-GOLD/McALILEY, U.S. District Court, Southern District of Florida. Order Granting Summary Judgment; Closing Case. 29 July 2008.

Miccosukee Tribe of Indians of Florida v. United States of America, et al. Lead Case No. 04-21448-CIV-GOLD/McALILEY, U.S. District Court, Southern District of Florida. Order Granting Plaintiffs' Motions [DE 357]; [DE 364] in Part; Granting Equitable Relief; Requiring Parties to Take Action by Dates Certain. 14 April 2010.

United States of America v. South Florida Water Management District, et al. Case No. 88-1886-CIV-MORENO, U.S. District Court for the Southern District of Florida, Miami Division. Order Granting Motion to Adopt the Special Master's Report, Motion Seeking Declaration of Violations, and Motion for Declaration of Breach of Commitments. 31 March 2010.

United States of America v. South Florida Water Management District, et al. Case No. 88-1886-CIV-MORENO, U.S. District Court for the Southern District of Florida, Miami Division. Order Requiring Special Master to Hold a Hearing on the Issue of Remedies and Submit a Report to the Court.

United States of America and Miccosukee Tribe of Indians v. South Florida Water Management District. Case No. 88-1886-CIV-MORENO. Miccosukee Tribe of Indians' Emergency Motion for Injunctive Relief and Supporting Memorandum of Law.

Articles and Papers

Alexander, Taylor R. "Effect of Hurricane Betsy on the Southeastern Everglades." *Quarterly Journal of the Florida Academy of Science* 30 (1967): 10-24.

Allison, R. V. "The Soil and Water Conservation Problem in the Everglades." *Soil Science Society of Florida Proceedings* 1 (1939): 35-42.

Alpert, Mark. "Replumbing the Everglades." *Scientific American* 281 (August 1999): 16-18.

Angier, Natalie. "Now You See It, Now You Don't." *Time* 124 (6 August 1984): 56.

Armistead, Thomas F. "Cash and Commitment Rescue Everglades Replumbing: Proof of Program's Success Remains Decades Away." *Engineering News-Record* 261 (31 March 2008): 24-27.

_____. "Feds Award First Everglades Contract." *Engineering News-Record* 262 (23 November 2009): 1.

_____. "Hope for Funding Remains Strong." *Engineering News-Record* 262, no. 3 (19 January 2009): 14.

_____. "Okeechobee Protection Plan Goes to Florida Legislature." *Engineering News-Record* 260, no. 4 (4 February 2008): 18.

Atlantis Scientific. "An Assessment of Water Resource Management in the Central and Southern Florida Flood Control District: A Review and Evaluation of Environmental Reports on the Kissimmee River and Lake Okeechobee." Copy in Library, Jacksonville District, U.S. Army Corps of Engineers, Jacksonville, Florida.

Baker, John H. "Time Is Running Out on the Everglades." *Audubon Magazine* 45 (May-June 1943): 176-179.

Ball, S. Mays. "Reclaiming the Everglades: Reversing the Far Western Irrigation Problem." *Putnam's Magazine* 7 (April 1910): 796-802.

Barlow, Gary. "A Sweet Deal for the Sugar Industry." *In These Times* 20 (14 October 1996): 10.

Becerra, Cesar A. "Birth of Everglades National Park." *South Florida History* 25-26 (Fall 1997/Winter 1998): 10-17.

Bennett, Charles E. "Early History of the Cross-Florida Barge Canal." *The Florida Historical Quarterly* 45, no. 2 (1966): 132-144.

Bestor, H. A. "Reclamation Problems of Sub-Drainage Districts Adjacent to Lake Okeechobee." *Soil Science Society of Florida Proceedings* 5-A (1943): 157-165.

Biemiller, Carl L. "The Water Wilderness—The Everglades." *Holiday* 10 (November 1951): 102-107, 114-119.

"Biennial Report Finds Little Progress, Many Hindrances." *Engineering News-Record* 261 (13 October 2008): 16.

"Bill Grants Land Inside Park to Miccosukee Tribe." *National Parks Magazine* 72 (September/October 1998): 11-12.

Black, Crow & Eidsness, Inc. "Report to Florida Cane Sugar League on Eutrophication of Lake Okeechobee." Copy in South Florida Water Management District Reference Center, West Palm Beach, Florida.

Boucher, Norman. "Smart as Gods: Can We Put the Everglades Back Together Again?" *Wilderness* 55 (Winter 1991): 10-21.

Boyle, Robert H. and Rose Mary Mechem. "Anatomy of a Man-Made Drought." *Sports Illustrated* 56 (15 March 1982): 46-54.

_____. "There's Trouble in Paradise." *Sports Illustrated* 54 (9 February 1981): 82-93.

Brooks, Paul. "Superjetport or Everglades Park?" *Audubon* 71 (July 1969): 4-11.

Broward, N. P. "Draining the Everglades." *The Independent* 64 (25 June 1908): 1448-1449.

Broward, Napoleon B. "Homes for Millions: Draining the Everglades." *Collier's* 44 (22 January 1910): 19.

Buckow, Ed. "Unraveling the Everglades Furor." *Field & Stream* 71 (October 1966): 12-16, 32-33.

Callison, Charles H. "National Outlook." *Audubon Magazine* 69 (May/June 1967): 56-57.

Canfield, Daniel E., Jr., and Mark V. Hoyer. "The Eutrophication of Lake Okeechobee." *Lake and Reservoir Management* 4, no. 2 (1988): 91-99.

Carney, James. "Last Gasp for the Everglades." *Time* 134 (25 September 1989): 26-27.

"Clinton Signs Bill Endorsing $7.8-Billion Everglades Job." *Engineering News-Record* 245 (18 December 2000): 9.

"Coalition Forms to Fight Florida Jetport." *National Parks Magazine* 43 (May 1969): 28-29.

"Coalition Outlines Plan to Halt Everglades Decline." *National Parks Magazine* 62 (March/April 1988): 9.

"Conservation: Jets v. Everglades." *Time* (22 August 1969): 42-43.

"Corps Unveils $4.6-Billion Stimulus Lineup." *Engineering News-Record* 262, no. 14 (4 May 2009): 11.

Culotta, Elizabeth. "Bringing Back the Everglades." *Science* 268 (23 June 1995): 1688-1690.

Davis, Jack E. "'Conservation is Now a Dead Word': Marjory Stoneman Douglas and the Transformation of American Environmentalism." *Environmental History* 8 (January 2003): 53-76.

"The Defense of the Everglades." *National Parks Magazine* 41 (August 1967): 2.

DeGeorge, Gail. "Big Sugar is Bitter over the Everglades." *Business Week* (4 November 1996): 192-193.

Derr, Mark. "Redeeming the Everglades." *Audubon* 95 (September/October 1993): 48-56, 128-131.

"Discord Swamps Everglades Talks." *Engineering News-Record* 232 (3 January 1994): 9.

"Dispute Over Big Reservoir Heads to Mediation this Month." *Engineering News-Record* 260, no. 19 (9 June 2008): 19.

Dougherty, Ryan. "Sugar Deal Sours 'Glades Restoration." *National Parks* 77 (September/October 2003): 10-11.

Douglas, Sue. "Save the Everglades." *Oceans* 18 (March/April 1985): 3-9.

Doyle, Mary. "Implementing Everglades Restoration." *Journal of Land Use and Environmental Law* 17 (Fall 2001): 59-66.

Elmer-Dewitt, Philip. "Facing a Deadline to Save the Everglades." *Time* 141 (21 June 1993): 56-57.

Engineering Department of Central and Southern Florida Flood Control District. "A Report on Water Resources of Everglades National Park, Florida." 22 May 1950. Copy in South Florida Water Management District Reference Center, West Palm Beach, Florida.

_____. "Review of the Plan of Flood Control for Central and Southern Florida in connection with the proposed development of the Everglades area and the operation of the conservation areas." November 1949. Copy in South Florida Water Management District Reference Center, West Palm Beach, Florida.

"Everglades Case Heads Back to Court." *National Parks Magazine* 68 (March/April 1994): 11-12.

"Everglades Hunt: The Deer Can't Win." *Newsweek* 100 (17 October 1982): 27.

"Everglades Land Swap Imperiled." *National Parks Magazine* 65 (September/October 1991): 10-11.

"Everglades Leader Appointed to Pentagon Post." *Everglades Restoration* (March/April 2009): 1, 4.

"Everglades Plan Comes Under Fire." *National Parks Magazine* 67 (September/October 1993): 10.

"Everglades Projects Receive $103 Million in Recovery Act Funds." *Everglades Report* (May/June 2009): 1, 3.

"Everglades: Render Back to Nature." *The Economist* 298 (9 December 1989): 50.

"Everglades Water Threatened." *Engineering News-Record* 221 (27 October 1988): 16.

"Five Major Problems Listed By Governor." *Water Management Bulletin* 5 (October-November 1971): 1-2.

"The Florida Everglades, River of Money." *The Economist* 362 (30 March 1996): 32.

"Florida Fairyland." *Reader's Digest* 28 (June 1936): 32.

"Florida's Battle of the Swamp." *Time* 118 (24 August 1981): 41.

"Florida's Growth Straining Fragile Groundwater." *Engineering News-Record* 216 (3 January 1985): 26.

"Florida's Land Boom Gets Down to Earth." *Business Week* (6 April 1968): 136-142.

"Folly in Florida." *National Parks Magazine* 43 (January 1969): 2, 21.

Fourqurean, James W. and Michael B. Robblee. "Florida Bay: A History of Recent Ecological Changes." *Estuaries* 22 (June 1999): 345-357.

Fumero, John J. and Keith W. Rizzardi. "The Everglades Ecosystem: From Engineering to Litigation to Consensus-Based Restoration." *St. Thomas Law Review* 13 (Spring 2001): 667-696.

Gilmour, Robert S. and John A. McCauley. "Environmental Preservation and Politics: The Significance of 'Everglades Jetport.'" *Political Science Quarterly* 90 (Winter 1975-1976): 719-738.

Glass, Stephen. "Rebirth of a River." *Restoration & Management Notes* 5 (Summer 1987): 7.

Graham, D. Robert. "A Quiet Revolution: Florida's Future on Trial." *The Florida Naturalist* 45 (October 1972): 145-151.

Griebenow, George W. "A Team Called GREAT." *Water Spectrum* 9 (Winter 1976-1977): 18-25.

Grumbine, R. Edward. "Reflections on 'What is Ecosystem Management?'" *Conservation Biology* 11 (February 1997): 41-47.

_____. "What is Ecosystem Management?" *Conservation Biology* 8 (March 1994): 27-38.

Grunwald, Michael. "Everglades." *Smithsonian* 36 (March 2006): 46-57.

_____. "Sugar Plum: Jeb v. the Everglades." *The New Republic* 228 (12 May 2003): 15-17.

_____. "Water World." *The New Republic* 230 (1 March 2004): 23.

Gunderon, Lance. "Resilience, Flexibility and Adaptive Management—Antidotes for Spurious Certitude?" *Conservation Ecology* 3, no. 1 (1999): 7. Copy available at <http://www.consecol.org/vol3/iss1/art7> (28 April 2006).

Hansen, Kevin. "South Florida's Water Dilemma: A Trickle of Hope for the Everglades." *Environment* 26 (June 1984): 14-20, 40-42.

Harwell, Mark A. "Ecosystem Management of South Florida." *BioScience* 47 (September 1997): 499-512.

Havens, Karl E. "Water Levels and Total Phosphorus in Lake Okeechobee." *Journal of Lake and Reservoir Management* 13, no. 1 (1997): 16-25.

Heitmann, John A. "The Beginnings of Big Sugar in Florida, 1920-1945." *The Florida Historical Quarterly* 77 (Summer 1998): 39-61.

Helvarg, David. "Destruction to Reconstruction: Restoring the Everglades." *National Parks Magazine* 72 (March/April 1998): 22-27.

Hilder, John Chapman. "America's Last Wilderness: The Florida Everglades Are Going Dry." *World's Work* 60 (February 1931): 54-56.

Holcombe, Randall G. "Why Has Florida's Growth Management Act Been Ineffective?" *Journal of the James Madison Institute* 28 (Spring/Summer 2004): 13-16.

"How the Florida Boom is Changing." *U.S. News & World Report* 46 (5 May 1969): 88-89.

Iannotta, Ben. "Mystery of the Everglades." *New Scientist* (9 November 1996): 3535.

Ichniowski, Tom. "Everglades Plan Sent to Hill." *Engineering News-Record* 243 (12 July 1999): 12.

James, R. Thomas, Val H. Smith, and Bradley L. Jones. "Historical Trends in the Lake Okeechobee Ecosystem, III. Water Quality." *Arch. Hydrobiology* 107 (January 1995): 52-53.

Katel, Peter. "Letting the Water Run into 'Big Sugar's' Bowl." *Newsweek* 116 (4 March 1996): 56-57.

Kersey, Harry A., Jr. "The East Big Cypress Case, 1948-1987: Environmental Politics, Law, and Florida Seminole Tribal Sovereignty." *The Florida Historical Quarterly* 69 (April 1991): 457-477.

Knetsch, Joe. "Hamilton Disston and the Development of Florida." *Sunland Tribune* 24, no. 1 (1998): 5-19.

Koebel, Joseph W., Jr. "An Historical Perspective on the Kissimmee River Restoration Project." *Restoration Ecology* 3, no. 3 (1995): 149-159.

Kohn, David. "Polishing an Ecological Jewel." *Engineering News-Record* 231 (9 August 1993): 29.

Kolter, Steven. "Reengineering the Everglades." *Wired* 10, no. 2 (February 2002): 104-111.

Kraft, Michael E. "U.S. Environmental Policy and Politics: From the 1960s to the 1990s." *Journal of Policy History* 12, no. 1 (2000): 17-39.

"A Last-Ditch Attempt to Save the Everglades." *Engineering News-Record* 240 (8 June 1998): 36-41.

Lee, Kai N. "Appraising Adaptive Management." *Conservation Ecology* 3, no. 2 (1999): 3. Copy at <http://www.consecol.org/vol3/iss2/art3> (28 April 2006).

"A Legal Ruling Needed on Everglades Water Rights." *Audubon* 69 (July/August 1967): 5.

Levin, Ted. "Bitter Sweets: A Politically Connected Industry Devastates the Everglades." *E: The Environmental Magazine* 14, no. 4 (July/August 2003): 34-39.

Light, Alfred R. "Miccosukee Wars in the Everglades: Settlement, Litigation, and Regulation to Restore an Ecosystem." *St. Thomas Law Review* 13 (Spring 2001): 729-742.

Light, Steven. "Key Principles for Adaptive Management in the Context of Everglades History and Restoration." Paper presented at Everglades Restoration Adaptive Management Strategy Workshop, October 2003.

Loftin, Kent. "Restoring the Kissimmee River." *Geotimes* 35 (December 1991): 15-16.

Luoma, Jon R. "Blueprint for the Future." *Audubon* 103 (July/August 2001): 66-72.

MacDonald, John D. "Threatened America—Last Chance to Save the Everglades." *Life* (5 September 1969): 58-66.

Marshall, Arthur R. "Are The Everglades Nearing Extinction?" *The Florida Naturalist* 44 (July 1971): 80-81, 87.

————. "Repairing the Florida Everglades Basin," n.d. Copy in South Florida Water Management District Reference Center, West Palm Beach, Florida.

McCormick, Robert. "Lavish Land." *Collier's* 110 (8 August 1942): 64-66.

McIntosh, Phyllis. "Reviving the Everglades." *National Parks* 76 (February 2002): 30-34.

McIver, Stuart A. "Death of a Bird Warden." *South Florida History* 29 (Fall 2001): 20-27.

Meindl, Christopher F. "Past Perceptions of the Great American Wetland: Florida's Everglades during the Early Twentieth Century." *Environmental History* 5 (July 2000): 378-395.

Merton, Robert K. "The Unanticipated Consequences of Purposive Social Action." *American Sociological Review* 1 (December 1936): 894-904.

Miller, James Nathan. "Rape on the Oklawaha." *Reader's Digest* 96 (January 1970): 54-60.

Mitchell, John G. "The Bitter Struggle for a National Park." *American Heritage* 22, no. 3 (1970): 97-109.

Moreau, Ron. "Everglades Forever?" *Newsweek* 107 (7 April 1986): 72-74.

Mormino, Gary R. "Sunbelt Dreams and Altered States: A Social and Cultural History of Florida, 1950-2000." *The Florida Historical Quarterly* 81 (Summer 2002): 3-21.

Morrison, Kenneth D. "Oil in the Everglades." *Natural History* 53 (June 1944): 282-283.

Munzer, Martha. "The Everglades and a Few Friends." *South Florida History Magazine* 23 (Winter 1995): 9-13.

Myers, Victoria. "The Everglades: Researchers Take a New Approach to an Old Problem." *Sea Frontiers* 40 (December 1994): 15-17.

Nelkin, Dorothy. "Scientists and Professional Responsibility: The Experience of American Ecologists." *Social Studies of Science* 7 (1977): 75-95.

"NPA Urges Protection from Everglades 'Salting.'" *National Parks Magazine* 40 (July 1966): 19.

O'Connell, Kim. "Gore Unveils Everglades Plan." *National Parks Magazine* 70 (May/June 1996): 13-14.

"Official Commitments Buoy Everglades Restoration Hope." *Southeast Construction* 8, no. 5 (1 March 2008): 9.

Ogden, Laura. "The Everglades Ecosystem and the Politics of Nature." *American Anthropologist* 110, no. 1 (March 2008): 21.

O'Keefe, M. Timothy. "Cows, Crackers and Cades: Bloody, Muddy and 'Unbent,' The Kissimmee River Has Seen Some Weird Goings-On." *Florida Sportsman* (August 1978): 18-21.

————. "The Kissimmee Problems: Trying to Unmuddle Man's Meddling." *Florida Sportsman* (August 1978): 28-34

"Opening of Canal 111 Is Delayed for Study." *National Parks Magazine* 40 (August 1966): 24.

O'Reilly, John. "Wildlife Protection in South Florida." *Bird-Lore* 41 (May-June 1939): 128-140.

Padgett, Tim. "Last Stand." *Time* 154 (5 July 1999): 60-61.

Partington, William M. "Oklawaha—The Fight Is On Again!" *Living Wilderness* 33 (Autumn 1969): 19-23.

Pearcy, Matthew T. "A History of the Ransdell-Humphreys Flood Control Act of 1917." *Louisiana History* 41 (Spring 2000): 133-159.

Pierce, Robert. "South Florida's Land Puzzle: Federal, State, and Private Agencies Purchase Protection." *National Parks Magazine* 59 (July/August 1985): 17.

"Political Fix For Wetland Woes." *ENR* 186 (2 September 1982): 33.

Pound, Edward T. "A Real Swampy Deal." *U.S. News and World Report* 138 (13 June 2005): 36.

Pratt, Theodore. "Papa of the Everglades National Park." *The Saturday Evening Post* 220 (9 August 1947): 32-33.

"The President's Report to You." *Audubon Magazine* 47 (January-February 1945): 45-50.

"The Problem & The Plan." *For the Future of Florida: Repair the Everglades* 2 (1981): 2.

"'Progress' Menaces the Everglades." *National Parks Magazine* 43 (July 1969): 8-10.

"The Proposed Everglades National Park." *Science* 77 (17 February 1933): 185.

Purcell, Aaron D. "Plumb Lines, Politics, and Projections: The Florida Everglades and the Wright Report Controversy." *The Florida Historical Quarterly* 80 (Fall 2001): 161-197.

Reiger, George. "The River of Grass is Drying Up!" *National Wildlife* 12 (December/January 1974): 55-62.

"Report of the Special Study Team on the Florida Everglades, August 1970." Copy in South Florida Water Management District Reference Center, West Palm Beach, Florida.

"Restudy Plan Shaped by Public Input," *C&SF Restudy Update* 3 (March 1999): 1-2.

Ridgley, Heidi. "Second Chance for a Dying Estuary." *National Wildlife* 40 (August/September 2002): 38-43.

Rizzardi, Keith W. "Alligators and Litigators: A Recent History of Everglades Regulation and Litigation." *The Florida Bar Journal* 18 (March 2001): n.p.

————. "Translating Science into Law: Phosphorus Standards in the Everglades." *Journal of Land Use and Environmental Law* 17 (Fall 2001): 149-168.

Roberts, Paul. "The Sweet Hereafter." *Harper's* 299 (November 1999): 54-68.

Rodgers, William H., Jr. "The Miccosukee Indians and Environmental Law: A Confederacy of Hope." *Environmental Law Reporter* 31 (August 2001): 10918-10927.

Rome, Adam. "'Give Earth a Chance': The Environmental Movement and the Sixties." *Journal of American History* 90 (September 2003): 525-554.

Ruddy, T. Michael. "Damming the Meramec: The Elusive Public Interest, 1927-1949." *Gateway Heritage* 10 (Winter 1989-1990): 36-45.

Russell, Edmund P. "Lost Among the Parts Per Billion: Ecological Protection at the United States Environmental Protection Agency, 1970-1993." *Environmental History* 2 (January 1997): 29-51.

Satchell, Michael. "Can the Everglades Still Be Saved?" *U.S. News & World Report* 108 (2 April 1990): 24.

"Saving the Everglades: Water In, Water Out." *The Economist* 370 (6 February 1999): 30.

Schneider, William J. and James H. Hartwell. "Troubled Waters of the Everglades." *Natural History* 93 (November 1984): 46-57.

Schulte, Bret. "Trouble in Swamplands." *U.S. News and World Report* 140 (13 March 2006): 26.

Scott, Harold A. "Distribution of Water in the Central and Southern Florida Project," n.d. Copy in South Florida Water Management District Reference Center, West Palm Beach, Florida.

Scott, Judy. "Good News from the Everglades: $160 Million STA 3/4 Project Winner of ACEC's Grand Conceptor Award." *Southeast Construction* 5, no. 13 (1 November 2005): 39.

Sewell, J. Richard. "Cross-Florida Barge Canal, 1927-1968." *The Florida Historical Quarterly* 46 (April 1968): 369-383.

"The 'Sewer Ditch' Undone." *Audubon* 89 (March 1987): 114-115.

"SFWMD Halts Work on Everglades Reservoir Project." *Southeast Construction* 8, no. 10 (1 August 2008): 9.

Sharp, Howard. "Farming the Muck Soil of the Everglades." *The Florida Grower* 32 (7 November 1925): 3-4, 22.

Shore, Jim and Jerry C. Straus. "The Seminole Water Rights Compact and the Seminole Indian Land Claims Settlement Act of 1987." *Journal of Land Use and Environmental Law* 6 (Winter 1990): 1-24.

Sirgo, Henry B. "Water Policy Decision-Making and Implementation in the Johnson Administration." *Journal of Political Science* 12, nos. 1-2 (1985): 53-63.

Stegner, Wallace. "Last Chance for the Everglades." *Saturday Review* (6 May 1967): 22-23, 72-73.

Stephens, John C. "The Cooperative Water Control Program for Central and Southern Florida." Paper presented at the Annual Winter Meeting of the American Society of Agricultural Engineers, 17 December 1958. Copy in Library, Jacksonville District, U.S. Army Corps of Engineers, Jacksonville, Florida.

Stine, Jeffrey K. "Regulating Wetlands in the 1970s: U.S. Army Corps of Engineers and the Environmental Organizations." *Journal of Forest History* 27 (April 1983): 60-75.

Stokstad, Erik. "Big Land Purchase Triggers Review of Plans to Restore Everglades." *Science* 321 (4 July 2008): 22.

Straight, Michael. "The Water Picture in Everglades National Park." *National Parks Magazine* 39 (August 1965): 4-9.

"Sugar Deal Sours 'Glades Restoration." *National Parks Magazine* 77 (September/October 2003): 10.

Tabb, D. C. and T. M. Thomas. "Prediction of Freshwater Requirements of Everglades National Park," n.d. Copy in South Florida Water Management District Reference Center, West Palm Beach, Florida.

Taylor, Ronald A. "Saving a Fountain of Life." *U.S. News and World Report* 100 (24 February 1986): 63-64.

Tilden, Paul M. "The Water Problem in Everglades National Park, Part II." *National Parks Magazine* 38 (March 1964): 8-11.

Todd, Kim. "Who Gets the Water?" *Sierra* 88 (January 2003): 10-11.

Toth, Louis A. "The Ecological Basis of the Kissimmee River Restoration Plan." *Florida Scientist* 56, no. 1 (1993): 25-32.

Toth, Louis A., Jayantha T. B. Obeysekera, William A. Perkins, and M. Kent Loftin. "Flow Regulation and Restoration of Florida's Kissimmee River." *Regulated Rivers: Research & Management* 8 (1993): 155-166.

Trimble, Paul J. and Jorge A. Marban. "A Proposed Modification to Regulation of Lake Okeechobee." *Water Resources Bulletin* 25 (December 1989): 1249-1257.

Trumbull, Stephen. "The River Spoilers." *Audubon* 68 (March-April 1966): 102-110.

Voss, Michael. "The Central and Southern Florida Project Comprehensive Review Study: Restoring the Everglades." *Ecology Law Quarterly* 27 (August 2000): 751-770.

Wallis, W. Turner. "The History of Everglades Drainage and Its Present Status." *Soil Science Society of Florida Proceedings*, 4-A (1942): 29-33.

————. "The Interests of Dade County in relation to the Cooperative Water Control Program for Central and Southern Florida," n.d. Copy in South Florida Water Management District Reference Center, West Palm Beach, Florida.

Wasserman, Harvey. "Cane Mutiny." *The Nation* 262 (11 March 1996): 6-7.

"Water, Bird and Man—The Everglades," *The Economist* 371 (8 October 2005): 32-34.

"Water for Everglades National Park." *National Parks Magazine* 41 (December 1967): 2.

Weiss, John. "Everglades Deer in Trouble." *Outdoor Life* 165 (April 1980): 32, 36.

"Welcome to *CERP Perspectives*," *CERP Perspectives* 1 (Fall 2001): 2.

Wheeler, Keith. "Florida's Never-Never Land." *Science Digest* 29 (May 1951): 28-32.

Williams, Verne O. "Man-Made Drouth Threatens Everglades National Park." *Audubon Magazine* 65 (September-October 1963): 290-294.

"WMD Seeks Permits for River Restoration." *South Florida Water Management District Bulletin* 9 (Winter 1983): 1.

Wolff, Anthony. "The Assault on the Everglades." *Look* (9 September 1969): 44-52.

Wood, Debra. "Everglades Restoration: What's Going On?" *Southeast Construction* 6 (1 April 2006): 74-76.

_____. "Life Support: Is the Everglades Restoration Dying?" *Southeast Construction* 7 (1 November 2007): 28-30.

_____. "S. Florida Water Management District Commits to Land Deal, Not Reservoir." *Engineering News-Record* (23 August 2010): 1.

Woody, Theresa. "Grassroots in Action: The Sierra Club's Role in the Campaign to Restore the Kissimmee River." *Journal of North American Benthological Society* 12, no. 2 (1993): 201-205.

Wright, Andrew and Tom Ichniowski. "GAO Hits Everglades Program." *Engineering-News Record* 242 (10 May 1999): 13.

Yates, Steve. "Marjory Stoneman Douglas and the Glades Crusade." *Audubon* 85 (March 1983): 112-127.

_____. "Saga of the Glades Continues." *Audubon* 87 (January 1985): 34-39.

Zaneski, Cyril T. "Anatomy of a Deal." *Audubon* 103 (July/August 2001): 48-53.

_____. "The Players: The Politicians." *Audubon* 103 (July/August 2001): 74-75.

Zimmer, Gale Koschmann. "Unless the Rains Come Soon . . . " *National Parks Magazine* 36 (June 1962): 4-7.

Dissertations and Theses

Dovell, Junius Elmore. "A History of the Everglades of Florida." Ph.D. diss., University of North Carolina at Chapel Hill, 1947.

Gannon, Patrick Thomas, Sr. "On the Influence of Surface Thermal Properties and Clouds on the South Florida Sea Breeze." Ph.D. diss., University of Miami, 1977.

Stine, Jeffrey K. "Environmental Politics and Water Resources Development: The Case of the Army Corps of Engineers during the 1970s." Ph.D. diss., University of California at Santa Barbara, 1984.

Strickland, Jeffrey Glenn. "The Origins of Everglades Drainage in the Progressive Era: Local, State and Federal Cooperation and Conflict." M.A. thesis, Florida Atlantic University, 1999.

Books

Arnold, Joseph L. *The Evolution of the 1936 Flood Control Act*. Fort Belvoir, Va.: Office of History, United States Army Corps of Engineers, 1988.

Blake, Nelson Manfred. *Land Into Water—Water Into Land: A History of Water Management in Florida*. Tallahassee: University of Florida Presses, 1980.

Bottcher, A. B. and F. T. Izuno, eds. *Everglades Agricultural Area (EAA): Water, Soil, Crop, and Environmental Management*. Gainesville: University Press of Florida, 1994.

Bryson, Michael A. *Visions of the Land: Science, Literature, and the American Environment from the Era of Exploration to the Age of Ecology*. Charlottesville: University Press of Virginia, 2002.

Buffalo Tiger and Harry A. Kersey, Jr. *Buffalo Tiger: A Life in the Everglades*. Lincoln: University of Nebraska Press, 2002.

Buker, George E. *Sun, Sand and Water: A History of the Jacksonville District, U.S. Army Corps of Engineers, 1821-1975*. Fort Belvoir, Va.: U.S. Army Corps of Engineers, 1981.

_____. *The Third E: A History of the Jacksonville District, U.S. Army Corps of Engineers, 1975-1998*. Jacksonville, Fla.: U.S. Army Corps of Engineers, 1998.

Cahn, Matthew Alan. *Environmental Deceptions: The Tension Between Liberalism and Environmental Policymaking in the United States*. Albany: State University of New York Press, 1995.

Cairns, John, Jr., Todd V. Crawford, and Hal Salwasser, eds. *Implementing Integrated Environmental Management*. Blacksburg, Va.: University Center for Environmental and Hazardous Material Studies, Virginia Polytechnic Institute and State University, 1994.

Caldwell, Lynton Keith. *The National Environmental Policy Act: An Agenda for the Future*. Bloomington: Indiana University Press, 1998.

Camillo, Charles A. and Matthew T. Pearcy. *Upon Their Shoulders: A History of the Mississippi River Commission from Its Inception Through the Advent of the Modern Mississippi River and Tributaries Project*. Vicksburg, Ms.: Mississippi River Commission, 2004.

Carter, Luther J. *The Florida Experience: Land and Water Policy in A Growth State*. Baltimore, Md.: Johns Hopkins University Press, 1974.

Central and Southern Florida Flood Control District. *Facts About F.C.D.* West Palm Beach, Fla.: Central and Southern Florida Flood Control District, 1955.

_____. *Ten Years of Progress: 1949-1959*. West Palm Beach, Fla.: Central and Southern Florida Flood Control District, 1959.

Davis, Frederick E. and Michael L. Marshall. *Chemical and Biological Investigations of Lake Okeechobee, January 1973-June 1974, Interim Report*. Technical Publication No. 75-1. West Palm Beach, Fla.: Central and Southern Florida Flood Control District, 1975.

Davis, Jack E. *An Everglades Providence: Marjory Stoneman Douglas and the American Environmental Century* . Athens: The University of Georgia Press, 2009.

Davis, Steven M. and John C. Ogden, eds. *Everglades: The Ecosystem and Its Restoration*. Delray Beach, Fla.: St. Lucie Press, 1994.

Douglas, Marjory Stoneman. *The Everglades: River of Grass*. 50th anniversary edition. Sarasota, Fla.: Pineapple Press, 1997.

Douglas, Marjory Stoneman with John Rothchild. *Voice of the River*. Sarasota, Fla.: Pineapple Press, 1987.

Dunlap, Thomas R. *DDT: Scientists, Citizens, and Public Policy*. Princeton, N.J.: Princeton University Press, 1981.

Everglades Engineering Board of Review. *Report of Everglades Engineering Board of Review to Board of Commissioners of Everglades Drainage District*. Tallahassee, Fla.: T. J. Appleyard, 1927.

Federico, Anthony C., Kevin G. Dickson, Charles R. Kratzer, and Frederick E. Davis. *Lake Okeechobee Water Quality Studies and Eutrophication Assessment*. Technical Publication 81-2. West Palm Beach, Fla.: South Florida Water Management District, Resource Planning Department, 1981.

Fernald, Edward A. and Elizabeth D. Purdum, eds. *Water Resources Atlas of Florida*. Institute of Science and Public Affairs, Florida State University, Tallahassee, 1998.

Flippen, J. Brooks. *Nixon and the Environment*. Albuquerque: University of New Mexico Press, 2000.

Florida Defenders of the Environment. *Environmental Impact of the Cross-Florida Barge Canal.* Gainesville, Fla.: Florida Defenders of the Environment, 1970.

Florida State Board of Conservation. *Fourth Biennial Report, Biennium Ending December 31, 1940.* Tallahassee: Florida State Board of Conservation, 1941.

Golley, Frank Benjamin. *A History of the Ecosystem Concept in Ecology: More Than the Sum of the Parts.* New Haven, Conn.: Yale University Press, 1993.

Goodrick, Robert L. and James F. Milleson. *Studies of Floodplain Vegetation and Water Level Fluctuation in the Kissimmee River Valley.* Technical Publication No. 74-2. West Palm Beach, Fla.: Central and Southern Florida Flood Control District, 1974.

Gore, Albert. *Earth in the Balance: Ecology and the Human Spirit.* New York: Houghton Mifflin Company, 1992.

Grunwald, Michael. *The Swamp: The Everglades, Florida, and the Politics of Paradise.* New York: Simon & Schuster, 2006.

Handbook of North American Indians. William Sturtevant, ed. Vol. 14, *Southeast.* Raymond D. Fogelson, ed. Washington, D.C.: Smithsonian Institution, 2004.

Hanna, Alfred Jackson and Kathryn Abbey Hanna. *Lake Okeechobee: Wellspring of the Everglades.* Indianapolis: The Bobbs-Merrill Company, 1948.

Hartzog, George B., Jr. *Battling for the National Parks.* Mt. Kisco, N.Y.: Moyer Bell Limited, 1988.

Hays, Samuel P. *Beauty, Health, and Permanence: Environmental Politics in the United States, 1955-1985.* Cambridge: Cambridge University Press, 1987.

_____. *Conservation and the Gospel of Efficiency: The Progressive Conservation Movement, 1890-1920.* Cambridge: Harvard University Press, 1959; reprint, New York: Atheneum, 1969.

_____. *A History of Environmental Politics Since 1945.* Pittsburgh, Penn.: University of Pittsburgh Press, 2000.

Healy, Robert G. *Land Use and the States.* Baltimore, Md.: The Johns Hopkins University Press, 1976.

Hickel, Walter J. *Who Owns America?* Englewood Cliffs, N.J.: Prentice-Hall, 1971.

Interim Report on the Special Project to Prevent the Eutrophication of Lake Okeechobee. Tallahassee: Florida Department of Administration, Division of State Planning, 1975.

Ives, Lieutenant J. C. *Memoir to Accompany a Military Map of the Peninsula of Florida, South of Tampa Bay.* New York: M. B. Wynkoop, 1856.

Johnson, Lamar. *Beyond the Fourth Generation.* Gainesville: The University Presses of Florida, 1974.

Keller, Robert H. and Michael F. Turek. *American Indians & National Parks.* Tucson: The University of Arizona Press, 1998.

The Kissimmee-Okeechobee Basin: A Report to the Cabinet of Florida. Miami, Fla.: Division of Applied Ecology, Center for Urban and Regional Studies, University of Miami, 1972.

Kleinberg, Eliot. *Black Cloud: The Great Florida Hurricane of 1928.* New York: Carroll & Graf Publishers, 2003.

Levin, Ted. *Liquid Land: A Journey Through the Florida Everglades.* Athens: The University of Georgia Press, 2003.

Marks, Stephen V. and Keith E. Maskus, eds. *The Economics and Politics of World Sugar Policies.* Ann Arbor: The University of Michigan Press, 1993.

McCally, David. *The Everglades: An Environmental History.* Gainesville: University Press of Florida, 1999.

McIntosh, Robert P. *The Background of Ecology: Concept and Theory.* Cambridge: Cambridge University Press, 1985.

McIver, Stuart A. *Death in the Everglades: The Murder of Guy Bradley, America's First Martyr to Environmentalism.* Gainesville: University Press of Florida, 2003.

Mead, Daniel W., Allen Hazen, and Leonard Metcalf. *Report on the Drainage of the Everglades of Florida.* Chicago: Board of Consulting Engineers, 1912.

Milleson, James F. *Vegetation Changes in the Lake Okeechobee Littoral Zone, 1972 to 1982.* Technical Publication 87-3. West Palm Beach, Fla.: South Florida Water Management District, Environmental Sciences Division, Resource Planning Department, 1987.

Morgan, Arthur E. *Dams and Other Disasters: A Century of the Army Corps of Engineers in Civil Works.* Boston: Porter Sargent Publisher, 1971.

Muschett, F. Douglas, ed. *Principles of Sustainable Development.* Delray Beach, Fla.: St. Lucie Press, 1997.

Mykle, Robert. *Killer 'Cane: The Deadly Hurricane of 1928.* New York: Cooper Square Press, 2002.

National Research Council of the National Academies. *Science and the Greater Everglades Ecosystem Restoration.* Washington, D.C.: The National Academies Press, 2003.

Okeechobee Flood Control District. *A Report to the Board of Commissioners of Okeechobee Flood Control District on the Activities of the District and on Lake Okeechobee.* Copy in Library, Jacksonville District, U.S. Army Corps of Engineers, Jacksonville, Florida.

Orsi, Jared. *Hazardous Metropolis: Flooding and Urban Ecology in Los Angeles.* Berkeley: University of California Press, 2004.

Porter, Douglas R. and David A. Salvesen, eds. *Collaborative Planning for Wetlands and Wildlife: Issues and Examples.* Washington, D.C.: Island Press, 1995.

Proctor, Samuel. *Napoleon Bonaparte Broward: Florida's Fighting Democrat.* Gainesville: University of Florida Press, 1950.

Reuss, Martin. *Designing the Bayous: The Control of Water in the Atchafalaya Basin, 1800-1995.* College Station: Texas A&M University Press, 2004.

_____. *Reshaping National Water Politics: The Emergence of the Water Resources Development Act of 1986.* IWR Policy Study 91-PS-1. Fort Belvoir, Va.: U.S. Army Corps of Engineers, Institute for Water Resources, 1991.

_____. *Shaping Environmental Awareness: The United States Army Corps of Engineers Environmental Advisory Board, 1970-1980.* Alexandria, Va.: Historical Division, Office of Administrative Services, Office of the Chief of Engineers, 1983.

Ritter, William F. and Adel Schirmohammadi, eds. *Agricultural Nonpoint Source Pollution: Watershed Management and Hydrology.* (Boca Raton, Fla.: Lewis Publishers, 2001).

Rosen, Howard and Martin Reuss, eds. *The Flood Control Challenge: Past, Present, and Future.* Chicago: Public Works Historical Society, 1988.

Runte, Alfred. *National Parks: The American Experience.* 3rd edition. Lincoln: University of Nebraska Press, 1997.

Ryding, Sven-Olof and Walter Rast, eds. *The Control of Eutrophication of Lakes and Reservoirs.* Park Ridge, N.J.: Parthenon Publishing Group, 1989.

Sale, Kirkpatrick. *The Green Revolution: The American Environmental Movement, 1962-1992.* New York: Hill and Wang, 1993.

Sellars, Richard West. *Preserving Nature in the National Parks: A History.* New Haven, Conn.: Yale University Press, 1997.

Short, C. Brant. *Ronald Reagan and the Public Lands: America's Conservation Debate, 1979-1984.* College Station: Texas A&M University Press, 1989.

Sitterson, J. Carlyle. *Sugar Country: The Cane Sugar Industry in the South, 1753-1950*. Lexington: Univeristy of Kentucky Press, 1953; reprint, Westport, Conn.: Greenwood Press, 1973.

Small, John Kunkel. *From Eden to Sahara: Florida's Tragedy*. Lancaster, Penn.: The Science Press Printing Company, 1929.

Soden, Dennis L., ed. *The Environmental Presidency*. Albany: State University of New York, 1999.

A Study of the Central and Southern Florida Flood Control Project. Washington, D.C.: The Library of Congress, 1953.

Tebeau, Charlton W. *A History of Florida*. Coral Gables, Fla.: University of Miami Press, 1971.

Vig, Norman J. and Michael E. Kraft, eds. *Environmental Policy in the 1980s: Reagan's New Agenda*. Washington, D.C.: Congressional Quarterly, 1984.

————. *Environmental Policy in the 1990s*. 3rd ed. Washington, D.C.: Congressional Quarterly, 1997.

Vileisis, Ann. *Discovering the Unknown Landscape: A History of America's Wetlands*. Washington, D.C.: Island Press, 1997.

Weisman, Brent Richards. *Unconquered People: Florida's Seminole and Miccosukee Indians*. Gainesville: University Press of Florida, 1999.

Wilkinson, Alec. *Big Sugar: Seasons in the Cane Fields of Florida*. New York: Alfred A. Knopf, 1989.

Williams, Jack E., Christopher A. Wood, and Michael P. Dombeck, eds. *Watershed Restoration: Principles and Practices*. Bethesda, Md.: American Fisheries Society, 1997.

Zaffke, Michael. *Wading Bird Utilization of Lake Okeechobee Marshes, 1977-1981*. Technical Publication 84-9. West Palm Beach, Fla.: South Florida Water Management District, Environmental Sciences Division, Resource Planning Department, 1984.

MATTHEW C. GODFREY is an associate historian with Historical Research Associates, Inc. He received a Ph.D. in American and Public History from Washington State University, and has completed several research projects and reports on water issues in the United States, including a study of the St. Paul District of the U.S. Army Corps of Engineers, co-authored with Theodore Catton. He has published articles and book reviews in *Agricultural History*, *Pacific Northwest Quarterly*, *Utah Historical Quarterly*, *Western Historical Quarterly*, and *Columbia: The Magazine of the Pacific Northwest*. He has a book forthcoming from Utah State University Press on the beet sugar industry in the Intermountain West.

THEODORE CATTON is an environmental historian based in Missoula, Montana. In his 15 years with Historical Research Associates, he specialized in the conservation mandates and history of various federal land management agencies including the Corps of Engineers, National Park Service, USDA Forest Service, and Bureau of Indian Affairs. His publications include *Inhabited Wilderness: Indians, Eskimos, and National Parks in Alaska*, and *National Park, City Playground: Mount Rainier in the Twentieth Century*.